CHIRAL DYNAMICS 2006

CHIRAL DYNAMICS 2006

Proceedings of the 5th International Workshop on
Chiral Dynamics, Theory and Experiment

Durham/Chapel Hill, North Carolina, USA   18 – 22 September 2006

editors

Mohammad W. Ahmed
Triangle Universities Nuclear Laboratory & Duke University, USA

Haiyan Gao
Triangle Universities Nuclear Laboratory & Duke University, USA

Barry Holstein
University of Massachusetts, USA

Henry R. Weller
Triangle Universities Nuclear Laboratory & Duke University, USA

World Scientific

NEW JERSEY • LONDON • SINGAPORE • BEIJING • SHANGHAI • HONG KONG • TAIPEI • CHENNAI

Published by

World Scientific Publishing Co. Pte. Ltd.

5 Toh Tuck Link, Singapore 596224

USA office: 27 Warren Street, Suite 401-402, Hackensack, NJ 07601

UK office: 57 Shelton Street, Covent Garden, London WC2H 9HE

British Library Cataloguing-in-Publication Data
A catalogue record for this book is available from the British Library.

CHIRAL DYNAMICS 2006
Proceedings of the 5th International Workshop on Chiral Dynamics, Theory and Experiment

ISBN-13 978-981-270-566-2
ISBN-10 981-270-566-X

Printed in Singapore by B & JO Enterprise

PREFACE

This volume presents the proceedings of the *Fifth International Workshop on Chiral Dynamics: Theory and Experiment*, which was hosted by Duke University in association with the Triangle Universities Nuclear Laboratory (TUNL). The location of the workshop was The William and Ida Friday Center for Continuing Education, located in Chapel Hill, North Carolina. Sponsors for the workshop included the School of Arts and Sciences at Duke University, TUNL, the Physics Departments of Duke, North Carolina State University and the University of North Carolina, Jefferson Lab, the Laboratory for Nuclear Science at MIT, Institute of Physics (IoP), CAEN Technologies Inc., and Wiener Plein & Baus, Ltd. The fourth workshop was held in Juelich in 2003, while the sixth workshop will be organized in Bern, Switzerland in 2009.

The main purpose of these workshops has been and continues to be the bringing together of physicists working in the field of Chiral Dynamics, so that they can discuss and debate the most important new developments as well as to explore the most promising new directions for both theory and experiment. One of the keys to the success of previous workshops has been perceived to be the excellent balance between theorists and experimentalists attending these workshops. This balance was, we believe, achieved again.

The workshop was organized along the traditional lines of the previous workshops in this series. There were four mornings of Plenary talks, and three afternoons of Working Group Sessions. The final session (on Friday morning, September 22) consisted of summary talks from the working group leaders.

The Plenary talks emphasized progress and exciting new results in areas including Chiral Perturbation Theory, tests of the Standard Model, Chiral Nuclear Forces and very recent Lattice QCD calculations. New experimental results from Mainz, JLAB, MIT-Bates, RIKEN and HIγS were also discussed. There were three working groups: Goldstone Boson Dynamics (G. Colangelo and S. Giovannella); Hadron Structure and Meson-Baryon Interactions (T. Hemmert, J. Feldman and A. Nathan); and Few-Body Physics (H. W. Hammer, D. Phillips, and N. Kalantur). These working groups comprised one of the most dynamic and productive aspects of the workshop, and were usually the venue for some of the most intense and informative discussions. The Working Group Summary Talks, among other things, addressed the important question of "what new experiments and calculations are needed in order for the field to make significant progress".

The location of the Workshop is connected to the commissioning of the newly Upgraded High Intensity Gamma-Ray Source (HIγS) at TUNL/Duke University. This facility will allow for detailed precision studies at energies just above and below pion-production threshold using intense, nearly monoenergetic, 100% polarized gamma-ray beams and polarized targets. In keeping with this setting, we emphasized some of the relevant physics which will be studied at HIγS including few-body physics, nucleon spin-polarizabilities, the GDH integrals, and near-threshold pion photoproduction.

The organizers are especially grateful to Mohammad W. Ahmed, the Scientific Secretary, for his hard work in making the workshop a success. His attention to the details of all of the arrangements before, during and after the workshop, was crucial to this success. The success of this workshop was also the result of the enthusiastic engagement of the participants. The excellent talks by all of the speakers and the hard work of the working group convenors were conspicuous. We are also very grateful to the International Advisory and Program Committees for their help in designing the program. The advice, support and engagement of the local organizing committee members was also very much appreciated. Finally, we wish to thank all of the sponsors of the workshop for their generous financial contributions, the staff of the Friday Center for their excellent running of the facility during our stay, and the graduate students at TUNL for their help before and during this workshop.

Haiyan Gao,
Barry Holstein, and
Henry R. Weller

22 September, 2006

ORGANIZING COMMITTEES

EDITORIAL BOARD

for the 5th International Workshop on Chiral Dynamics: Theory and Experiment

M. W. Ahmed	– Duke University, Durham, NC, USA
H. Gao	– Duke University, Durham, NC, USA
B. Holstein	– University of Massachusetts, Amherst, MA, USA
H. R. Weller	– Duke University, Durham, NC, USA

INTERNATIONAL ADVISORY COMMITTEE

for the 5th International Workshop on Chiral Dynamics: Theory and Experiment

Aron Bernstein	– Massachusetts Institute of Technology
Hans Bijnens	– MAX-lab, Lund University
Kees de Jager	– Jefferson Laboratory
Xiangdong Ji	– University of Maryland
Bira van Kolck	– University of Arizona
Kuniharu Kubodera	– University of South Carolina
Heiri Leutwyler	– University of Bern
Aneesh Manohar	– University of California
Judith McGovern	– University of Manchester
Ulf Meißner	– University of Bonn & FZ Juelich
Al Nathan	– University of Illinois
Andy Sandorfi	– Brookhaven National Laboratory
Martin Savage	– University of Washington
Carlo Schaerf	– Istituto Nazionale di Fisica Nucleare
Tony Thomas	– Jefferson Laboratory
Marc Vanderhaeghen	– Jefferson Laboratory
Thomas Walcher	– University of Mainz

PROGRAM COMMITTEE

for the 5th International Workshop on Chiral Dynamics: Theory and Experiment

Mike Birse	– University of Manchester
Reinhard Beck	– University of Mainz
Veronique Bernard	– Universite Louis Pasteur
Aron Bernstein	– Massachusetts Institute of Technology
Elizabeth Beise	– University of Maryland
Stephen Cotanch	– North Carolina State University
Evgeni Epelbaum	– Jefferson Laboratory
Ada Farilla	– Istituto Nazionale di Fisica Nucleare
Paul Frampton	– University of North Carolina
Haiyan Gao	– Duke University
Juerg Gasser	– University of Bern
Jose Goity	– Jefferson Laboratory
Hans-Werner Hammer	– Institute of Nuclear Theory
Thomas Hemmert	– TU Munchen
Barry Holstein	– University of Massachusetts
Marc Knecht	– Marseillei, CPT
Bira van Kolck	– University of Arizona
Ulf Meissner	– University of Bonn & FZ Juelich
Rory Miskimen	– University of Massachusetts
Blaine Norum	– University of Virginia
Dan Phillips	– Ohio University
Roxanne Springer	– Duke University
Tony Thomas	– Jefferson Laboratory
Thomas Walcher	– University of Mainz
Henry Weller	– Duke University

LOCAL ORGANIZING COMMITTEE

for the 5th International Workshop on Chiral Dynamics: Theory and Experiment

Mohammad Ahmed	– Duke University
Art Champagne	– University of North Carolina, Chapel Hill
Alex Crowell	– Duke University
Dipangkar Dutta	– Duke University
Jon Engel	– University of North Carolina, Chapel Hill
Haiyan Gao	– Duke University
Calvin Howell	– Duke University
John Kelley	– North Carolina State University
Dean Lee	– North Carolina State University
Diane Markoff	– North Carolina Central University
Tom Mehen	– Duke University
Gary Mitchell	– North Carolina State University
Werner Tornow	– Duke University
Ying Wu	– Duke University
Henry Weller	– Duke University

WORKING GROUP LEADERS

for the 5th International Workshop on Chiral Dynamics: Theory and Experiment

Goldstone Boson Dynamics

Simona Giovannella	– Laboratori Nazionali di Frascati
Gilberto Colangelo	– University of Bern

Few-Body Dynamics

Hans-Werner Hammer	– Institute of Nuclear Theory
Dan Phillips	– Ohio University
Nasser Kalantar	– Kernfysisch Versneller Instituut

Hadron Structure & Meson-Baryon Interactions

Thomas Hemmert	– TU Munchen
Gerald Feldman	– George Washington University
Alan Nathan	– University of Illinois

CONTENTS

PART A

PLENARY SESSION

OPENING REMARKS: EXPERIMENTAL TESTS OF CHIRAL SYMMETRY BREAKING

A.M.BERNSTEIN *

Department of Physics, LNS, MIT,
Cambridge, MA 02139, USA

A physical introduction to the basics of chiral dynamics is presented. Emphasis is placed on experimental tests which have generally demonstrated a strong confirmation of the predictions of chiral perturbation theory, a low energy effective field theory of QCD. Special attention is paid to a few cases where discrepancies exist, requiring further work. Some desirable future tests are also recommended.

Keywords: Chiral Physics, QCD

1. Introduction: Brief History of this Workshop Series and an Introduction to Chiral Dynamics

This workshop is the fifth of a series dedicated to Chiral Dynamics: Theory and Experiment that Barry Holstein and I started at MIT in 1994.[1] At that time, the theory was generally far ahead of experiments, and we decided that it was important to start serious discussions in which the experimental aspects of the field would be treated on an equal footing with the theory. We also decided that it should be a workshop (not a conference) allowing plenty of time for active discussion. The format of having plenary talks in the mornings and working groups in the afternoons was established. Our goal was to evaluate what had happened in the field in the last few years and to discuss where we should go in the near future. In the early period of the workshops we decided to have them every three years and to alternate between Europe and the U.S. We have now had workshops in Mainz (1997[2]), Jefferson Lab (2000[3]), Bonn (2003[4]), and now at Duke. I am very proud that this series has become an important measure of the progress of this field.

*E-mail: bernstei@lns.mit.edu

As the first speaker of this workshop I thought that it was important to introduce some of the terminology and concepts of chiral dynamics and the experiments that test its theoretical predictions. As is well known the coupling between quarks in QCD increases at low energies (for an introduction to QCD and chiral dynamics see e.g.[5]) which leads to confinement. This has the consequence that normal perturbation theory does not work at low energies. However there is a low energy effective theory known as chiral perturbation theory(ChPT).[5–7] The QCD Lagrangian can be written as a sum of two terms, L_0 which is independent of the light quark masses (up, down, strange) and L_m which contains the masses of the three light quarks. Consider the chiral limit in which the three light quark masses $m_q \rightarrow 0$. The solutions to the Dirac equation for massless quarks have a definite chirality or equivalently helicity $= \hat{\sigma} \cdot \hat{p} = \pm 1$. The terminology is that when the quark spin $\hat{\sigma}$ is parallel (anti-parallel) to the momentum vector $\hat{p}$, the quark is labeled as right(left) handed. For massless quarks, the left and right handed solutions are independent and this is known as chiral symmetry. Another language to express this is that there are separate conservation laws for vector (left +right) currents and axial vector(left- right) currents.[5] As is well known, the vector current is conserved while the axial vector current is conserved only in the chiral limit (i.e. $m_q \rightarrow 0$) and slightly non-conserved in the real world. This is one of the approximate symmetries of QCD which was earlier known as current algebra/PCAC(partially conserved axial currents).[5]

Despite the fact that the light quark mass independent part of the QCD Lagrangian L_0, has chiral symmetry, matter does not seem to obey the rules. The chiral symmetry is expected to show up by the parity doubling of all hadronic states, i.e., the proton with $j^p = 1/2^+$ would have a $1/2^-$ partner(the Wigner-Weyl manifestation of the symmetry). Clearly this is not the case. This indicates that the symmetry is spontaneously hidden (often stated as spontaneously broken) and is manifested in the Nambu-Goldstone mode; parity violation occurs through the appearance of a massless pseudo scalar (0^-) meson. The opposite parity partner of the proton is a proton plus a "massless pion".

There are profound consequences of spontaneous chiral symmetry hiding which are subject to experimental tests. In the SU(2) version of the picture the up and down quark masses are considered small and the three π mesons are the Nambu-Goldstone Bosons. In the SU(3) version the up, down, and strange quark masses are considered small and there are eight Nambu- Goldstone Bosons ($\pi^{\pm,0}, \eta, K^{0,\pm}, \bar{K^0}$). Below the chiral symmetry breaking scale $\Lambda_x \simeq 1$ GeV all of the lightest hadrons are Nambu-Goldstone Bosons with the pions $\simeq$ 140 MeV, the $\eta \simeq$ 547 MeV and the Kaons $\simeq$ 496 MeV.[8] Clearly these masses are not zero, due to the explicit chiral symmetry breaking term L_m in the QCD Lagrangian. The lowest order estimate for the pion mass $m_\pi^2 = B_0(m_u + m_d)$ (Gell-Mann, Oakes, Renner relation) where B_0 is proportional to the scalar quark vacuum condensate $< 0 \mid \bar{q}q \mid 0 >$ and q represents up or down quarks. This is an order parameter of QCD which mixes left and right handed states. From this formula one can see that $m_\pi \rightarrow 0$ in the chiral limit $m_q \rightarrow 0$. This formula for m_π represents the strong but not the electromagnetic interaction contribution. The experimentally observed mass difference $m_{\pi^\pm} - m_{\pi^0} = 4.59$ MeV[8] is almost purely electromagnetic in origin. Since the lightest hadron is the π^0 meson its primary decay mode is $\pi^0 \rightarrow \gamma\gamma$. This allows us to perform a precision test of the predictions of the QCD axial anomaly which is the principal mechanism for this decay[5] (see Sec.4).

The formulas for the masses of the K and η mesons contain the strange quark mass m_s. From the masses of these mesons (subtracting the electromagnetic contribution) the ratios of the light quark masses can be accurately obtained. If in addition QCD sum rules are also invoked the absolute values of $m_u = 5.1 \pm 0.9, m_d = 9.3 \pm 1.4, m_s = 175 \pm 25$ MeV have been obtained.[9] Since the quarks are confined these absolute values are quoted in the $\bar{M}\bar{S}$ scheme at a scale of 1 GeV[a] . It should be mentioned that there are other empirical and lattice values in the literature (see[8] for a summary) indicating that the absolute values might change in the future. However the ratios are more accurately determined and are less likely to change significantly. The large strange quark mass compared to the up and down quark comes from the fact that the kaon and η are much more massive than the pion. Physically this means that the most accurate tests of chiral dynamics are in the pion sector. The fact that the up and down quark masses differ by almost a factor of two means that there is strong isospin (SU(2)) breaking in addition to electromagnetic effects.[10] However since both of these masses are small the magnitude of this will be typically $\leq 1\%$[11] [Meissner][b]. The pseudoscalar η' meson with a mass of $\simeq$ 958 MeV is very interesting. In the large N_c(number of colors) limit it would be the ninth Nambu-Goldstone

[a]In QCD the coupling constant α_s and quark masses are scale and renormalization scheme dependent (see e.g. the QCD section of the particle data book[8]). It is customary to use the "modified minimum subtraction scheme" $\bar{M}\bar{S}$, to quote the value of α_S at the mass of the Z meson, and the quark masses at a scale of 1 or 2 GeV.

[b]references to talks at this meeting will be presented in brackets

Boson. Its large mass is due to the QCD axial anomaly[5] and therefore has a large component which does not vanish in the chiral limit.

The effective field theory that utilizes the concept of spontaneously hidden chiral symmetry is called chiral perturbation theory (ChPT)[c]. This is an effective (low energy) theory of QCD in which the quark and gluon fields are replaced by a set of fields $U(x)$ describing the degrees of freedom of the observed hadrons. For the Nambu-Goldstone Boson sector this is usually taken to be of the non-linear exponential form $U(x) = exp[i\phi(x)/F_\pi]$ where F_π is the pion decay constant $\simeq$ 92 MeV and ϕ represents the Nambu-Goldstone fields, a 2×2 matrix for the pion fields if we assume only that the up and down quarks are active, and a 3×3 matrix representing pion, η, and kaon fields when we take the strange quark into account. The Lagrangian of QCD is replaced by an effective Lagrangian, which only involves the field $U(x)$, and even powers of its derivatives $L_{QCD} \to L_{eff}(U, \partial U, \partial^2 U, \ldots) = L^2_{eff} + L^4_{eff} + L^6_{eff} + \ldots$ where the superscript n on L^n_{eff} represents the number of derivatives. The form of the terms are fully determined by the requirements of chiral symmetry. However the magnitudes of the terms are not determined by the symmetries and must be determined empirically or on the lattice. These are in reasonable agreement with model estimates[7,14]which shows that the physics is understood. For the SU(2) case the lowest (n=2) term is: $L^2_{eff} = (F^2_\pi/4)[\mathrm{tr}\{\partial_\mu U^+ \partial^\mu U\} + m^2_\pi \mathrm{tr}\{U + U^+\}]$ which contains the well known pion decay constant and mass. The derivative term predicts that Nambu-Goldstone Bosons are emitted and absorbed in p waves and have no interaction as the momentum $\to$ 0 in accordance with Goldstone's theorem. The mass term explicitly breaks chiral symmetry and causes a small interaction at zero momentum.

ChPT represents a systematic expansion with definite counting rules which are governed by the order of which one chooses to work. The predictions are expansions in the Nambu-Goldstone Boson masses and momenta. To converge they must be small compared to the chiral symmetry breaking scale $\Lambda_x = 4\pi F_\pi \simeq 1$ GeV. They must also be in a region below any resonances or branch cuts. In πN scattering the energies must be significantly below the Δ resonance unless it is included as a dynamical degree of freedom in the calculations [Pascalutsa]. In their domain of validity they represent the predictions of QCD subject to the errors which are imposed by uncertainties in the low energy constants and by the neglect of higher order terms. As such they are worthy of great experimental effort in or-

---

[c]For an introduction to ChPT see[5,12] and for more complete reviews see[1–3,13,14]

der to check them. Any discrepancy which is significantly larger than the combined experimental and theoretical errors must be taken seriously!

2. Chiral Dynamics Phenomena and Experiments

Following this brief introduction to the basic ideas of spontaneous chiral symmetry hiding in QCD and ChPT the phenomena that are associated with this subject and the associated experimental possibilities will be outlined.

- Nambu-Goldstone Bosons at Low Energies
  interactions: $\pi - \pi, \pi - K, \pi - \eta, \pi - \eta^{'}, K - \eta, \ldots$
  properties: RMS radii, polarizabilities
  electromagnetic and hadronic decays: $\pi^0, \eta, \eta^{'} \to \gamma\gamma$,
  $\gamma\gamma \to \pi\pi, \eta\eta, \eta \to \pi\pi\gamma, \eta \to 3\pi, \ldots$
  leptonic and semi-leptonic decays: $\pi, K \to e\nu\gamma, K^+ \to \pi^+ l\nu_l \ldots$
- Nambu-Goldstone Boson-hadron scattering: $\pi - N, K - N, \ldots$
- photo and electro-production of Nambu-Goldstone Bosons:
  $\gamma^* N \to \pi N, K\Lambda, \gamma^*\pi \to \pi\pi \ldots$
- Hadron structure at low Q^2
  Nucleon EM,axial, strange form factors:
  RMS radii, magnetic moments
  quadrupole amplitudes in $\gamma^* N \to \Delta$
  Electric and magnetic polarizabilities, $\pi N - \sigma$ term
- long range part of N-N interaction. nuclear physics at low energies, nuclear astrophysics

This incomplete list shows the broad range of phenomena included in chiral dynamics. In this short presentation it is only possible to discuss a small fraction of these topics. The most pristine testing grounds for chiral dynamics is in the Nambu-Goldstone Boson section and $\pi - \pi$ scattering is the best case, both theoretically and experimentally. However experiments are difficult since precision is vital and there are no free pion targets. The best experimental method has been to study the final state interactions in $K \to \pi\pi l\nu$ and more recently by the cusp in $K \to 3\pi$. Since this subject was extensively covered in this workshop[Leutwyler, Goy-Lopez, Pelaez, Bedaque] I will say only that it is still very interesting and evolving. More open is the study of $\pi - K$ interactions for which values of the scattering lengths have been extracted from higher energy data using dispersion theory[15] based on data from the KN $\to \pi$ KN reaction. Lower energy data has been taken in the decay $D^+ \to \pi^- K^+ \mu^+ \nu$ from the FOCUS collaboration at Fermilab,[16] but have not yet been incorporated into this analysis. As far as I know, there are no data on the interactions with the η or η' ; the latter is particularly interesting since it is a Nambu-Goldstone Boson in the large N_c limit and it is not clear how different its dynamics (e.g. scattering lengths) will be from, say, the η.

The polarizability of the pion has proven to be difficult to measure (for a summary see[17] [Walcher, Kashevarov]). A recent experiment at Mainz[18] is in serious disagreement with the result extracted from $e^+e^- \rightarrow \gamma\gamma \rightarrow \pi^+\pi^-$ experiments.[19] In my view this is a potentially serious situation which urgently needs additional experimental effort. I think that an improvement in the accuracy of the e^+e^- experiments is needed since the sensitivity to the pion polarizability is not large. It would also be highly desirable to perform a modern radiative Primakoff experiment ($\pi^{\pm}\gamma \rightarrow \pi^{\pm}\gamma$ with pion beams.

The phenomena that are associated with the πN system are at the heart of nuclear physics. The π meson has a special role in the universe. The exchange of pions between nucleons (Yukawa interaction) is the long range part of the nucleon-nucleon potential, and governs low energy nuclear interactions and stellar formation. Indeed the effective theory of few nucleon systems[Nogga, Hahnhart] has become an integral part of chiral dynamics and of these workshops. The pion cloud which surrounds hadrons plays a major role in their structure, e.g. their form factors and polarizeabilities[McGovern, Miskimen]. The Δ resonance (the first excited state of the nucleon) plays a central role in πN dynamics, a topic discussed below.

The Nambu-Goldstone Boson- Fermion sector is more difficult theoretically (compared to the purely Nambu-Goldstone Boson physics) but parts of it, e.g. πN physics, including low energy electromagnetic meson production, are much more easily accessible experimentally which has lead to extensive and accurate experimental data. This does not mean that all problems are solved. For example there is an old open problem that was not even discussed at this workshop, namely the value of the $\pi N - \sigma$ term. Generally its interpretation leads to a greater than 20% contribution of the strange quark to the nucleon mass. Yet, as we learn from parity violating electron scattering [Michaels], the strangeness magnetic moment is close to zero. There is no simple connection between these quantities since they are the expectation values of different operators. However, the fact that one is large and the other is small needs explanation.

3. Pion-Nucleon Interactions and Electromagnetic Pion Production

In this and the next section I shall give a few specific examples to illustrate some of the applications of chiral dynamics. From the large number of possibilities I have chosen ones that I have worked on, and are close to my heart.

The πN interaction in momentum space $= g_{\pi N}\sigma \cdot \mathbf{p}_\pi$ where σ is the nucleon spin. In accordance with Goldstone's theorem, this interaction $\to 0$ as the pion momentum $\to 0$. Furthermore $g_{\pi N}$ can be computed from the Goldberger-Treiman relation[5] and chiral corrections,[20] and is accurate to the few % level. The πN interaction is very weak in the s wave and strong in the p wave which leads to the Δ resonance, the tensor force between nucleons, and to long range non-spherical virtual pionic contributions to hadronic structure. For illustrative purposes consider the lowest order ChPT calculation $O(p^2)$ for $a(\pi, h)$, the s wave π hadron scattering length; $a^I(\pi, h) = -\mathbf{I}_\pi \cdot \mathbf{I_h} m_\pi/(\Lambda_x F_\pi)$ where $\mathbf{I} = \mathbf{I}_\pi + \mathbf{I_h}$ is the total isospin, and I_π, and I_h are the isospin of the pion and hadron respectively, F_π is the pion decay constant, and $\Lambda_x = 4\pi F_\pi \simeq 1$ GeV is the chiral symmetry breaking scale.[21] Note that $a(\pi, h) \to 0$ in the chiral limit $m_\pi \to 0$ as it must to obey Goldstone's theorem. Also note that $a(\pi, h) \simeq 1/\Lambda_x \simeq 0.1$ fm, which is small compared to a typical strong interaction scattering length of $\simeq 1$ fm. This small scattering length is obtained from the explicit chiral symmetry breaking due to the finite quark masses. The predictions of ChPT for πN scattering lengths have been verified in detail in a beautiful series of experiments on pionic hydrogen and deuterium at PSI;[22] this includes the isospin breaking due to the difference in $m_d - m_u$ mentioned previously [Meissner].

Low energy electromagnetic production of Goldstone Bosons is as fundamental as Goldstone Boson scattering for two reasons: 1) the production amplitudes vanish in the chiral limit (as in scattering); and 2) the phase of the production amplitude is linked to scattering in the final state by unitarity or final state interaction (Fermi- Watson) theorem suitably modified to take the up, down quark masses into account.[23] First consider the low energy limit of the electric dipole E_{0+} for s wave photo-pion production:[24]

$$\begin{aligned} E_{0+}(\gamma p \to \pi^0 p) &= -D_0 \mu (1 + O(\mu) + ..) \to 0 \\ E_{0+}(\gamma p \to \pi^+ n) &= \sqrt{2} D_0/(1 + \mu + ...)^{3/2} \to \sqrt{2} D_0 \\ \mu &= m_\pi/M \to 0 \\ D_0 = e \cdot g_{\pi N}/8\pi M &= 24 \cdot 10^{-3}(1/m_\pi) \end{aligned} \tag{1}$$

where M is the nucleon mass and the right arrow denotes the chiral limit ($m_u, m_d, m_\pi \rightarrow 0$). Eq. 1 shows that for neutral pion production the amplitude vanishes in the chiral limit. For charged pion production, there is a different low energy theorem.[24] Therefore the amplitude that is most sensitive to explicit chiral symmetry breaking is neutral pion production and most of the modern experiments have concentrated on this channel. In general, ChPT to one loop calculated in the heavy Fermion approximation, has been highly successful in calculating the observed cross sections and linearly polarized photon asymmetry.[24]

The application of these ideas to data from low energy πN scattering and electromagnetic pion production from the nucleon is instructive. The left panel of Fig.1 shows the shape of the Δ resonance from fits to the total cross sections for $\pi^+ p$ (scaled) scattering and and for the $\gamma p \rightarrow \pi^0 p, \pi^+ n$ reactions versus W (the center of mass energy).[25] All of the these reactions have a strong Δ resonance. The $\pi^+ p$ and $\gamma p \rightarrow \pi^0 p$ reactions have small cross sections near threshold and therefore clearly show the Δ resonance without any interference (the small shift between them is due to the mass difference of the Δ^0 and Δ^+). Indeed these cross sections are text book example of an isolated resonance. Although not usually mentioned in text books it is the combination of a strong resonance and a small cross section at threshold that produces this beautiful example (as predicted by chiral dynamics)! In the case of the $\gamma p \rightarrow \pi^+ n$ reaction there is strong s wave production starting at threshold due to the Kroll-Ruderman low energy theorem (see Eq.1. In this case the Δ resonance curve is superimposed on the strong s wave amplitude and looks quite different!

The photo and electro-pion $\gamma^* p \rightarrow \Delta$ reactions have been extensively used to study non-spherical amplitudes (shape) in the nucleon and Δ structure.[28] This is studied by measuring the electric and Coulomb quadrupole amplitudes (E2,C2) in the predominantly magnetic dipole, quark spin flip (M1) amplitude. At low Q^2 the non-spherical pion cloud is a major contributor to this (for a review see[28]). Recently there have been chiral calculations of this process [Pascalutsa]. The right panel of Fig.1 shows our best estimate of the difference between the electro-excitation Δ for the spherical case(the relatively flat grey band) and the fit to the Bates data for the transverse-longitudinal interference cross section σ_{LT}[26] which shows the C2 magnitude which is primarily due to the pion cloud [Stave].[27,28] The evolution of the Coulomb quadrupole amplitude with Q^2 indicates that the quark models do not agree with experiment, but that models with pionic degrees of freedom do, demonstrating that the crucial ingredient in the

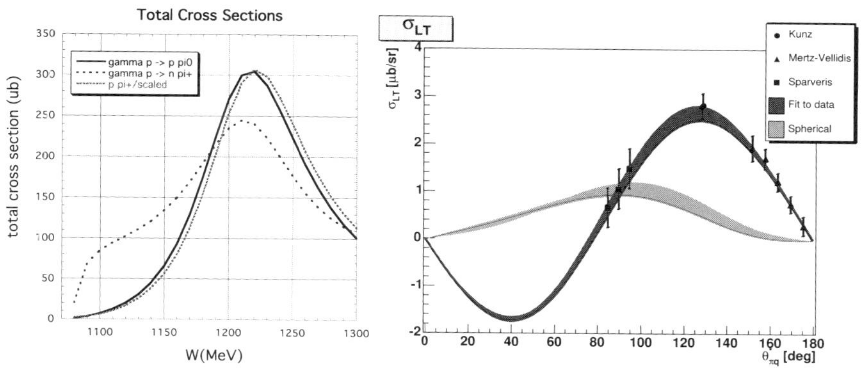

Fig. 1. Left panel: The shape of the Δ resonance from fits to the total cross sections for π^+p(scaled solid curve) scattering and and for the $\gamma p \longrightarrow \pi^0 p$(short red dots), and $\gamma p \longrightarrow \pi^+ n$ reactions(blue dotted curve which is larger at low W) versus W, the center of mass energy.the curves represent fits to the data.[25] Right panel: σ_{LT} from the $ep \longrightarrow e' \pi^0 p$ reaction at $Q^2 = 0.126 GeV^2$, W=1232 MeV(see text). The blue curve is fit to the data[26] and the relatively flat grey curve shows the calculation for the spherical case, i.e. when the quadrupole transition amplitudes are set to zero.[27,28]

non-spherical amplitude at long range is the pion cloud[Stave].

A great deal of effort has gone into the study of the near threshold $\gamma p \longrightarrow \pi^0 p$ reaction experimentally at Mainz[29] and Saskatoon[30] and with ChPT calculations.[24] In addition we are planning to conduct future experiments at HIγS, a new photon source being constructed at Duke[Weller]. These experiments will have full photon and target polarization and will be a significant extension of the results we have at present. The unpolarized cross sections were accurately measured despite their small size and the results from Mainz and Saskatoon are in reasonable agreement. The p wave amplitudes tend to dominate even close to threshold. The real part of the s wave electric dipole amplitude ReE_{0+} is extracted from the data using the interference between s and p waves which goes as $cos(\theta_\pi)$ in the differential cross section and leads to larger errors. The results for ReE_{0+} versus photon energy are plotted in the left panel of Fig. 2. There is reasonable agreement between the Mainz and Saskatoon points as well as with ChPT[24] and the unitary model calculations.[23] The sharp downturn in ReE_{0+} between the threshold at 144.7 MeV and the π^+ n threshold at 151.4 MeV is due to a unitary cusp caused by the interference between the $\gamma p \longrightarrow \pi^0 p$ and $\gamma p \longrightarrow \pi^+ n$ channels. The magnitude of the cusp is $\beta = ReE_{0+}(\gamma p \longrightarrow \pi^+ n) \cdot a_{cex}(\pi^+ n \longrightarrow \pi^0 p)$ which is measured to an accuracy of $\simeq 30\%$ from the data shown. The reason for this accuracy limitation is due to the fact that the ReE_{0+} is a sum of a smooth and cusp functions and the smooth function is not known precisely.[23] Therefore it is important to measure ImE_{0+} which starts from close to zero at the π^+n threshold energy and rises rapidly as βp_{π^+}. This makes the extraction of β as accurate as the measured asymmetry for π^0 photoproduction from a polarized target normal to the reaction plane. The estimated error for such an experiment running at HIγS for $\simeq$ 400 hours of anticipated operation of the accelerator is presented in the right panel of Fig.2. This experiment, along with an independent measurement of the $\gamma p \to \pi^+ n$ cross section will allow us to extract β at the few % level and measure the charge exchange scattering length $a_{cex}(\pi^+ n \to \pi^0 p)$ for the first time. We will be able to compare this to the measured value of $a_{cex}(\pi^- p \to \pi^0 n)$[22] as an isospin conservation test. This illustrates the power of photopion reaction studies with transversely polarized targets to measure πN phase shifts in completely neutral charge channels which are not accessible to pion beam experiments! This is potentially valuable to help pin down experimentally the value of the $\pi N - \sigma$ term which has had a long, difficult measurement history.

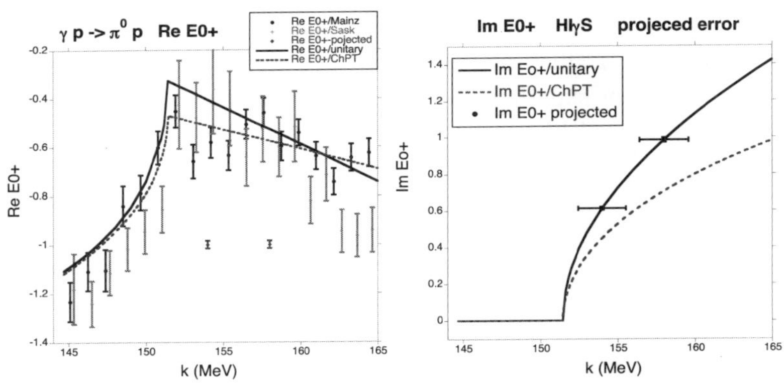

Fig. 2. The $\gamma p \to \pi^0 p$ Reaction. Left panel: ReE_{0+} versus photon energy. The data points are from Mainz[29] and Saskatoon.[30] The curves are from ChPT[24] and a unitary fit to the data.[23] The two projected points from HIγS are plotted at an arbitrary value (ReE_{0+} = -1) to show the anticipated errors . Right panel: Im E_{0+} versus photon energy. The curves are the same as in the left panel and the projected HIγS points are arbitrarily plotted on the unitary curve. There are no experimental points(see text).

ChPT has been extremely successful in predicting the cross sections and the linearly polarized photon asymmetry in the $\gamma p \to \pi^0 p$ reaction. However

I would like to point out a significant discrepancy with the $ep \rightarrow e^{'}p\pi^0$ reaction data at $Q^2 = 0.05GeV^2$ taken at Mainz[31] shown in Fig.3. It can be seen that the ChPT calculations[32] do not agree with the data although the DMT dynamical model does.[33] This discrepancy is a potentially serious problem which needs to be resolved!

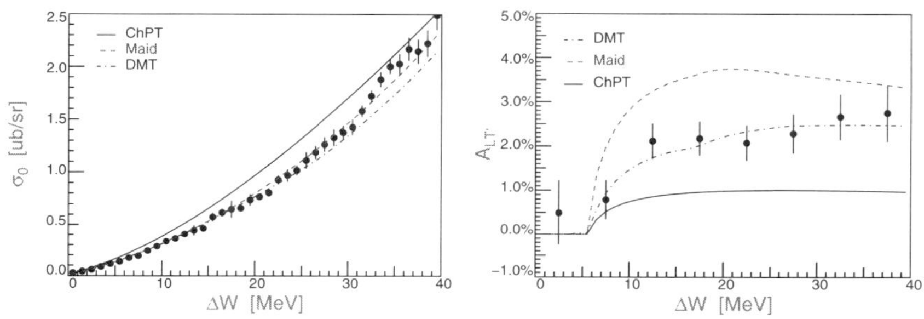

Fig. 3. Cross Section(Left panel) and $LT^{'}$ Asymmetrry (right panel) for the $ep \rightarrow e^{'}\pi^0 p$ Reaction at $Q^2 = 0.05GeV^2$ versus dW, the center of mass energy above threshold.[31] See text for discussion.

4. The $\pi^0 \rightarrow \gamma\gamma$ Decay Width and the QCD Axial Anomaly

As the final special topic I would like to discuss a test of the axial anomaly by an accurate measurement of the π^0 lifetime. As was discussed in the introduction, due to the spontaneous breaking of chiral symmetry, the π^0 is the lightest hadron and its primary decay mode is $\pi^0 \rightarrow \gamma\gamma$. This decay rate is exactly predicted in the chiral limit by the QCD axial anomaly. As is quoted in most textbooks on QCD (see e.g.[5]) this prediction is in agreement with the average in the particle data book[8] which has a error $\simeq$ 10%. However this oversimplifies the experimental situation which is shown in Fig.4. In my opinion, almost all of the errors quoted in the literature are underestimates. This is indicated by the spread in the experimental values. Also at issue are the chiral corrections to the decay rate. These have been worked out to next to leading order. They primarily involve $\pi - \eta, \eta^{'}$ mixing which is isospin breaking and therefore proportional to $m_d - m_u$; they increase the predicted decay width by 4± 1%.[34]

This experiment has been performed by the Primex collaboration at JLab and the data analysis is in the final stages[McNulty]. The experiment measures the photo-production of π^0 mesons from C and Pb at an average photon energy $\simeq$ 5.2 GeV and has the goal of achieving an accuracy of a few %. It is the first Primakoff measurement to use a tagged photon beam. Prelilminary results are shown in Fig.4. The large forward peak is due to the Primakoff effect which is the production of π^0's in the Coulomb field of the target. The larger peak at a few degrees in C is due to coherent nuclear production and there is a small quantum interference amplitude. The fits to these processes are shown. For the Pb target (not shown) the nuclear coherent peak is small compared to the Primakoff peak. This is due to final state absorption causing the coherent nuclear cross section to scale $\simeq$ A, while the Primakoff cross section scales $\simeq Z^2$. Therefore the relative coherent to Primakoff peak decreases with heavier targets. As can be seen the preliminary data look good and our collaboration expects to release preliminary lifetime results in the next half year.

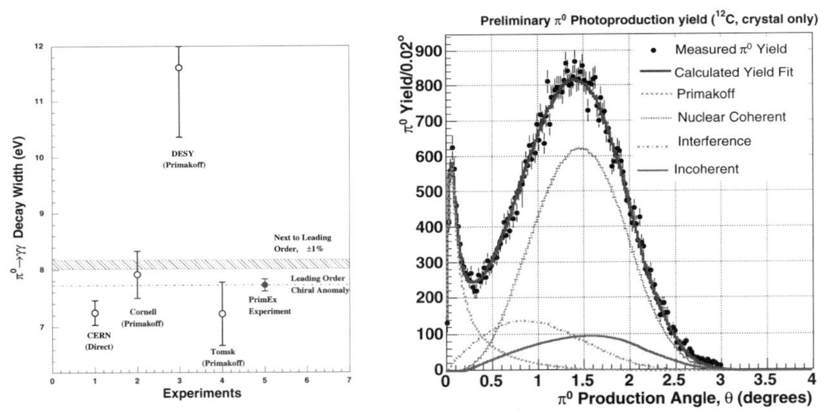

Fig. 4. Left panel: the $\pi^0 \rightarrow \gamma\gamma$ decay width in eV. These include the experimental points,[8] the projected error of the Primex collaboration (arbitrarily plotted to agree with the predictions of the axial anomaly), and the next to leading order chiral correction.[34] Right panel: Preliminary yield and fit for the Primakoff effect on Carbon versus pion angle with the individual contributions to the total yield exhibited (see text for discussion).

5. Conclusions

It is clear that chiral physics includes an impressive array of reactions and particle properties. The general concepts and detailed calculations of ChPT are generally verified for the Nambu-Goldstone meson sector where they have been tested. The physics is more complicated when Fermions are also included. This sector is more closely related to the observable world, e.g. the properties of nucleons, and the formation of the stars. In this sector there are many areas of good agreement between ChPT and experiment but there are also a few outstanding problems

Finally here is a short wish list that I would like to see granted by CD2009:

- measurements of the pion polarizabilities by different techniques to address the existing discrepancy
- further calculations and measurements of the $ep \to e^{'} p\pi^0$ reaction to resolve the existing discrepancy
- accurate measurements of the decay $\pi^0, \eta, \eta^{'} \to \gamma\gamma$ decay widths
- further isospin tests in the πN scattering and $\gamma N \to \pi N$
- progress on the $\pi N - \sigma$ term
- further calculations and experiments on the nature of $\eta^{'}$ Boson; what does it mean physically that this is a Nambu-Goldstone Boson in the large N_c limit?
- more data on the interactions between the heavier Nambu-Goldstone Bosons

It was a pleasure for me to experience this workshop that was so stimulating and well organized. Personally I learned a great deal in a collegial and enthusiastic atmosphere. I would like to thank that organizers and Duke University for their wonderful organization and hospitality. I would also like to thank M. Kohl and D. McNulty for their careful reading of this manuscript. This work has been supported in part by the U.S. Department of Energy under Grant No. DEFG02-94ER40818.

References

1. A. M. Bernstein and B. R. Holstein (eds.), *Chiral dynamics: Theory and experiment. Proceedings of the MIT Workshop, Cambridge, MA, USA, July 25-29, 1994* (Springer Lecture notes in physics, Vol. 452).
2. A. M. Bernstein, D. Drechsel and T. Walcher (eds.), *Chiral dynamics: Theory and experiment. Proceedings, Workshop, Mainz, Germany, September 1-5, 1997* (Springer Lecture notes in physics, Vol. 513).
3. A. M. Bernstein, J. L. Goity and U. G. Meissner (eds.), *Chiral dynamics: Theory and experiment. Proceedings, 3rd Workshop, Newport News, USA, July 17-22, 2000* (World Scientific, 2001).
4. U. G. Meissner, Fourth Workshop on Chiral Dynamics - Chiral Dynamics 2003: Theory and Experiment, Bonn, Germany, Sept. 8-13,2003, http://www.itkp.uni-bonn.de/ cd2003/.

5. J. F. Donoghue, E. Golowich and B. R. Holstein, *Dynamics of the standard model* (Camb. Monogr. Part. Phys. Nucl. Phys. Cosmol., 1992).
6. S. Weinberg, *Physica* **A96**, p. 327 (1979).
7. J. Gasser and H. Leutwyler, Annals Phys.,158, 142 (1984) , Phys. Lett. B, 125, 321 (1983), Phys. Lett. B,125, 325 (1983).
8. W. M. Yao *et al.*, *J. Phys.* **G33**, 1 (2006).
9. H. Leutwyler, hep-ph/9609467; J. Gasser and H.Leutwyler, Phys. Rept.87,77(1982).
10. S. Weinberg, *Trans. New York Acad. Sci.* **38**, 185 (1977).
11. B. Kubis and U.-G. Meissner, Phys. Lett. B529, 69(2002); Fettes, Nadia and Meissner, Ulf-G, Phys. Rev.C63, 045201(2001).
12. H. Leutwyler, hep-ph/9409423 (article in[1]).
13. H. Leutwyler, hep-ph/0008124, hep-ph/9409422.
14. S. Scherer, Adv. Nucl. Phys.,27(2003),G. Ecker, Prog. Part. Nucl. Phys. **35**, 1 (1995); A. Pich, Rept. Prog. Phys. **58**, 563 (1995), U. G. Meissner, Rept. Prog. Phys. **56**, 903 (1993).
15. P. Buettiker, S. Descotes-Genon and B. Moussallam, *Eur. Phys. J.* **C33**, 409 (2004).
16. J. M. Link *et al.*, *Phys. Lett.* **B535**, 43 (2002).
17. L. V. Fil'kov and V. L. Kashevarov, *Phys. Rev.* **C73**, p. 035210 (2006).
18. J. Ahrens *et al.*, *Eur. Phys. J.* **A23**, 113 (2005).
19. J. Gasser, M. A. Ivanov and M. E. Sainio, *Nucl. Phys.* **B745**, 84 (2006).
20. J. L. Goity, R. Lewis, M. Schvellinger and L.-Z. Zhang, *Phys. Lett.* **B454**, 115 (1999).
21. S. Weinberg, *Phys. Rev. Lett.* **17**, 616 (1966).
22. D. Gotta, *Int. J. Mod. Phys.* **A20**, 349 (2005).
23. A. M. Bernstein, *Phys. Lett.* **B442**, 20 (1998).
24. V. Bernard, N. Kaiser and U.-G. Meissner, Eur. Phys. J., A11, 209(2001); Z. Phys.,C70, 483(1996); Phys. Lett.B383,116(1996)
25. R. A. Arndt, W. J. Briscoe, I. I. Strakovsky and R. L. Workman, Phys. Rev. C74, 045205(2006),nucl-th/0607017 , http://gwdac.phys.gwu.edu/.
26. N. F. Sparveris *et al.*, *Phys. Rev. Lett.* **94**, p. 022003 (2005).
27. C.N.Papanicolas, Eur. Phys. J. A, 18, 141 (2003).
28. C. Papanicolas and A. Bernstein, Shape of Hadrons Workshop, Athens, Greece(2006); http://microtron.iasa.gr/hadrons/index.html, to be published.
29. A. Schmidt *et al.*, *Phys. Rev. Lett.* **87**, p. 232501 (2001).
30. J. C. Bergstrom *et al.*, *Phys. Rev.* **C53**, 1052 (1996).
31. M. Weiss, Ph.D. thesis, Johannes Gutenberg-univ. Mainz (2003).
32. V. Bernard, N. Kaiser and U.-G. Meissner, Nucl. Phys.A607, 379(1996); Phys. Rev. Lett., 74, 3752(1995).
33. S. S. Kamalov, G.-Y. Chen, S.-N. Yang, D. Drechsel and L. Tiator, *Phys. Lett.* **B522**, 27 (2001).
34. J. L. Goity, A. M. Bernstein and B. R. Holstein, Phys. Rev.D66, 076014(2002); B. Ananthanarayan and B. Moussallam, JHEP **0205**, 052 (2002).

$\pi\pi$ SCATTERING

H. LEUTWYLER
Institute for Theoretical Physics, University of Bern,
Sidlerstr. 5, CH-3012 Bern, Switzerland
**E-mail: leutwyler@itp.unibe.ch*

Recent work in low energy pion physics is reviewed. One of the exciting new developments in this field is that simulations of QCD on a lattice now start providing information about the low energy structure of the continuum theory, for physical values of the quark masses. Although the various sources of systematic error yet need to be explored more thoroughly, the results obtained for the correlation function of the axial current with the quantum numbers of the pion already have important implications for the effective Lagrangian of QCD. The consequences for $\pi\pi$ scattering are discussed in some detail. The second part of the report briefly reviews recent developments in the dispersion theory of the scattering amplitude. One of the important results here is that the position of the lowest resonances of QCD can now be determined in a model independent manner and rather precisely. Beyond any doubt, the partial wave with $I = \ell = 0$ contains a pole on the second sheet, not far from the threshold: the lowest resonance of QCD carries the quantum numbers of the vacuum.

1. Introduction

The pions are the lightest hadrons and we know why they are so light. Since the underlying approximate symmetry also determines their basic properties at low energy, the interaction among the pions is understood very well. In fact, in the threshold region, the $\pi\pi$ scattering amplitude is now known to an amazing degree of accuracy.[1] In particular, we know how to calculate the mass and width of the lowest resonance of QCD in a model independent manner.[2] The actual uncertainty in the pole position is smaller than the estimate given in the 2006 edition of the Review of Particle Physics,[3] by more than an order of magnitude.

The progress made in this field heavily relies on the fact that the dispersion theory of $\pi\pi$ scattering is particularly simple: the s-, t- and u-channels represent the same physical process. As a consequence, the scattering amplitude can be represented as a dispersion integral over the imaginary part and the integral exclusively extends over the physical region.[4] Throughout the following, I work in the isospin limit ($m_u = m_d = m$ and $e = 0$), where the representation involves only two subtraction constants.[a] These may be identified with the S-wave scattering lengths a_0^0, a_0^2. The projection of the amplitude on the partial waves leads to a dispersive representation for these, the Roy equations. For a thorough discussion, I refer to ACGL.[5]

The pioneering work on the physics of the Roy equations was carried out more than 30 years ago.[6] The main problem encountered at that time was that the two subtraction constants occurring in these equations were not known: if the values of a_0^0, a_0^2 are contained in the so-called universal band – the region spanned by the two thick lines in figure 1 below – the Roy equations admit a solution. Since the data available at the time were consistent with a very broad range of S-wave scattering lengths, the Roy equation analysis was not conclusive.

2. Low energy theorems for the S-wave scattering lengths

The insights gained by means of chiral perturbation theory (χPT) thoroughly changed the situation.[7] The corrections to Weinberg's low energy theorems[8] for a_0^0, a_0^2 (left dot in figure 1) have been worked out to first non-leading order[9] (middle dot) and those of next-to-next-to leading order are also known[10] (dot on the right). As demonstrated in CGL,[1] the chiral perturbation series converges particularly rapidly near the center of the Mandelstam triangle, so that very accurate predictions for the scattering lengths are obtained by matching the chiral and dispersive representations there. Using this method, the low energy theorems for a_0^0, a_0^2 may be brought to the form

$$2a_0^0 + 7a_0^2\alpha = \alpha - 6\pi\left(\frac{M_\pi}{4\pi F_\pi}\right)^4\left(\bar{\ell}_3 - \frac{193}{210}\right) + M_\pi^4\alpha_3 + O(M_\pi^6)\,, \qquad \text{(1a)}$$

$$2a_0^0 - 5a_0^2\alpha = \alpha\frac{3M_\pi^2}{4\pi F_\pi^2} + 24\pi\left(\frac{M_\pi}{4\pi F_\pi}\right)^4\left(\bar{\ell}_4 - \frac{887}{840}\right) + M_\pi^4\alpha_4 + O(M_\pi^6)\,. \quad \text{(1b)}$$

While the chiral expansion of a_0^0, a_0^2 starts at order M_π^2, with a term that is fixed by the pion decay constant, the combination $2a_0^0 + 7a_0^2$ vanishes at leading order. Two types of corrections occur at $O(M_\pi^4)$: a contribution

[a] The value used for the basic QCD parameters in this theoretical limit is a matter of convention. It is convenient to choose Λ_{QCD} and m such that F_π and M_π agree with the observed values of F_{π^+} and M_{π^+}. The result quoted by the PDG[3] corresponds to $F_{\pi^+} = 92.42 \pm 0.10 \pm 0.26$ MeV. The scattering lengths are given in units of M_{π^+}.

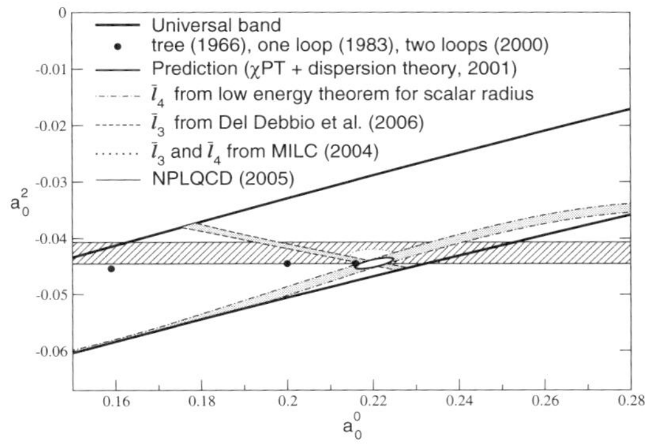

Fig. 1. S wave scattering lengths: theoretical results.

from the coupling constants $\bar{\ell}_3, \bar{\ell}_4$ of the effective Lagrangian and one involving dispersion integrals, which I denote by $M_\pi^4\alpha_3$ and $M_\pi^4\alpha_4$, respectively[b] (these also contain higher order contributions).

A representation similar to (1) was given earlier,[9] but there, the dispersion integrals $M_\pi^4\alpha_3, M_\pi^4\alpha_4$ were evaluated at leading order, where they can be expressed in terms of the D-wave scattering lengths. The virtue of the above form of the low energy theorems is that (a) the neglected terms of order M_π^6 are much smaller than in the straightforward chiral perturbation series and (b) the dispersion integrals can be evaluated quite accurately, on the basis of the Roy equations. Since the integrals converge very rapidly, they depend almost exclusively on the scattering lengths a_0^0, a_0^2. In the vicinity of the physical values, the linear representations

$$M_\pi^4\alpha_3\alpha = \alpha 0.135 + 0.77\,(a_0^0 - 0.220) - 1.50\,(a_0^2 + 0.0444)\,, \qquad (2a)$$
$$M_\pi^4\alpha_4\alpha = \alpha 0.061 + 0.48\,(a_0^0 - 0.220) - 0.26\,(a_0^2 + 0.0444)\,, \qquad (2b)$$

provide an excellent approximation. For a detailed discussion, in particular also of the uncertainties, I refer to CGL. The estimates given there show that the contributions of order M_π^6 in (1a) generate a correction of order 0.002, while those in equation (1b) are of order 0.003.

The low energy theorems (1) show that the S-wave scattering lengths are related to the coupling constants $\bar{\ell}_3$ and $\bar{\ell}_4$. Indeed, these play a central role in the effective theory, because they determine the dependence of M_π and F_π on the quark mass at first non-leading order: the expansion of M_π^2

[b]In the notation of CGL: $\alpha_3 = 2\alpha_0 + 7\alpha_2$, $\alpha_4 = 2\alpha_0 - 5\alpha_2$.

and F_π in powers of the quark mass starts with

$$M_\pi^2 = M^2\{1 + \tfrac{1}{2}\,x\,\bar{\ell}_3 + O(x^2)\}\,, \qquad F_\pi = F\{1 + x\,\bar{\ell}_4 + O(x^2)\}\,, \qquad (3)$$

where $M^2 = 2Bm$ is proportional to the quark mass, F is the value of the pion decay constant in the chiral limit, and $x \equiv M^2/(16\pi^2F^2)$. A crude estimate for $\bar{\ell}_3$ can be obtained from the mass spectrum of the pseudoscalar octet: $\bar{\ell}_3 = 2.9 \pm 2.4$.[9] For $\bar{\ell}_4$, an analogous estimate ($\bar{\ell}_4 = 4.3 \pm 0.9$) follows from the experimental value of the ratio F_K/F_π. The low energy theorem for the radius of the scalar pion form factor yields a more accurate result: $\bar{\ell}_4 = 4.4 \pm 0.2$.[11] The error is smaller here, because the theorem holds within SU(2)×SU(2), so that an expansion in the mass of the strange quark is not necessary. The small ellipse in figure 1 shows the prediction for the scattering lengths obtained in CGL with these values of $\bar{\ell}_3$ and $\bar{\ell}_4$ (since the residual errors are small, the corrections of order M_π^6 are not entirely negligible – the curve shown includes an estimate for these).

The narrow strip that runs near the lower edge of the universal band indicates the region allowed if the coupling constant $\bar{\ell}_3$ is treated as a free parameter, while $\bar{\ell}_4$ is fixed with the scalar radius, like for the ellipse. By construction, the upper and lower edges of the strip are tangent to the ellipse. As implied by equations (1b) and (2b), the strip imposes an approximately linear correlation between the two scattering lengths – a curvature becomes visible only in the region where the theoretical prediction would be totally wrong.

3. Lattice results

On the lattice, dynamical quarks can now be made sufficiently light to establish contact with the effective low energy theory of QCD. In particular, the MILC collaboration obtained an estimate for the coupling constants L_4, L_5, L_6, L_8 of the effective chiral SU(3)×SU(3) Lagrangian.[12] Using standard one loop formulae,[13] these results lead to $\bar{\ell}_3 = 0.8 \pm 2.3$ and $\bar{\ell}_4 = 4.0 \pm 0.6$. In the plane of the two S-wave scattering lengths, the MILC results select the region indicated by the second ellipse shown in figure 1, which is slightly larger. It would be worthwhile to analyze these data within SU(2)×SU(2), in order to extract the constants $\bar{\ell}_3$ and $\bar{\ell}_4$ more directly. The one loop formulae of SU(3)×SU(3) are subject to inherently larger corrections from higher orders, so that the results necessarily come with a larger error (note that, at the present state of the art, the analysis of the lattice data relies on the one loop formulae).

The second narrow strip shown in the figure is obtained by treating $\bar{\ell}_4$ as an unknown. Instead of showing the region that belongs to the value of $\bar{\ell}_3$ used for the ellipse, I am taking $\bar{\ell}_3$ from a lattice study, which appeared very recently.[14] In that work, the dependence of M_π^2 and F_π on the quark mass is investigated for two flavours of light Wilson quarks. The results for M_π^2 are consistent with one loop χPT . With $M_{K_{\rm ref}} = 495$ MeV, the result $\hat{\ell}_3 = 0.5(5)(1)$ implies $\bar{\ell}_3 = 3.0 \pm 0.5 \pm 0.1$. Since this is close to the center of the estimated range, the strip runs through the middle of the ellipse. Taken at face value, the uncertainty in the lattice result for $\bar{\ell}_3$ is four times smaller than the one in the estimate $\bar{\ell}_3 = 2.9 \pm 2.4$, obtained from the SU(3) mass formulae for the pseudoscalars, more than 20 years ago.[9] While the width of the ellipse is dominated by the uncertainty in the value used for $\bar{\ell}_3$, those associated with the higher order contributions and with the phenomenological uncertainties in the dispersion integrals do affect the width of the strip belonging to the value of Del Debbio et al. – this is why the higher precision for $\bar{\ell}_3$ reduces the width by less than a factor of four.

The coupling constants depend logarithmically on the pion mass:

$$\bar{\ell}_3 = \ln \frac{\Lambda_3^2}{M_\pi^2}\,, \qquad \bar{\ell}_4 = \ln \frac{\Lambda_4^2}{M_\pi^2}\,, \tag{4}$$

so that their value changes if the quark mass is varied. The quoted value for $\bar{\ell}_3$ corresponds to $\Lambda_3 \simeq 560$ MeV. For $M_\pi < \Lambda_3$, the curvature term is positive and reaches a maximum at $M_\pi = \Lambda_3/\sqrt{e} \simeq 340$ MeV. Even there, $\frac{1}{2}x\,\bar{\ell}_3$ is less than 0.05, so that M_π^2/m is indeed nearly a constant. Although more work is needed to clarify all sources of uncertainty, the calculation beautifully confirms that, to a very good approximation, M_π^2 is linear in the quark mass. Hence the quark condensate indeed represents the leading order parameter of the spontaneously broken symmetry.[15] The same lattice data also yield information about the dependence of F_π on the quark mass, but the comparison with the one loop formula of χPT is not conclusive in this case: the result for $\bar{\ell}_4$ depends on how the data are analyzed.[14]

The quark mass dependence of M_π and F_π is also being studied by the European Twisted Mass Collaboration. The preliminary results for the relevant effective coupling constants read $\bar{\ell}_3 = 3.5$ and $\bar{\ell}_4 = 4.4$. A thorough analysis of the uncertainties is under way.[16]

Finally, I mention that the scattering length of the exotic S-wave, a_0^2, can be determined directly, from the volume dependence of the energy levels occurring on the lattice. The horizontal band shown in figure 1 indicates the result obtained in this way by NPLQCD.[17] It is also consistent with the other pieces of information shown in the figure. Although possible in

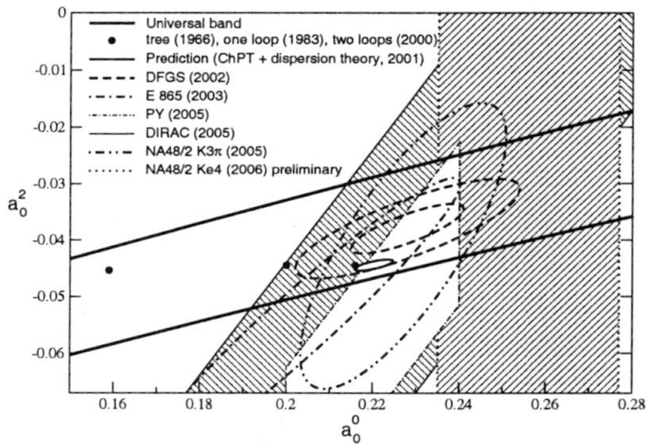

Fig. 2. *S* wave scattering lengths. Experimental results

principle, it is difficult to extend this method to the isoscalar channel.

4. Experiment

Since the pions are not stable, they first need to be produced before they can be studied, so that the experimental information is of limited accuracy. Moreover, there are inconsistencies among the various data sets. One of the problems is that, in most production experiments, the two pions in the final state are accompanied by other hadrons. I know of three exceptions: production via photons in e^+e^- collisions or via W-bosons in the decays $\tau^\pm \to \nu\,\pi^\pm\pi^0$ and $K \to e\nu\pi\pi$. The first two yield excellent information about the electromagnetic and weak vector form factors of the pion and hence also about the P-wave $\pi\pi$ phase shift. The electromagnetic form factor is of particular interest in view of the Standard Model prediction for the magnetic moment of the muon. In this connection, the theoretical understanding of the final state interaction among the pions achieved in recent years can be used to reduce the experimental uncertainties, in particular in the region below 600 MeV, where the data are meager.[18] The K to $\pi\pi$ transition form factors relevant for the third process allow a measurement of the phase difference between the S- and P-waves, $\delta_0^0 - \delta_1^1$. The ellipse labeled E865[19] in figure 2 shows the constraint imposed on a_0^0 and a_0^2 by the K_{e4} data collected at Brookhaven. The two ellipses denoted DFGS[20] represent the 1σ and 2σ contours obtained by combining these with other $\pi\pi$ data. The region labeled PY[21] has roughly the same experimental basis.

A very interesting proposal due to Cabibbo[22] has recently been explored in the NA48/2 experiment at CERN: the cusp seen near threshold in the decay $K^{\pm} \to \pi^{\pm}\pi^{0}\pi^{0}$ allows a measurement of the difference $a_0^0 - a_0^2$ of scattering lengths.[23] In this case, the final state involves three pions, so that the analysis of the data is not a trivial matter.[24–26] As seen in the figure, the result is in good agreement with the Brookhaven data, as well as with the theoretical predictions. The same group has also investigated a very large sample of K_{e4} decays.[27] The result for the scattering length a_0^0 is indicated by the broad vertical band. It is not in good agreement with E865, nor with their $K \to 3\pi$ result. The discrepancy calls for clarification.

An ideal laboratory for exploring the low energy properties of the pions is the atom consisting of a pair of charged pions, also referred to as pionium. The DIRAC collaboration at CERN has demonstrated that it is possible to generate such atoms and to measure their lifetime.[28] Since the physics of the bound state is well understood,[29] there is a sharp theoretical prediction for the lifetime.[1] The band labeled DIRAC in the figure shows that the observed lifetime confirms this prediction. Pionium level splittings would offer a clean and direct measurement of the second subtraction constant. Data on πK atoms would also be very valuable, as they would allow to explore the role played by the strange quarks in the QCD vacuum.[29,30]

5. Dispersion relations

After the above extensive discussion of our current knowledge of the two subtraction constants, I now wish to discuss their role in the low energy analysis of the $\pi\pi$ scattering amplitude, avoiding technical machinery as much as possible. Although the Roy equations represent an optimal and comprehensive framework for that analysis, the main points can be seen in a simpler context: forward dispersion relations.[31] More specifically, I consider the component of the scattering amplitude with s-channel isospin $I = 0$, which I denote by $T^0(s,t)$. It satisfies a twice subtracted fixed-t dispersion relation in the variable s. In the forward direction, $t = 0$, this relation reads

$$\mathrm{Re}\, T^0(s,0)\alpha = \alpha c_0 + c_1\, s + \frac{s(s-4M_\pi^2)}{\pi}\, P\!\int_{4M_\pi^2}^{\infty} \frac{dx\, \mathrm{Im}\, T^0(x,0)}{x\,(x-4M_\pi^2)\,(x-s)} + \quad (5)$$
$$+ \frac{s(s-4M_\pi^2)\,\cancel{\alpha\alpha}}{\pi} \int_{4M_\pi^2}^{\infty} \frac{dx\, \{\mathrm{Im}\, T^0(x,0) - 3\, \mathrm{Im}\, T^1(x,0) + 5\, \mathrm{Im}\, T^2(x,0)\}}{3\,x\,(x-4M_\pi^2)\,(x+s-4M_\pi^2)}\,.$$

The symbol P indicates that the principal value must be taken. The first integral accounts for the discontinuity across the right hand cut, while the second represents the analogous contribution from the left hand cut, where the components of the scattering amplitude with $I = 1, 2$ also show up.

According to the optical theorem, the imaginary part of the forward scattering amplitude represents the total cross section: in the normalization of ACGL,[5] we have $\mathrm{Im}\, T^I(s,0) = \sqrt{s(s-4M_\pi^2)}\;\sigma^I_{tot}(s)$. In this notation, the physical total cross sections are given by

$$\begin{aligned} \sigma^{\pi^\pm\pi^\pm}_{tot}\alpha &= \alpha\sigma^2_{tot} \\ \sigma^{\pi^\pm\pi^0}_{tot} &= \tfrac{1}{2}\sigma^1_{tot} + \tfrac{1}{2}\sigma^2_{tot} \\ \sigma^{\pi^\mp\pi^\pm}_{tot}\alpha &= \alpha\tfrac{1}{3}\sigma^0_{tot} + \tfrac{1}{2}\sigma^1_{tot} + \tfrac{1}{6}\sigma^2_{tot} \\ \sigma^{\pi^0\pi^0}_{tot} &= \tfrac{1}{3}\sigma^0_{tot} + \tfrac{2}{3}\sigma^2_{tot} \end{aligned} \tag{6}$$

As mentioned already, the subtraction term is determined by a_0^0, a_0^2:

$$c_0 + c_1\, s = 32\,\pi \left\{ a_0^0 + (2a_0^0 - 5a_0^2)\,\frac{s-4M_\pi^2}{12M_\pi^2} \right\}\,. \tag{7}$$

A dispersion relation of the above type also holds for other processes. What is special about $\pi\pi$ is that the contribution from the crossed channels can be expressed in terms of observable quantities – total cross sections in the case of forward scattering. The contribution from the left hand cut is dominated by the ρ-meson, which generates a pronounced peak in the total cross section with $I = 1$. This contribution is known very accurately from the process $e^+e^- \rightarrow \pi^+\pi^-$. In the physical region, $s > 4M_\pi^2$, the entire contribution from the crossed channels is a smooth function that varies only slowly with the energy. Note, however, that this contribution is by no means small.[32]

The angular momentum barrier suppresses the higher partial waves: at low energies, the first term in the partial wave decomposition

$$\begin{aligned} \mathrm{Re}\, T^0(s,t)\alpha &= \alpha 32\pi \left\{ \mathrm{Re}\, t_0^0(s) + 5\, P_2(z)\, \mathrm{Re}\, t_2^0(s) + \ldots \right\} \\ t\alpha &= \alpha\tfrac{1}{2}(4M_\pi^2 - s)(1-z) \end{aligned} \tag{8}$$

represents the most important contribution. In the vicinity of the threshold, where the contribution from the higher angular momenta is negligibly small, the dispersion relation (5) thus amounts to an expression for the real part of the isoscalar S-wave. For brevity, I refer to this partial wave as S^0.

Equation (5) imposes a very strong constraint on S^0, for the following reason. Suppose that all partial waves except this one are known. The relation then determines $\mathrm{Re}\, t_0^0(s)$ as an integral over $\mathrm{Im}\, t_0^0(x)$. In the elastic region, unitarity already fixes the real part in terms of the imaginary part – we thus have two equations for the two unknowns $\mathrm{Re}\, t_0^0(s)$ and $\mathrm{Im}\, t_0^0(s)$.

Accordingly, we may expect that, in the elastic region, the f.d.r. unambiguously fixes the S^0-wave. Indeed, this is borne out by the calculation, which is described in some detail elsewhere.[33]

One of the main results established in CGL is that the Roy equations fix the behaviour of the S^0-wave below 800 MeV almost entirely in terms of three parameters: the two subtraction constants a_0^0, a_0^2 and the value of the phase at 800 MeV. In particular, the behaviour of the scattering amplitude at high energies is not important. This can also be seen on the basis of equation (5): replacing our representation of the scattering amplitude above $K\bar{K}$ threshold by the one proposed in KPY,[31] for instance, the solution of the f.d.r. very closely follows the solution of the Roy equations that belongs to the same value of the phase at 800 MeV.[33]

6. The lowest resonances of QCD

The positions of the poles in the S-matrix represent universal properties of the strong interaction, which are unambiguous even if the width of the corresponding resonance turns out to be large,[34] but they concern the non-perturbative domain, where an analysis in terms of the local degrees of freedom of QCD – quarks and gluons – is not in sight. First quenched lattice explorations of the pole from the σ have appeared,[35] but in view of the strong final state interaction in this channel, it will take some time before this state can reliably be reached on the lattice.

One of the reasons why the values for the pole position of the σ quoted by the Particle Data Group cover a very broad range is that all of these rely on the extrapolation of hand made parametrizations: the data are represented in terms of suitable functions on the real axis and the position of the pole is determined by continuing this representation into the complex plane. If the width of the resonance is small, the ambiguities inherent in the choice of the parametrization do not significantly affect the result, but the width of the σ is not small.

A popular approach to the problem is based on the so-called *inverse amplitude method.* Applying it to the χPT representation of the scattering amplitude[36] invariably produces a pole in the right ball park. The procedure definitely improves the quality of the two-loop approximation of χPT on the real axis, because it respects unitarity in the elastic region. The extrapolation into the complex plane, however, also contains a number of fake singularities. Like all other parametrizations, this approach relies on a model. I do not know of a way to estimate the systematic uncertainties generated if QCD is replaced by one model or the other – if error estimates are given at all, these necessarily rely on guesswork. A thorough discussion of the problems inherent in the IAM approach appeared several years ago.[37]

We have found a method that does not require a parametrization of the data at all. It relies on the fact that (a) the S-matrix has a pole on the second sheet if and only if it has a zero on the first sheet, (b) the Roy equations are valid not only on the real axis, but in a limited domain of the first sheet, (c) the poles from the lowest resonances, σ, ρ, $f_0(980)$, all occur in that domain. The numerical evaluation of the pole position is straightforward. The one closest to the origin occurs at[2]

$$M_\sigma - \frac{i}{2}\,\Gamma_\sigma = \sqrt{s_\sigma} = 441\,{}^{+16}_{-\,8} - i\,272\,{}^{+\,9}_{-12.5}\text{ MeV}\,, \tag{9}$$

where the error accounts for all sources of uncertainty. We may, for instance, replace our representation for S^0 by the one proposed by Bugg[38] or replace the entire scattering amplitude by the parametrization in KPY.[31] In either case, the outcome for the pole position is in the above range. For more details, I refer to CCL[2] and to a recent conference report.[39]

In the meantime, the method described above was applied to the case of πK scattering, with the result that the lowest resonance in that channel occurs at $m_\kappa = (658 \pm 13) - i\,(278.5 \pm 12)$ MeV.[40] Evidently, the physics of the κ is very similar to the one of the σ.

7. On the working bench

We are currently extending the work described in CGL to higher energies. The price to pay is that the contributions from the high energy region then become more important. Since the first few terms of the partial wave expansion do not represent a decent approximation there, one instead uses a Regge representation for the asymptotic domain. In CGL, we borrowed that from Pennington.[41] In the meantime, we have performed a new Regge analysis, invoking experimental information as well as sum rules to pin down the residue functions. Brief accounts of this work were given elsewhere.[42] Figure 3 shows the preliminary results obtained for the total cross sections with t-channel isospin 0 and 1:

$$\sigma^{(0)}_{tot} = \tfrac{1}{3}\sigma^0_{tot} + \sigma^1_{tot} + \tfrac{5}{3}\sigma^2_{tot}\,,\quad \sigma^{(1)}_{tot} = \tfrac{1}{3}\sigma^0_{tot} + \tfrac{1}{2}\sigma^1_{tot} - \tfrac{5}{6}\sigma^2_{tot}\,. \tag{10}$$

We include pre-asymptotic contributions and assume that the Regge parametrization yields a decent approximation above 1.7 GeV. At lower energies, the uncertainties in that representation become larger than those in the sum over the partial waves, which is also shown. The comparison with

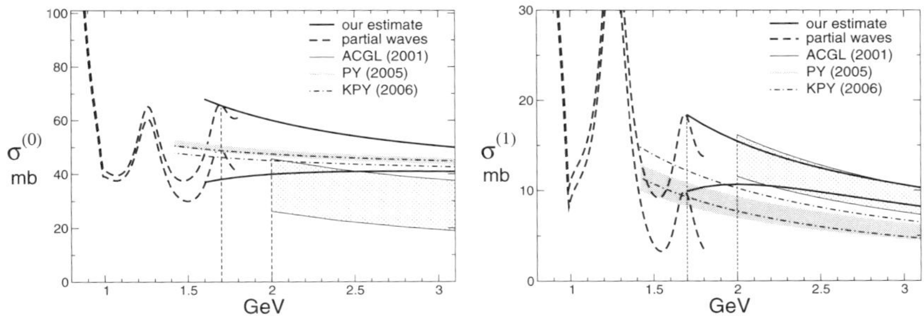

Fig. 3. Asymptotic behaviour of the total cross sections with $I_t = 0$ and $I_t = 1$.

the representation used in ACGL confirms the cross section with $I_t = 1$, while the one for $I_t = 0$ comes out larger by 1 or $2\,\sigma$. In the solutions of the Roy equations, the effects generated by such a shift are barely visible below 800 MeV. We hope to complete the analysis soon, as well as the application to the electromagnetic form factor of the pion, for which an accurate representation is needed in connection with the Standard Model prediction for the magnetic moment of the muon.

It is a pleasure to thank the organizers of the meeting for their kind hospitality during a very pleasant stay at Chapel Hill and Claude Bernard, Leonardo Giusti and Steve Sharpe for useful comments about the text. Also, I acknowledge Balasubramanian Ananthanarayan, Irinel Caprini, Gilberto Colangelo and Jürg Gasser for a most enjoyable and fruitful collaboration – the present report is based on our common work.

References

1. G. Colangelo, J. Gasser and H. Leutwyler, Nucl. Phys. B **603** (2001) 125.
2. I. Caprini, G. Colangelo and H. Leutwyler, Phys. Rev. Lett. **96** (2006) 132001.
3. W.-M. Yao et al. [Particle Data Group], Journal of Physics G **33** (2006) 1.
4. S. M. Roy, Phys. Lett. B **36** (1971) 353.
5. B. Ananthanarayan et al., Phys. Rept. **353** (2001) 207.
6. J. L. Basdevant, C. D. Froggatt and J. L. Petersen, Nucl. Phys. B **72** (1974) 413.
7. V. Bernard and U. G. Meissner, hep-ph/0611231, give an excellent overview of recent developments in χPT , with many references to ongoing work.
8. S. Weinberg, Phys. Rev. Lett. **17** (1966) 616.
9. J. Gasser and H. Leutwyler, Phys. Lett. B **125** (1983) 325; Annals Phys. **158** (1984) 142.
10. J. Bijnens, G. Colangelo, G. Ecker, J. Gasser and M. E. Sainio, Phys. Lett. B **374** (1996) 210; Nucl. Phys. B **508** (1997) 263; ibid. B **517** (1998) 639 (E).

11. For a detailed discussion of the scalar form factor and references to related work, see B. Ananthanarayan et al., Phys. Lett. **B602**, 218 (2004).
12. C. Aubin *et al.*, [MILC Collaboration], Phys. Rev. D **70** (2004) 114501.
13. J. Gasser and H. Leutwyler, Nucl. Phys. B **250** (1985) 465, section 11.
14. L. Del Debbio, L. Giusti, M. Lüscher, R. Petronzio and N. Tantalo, hep-lat/0610059.
15. G. Colangelo, J. Gasser and H. Leutwyler, Phys. Rev. Lett. **86** (2001) 5008.
16. I thank Karl Jansen and Andrea Shindler for this information. The method used is described in K. Jansen and C. Urbach [ETM Collaboration], hep-lat/0610015; A. Shindler [ETM Collaboration], hep-ph/0611264.
17. S. R. Beane, P. F. Bedaque, K. Orginos and M. J. Savage [NPLQCD Collaboration], Phys. Rev. D **73** (2006) 054503.
18. G. Colangelo, AIP Conf. Proc. **756** (2005) 60.
19. S. Pislak *et al.* [BNL-E865 Collaboration], Phys. Rev. Lett. **87** (2001) 221801; Phys. Rev. D **67** (2003) 072004.
20. S. Descotes-Genon, N. H. Fuchs, L. Girlanda and J. Stern, Eur. Phys. J. C **24** (2002) 469.
21. J. R. Pelaez and F. J. Yndurain, Phys. Rev. D **71** (2005) 074016.
22. N. Cabibbo, Phys. Rev. Lett. **93** (2004) 121801.
23. J. R. Batley *et al.* [NA48/2 Collaboration], Phys. Lett. B **633**, 173 (2006).
24. N. Cabibbo and G. Isidori, JHEP **0503** (2005) 021.
25. G. Colangelo, J. Gasser, B. Kubis and A. Rusetsky, Phys. Lett. B **638** (2006) 187.
26. E. Gamiz, J. Prades and I. Scimemi, hep-ph/0602023.
27. L. Masetti, Proc. ICHEP06, hep-ex/0610071.
28. B. Adeva *et al.* [DIRAC Collaboration], Phys. Lett. B **619** (2005) 50.
29. A thorough review is contained in *Proc. Workshop on Hadronic Atoms 2005,* eds. L. Afanasyev, G. Colangelo and J. Schacher, hep-ph/0508193.
30. J. Schweizer, Eur. Phys. J. C **36** (2004) 483; Phys. Lett. B **587** (2004) 33.
31. R. Kaminski, J. R. Pelaez and F. J. Yndurain, Phys. Rev. D **74** (2006) 014001 [Erratum-ibid. D **74** (2006) 079903].
32. The significance of the contributions from the left hand cut is thoroughly discussed in Z. Y. Zhou, G. Y. Qin, P. Zhang, Z. G. Xiao, H. Q. Zheng and N. Wu, JHEP **0502** (2005) 043.
33. H. Leutwyler, in *Proc. Quark Confinement and the Hadron Spectrum VII*, Ponta Delgada, Azores islands, Portugal (2006), hep-ph/0612111.
34. See for example R.J. Eden, P.V. Landshoff, D.I. Olive, J.C. Polkinghorne, *The analytic S-matrix*, Cambridge University Press (1966) or D. Zwanziger, Phys. Rev. 131 (1963) 888.
35. H. Y. Cheng, C. K. Chua and K. F. Liu, Phys. Rev. D **74** (2006) 094005.
36. T. N. Truong, Phys. Rev. Lett. **61**, 2526 (1988); A. Dobado, M. J. Herrero and T. N. Truong, Phys. Lett. B **235**, 134 (1990); A. Dobado and J. R. Pelaez Phys. Rev. D **56** (1997) 3057; J. A. Oller, E. Oset and J. R. Pelaez, Phys. Rev. D **59** (1999) 074001, D **60** (1999) 099906 (E); T. Hannah, Phys. Rev. D **60** (1999) 017502.
37. G. Y. Qin, W. Z. Deng, Z. G. Xiao and H. Q. Zheng, Phys. Lett. B **542**

(2002) 89.
38. D. V. Bugg, hep-ph/0608081.
39. H. Leutwyler, in *Proc. MESON 2006*, Krakow, Poland, hep-ph/0608218.
40. S. Descotes-Genon and B. Moussallam, arXiv:hep-ph/0607133.
41. M. R. Pennington, Annals Phys. **92** (1975) 164.
42. I. Caprini, in these proceedings; I. Caprini, G. Colangelo and H. Leutwyler, Int. J. Mod. Phys. A **21** (2006) 954.

CHIRAL EFFECTIVE FIELD THEORY IN THE Δ-RESONANCE REGION

VLADIMIR PASCALUTSA*

Physics Department, The College of William & Mary, Williamsburg, VA 23187, USA
Theory Center, Jefferson Lab, 12000 Jefferson Ave, Newport News, VA 23606, USA

I discuss the problem of constructing an effective low-energy theory in the vicinity of a resonance or a bound state. The focus is on the example of the $\Delta(1232)$, the lightest resonance in the nucleon sector. Recent developments of the chiral effective-field theory in the Δ-resonance region are briefly reviewed. I conclude with a comment on the merits of the manifestly covariant formulation of chiral EFT in the baryon sector.

Keywords: Chiral Lagrangians, power counting, resonances

1. Introduction

Chiral Perturbation Theory (χPT) provides a systematic field-theoretic framework for the description of the low-energy strong interaction. The fundamental degrees of freedom in χPT are therefore the low-energy hadron excitations, such as the Goldstone bosons (GBs) of chiral symmetry breaking, nucleons and a few others. The corresponding hadron fields appear in the effective Lagrangian which can be organized in powers of derivatives of the GB fields, or schematically as:

$$\mathcal{L}(\pi, N, \ldots) = \sum_n \mathcal{L}^{(n)} = \sum_n \mathcal{O}_n(c_i) \frac{(\partial \pi)^n}{\Lambda^n} \tag{1}$$

where O_n are some field operators which may contain GB fields but not their derivatives. All possible field operators, constrained by chiral and other symmetries, appear with the free parameters, c_i, the so-called low energy constants (LECs). The mass scale Λ is the heavy scale which sets the upper limit of applicability of χPT and is believed to be of order of 1 GeV — the

*Present address: *ECT*, Villa Tambosi, Strada delle Tabarelle 286, I-38050 Villazzano-Trento, Italy*. E-mail: *vlad@ect.it* .

scale of spontaneous chiral symmetry breaking that led to the appearance of the GBs.

This expansion of the Lagrangian translates into a low-energy expansion of the S-matrix, schematically:

$$S = \sum_n A_n(c_i)\, \frac{p^n}{\Lambda^n} \tag{2}$$

where A's are amplitudes which depend on LECs, and p denotes the momentum of the GBs. The anzatz[1] is that the same expansion can (one day) be obtained directly from QCD, provided the LECs are matched onto the QCD parameter: $c_n = c_n(\Lambda_{QCD})$. In the absence of the correspondent calculation in QCD, the best one can do is to match (or, fit) the LECs to experimental data, making sure that they take reasonable (or, natural) values such that the above expansion is convergent.

One case where the convergence of the χPT expansion is immediately questioned is the case of hadronic bound states and resonances. In the presence of a bound state or a resonance the low-energy expansion of the S-matrix goes as

$$S \sim \sum_n A_n \left(\frac{p}{\Delta E}\right)^n, \tag{3}$$

where ΔE is the excitation (binding) energy of the resonance (bound state). Thus, the limit of applicability of χPT is limited not by $\Lambda \sim 1$ GeV but by the characteristic energy scale ΔE of the closest bound or excited state. Furthermore, in the vicinity of a bound state or a resonance the S-matrix has a pole, which cannot be reproduced in a purely perturbative expansion in energy that is utilized in χPT.

This problem arises in various contexts, ranging from pion-pion scattering[2] to halo nuclei.[3] Some are being discussed at this meeting (e.g., resonances in the $\pi\pi$ system,[4] or bound states and resonances in the few-nucleon system[5]). In this contribution I focus on the πN system where the first resonance is the $\Delta(1232)$.

The Δ resonance is an ideal study case for the problem of resonances in χPT. It is relatively light, with the excitation energy of $\varDelta \equiv M_\Delta - M_N \approx$ 300 MeV, elastic, and well separated from the other nucleon resonances. It is also a very prominent resonance and plays an important role in many processes, including astrophysical ones. It is, for instance, responsible for the damping of the high-energy cosmic rays by the cosmic microwave background, the so-called GZK cutoff. Let us therefore look at the description of this resonance in the framework of chiral effective-field theory (χEFT).

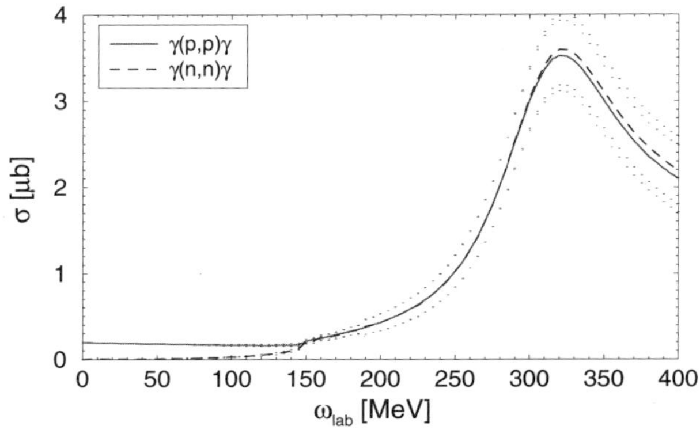

Fig. 1. (Color online) Total cross-section of the Compton scattering on the nucleon (proton – red solid curve, neutron – blue dashed curve), as the function of the incident photon lab energy. The curves are obtained in a χEFT calculation.[6]

2. Power counting(s) for the Δ resonance

Imagine Compton scattering on the nucleon. The total cross-section of this process, as a function of photon energy ω, is shown in Fig. 1. In this case we are able to examine the entire energy range, starting with $\omega = 0$, through the pion production threshold $\omega \simeq m_\pi$ and into the resonance region $\omega \sim \Delta$.

At energies up to around the pion production threshold the cross section shows a smooth behavior which can reproduced by a low energy expansion. In this region the Δ-resonance can be "integrated out", as its tail contribution can be mimicked by the terms already present in the χPT Lagrangian with nucleons only.[7]

Higher in energy, however, the rapid energy variation induced by the resonance pole is not reproducible by a naive low-energy expansion. Obviously, to describe this behavior it is necessary to introduce the Δ as an explicit degree of freedom,[8] hence include a corresponding field in the effective chiral Lagrangian. The details of how this is done have recently been reviewed in.[9]

Once the Δ appears in the Lagrangian the question is how to power-count its contributions. In χEFT with pions and nucleons alone the power-counting index of a graph with L loops, N_π (N_N) internal pion (nucleon) lines, and V_k vertices from kth-order Lagrangian is found as

$$n_{\chi\mathrm{PT}} = 4L - 2N_\pi - N_N + \sum_k kV_k\,. \tag{4}$$

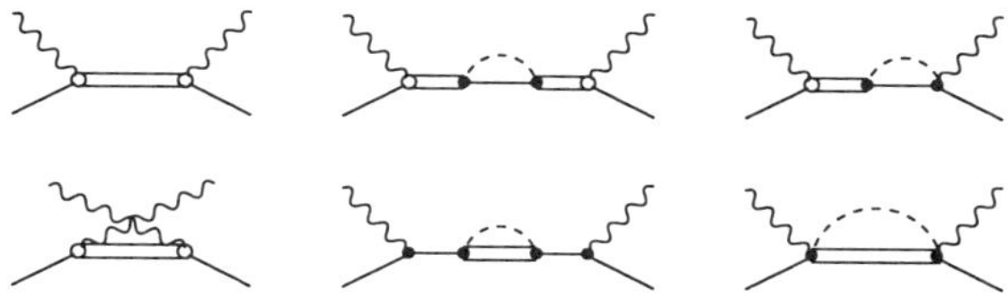

Fig. 2. Examples of the one-Delta-reducible (1st row) and the one-Delta-irreducible (2nd row) graphs in Compton scattering.

What about the graphs with the Δ, such as those depicted in Fig. 2 ? Their power counting turns out to be dependent on how one weighs the excitation energy $\varDelta$ in comparison with the other mass scales of the theory. In this case we have the soft momentum p (or, ω), the pion mass m_π, and heavy scales which we collectively denote Λ.

The Small Scale Expansion[10] (SSE) counts all light scales equally: $p \sim m_\pi \sim \varDelta$. The small parameter is then:

$$\epsilon = \left\{ \frac{p}{\Lambda}, \frac{m_\pi}{\Lambda}, \frac{\varDelta}{\Lambda} \right\} . \tag{5}$$

An unsatisfactory feature of such a democratic counting (ϵ-expansion) is that the Δ-resonance contributions are always estimated to be of the same size as the nucleon contributions. As we have seen from Fig. 1, in reality the resonance contributions are suppressed at low energies while being dominant in the resonance region. Therefore the power counting *overestimates* the Δ-contributions at lower energies and *underestimates* them at the resonance energies. Despite this flaw, the SSE has been widely and quite successfully used to account the Δ contributions below the resonance.[11–16] No applications of this scheme to calculation of observables in the resonance region have been reported yet.

A more adequate power counting is achieved by separating out the resonance energy, e.g., by maintaining the following scale hierarchy $m_\pi \ll \varDelta \ll \Lambda$ in the power-counting scheme.[6,17] In the so-called *"δ expansion"*[6] this is done by introducing a small parameter $\delta = \varDelta/\Lambda$, and then counting m_π/Λ as δ^2. The power 2 is chosen here because it is the closest integer representing the ratio of these scales in the real world.

Obviously, the power counting of the Δ contributions then becomes dependent on the energy domain: in the *low-energy region* ($p \sim m_\pi$) and the *resonance region* ($p \sim \varDelta$), the momentum counts differently, see Table 1. This dependence most significantly affects the counting of the one-Delta-

Table 1. The counting of momenta in the three different χEFT expansions.

EFT	$p \sim m_\pi$	$p \sim \Delta$
$\not\Delta$-χPT	$\mathcal{O}(p)$	$\mathcal{O}(1)$
ϵ-expansion	$\mathcal{O}(\epsilon)$	$\mathcal{O}(\epsilon)$
δ-expansion	$\mathcal{O}(\delta^2)$	$\mathcal{O}(\delta)$

reducible (ODR) graphs. The 1st row of graphs in Fig. 2 illustrates examples of the ODR graphs for the Compton scattering case. These graphs are all characterized by having a number of ODR propagators, each going as

$$S_{ODR} \sim \frac{1}{s - M_\Delta^2} \sim \frac{1}{2M_\Delta} \frac{1}{p - \varDelta}, \tag{6}$$

where $s = M_N^2 + 2M_N\omega$ is the Mandelstam variable, and the soft momentum p in this case given by the photon energy. In contrast, the nucleon propagator in analogous graphs would go simply as $S_N \sim 1/p$. Therefore, in the low-energy region, the Δ and nucleon propagators would count respectively as $\mathcal{O}(1/\delta)$ and $\mathcal{O}(1/\delta^2)$, the Δ being suppressed by one power of the small parameter as compared to the nucleon. In the resonance region, the ODR graphs obviously all become large. Fortunately they all can be subsumed, leading to "dressed" ODR graphs with a definite power-counting index. Namely, it is not difficult to see that the resummation of the classes of ODR graphs results in ODR graphs with only a single ODR propagator of the form

$$S^*_{ODR} = \frac{1}{S^{-1}_{ODR} - \Sigma} \sim \frac{1}{p - \varDelta - \Sigma}, \tag{7}$$

where Σ is the Δ self-energy. The expansion of the self-energy begins with p^3, and hence in the low-energy region does not affect the counting of the Δ contributions. However, in the resonance region the self-energy not only ameliorates the divergence of the ODR propagator at $s = M_\Delta^2$ but also determines power-counting index of the propagator. Defining the Δ-resonance region formally as the region of p where

$$|p - \varDelta| \leq \delta^3 \Lambda, \tag{8}$$

we deduce that an ODR propagator, in this region, counts as $\mathcal{O}(1/\delta^3)$. Note that the nucleon propagator in this region counts as $\mathcal{O}(1/\delta)$, hence is suppressed by two powers as compared to ODR propagators. Thus, within this power-counting scheme we have the mechanism for estimating correctly the relative size of the nucleon and Δ contributions in the two energy domains. In Table 2 we summarize the counting of the nucleon, ODR, and

Table 2. The counting for the nucleon, one-Delta-reducible (ODR), and one-Delta-irreducible (ODI) propagators in the two different expansion schemes. The counting in the δ-expansion depends on the energy domain.

	ϵ-expansion	δ-expansion	
	$p/\Lambda_{\chi SB} \sim \epsilon$	$p \sim m_\pi$	$p \sim \Delta$
S_N	$1/\epsilon$	$1/\delta^2$	$1/\delta$
S_{ODR}	$1/\epsilon$	$1/\delta$	$1/\delta^3$
S_{ODI}	$1/\epsilon$	$1/\delta$	$1/\delta$

one-Delta-irreducible (ODI) propagators in both the ϵ- and δ-expansion.

We conclude this discussion by giving the general formula for the power-counting index in the δ-expansion. The power-counting index, n, of a given graph simply tells us that the graph is of the size of $\mathcal{O}(\delta^n)$. For a graph with L loops, V_k vertices of dimension k, N_π pion propagators, N_N nucleon propagators, N_Δ Delta propagators, N_{ODR} ODR propagators and N_{ODI} ODI propagators (such that $N_\Delta = N_{ODR} + N_{ODI}$) the index is

$$n = \begin{cases} 2n_{\chi\mathrm{PT}} - N_\Delta\,, & p \sim m_\pi; \\ n_{\chi\mathrm{PT}} - 3N_{ODR} - N_{ODI}\,, & p \sim \Delta, \end{cases}$$

where $n_{\chi\mathrm{PT}}$, given by Eq. (4), is the index of the graph in χPT with no Δ's.

In the following I show a few applications of the δ expansion to the calculation of processes in the Δ resonance region.

3. Pion-nucleon scattering

The pion-nucleon (πN) scattering amplitude at leading order in the δ-expansion in the resonance region, is given by the graph (*LO*) in Fig. 3. This is an example of an ODR graph and thus the Δ-propagator counts as δ^{-3}. The leading-order vertices are from $\mathcal{L}^{(1)}$ and since $p \sim \delta$, the whole graph is $\mathcal{O}(\delta^{-1})$.

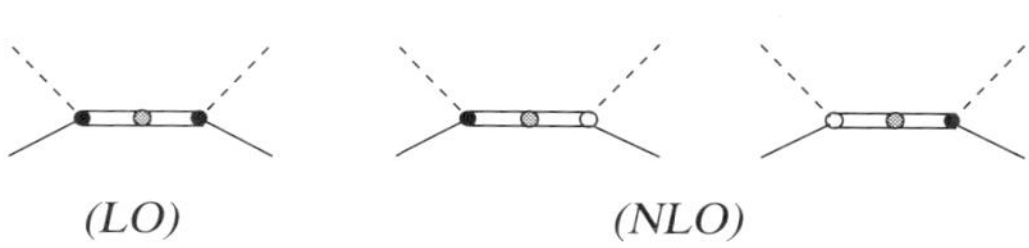

Fig. 3. The leading and next-to-leading order graphs of the πN-scattering amplitude.

At the NLO, the graphs labeled (NLO) Fig. 3 begin to contribute. The $\pi N\Delta$ vertices denoted by dots stand for the h_A coupling from $\mathcal{L}^{(1)}$ and the circles for the h_1 $\pi N\Delta$ coupling[9] from $\mathcal{L}^{(2)}$. The NLO graphs are thus $\mathcal{O}(\delta^0)$. The graphs containing the loop correction to the vertex, as well as the nucleon-exchange graphs, begin to contribute at N^{2}LO [$\mathcal{O}(\delta)$].

The ODR graphs at NLO contribute only to the P_{33} partial wave and this contribution can conveniently be written in terms of the following partial-wave 'K-matrix':

$$K_{P33} = -\frac{1}{2}\frac{\Gamma(W)}{W - M_\Delta}\,, \tag{10}$$

where $W = \sqrt{s}$ is the total energy and Γ is an energy-dependent width, which arises from the Δ self-energy. At this stage it is already taken into account that the real part of the self-energy will lead to the mass and field renormalization and otherwise are of N^{2}LO. Thus, only the imaginary part of the self-energy affects the NLO calculation.

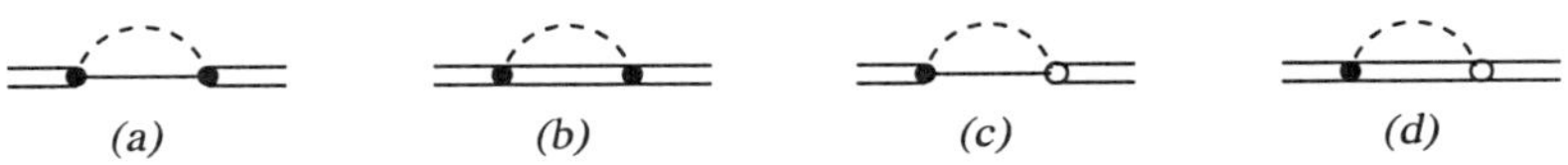

Fig. 4. The leading and next-to-leading order graphs of the Δ self-energy.

In the ODR graphs of Fig. 3, the Δ-propagator is dressed by the self-energy given to NLO by the graphs in Fig. 4, which give rise to the energy-dependent width:

$$\Gamma(W) = -2\,\mathrm{Im}\,[\Sigma(M_\Delta) + (W - M_\Delta)\,\Sigma'(M_\Delta)]\;. \tag{11}$$

Therefore the expression for the K-matrix Eq. (10) becomes

$$K_{P33} = \frac{\mathrm{Im}\Sigma(M_\Delta)}{W - M_\Delta} + \mathrm{Im}\Sigma'(M_\Delta)\;. \tag{12}$$

The πN scattering phase-shift is related to the partial-wave K-matrix simply as

$$\delta_l = \arctan K_l\,, \tag{13}$$

where l stands for the conserved quantum numbers: spin (J), isospin (I) and parity (P). The P_{33} phase (corresponding to $J = 3/2 = I$, $P = +$) is the only nonvanishing one at NLO in the resonance region. One can then fix the LECs h_A and h_1 by fitting the result to the well-established empirical

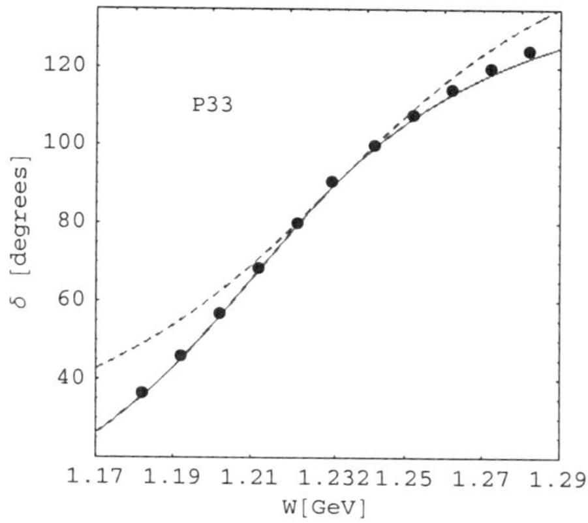

Fig. 5. (Color online) The energy-dependence of the P33 phase-shift of elastic pion-nucleon scattering in the Δ-resonance region. The red solid (blue dashed) curve represents the NLO (LO) result. The data points are from the SP06 SAID analysis.[18]

information about this phase-shift.

In Fig. 5 the red solid curve shows the NLO description of the empirical P_{33} phase-shift represented by the data points. The blue dashed line in Fig. 5 shows the LO result, obtained by neglecting $\mathrm{Im}\Sigma'$ and h_1. This corresponds with the so-called "constant width approximation". At both LO and NLO, the resonance width takes the value $\Gamma(M_\Delta) \simeq 115$ MeV.

One can conclude that the resonant phase-shift is remarkably well reproduced at NLO in the δ expansion. The comparison of the LO and NLO shows a very good convergence of this expansion in the broad energy window around the resonance position.

4. Pion electroproduction

The pion electroproduction on the proton in the Δ-resonance region has been under intense study at many electron beam facilities, most notably at MIT-Bates, MAMI, and Jefferson Lab. The primary goal of these recent experiments is to map out the three electromagnetic $N \to \Delta$ transition form factors. On the theory side, these form factors have been studied in both the SSE[14,15] and the δ-expansion.[19,20] They both have been reviewed very recently[21,22] and I will therefore skip to the next topic.

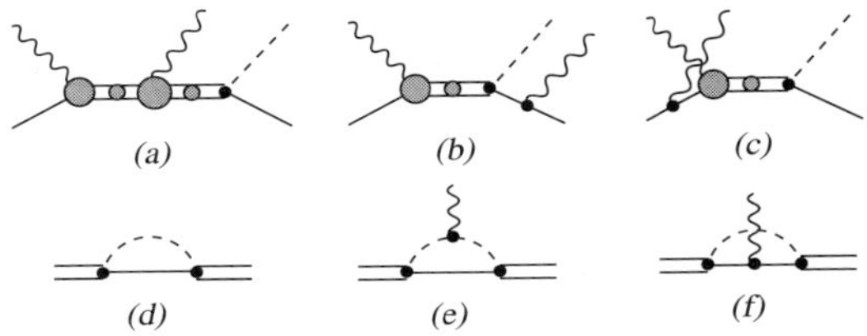

Fig. 6. Diagrams for the $\gamma p \to \pi^0 p \gamma'$ reaction at NLO in the δ-expansion.

5. Radiative pion photoproduction

Radiative pion photoproduction ($\gamma N \to \pi N \gamma'$) in the Δ-resonance region is used to access the Δ^+ magnetic dipole moment (MDM).[23,24] The pioneering experiment[25] was carried out at MAMI in 2002 and a series of dedicated experiments were run in 2005 by the Crystal Ball Collaboration with preliminary results announced this year.[26]

The first, and thusfar the only, study of this process within χEFT had been performed using the δ expansion.[27] This case is particularly interesting from the viewpoint of δ expansion, because the kinematics is such (for the optimal sensitivity to the MDM) that the incident photon energy ω is in the vicinity of $\varDelta$, while the outgoing photon energy ω' is of order of m_π. In this case the $\gamma p \to \pi^0 p \gamma'$ amplitude to NLO in the δ-expansion is given by the diagrams Fig. 6(a),(b), and (c), where the shaded blobs, in addition to the couplings from the chiral Lagrangian, contain the one-loop corrections shown in Fig. 6(e), (f).

Figure 7 shows the pion-mass dependence of real and imaginary parts of the Δ^+ and Δ^{++} MDMs, according to the calculation of Ref.[27] Each of the two solid curves has a free parameter, a counterterm κ_Δ from $\mathcal{L}_\Delta^{(2)}$, adjusted to agree with the lattice data at larger values of m_π. As can be seen from Fig. 7, the Δ MDM develops an imaginary part when $m_\pi < \varDelta$, whereas the real part has a pronounced cusp at $m_\pi = \varDelta$. The dashed-dotted curve in Fig. 7 shows the result[28] for the magnetic moment of the proton. One can see that μ_{Δ^+} and μ_p, while having very distinct behavior for small m_π, are approximately equal for larger values of m_π.

The NLO calculation, completely fixes the imaginary part of the $\gamma\Delta\Delta$ vertex. (For an alternative recent calculation of the imaginary part of the Δ MDM see Ref.[31]).

The expansion for the real part of the $\gamma\Delta\Delta$ begins with LECs from $\mathcal{L}^{(2)}$ which represent the isoscalar and isovector MDM couplings:[9] $\kappa_\Delta^{(S)}$ and $\kappa_\Delta^{(V)}$. A linear combination of these parameters, $\mu_{\Delta^+} = [3 + (\kappa_\Delta^{(S)} +$

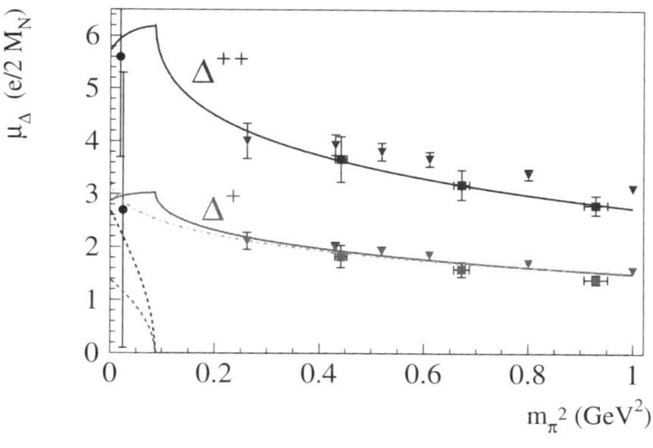

Fig. 7. Pion mass dependence of the real (solid curves) and imaginary (dashed curves) parts of Δ^{++} and Δ^{+} MDMs [in nuclear magnetons]. Dashed-dotted curve is the result for the proton magnetic moment from Ref.[28] The experimental data points for Δ^{++} and Δ^{+} (circles) are the values quoted by the PDG. Quenched lattice data are from Refs.[29] (squares) and from Ref.[30] (triangles).

$\kappa_\Delta^{(V)})/2](e/2M_\Delta)$, is to be extracted from the $\gamma p \to \pi^0 p \gamma'$ observables. For further discussion of how this is done I, for the reason of space, have to refer to the original paper.[27]

6. HBχPT *vs* manifestly covariant baryon χPT

The original formulation of chiral perturbation theory with nucleons is done in the manifestly Lorentz-invariant fashion.[7] However, essentially because of the use of the $\overline{MS}$ renormalization scheme, it was found violate the chiral power counting. The heavy-baryon chiral perturbation theory (HBχPT), which treats nucleons semi-relativistically, was developed to cure the power-counting problem.[32] The heavy-baryon formalism was extensively used in the previous decade and is still used sometimes. More recently, Becher and Leutwyler[33] proposed a manifestly Lorentz-invariant formulation supplemented with so-called "infrared regularization" (IR) of loops in which the chiral power-counting is manifest. The IR, however, alters the analytic structure of the loop amplitude – a very undesirable feature. At about the same time it was realized[34] that power-counting can be maintained in a manifestly covariant formalism without the IR or the heavy-baryon expansions. The original formulation[7] satisfies the power-counting provided the appropriate renormalizations of available low-energy constants are performed, such that the renormalization scale is set at the nucleon mass scale.

It since has been argued that the manifestly covariant calculations have, at least in some cases, better convergence than their heavy-baryon counterparts. This means that the convergence is improved by a resummation of terms which are required by analyticity, but are formally higher-order relativistic corrections to the non-analytic terms.

In the Δ-resonance region calculations, reviewed here, the use of the manifestly covariant formalism has been seen to be of utmost importance from the viewpoint of convergence and naturalness of the theory. The same statement can be for the studies of the chiral behavior for larger than physical values of pion mass.[28,35]

7. Summary

In the single-nucleon sector the limit of applicability of chiral perturbation theory is set by the excitation energy of the first nucleon resonance – the Δ(1232). Inclusion of the Δ in the chiral Lagrangian extends the limit of applicability into the resonance energy region. The power counting of the Δ contribution depends crucially on how the $\varDelta = M_\Delta - M_N$, weighted in comparison to the other mass scales in the problem, in this case the pion mass m_π and the scale of chiral symmetry breaking Λ.

Two different schemes exist in the literature. In the Small Scale Expansion $\varDelta \sim m_\pi \ll \Lambda$, while in the "$\delta$-expansion" $m_\pi \ll \varDelta \ll \Lambda$. I have argued that the hierarchy of scales used in the δ expansion provides a more adequate power-counting of the Δ-resonance contributions. It provides a justification for "integrating out" the resonance contribution at very low energies, as well as for resummation and dominance of resonant contributions in the resonance region.

The δ expansion has already been successfully applied (at NLO) to the calculation of observables for processes such as pion-nucleon and Compton scattering, pion electroproduction and radiative pion photoproduction in the Δ-resonance region. This applications show good convergence properties of this chiral EFT expansion. The use of the manifestly Lorentz-invariant formalism is seen to play an important role in naturalness of the theory.

I have also given examples of how the chiral EFT plays here a dual role in that it allows for an extraction of resonance properties from observables *and* predicts their pion-mass dependence. In this way it may provide a crucial connection of present lattice QCD results to the experiment.

Acknowledgments

This work was partially supported by DOE grant no. DE-FG02-04ER41302 and contract DE-AC05-06OR23177 under which Jefferson Science Associates operates the Jefferson Laboratory.

References

1. S. Weinberg, Physica A **96**, 327 (1979); J. Gasser and H. Leutwyler, Annals Phys. **158**, 142 (1984); Nucl. Phys. B **250**, 465 (1985).
2. I. Caprini, G. Colangelo and H. Leutwyler, Phys. Rev. Lett. **96**, 132001 (2006).
3. P. F. Bedaque, H. W. Hammer and U. van Kolck, Phys. Lett. B **569**, 159 (2003); U. van Kolck, Nucl. Phys. A **752**, 145 (2005).
4. H. Leutwyler, "pi pi scattering," in these Proceedings [arXiv:hep-ph/0612112].
5. H. W. Hammer, N. Kalantar-Nayestanaki and D. R. Phillips, a Working Group Summary in these Proceedings [arXiv:nucl-th/0611084].
6. V. Pascalutsa and D. R. Phillips, Phys. Rev. C **67**, 055202 (2003).
7. J. Gasser, M. E. Sainio and A. Svarc, Nucl. Phys. B **307**, 779 (1988).
8. E. Jenkins and A. V. Manohar, Phys. Lett. B **259**, 353 (1991).
9. V. Pascalutsa, M. Vanderhaeghen and S. N. Yang, Phys. Rep. (in press) [arXiv:hep-ph/0609004].
10. T. Hemmert, B. R. Holstein and J. Kambor, Phys. Lett. B **395**, 89 (1997); J. Phys. G **24**, 1831 (1998).
11. T. R. Hemmert, M. Procura and W. Weise, Phys. Rev. D **68**, 075009 (2003).
12. V. Bernard, Th. Hemmert and U. G. Meißner, Phys. Lett. B **622**, 141 (2005).
13. C. Hacker, N. Wies, J. Gegelia, S. Scherer, Phys. Rev. C **72**, 055203 (2005).
14. G. C. Gellas *et al.*, Phys. Rev. D **60**, 054022 (1999).
15. T. A. Gail and T. R. Hemmert, arXiv:nucl-th/0512082.
16. R. P. Hildebrandt *et al.*, Eur. Phys. J. A **20**, 293 (2004); R. P. Hildebrandt, PhD Thesis (University of Munich, 2005) [arXiv:nucl-th/0512064].
17. C. Hanhart and N. Kaiser, Phys. Rev. C **66**, 054005 (2002).
18. R. A. Arndt, I. I. Strakovsky, R. L. Workman, Phys. Rev. C **53**, 430 (1996).
19. V. Pascalutsa and M. Vanderhaeghen, Phys. Rev. Lett. **95**, 232001 (2005).
20. V. Pascalutsa and M. Vanderhaeghen, Phys. Rev. D **73**, 034003 (2006).
21. T. A. Gail and T. R. Hemmert, in Proc. of *Shape of Hadrons*, eds. C. N. Papanicolas and A. M. Bernstein, AIP (2007) [arXiv:nucl-th/0610081].
22. V. Pascalutsa and M. Vanderhaeghen, in Proc. of *Shape of Hadrons*, eds. C. N. Papanicolas and A. M. Bernstein, AIP (2007) [arXiv:hep-ph/0611317].
23. D. Drechsel *et al.*, Phys. Lett. B **484**, 236 (2000).
24. D. Drechsel and M. Vanderhaeghen, Phys. Rev. C **64**, 065202 (2001).
25. M. Kotulla *et al.*, Phys. Rev. Lett. **89**, 272001 (2002).
26. M. Kotulla, Talk at the Workshop *Shape of Hadrons, Athens, 2006.*
27. V. Pascalutsa and M. Vanderhaeghen, Phys. Rev. Lett. **94**, 102003 (2005).
28. V. Pascalutsa, B. R. Holstein and M. Vanderhaeghen, Phys. Lett. B **600**, 239 (2004); Phys. Rev. D **72**, 094014 (2005).
29. D. B. Leinweber, T. Draper, R. M. Woloshyn, Phys. Rev. D **46**, 3067 (1992);

I. C. Cloet, D. B. Leinweber, A. W. Thomas, Phys. Lett. B **563**, 157 (2003).
30. F. X. Lee, R. Kelly, L. Zhou and W. Wilcox, Phys. Lett. B **627**, 71 (2005).
31. C. Hacker, N. Wies, J. Gegelia, and S. Scherer, arXiv:hep-ph/0603267.
32. E. Jenkins and A. V. Manohar, Phys. Lett. B **255**, 558 (1991).
33. T. Becher and H. Leutwyler, Eur. Phys. J. C **9**, 643 (1999).
34. J. Gegelia and G. Japaridze, Phys. Rev. D **60**, 114038 (1999).
35. V. Pascalutsa and M. Vanderhaeghen, Phys. Lett. B **636**, 31 (2006).

SOME RECENT DEVELOPMENTS IN CHIRAL PERTURBATION THEORY*

ULF-G. MEIßNER[‡]
HISKP, Universität Bonn, D-53115 Bonn, Germany
and
IKP, Forschungszentrum Jülich, D-52425 Jülich, Germany
[‡] *E-mail: meissner@itkp.uni-bonn.de*

In this talk, I address some recent developments in chiral perturbation theory at unphysical and physical quark masses.

Keywords: Chiral Lagrangians, chiral extrapolations, isospin violation

1. Introduction I: Remarks on chiral extrapolations

The first part of this talk concerns the application of chiral perturbation theory (CHPT) at **unphysical** quark masses. More precisely, lattice QCD (LQCD) allows one in principle to calculate hadronic matrix elements ab initio using capability computing on a discretized space-time. To connect to the real world, various extrapolations are necessary: LQCD operates at a finite volume V, at a finite lattice spacing a and at large (unphysical) quark masses m_q. All these effects can be treated in suitably tailored effective field theories (EFTs) (for a recent review, see[1]). Here, I consider the quark mass expansion of certain baryon observables. Various nucleon (baryon) observables have already been computed on the lattice, like e.g. masses of ground and excited states, magnetic moments, nucleon electromagnetic radii, the nucleon axial-vector coupling, and so on. CHPT in principle provides extrapolation functions for all these observables, parameterized in terms of a number of low-energy constants (LECs). These LECs relate many observables, they are *not* dependent on the process one considers. Given the present situation with only a few lattice results at reasonably small quark masses available, it is mandatory to incorporate as much phenomenological

*Work supported in part by DFG (TR-16), by BMBF (06BN411) and by EU I3HP (RII3-CT-2004-506078).

input as is available for the LECs from studies of pion-nucleon scattering, pion production and so on. Also, since many LECs appear in various observables, a true check of our understanding of the chiral symmetry breaking of QCD requires global fits at sufficiently small quark masses. It is absolutely necessary for such extrapolations to make sense that one is in a regime where higher order terms stay sufficiently small. Consequently, results must be independent of the regularization scheme, these can differ by higher order terms (e.g. comparing results based on heavy baryon CHPT to ones obtained employing e.g. infrared regularization). For this interplay of CHPT and LQCD to make sense, the lattice "data" should be in the *true* chiral regime. I will illustrate these issues for two very different examples: the axial-vector coupling g_A and the Roper mass m_R.

2. Application I: The nucleon axial-vector coupling

The nucleon axial-vector coupling g_A is a fundamental quantity in hadron physics as it appears prominently in the Goldberger-Treiman relation. Lattice results for g_A obtained by various collaborations for pion masses between 300 and 1000 MeV show a very flat quark (pion) mass dependence. On the the other hand, it is long since known that the one–loop representation of g_A is not converging well and is dominated by the M_π^3 term with increasing pion mass,[2] thus $g_A(M_\pi)$ rises steeply as the pion mass increases. The large coefficient of this term is, however, understood in terms of the large values of the dimension two LECs c_3 and c_4 combined with some large numerical prefactors. In Ref.[3] we have therefore worked out the two–loop representation of g_A,

$$\begin{aligned} g_A &= g_0 \left\{ 1 + \left(\frac{\alpha_2}{(4\pi F)^2} \ln \frac{M_\pi}{\lambda} + \beta_2 \right) M_\pi^2 + \alpha_3 \, M_\pi^3 \right. \\ &+ \left. \left(\frac{\alpha_4}{(4\pi F)^4} \ln^2 \frac{M_\pi}{\lambda} + \frac{\gamma_4}{(4\pi F)^2} \ln \frac{M_\pi}{\lambda} + \beta_4 \right) M_\pi^4 + \alpha_5 \, M_\pi^5 \right\} + \mathcal{O}(M_\pi^6) \, , \\ &= g_0 \left\{ 1 + \Delta^{(2)} + \Delta^{(3)} + \Delta^{(4)} + \Delta^{(5)} \right\} + \mathcal{O}(M_\pi^6) \, , \end{aligned} \tag{1}$$

with g_0 the chiral limit value of g_A, λ is the scale of dimensional regularization, and the coefficients $\alpha_{2,3}, \beta_2$ encode the one-loop result. Further, F denotes the chiral limit value of the pion decay constant, $F \simeq 86\,\text{MeV}$, and $\Delta^{(n)}$ collects the corrections $\sim M_\pi^n$. At two-loop order, one has corrections of fourth and fifth order in the pion mass, given in terms of the coefficients $\alpha_{4,5}, \beta_4, \gamma_4$ (note that there is also a $M_\pi \ln M_\pi^5$ terms whose contribution we have absorbed in the uncertainty of α_5). The LEC α_4 can be analyzed

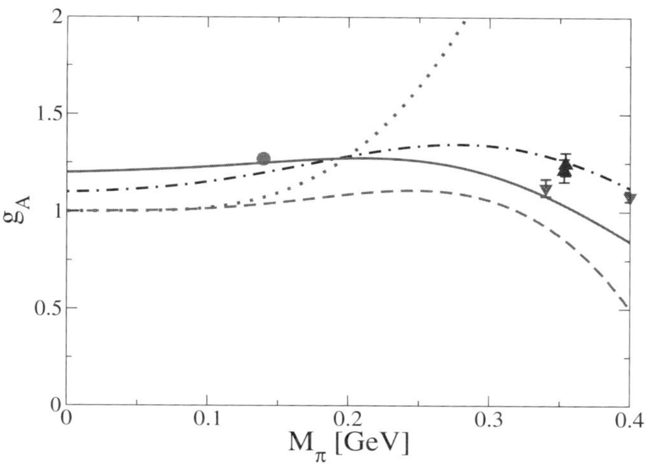

Fig. 1. The axial-vector coupling g_A as a function of the pion mass. The solid (red), the dot-dashed (black) and the dashed (green) line correspond to various input values for the LECs (see[3] for details). The (magenta) circle denotes the physical value of g_A at the physical pion mass, the triangles are the lowest mass data from Ref.[8] and the inverted triangles are recent results from QCDSF.[9]

using the renormalization group (as stressed long ago by Weinberg[4]) and is entirely given in terms of the dimension three coefficients of the one–loop generating functional. We have performed this calculation based on two existing versions of the dimension three pion-nucleon chiral Lagrangian without and with equation of motion terms (see[5] and,[6] respectively) and obtained

$$\alpha_4 = -\frac{16}{3} - \frac{11}{3}g_0^2 + 16g_0^4 \ . \qquad (2)$$

In[3] the dominant contributions to the LECs $\alpha_5, \beta_4, \gamma_4$ from $1/m_N$ corrections to dimension two and three insertions (with m_N the nucleon mass) as well as from the pion mass expansion of F_π were also worked out, for details see that paper. Setting the remaining contributions to zero, we find (in the notation of Eq. (1)) $g_A = g_0(1 - 0.15 + 0.26 - 0.06 - 0.001)$ and $g_0 = 1.21$, using $g_A = 1.267$ and central values for the LECs c_3, c_4, d_{16}, see e.g.[7] This shows that for the physical pion mass, the higher order corrections are small and one thus has a convergent representation. Varying the LECs $\alpha_5, \beta_4, \gamma_4$ within bounds given by naturalness, one finds that the pion mass dependence of g_A stays flat for $M_\pi \lesssim 350\,\text{MeV}$, see Fig. 1. From this figure one also sees that there is just a little overlap between the lattice results and the chiral representation, where it can be applied with a tolerable uncertainty. We remark that another solution to this problem was offered in Ref.[10] where an effective field theory with explicit delta degrees of freedom at leading one–loop order could lead to a flat pion mass dependence of g_A, requiring, however, a fine tuning of certain low–energy constants. For a recent update, see.[11] It should also be noted that most of the lattice results analyzed in these papers are far outside the range of applicability of that particular EFT evaluated only to leading one-loop order – that such a representation works at such large pion masses is an interesting observation but certainly does not support claims of a controlled and precise determination of g_A from LQCD.

3. Application II: Chiral corrections to the Roper mass

Understanding the (ir)regularities of the light quark baryon spectrum poses an important challenge for lattice QCD. In particular, the first even-parity excited state of the nucleon, the Roper $N^*(1440)$ (from here on called the Roper) is very intriguing—it is lighter than the first odd-parity nucleon excitation, the $S_{11}(1535)$, and also has a significant branching ratio into two pions. Recent lattice studies have not offered a clear picture about the nucleon resonance spectrum. In particular, in Ref.[12] an indication of a rapid cross over of the first positive and negative excited nucleon states close to the chiral limit was reported – so far not seen in other simulations at higher quark masses. Note also that so far very simple chiral extrapolation functions have been employed in most approaches, e.g., a linear extrapolation in the quark masses, thus $\sim M_\pi^2$, was applied in.[13] It is therefore important to provide the lattice practitioners with improved chiral extrapolation functions. A complete one–loop representation for the pion mass dependence of the Roper mass was recently given in.[14] Since the Roper is the first even-parity excited state of the nucleon, the construction of the chiral SU(2) effective Lagrangian follows standard procedures, see e.g.[15] The effective Lagrangian relevant for our calculation is (working in the isospin limit $m_u = m_d$ and neglecting electromagnetism)

$$\mathcal{L} = \mathcal{L}_0 + \mathcal{L}_R + \mathcal{L}_{NR}, \tag{3}$$

$$\mathcal{L}_0 = i\bar{N}\gamma_\mu D^\mu N - m_N \bar{N}N + i\bar{R}\gamma_\mu D^\mu R - m_R \bar{R}R \ , \tag{4}$$

$$\mathcal{L}_R^{(1)} = \frac{1}{2} g_R \bar{R}\gamma_\mu\gamma_5 u^\mu R \ , \ \mathcal{L}_{NR}^{(1)} = \frac{g_{NR}}{2} \bar{R}\gamma_\mu\gamma_5 u^\mu N + \text{h.c.} \ , \tag{5}$$

$$\mathcal{L}_R^{(2)} = c_1^* \langle \chi_+ \rangle \bar{R}R - \frac{c_2^*}{8m_R^2} \bar{R} \left(\langle u_\mu u_\nu \rangle \{D^\mu, D^\nu\} + \text{h.c.} \right) R + \frac{c_3^*}{2} \langle u_\mu u^\mu \rangle \bar{R}R \ , \tag{6}$$

$$\mathcal{L}_R^{(4)} = -\frac{e_1^*}{16} \langle \chi_+ \rangle^2 \bar{R}R \ , \tag{7}$$

where N, R are nucleon and Roper fields, respectively, and m_N, m_R the corresponding baryon masses in the chiral limit. The pion fields are collected in $u_\mu = -\partial_\mu \boldsymbol{\pi}/F_\pi + \mathcal{O}(\boldsymbol{\pi}^3)$. D_μ is the chiral covariant derivative, for our purpose we can set $D_\mu = \partial_\mu$, see e.g.[15] for definitions. Further, χ_+ is proportional to the pion mass and induces explicit chiral symmetry breaking, and $\langle \ \rangle$ denotes the trace in flavor space. The dimension two and four LECs c_i^* and e_i^* correspond to the c_i and e_i of the effective chiral pion–nucleon Lagrangian. The pion-Roper coupling is given to leading chiral order by $\mathcal{L}_R^{(1)}$, with a coupling g_R. This coupling is bounded by the nucleon axial coupling, $|g_R| < |g_A|$, in what follows we use $g_R = 1$. The leading interaction piece between nucleons and the Roper is given by $\mathcal{L}_{NR}^{(1)}$. The coupling g_{NR} can be determined from the strong decays of the Roper resonance, its actual value is $g_{NR} = 0.35$ using the Roper width extracted from the speed plot (and not from a Breit-Wigner fit). Further pion-Roper couplings are encoded in $\mathcal{L}_R^{(2)}$ and $\mathcal{L}_R^{(4)}$. To analyze the real part of the Roper self-energy, one has to calculate a) tree graphs with insertion $\sim c_1^*, e_1^*$, self-energy diagrams with intermediate b) nucleon and c) Roper states and d) tadpoles with vertices from $\mathcal{L}_R^{(2)}$. In fact, the graphs of type b) require a modification of the regularization scheme due to the appearance of the two large mass scales m_N and m_R. The solution to this problem – assuming $m_N^2/m_R^2 \ll 1$ (in nature, this ratio is $\simeq 1/2.4$) – is described in.[14] As discussed in that paper, the LECs c_i^* and e_i^* can be bounded assuming naturalness and by direct comparison with the corresponding pion-nucleon couplings: $|c_1^*| \lesssim 0.5\,\text{GeV}^{-1}$, $|c_{2,3}^*| \lesssim 1.0\,\text{GeV}^{-1}$ and $|e_1^*| \lesssim 0.5\,\text{GeV}^{-3}$. In

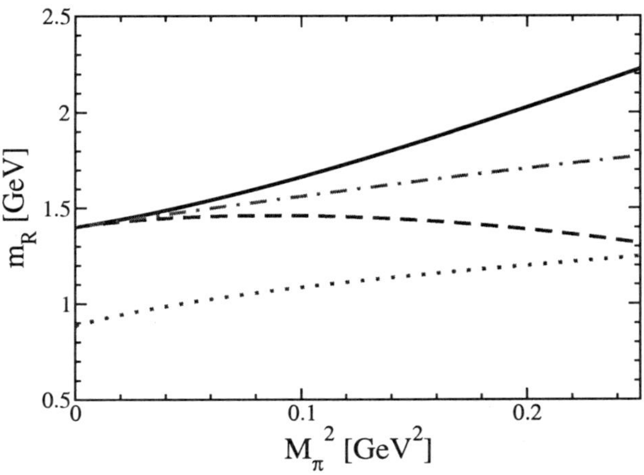

Fig. 2. Quark mass dependence of the Roper mass for different parameter sets $c_1^* = -0.5, c_{2,3}^*, e_1^*$. The c_i are in units of GeV^{-1} and e_1 is given in GeV^{-3}. and couplings $g_R = 1.0, g_{NR} = 0.35$. The solid curve corresponds to $c_2^* = 1.0, c_3^* = 1.0, e_1^* = 0.5$, the dashed one to $c_2^* = -1.0, c_3^* = -1.0, e_1^* = -0.5$ and the dot–dashed one to $c_2^* = c_3^* = e_1^* = 0$. The dotted curve represents the quark mass dependence of the nucleon, see Ref.[16] The values of the corresponding LECs are: $c_1 = -0.9, c_2 = 3.2, c_3 = -3.45, e_1 = -1.0$.

Fig. 2 an estimated range for the pion mass dependence of the Roper mass is presented by taking the extreme values for $c_{2,3}^*$ and e_1^*, while keeping $c_1^* = -0.5\,\mathrm{GeV}^{-1}, g_{NR} = 0.35, g_R = 1$ fixed. The masses of the baryons in the chiral limit are taken to be $m_N = 0.885\,\mathrm{GeV}$[16] and $m_R = 1.4\,\mathrm{GeV}$, respectively. The dash-dotted curve is obtained by setting the couplings $c_{2,3}^*, e_1^*$ all to zero, and exhibits up to an offset a similar quark mass dependence as the nucleon result (dotted curve, taken from Ref.[16]). It should be emphasized, however, that the one-loop formula cannot be trusted for pion masses much beyond 350 MeV. No sharp decrease of the Roper mass for small pion masses is observed for natural values of the couplings. Note that the important $\Delta\pi$ and $N\pi\pi$ channels are effectively included through the dimension two and four contact interactions, still it would be worthwhile to extend these considerations including the delta explicitely. Note further that the formalism developed in[14] is in general suited to study systems with two heavy mass scales in addition to a light mass scale. In this sense, it can be applied to other resonances as well, such as the $S_{11}(1535)$. In this case, however, an SU(3) calculation is necessary due to the important ηN decay channel.

4. Intoduction II: Hadronic atoms

Let us come back to the real world of physical quark masses. In what follows, I will discuss a spectacular effect of isospin violation in pionic deuterium, which is one particular hadronic atom. More generally, such atoms are made of certain hadrons bound by the static Coulomb force. There exist many species, e.g. $\pi^+\pi^-$, $\pi^\pm K^\mp$, $\pi^- p$, $\pi^- d$, $K^- p$, or $K^- d$. In these systems, the Bohr radii are much larger than any typical scale of strong interactions (QCD), so that their effects can be treated as perturbations. These are the energy shift ΔE from the Coulomb value and the decay width Γ (often combined in the complex valued energy shift). Since the average momenta in such systems are very small compared to any hadronic scale, hadronic atoms give access to scattering at zero energy and thus the pertinent S-wave scattering length(s). As it is well known, these scattering lengths are very sensitive to the chiral and isospin symmetry breaking in QCD. In fact, hadronic atoms offer may be the most precise method of determining these fundamental parameters since theory and experiment can in some cases be driven to an accuracy of one or a few percent. On the theoretical side, hadronic atoms can be analyzed systematically and consistently in the framework of non-relativistic low-energy EFT including virtual photons, see e.g.[17]

5. Isospin violation in pionic deuterium

Combined measurements of the energy shift and decay width of pionic hydrogen and the energy shift of pionic deuterium offer an excellent test of isospin symmetry and its breaking because these three quantities are expressed in terms of two scattering lengths, $\Delta E(\pi^- p) \sim a^+ + a^-$, $\Gamma(\pi^- p) \sim a^-$, and $\mathrm{Re}\,\Delta E(\pi^- d) \sim a^+ + \ldots$, where the ellipses denotes three-body effects such as multiple scattering within the deuteron. Here, a^+ and a^- are the isoscalar and the isovector S-wave πN scattering lengths, respectively. The Bern group has championed the EFT treatment of pionic hydrogen, the calculations including strong and electromagnetic isospin violation can be found in[18] and[19] for the ground state energy and the width, respectively. Using this formalism to analyze the data from PSI (as reviewed in[20]) and combining these with the EFT treatment of pionic deuterium in the isospin limit[21] (compare the bands denoted hydrogen energy, isospin breaking and hydrogen width, isospin breaking and deuteron, no isospin breaking in Fig. 3) one faces a problem - these bands do not intersect. Since the analysis of pionic deuterium in Ref.[21] was done in the isospin limit, the question naturally arises whether this is the source of the trouble? In principle, isospin violation (IV) leads to corrections in the bound-state as well as in the pertinent scattering amplitude. From experience with pionium, pionic hydrogen and πK atoms (see e.g. Ref.[22]), such bound-state corrections are expected to be small. On the other hand, already in 1977 Weinberg pointed out that IV effects can be unnaturally large if the isospin-conserving (IC) contribution is chirally suppressed.[23] In particular, such an effect is very pronounced in neutral pion scattering off nucleons (for an update, see[24]), but it is very hard to observe. On the other hand, the leading order contribution to πd scattering is chirally suppressed, $\mathrm{Re}\, a_{\pi d} \sim (a_{\pi^- p} + a_{\pi^- n}) = \mathcal{O}(p^2)$, despite the fact that $a_{\pi^- p}$ and $a_{\pi^- n}$ are individually of $\mathcal{O}(p)$. Here, p denotes collectively the small parameters of CHPT. While this is well-known, nobody has ever systematically investigated IV in $\pi^- d$. The leading order IV in pionic deuterium was only analyzed recently in Ref.[25] To be specific, consider the threshold pion-deuteron scattering amplitude:

$$\mathrm{Re}\ a_{\pi d}^{\mathrm{thr}} = \mathrm{Re}\ a_{\pi d}^{(0)} + \Delta a_{\pi d}\,, \tag{8}$$

where the IV piece $\sim \Delta a_{\pi d}$ appears at the same order as the leading IC piece $\sim \mathrm{Re}\ a_{\pi d}^{(0)}$. Also, to this order one has no dependence on the deuteron structure. The explicit calculation leads to

$$\Delta a_{\pi d}^{\mathrm{LO}} = (4\pi(1+\mu/2))^{-1}(\delta\mathcal{T}_p + \delta\mathcal{T}_n)\,. \tag{9}$$

Here, $\mu = M_{\pi^+}/m_p$ and $\mathcal{T}_{p,n}$ are the leading isospin breaking corrections to the $\pi^- p$ and $\pi^- n$ threshold scattering amplitudes. These are given by

$$\begin{aligned} \delta\mathcal{T}_p &= \frac{4(M_{\pi^+}^2 - M_{\pi^0}^2)}{F_\pi^2}\, c_1 - \frac{e^2}{2}\,(4f_1 + f_2) + \mathcal{O}(p^3)\,, \\ \delta\mathcal{T}_n &= \frac{4(M_{\pi^+}^2 - M_{\pi^0}^2)}{F_\pi^2}\, c_1 - \frac{e^2}{2}\,(4f_1 - f_2) + \mathcal{O}(p^3)\,, \end{aligned} \tag{10}$$

where $F_\pi = 92.4$ MeV is the pion decay constant, $g_A = 1.27$ denotes the axial-vector charge of the nucleon and c_1 is a strong and f_1, f_2 are electromagnetic $\mathcal{O}(p^2)$ LECs, respectively. Note also that at lowest order in CHPT c_1 is directly related to the value of the pion-nucleon σ-term and f_2 to the proton-neutron mass difference. In the numerical calculations we take $c_1 = -0.9^{+0.5}_{-0.2}$ GeV^{-1},[7] $f_2 = -(0.97 \pm 0.38)$ GeV^{-1}.[18,27] Note that the errors on the LEC c_1 are most conservative. The largest uncertainty in the results is introduced by the constant f_1, whose value at present is unknown and for which the dimensional estimate $|f_1| \leq 1.4$ GeV^{-1} has been used.

Note also, that the hydrogen energy band, which is shown in Fig. 3 corresponds to the new value of c_1 given above. From this, the leading order IV contribution to pionic deuterium follows as:

$$\begin{aligned}\Delta a_{\pi d}^{\mathrm{LO}} &= (4\pi(1+\mu/2))^{-1}(\delta\mathcal{T}_p+\delta\mathcal{T}_n)\\ &= \frac{1}{4\pi(1+\mu/2)}\left\{\frac{8\Delta M_\pi^2}{F_\pi^2}\,c_1-4e^2\,f_1\right\}+\mathcal{O}(p^3)\,,\end{aligned} \tag{11}$$

with $\Delta M_\pi^2 = M_{\pi^+}^2 - M_{\pi^0}^2$ the squared charged-to-neutral pion mass difference. Substituting numerical values for the various low-energy constants, which were specified above, one obtains that the correction at $\mathcal{O}(p^2)$ is extremely large

$$\Delta a_{\pi d}^{\mathrm{LO}} = -(0.0110^{+0.0081}_{-0.0058})\,M_\pi^{-1}\,, \tag{12}$$

that is $\Delta a_{\pi d}^{\mathrm{LO}}/\mathrm{Re}\,a_{\pi d}^{\mathrm{exp}} = 0.42$ (central values), using the experimental value $\mathrm{Re}\,a_{\pi d}^{\mathrm{exp}} = -(0.0261 \pm 0.0005)M_\pi^{-1}$.[26] Moreover, one can immediately see that the correction moves the deuteron band in Fig. 3 in the right direction: the isospin-breaking corrections amount for the bulk of the discrepancy between the experimental data on pionic hydrogen and deuterium. Including the corrections $\Delta a_{\pi d}^{\mathrm{LO}}$, all bands now have a common intersection area in the a^+, a^--plane, see Fig. 3. The resulting values for the πN scattering lengths are:

$$\begin{aligned}a^+ &= (0.0015 \pm 0.0022)\,M_\pi^{-1}\,,\\ a^- &= (0.0852 \pm 0.0018)\,M_\pi^{-1}\,.\end{aligned} \tag{13}$$

Further, using the hydrogen energy shift to estimate the LEC f_1, we obtain

$$f_1 = -2.1^{+3.2}_{-2.2}\ \mathrm{GeV}^{-1}\,, \tag{14}$$

which is consistent with the dimensional analysis and a recent evaluation based on a quark model.[28] Note that the error displayed here does not include the uncertainty coming from the higher orders in CHPT and should thus be considered preliminary.

As we see, the presence of the $\mathcal{O}(p^2)$ LECs in the expressions for the isospin-breaking corrections leads to a sizeable increase of the uncertainty in the output. In order to gain precision, in the fit one might also use those particular linear combination(s) of the experimental observables that do not contain f_1 and c_1. However, it should be pointed out that such a fit imposes much more severe constraints on the data than the uncorrelated fit considered above. In fact, applying isospin-breaking corrections only at $\mathcal{O}(p^2)$, we find that the data are still over-constrained in the combined fit.

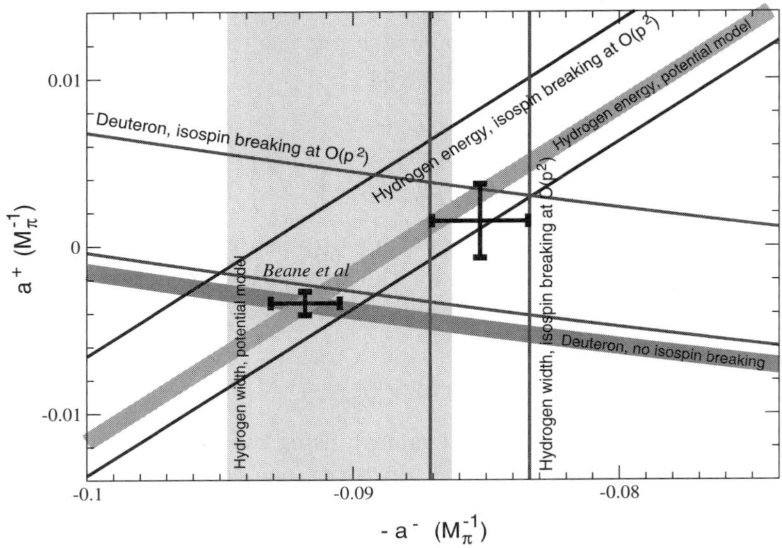

Fig. 3. Determination of the πN S-wave scattering lengths a^+ and a^- from the combined analysis of the experimental data on the pionic hydrogen energy shift and width, as well as the pionic deuterium energy shift (details in the text). The cross denoted as *Beane et al* is taken from Ref.[21] The second cross corresponds to the scattering lengths given in Eq. (13).

From this we finally conclude that to carry out such a combined analysis with the required precision, one would have e.g. first to evaluate the isospin-breaking corrections with a better accuracy.

Up to now, we have restricted ourselves to the leading-order isospin-breaking correction in CHPT. Calculations at $\mathcal{O}(p^3)$ exist only for the hydrogen energy shift and yield $\delta_\epsilon = (-7.2 \pm 2.9) \cdot 10^{-2}$ (using $c_1 = (-0.93 \pm 0.07)$ GeV^{-1}).[18] The corrections to $O(p^2)$ result are sizable (the energy band in Fig. 3 will be shifted further upwards), but the uncertainty, which is almost completely determined by the $\mathcal{O}(p^2)$ LECs, remains practically the same. On the other hand, consistent studies at $\mathcal{O}(p^3)$ imply the treatment of the scattering process in the three-body system in the effective field theory with virtual photons. To the best of our knowledge, such investigations have not been yet carried out, although certain three-body contributions at $\mathcal{O}(p^3)$ were calculated in the past.[29] It is natural to expect that generally three-body terms at $\mathcal{O}(p^3)$ should not depend on the additional LECs from the two-nucleon sector and hence the extraction of the

πN scattering lengths at a high precision is still possible. Of course, these arguments can not be a substitute for a rigorous proof in the framework of EFT, which in the light of the above discussion, is urgently called for. It was also shown in Ref.[25] that the $\mathcal{O}(p^4)$ correction which emerges from the double-scattering term in the multiple-scattering series is very small, $\Delta a_{\pi d}^{\text{double scat.}} = 0.003\,\text{Re}\,a_{\pi d}^{\text{exp}}$, using the scattering lengths from Eq. (13) as input. Of course, such a partial result can only be considered indicative. Evidently, the systematic analysis of all $\mathcal{O}(p^3)$ (and eventually $\mathcal{O}(p^4)$) corrections should be carried out.

6. Summary and outlook

In the first part of this talk, I have considered aspects related to baryon CHPT for light quark masses above their physical values. It should be stressed that baryon CHPT is a mature field in the up and down quark sector and provides unambiguous extrapolation functions for LQCD – certainly more work is needed for the three-flavor case. To carry out the required chiral extrapolations, one should keep in mind that different observables are linked by general operator structures and the appearing low-energy constants (LECs) are universal, which means that they are independent of the process considered. Given the present status of LQCD, it appears mandatory to perform global fits to observables and constrain the appearing LECs by input from phenomenology, whenever available. I have discussed two specific examples of the interplay between CHPT and LQCD. A chiral extrapolation for g_A exists now at two–loop accuracy. For pion masses $\lesssim 350$ MeV, the theoretical uncertainty related to it is reasonably small. I have also provided a chiral extrapolation function for the Roper mass - certainly much more work is needed for such excited states from both CHPT and LQCD. Evidently, we need more lattice "data" at low quark masses to really perform precision studies.

In the second part of this talk, I returned to the real world (physical quark masses) and considered hadronic atoms. These can be systematically analyzed in non-relativistic effective field theory including virtual photons. We have found a very large isospin-violating effect in pionic deuterium at leading order. That there is such an effect is not so surprising because the leading isospin-conserving contribution is chirally suppressed. What is surprising, however, is the actual size of the effect and that it was only found recently. Combining this with the information obtained from the analysis of the energy shift and width in pionic hydrogen, one is led to a consistent extraction of the S-wave πN scattering lengths and can furthermore determine the electromagnetic LEC f_1. Clearly, higher order calculations are necessary to reduce the theoretical uncertainty. In this context it is also important to stress that it was recently shown that there are only *tiny* dispersive corrections to Re $a_{\pi d}$, see[30] and Hanhart's talk at this conference.

Acknowledgements

I thank the organizers for a superb job and all my collaborators for sharing their insight into the topics discussed here.

References

1. S. R. Sharpe, arXiv:hep-lat/0607016.
2. J. Kambor and M. Mojžiš, JHEP **9904** (1999) 031 [arXiv:hep-ph/9901235].
3. V. Bernard and U.-G. Meißner, Phys. Lett. B **639** (2006) 278 [arXiv:hep-lat/0605010].
4. S. Weinberg, PhysicaA **96** (1979) 327.
5. G. Ecker and M. Mojžiš, Phys. Lett. B **365** (1996) 312 [arXiv:hep-ph/9508204].
6. N. Fettes, U.-G. Meißner and S. Steininger, Nucl. Phys. A **640** (1998) 199 [arXiv:hep-ph/9803266].
7. U.-G. Meißner, PoS **LATT2005** (2005) 009 [arXiv:hep-lat/0509029].
8. R. G. Edwards *et al.* [LHPC Collaboration], Phys. Rev. Lett. **96** (2006) 052001 [arXiv:hep-lat/0510062].
9. M. Göckeler, private communication; QCDSF collaboration, to be published.
10. T. R. Hemmert, M. Procura and W. Weise, Phys. Rev. D **68** (2003) 075009 [arXiv:hep-lat/0303002].
11. M. Procura, B. U. Musch, T. R. Hemmert and W. Weise, Nucl. Phys. Proc. Suppl. **153** (2006) 229 [arXiv:hep-lat/0512026].
12. N. Mathur *et al.*, Phys. Lett. B **605** (2005) 137 [arXiv:hep-ph/0306199].
13. T. Burch, C. Gattringer, L. Y. Glozman, C. Hagen, D. Hierl, C. B. Lang and A. Schäfer, Phys. Rev. D **74** (2006) 014504 [arXiv:hep-lat/0604019].
14. B. Borasoy, P. C. Bruns, U.-G. Meißner and R. Lewis, Phys. Lett. B **641** (2006) 294 [arXiv:hep-lat/0608001].
15. N. Fettes, U.-G. Meißner, M. Mojžiš and S. Steininger, Annals Phys. **283** (2000) 273 [Erratum-ibid. **288** (2001) 249] [arXiv:hep-ph/0001308].
16. V. Bernard, T. R. Hemmert and U.-G. Meißner, Nucl. Phys. A **732** (2004) 149 [arXiv:hep-ph/0307115].
17. A. Rusetsky, arXiv:hep-ph/0011039.
18. J. Gasser, M. A. Ivanov, E. Lipartia, M. Mojžiš and A. Rusetsky, Eur. Phys. J. C **26** (2002) 13 [arXiv:hep-ph/0206068].
19. P. Zemp, Ph.D. thesis, Berne University, 2004.
20. D. Gotta [Pionic Hydrogen Collaboration], Int. J. Mod. Phys. A **20** (2005) 349.
21. S. R. Beane, V. Bernard, E. Epelbaum, U.-G. Meißner and D. R. Phillips, Nucl. Phys. A **720** (2003) 399 [arXiv:hep-ph/0206219].

22. J. Schweizer, Eur. Phys. J. C **36** (2004) 483 [arXiv:hep-ph/0405034].
23. S. Weinberg, Trans. New York Acad. Sci. **38** (1977) 185.
24. N. Fettes, U.-G. Meißner and S. Steininger, Phys. Lett. B **451** (1999) 233 [arXiv:hep-ph/9811366].
25. U.-G. Meißner, U. Raha and A. Rusetsky, Phys. Lett. B **639** (2006) 478 [arXiv:nucl-th/0512035].
26. P. Hauser *et al.*, Phys. Rev. C **58** (1998) 1869.
27. U.-G. Meißner and S. Steininger, Phys. Lett. B **419** (1998) 403 [arXiv:hep-ph/9709453].
28. V. E. Lyubovitskij, T. Gutsche, A. Faessler and R. Vinh Mau, Phys. Rev. C **65** (2002) 025202 [arXiv:hep-ph/0109213].
29. R. M. Rockmore, Phys. Lett. B **356** (1995) 153.
30. V. Lensky, V. Baru, J. Haidenbauer, C. Hanhart, A. E. Kudryavtsev and U.-G. Meißner, arXiv:nucl-th/0608042.

CHIRAL EXTRAPOLATION AND NUCLEON STRUCTURE FROM THE LATTICE

R. D. YOUNG

Jefferson Lab, 12000 Jefferson Ave., Newport News, VA 23606, USA

The slow convergence of the quark-mass expansion necessitates some form of resummation of the chiral series to make connection with the results of dynamical lattice simulations. Here we outline the formulation of finite-range regularisation (FRR), demonstrating the mathematical equivalence with any other renormalised expansion. We use the nucleon mass as an example to demonstrate the improved convergence offered by FRR.

1. Introduction

With continuing advances in lattice QCD simulations with dynamical fermions, there is a significant effort to make genuine comparisons between nonperturbative QCD and modern experimental programs. Beyond ensuring the control of finite-volume and discretisation errors, the large computational demands restrict the simulated quark masses to a domain substantially larger than the physical light-quark masses. This necessitates an extrapolation of the lattice results to access the physical regime. In attempt to make connection of effective field theory (EFT) with lattice results, there has been recent interest in the regularisation.[1–5]

Here we review the advances in finite-range regularised (FRR) EFT that offers reliable extrapolations from beyond the realm of the perturbative chiral series. Further, even when the chiral regime is readily accessible within lattice simulations, implementation of FRR offers a natural check on the convergence of a given truncation.

We describe the renormalisation of the nucleon mass expansion in FRR. We prove the mathematical equivalence of the FRR approach to a minimal subtraction scheme, such as dimensional regularisation (DR). The FRR expansion is applied to the extrapolation of lattice results and contrasted with the DR form. FRR is found to offer superior stability between successive orders of the expansion.

2. Radius of convergence

The heart of the success of perturbative expansion techniques, such as chiral pertubation theory, is a small expansion parameter and the reliance of a series expansion of natural-sized coefficients. The perfect example of such a series is the geometric series

$$1 - x + x^2 - x^3 + \ldots = \frac{1}{1+x}, \tag{1}$$

where the radius of convergence for the series expansion about $x = 0$ is $|x| < 1$. With such an expansion one can trivially ascertain the error in truncation within a range of x by considering the relative magnitude of succesive terms in the series, without necessarily have knowledge of the whole function.

This simple description can provide insight into the natural convergence radius of the quark-mass expansion of the nucleon mass. Writing the renormalised nucleon mass expansion to leading nonanalytic order, we have

$$M_N = M_N^0 + c_2 M_\pi^2 + \chi_\pi M_\pi^3 + \ldots . \tag{2}$$

The model-independence of nonanalytic terms ($M_\pi \propto \sqrt{m_q}$) in the quark-mass expansion of hadronic observables is one of the celebrated features of EFT.[6] In the case of the nucleon mass, the coefficient of this leading nonanalytic term is given by

$$\chi_\pi = -\frac{3\, g_A^2}{32\pi f_\pi^2}, \tag{3}$$

where g_A and f_π are to be determined in the chiral limit. Nevertheless, inserting the physical values indicates that this coefficient takes a numerical value $\chi_\pi \simeq -5.6\,\mathrm{GeV}^{-2}$. Further, the term in M_π^2 is dominated by the pion-nucleon sigma term. Assuming this to take its canonical value $\sigma_N = 45\,\mathrm{MeV}$,[7] and neglecting the contribution of higher order terms, one arrives at $c_2 \simeq 3.6\,\mathrm{GeV}^{-1}$. Finally, the physical nucleon mass sets the overall scale, giving $M_N^0 \simeq 0.89\,\mathrm{GeV}$.

Inserting these estimate numerical values into the expansion, eq. (2), at a pion mass of $0.54\,\mathrm{GeV}$ gives

$$M_N = 0.89(1 + 1.2 - 1.0 + \ldots)\,\mathrm{GeV}, \tag{4}$$

and it is observed that each term contributes ~100% to the total mass. Thereby at a pion mass of the order $M_\pi \sim 0.54\,\mathrm{GeV}$ one is at the radius of convergence of this series, requiring and infinite number of terms to accurately reproduce the true curve. In a practical case, one may then ask how close to this radius of convergence can one get with a finite number of terms in the truncation.

The expansion in Eq. (4) is very suggestive of a geometric series in M_π (with vanishing linear term). Demanding that the series expansion in eq. (1) up to x^3 is accurate to the 1% level, limits one to $x < 0.34$. Simply scaling this to the nucleon mass case, with radius of convergence 0.54 GeV, indicates that the 1% precision at order M_π^3 is attained within

$$M_\pi < 0.34 * 0.54\,\mathrm{GeV} = 0.18\,\mathrm{GeV}\,. \tag{5}$$

Continuing the natural-size argument means that at order M_π^4 demanding this precision limits one to $M_\pi < 0.23\,\mathrm{GeV}$. We note that this is a model-independent result, relying only on the assumption of a natural perturbation parameter — the heart of effective field theory.

The only way to extrapolate lattice results to the chiral regime is to look to resummation schemes which can improve the convergence properties of the expansion.

3. Renormalisation

Here we review the regularisation and renormalisation that produced to the leading expansion of the nucleon mass shown in eq. (2). Before regularisation or renormalisation the nucleon mass expansion can be written as

$$M_N = a_0 + a_2 M_\pi^2 + \chi_\pi I_\pi\,, \tag{6}$$

where χ_π is the leading non-analytic (LNA) coefficient as defined in eq. (3) and I_π denotes the leading one-loop integral. In the heavy baryon limit, this integral over pion momentum can be written as

$$I_\pi = \frac{2}{\pi}\int_0^\infty dk\,\frac{k^4}{k^2 + M_\pi^2}\,. \tag{7}$$

This integral suffers from a cubic divergence for large momenta. The infrared behaviour of this integral gives the leading non-analytic correction to the nucleon mass. This arises from the pole in the pion propagator at complex momentum $k = iM_\pi$ and will be determined independent of how the ultraviolet behaviour of the integral is treated. Rearranging eq. (7) we see that the pole contribution can be isolated from the divergent part

$$I_\pi = \frac{2}{\pi}\int_0^\infty dk\,\left(k^2 - M_\pi^2\right) + \frac{2}{\pi}\int_0^\infty dk\,\frac{M_\pi^4}{k^2 + M_\pi^2}\,. \tag{8}$$

The final term converges and is given simply by

$$\frac{2}{\pi}\int_0^\infty dk\,\frac{M_\pi^4}{k^2+M_\pi^2}=M_\pi^3\,, \tag{9}$$

where we now recognise the choice of normalisation of the loop integral, defined such that the coefficient of the LNA term is set to unity. This choice is purely conventional and allows for a much more transparent presentation of the differences in the chiral expansion with various regularisation schemes.

In the most basic form of renormalisation we could simply imagine absorbing the infinite contributions arising from the first term in eq. (8) into a redefinition of the coefficients a_0 and a_2 in eq. (6). This solution is simply a minimal subtraction scheme and the renormalised expansion can be given without making reference to an explicit scale,

$$m_N=c_0+c_2M_\pi^2+\chi_\pi M_\pi^3\,, \tag{10}$$

and the renormalised coefficients indentified by

$$c_0=a_0+\chi_\pi\frac{2}{\pi}\int_0^\infty dk\,k^2\,, \tag{11}$$

$$c_2=a_2-\chi_\pi\frac{2}{\pi}\int_0^\infty dk\,. \tag{12}$$

Equation (10) therefore encodes the complete quark mass expansion of the nucleon mass to $\mathcal{O}(M_\pi^3)$. This result will be precisely equivalent to any form of minimal subtraction scheme, where all the ultraviolet behaviour will be absorbed into the two leading coefficients of the expansion.

We now describe the chiral expansion within finite-range regularisation, where the cut-off scale remains explicit. In particular, we highlight the mathematical equivalence of FRR and dimensional regularisation in the low energy regime. We introduce a functional cutoff, $u(k)$, defined such that the loop integral is ultraviolet finite,

$$I_\pi=\frac{2}{\pi}\int_0^\infty dk\,\frac{k^4u^2(k)}{k^2+M_\pi^2}\,. \tag{13}$$

To preserve the infrared behaviour of the loop integral, the regulator is defined to be unity as $k\to 0$. For demonstrative purposes, we choose a dipole regulator $u(k)=(1+k^2/\Lambda^2)^{-2}$, giving

$$I_\pi^{\rm DIP}=\frac{\Lambda^5(M_\pi^2+4M_\pi\Lambda+\Lambda^2)}{16(M_\pi+\Lambda)^4}\,. \tag{14}$$

A Taylor series expansion of this integral about vanishing M_π defines the renormalisation of the chiral expansion in FRR,

$$I_\pi^{\rm DIP} \sim \frac{\Lambda^3}{16} - \frac{5\Lambda}{16} M_\pi^2 + M_\pi^3 - \frac{35}{16\Lambda} M_\pi^4 + \ldots \tag{15}$$

The renormalised expansion in FRR is therefore precisely equivalent to eq. (10) up to $\mathcal{O}(M_\pi^3)$ where the leading coefficients are given by

$$c_0 = a_0 + \chi_\pi \frac{\Lambda^3}{16}\,, \tag{16}$$

$$c_2 = a_2 - \chi_\pi \frac{5\Lambda}{16}\,, \tag{17}$$

with a tower of higher-order terms. Retaining the full form of eq. (14) will therefore build in a resummation of higher-order terms in the chiral series.

A systematic study of the quark mass expansion in numerous regularisation schemes, particularly dimensional and finite-range regularisation has been reported in Ref.[1] By studying a one-to-one correspondence between individual schemes it was found that the DR forms could only reproduce the FRR curves over a limited range of pion masses. In contrast, all FRR forms demonstrated equivalence to DR over a much wider window of quark masses. The results found in Ref.[1] established quantitatively that there is minimal dependence on the choice of finite-range regulator.

4. Chiral Extrapolation

Using the tools developed in the previous section, we study chiral extrapolation with various regularisation prescriptions — the features of these results have previously been published in Ref.[8]

Figure 1 shows results of two-flavour dynamical QCD simulations of mean-field improved clover fermions on an improved gluon action.[9] Improvement in the lattice actions acts to remove ultraviolet cutoff effects associated with the finite lattice spacing. Results are demonstrated to represent a good approximation to the continuum theory.

The curves in Figure 1 are fits to the lattice data based upon the nucleon mass expansion to leading non-analytic order. Equation (6) is extended, adding an extra counter-term at M_π^4 to minimise any residual regulator dependence,

$$m_N = a_0 + a_2 M_\pi^2 + \chi_\pi I_\pi + a_4 M_\pi^4\,. \tag{18}$$

The dashed curve is that obtained with I_π regularised with a minimal subtraction scheme as described above. The four solid curves, which all overlap,

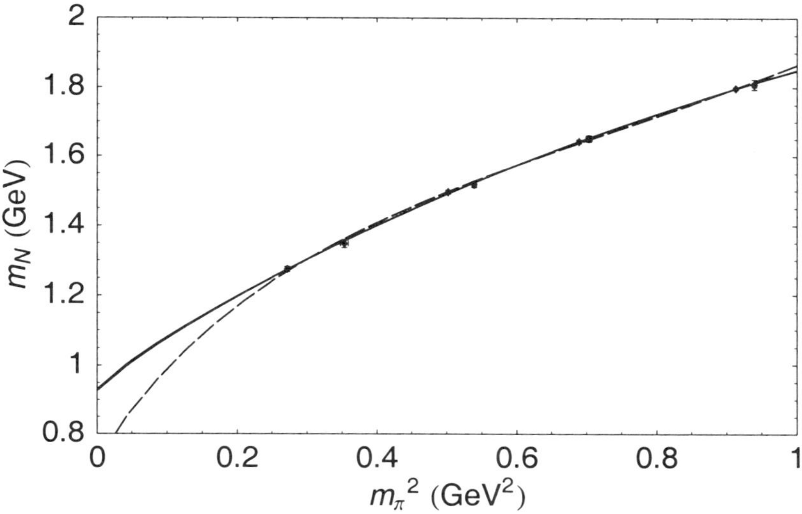

Fig. 1. Extrapolation of nucleon mass lattice data based on different regularisation schemes to order $m_q^{3/2}$. The four, indistinguishable, solid lines correspond to the finite-range regulators, the dashed curve is the dimensionally regularised fit. The dash-dot curve displays the DR form with the correct branch structure at $M_\pi = \Delta$.

correspond to I_π evaluated with different choices of FRR: a sharp cutoff, monopole, dipole and Gaussian forms.

The extrapolations based on the FRR forms display good agreement with the data and are consistent with the experimental nucleon mass. It appears that the leading order dimensionally regularised extrapolation is not reliably convergent at these lattice quark masses. We therefore extend our analysis to next-to-leading non-analytic order, explicitly including all terms up to $m_q^2 \sim M_\pi^4$. Most importantly, there are non-analytic contributions of order $M_\pi^4 \log M_\pi$ arising from the Δ-baryon and tadpole loop contributions.

There are two additional loop diagrams to consider at this order. Firstly, we incorporate a contribution arising from the nucleon coupling to $\Delta\pi$. This contribution is similar to that arising from the $N\pi$ self-energy, with modified coupling strength $\chi_{\pi\Delta}$ and and an explicit mass splitting between the octet and decuplet baryons,

$$I_{\pi\Delta} = \frac{2}{\pi} \int_0^\infty dk \, \frac{k^4}{\omega(k)[\Delta + \omega(k)]} \, . \tag{19}$$

In a minimal subtraction scheme, with $M_\pi \ll \Delta$, this contribution can be described simply as

$$I_{\pi\Delta}^{\rm DR} = -\frac{3}{4\pi\Delta} M_\pi^4 \log \frac{M_\pi}{\mu} \, . \tag{20}$$

In addition to this form, we also retain the integral which preserves the correct branch-point (BP) structure at $M_\pi = \Delta$.[10] In Ref.[1] the renormalsation details of the $\Delta\pi$ loop are given.

To this order, we also have a loop contribution arising from the expansion of the $\mathcal{O}(m_q)$ chiral Lagrangian. This comes as a tadpole diagram, with a coupling proportional to the renormalised m_q expansion coefficient, c_2. The corresponding loop integral can be expressed as

$$I_{\rm tad} = M_\pi^2 \int dk \frac{2k^2}{\sqrt{k^2 + M_\pi^2}} \, , \tag{21}$$

with coefficient given by

$$\chi_{\rm tad} = -c_2 \frac{3}{16\pi^2 f_\pi^2} \, . \tag{22}$$

We note that in evaluating $I_{\rm tad}$ with a FRR we must subtract the leading constant from the integral such that the renormalisation of c_2 remains linear. We denote this integral as $I_{\rm tad}^s$.

In summary, we now have our nucleon mass expansion to $\mathcal{O}(m_q^2)$ given by

$$\begin{aligned} m_N &= a_0 + a_2 M_\pi^2 + a_4 M_\pi^4 + a_6 M_\pi^6 \\ &\quad + \chi_\pi I_\pi + \chi_{\pi\Delta} I_{\pi\Delta} + \chi_{\rm tad} I_{\rm tad}^s \, . \end{aligned} \tag{23}$$

Once again, with appropriate renormalisation, the expansion is equivalent for all regularisation schemes

$$\begin{aligned} m_N &= c_0 + c_2 M_\pi^2 + + \chi_\pi M_\pi^4 + c_4 M_\pi^4 \\ &\quad (\chi_{\rm tad} - \chi_{\pi\Delta} \frac{3}{4\pi\Delta}) M_\pi^4 \log \frac{M_\pi}{\mu} + \ldots \end{aligned} \tag{24}$$

We now show the fit to the lattice data based on the NLNA order in Fig. 2. The FRR results display excellent mutual agreement. The dimensional regularisation forms, particularly the branch-point form, show significantly improved fits over the leading order result.

In Table 1 the extrapolated nucleon masses for the various fits are displayed. Remarkable agreement is observed between the different orders of the expansion for the FRR fits. This gives very strong support, even to the large pion masses considered in this study, that the series is sufficiently convergent. The NLNA dimensionally regularised forms are consistent with

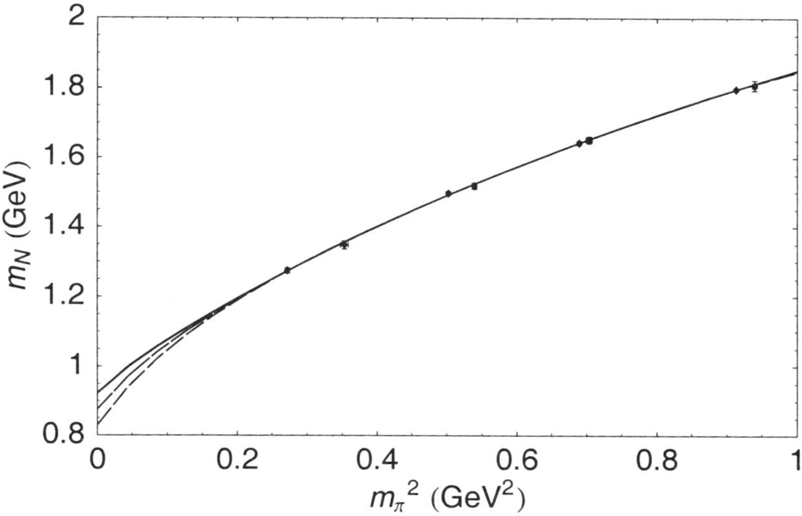

Fig. 2. Extrapolation of nucleon mass lattice data based on different regularisation schemes to order m_q^2. The four, indistinguishable, sold lines correspond to the finite-range regulators, the dashed curve is the dimensionally regularised fit. The dash-dot curve displays the DR form with the correct branch structure at $M_\pi = \Delta$.

those of FRR within statistical error, yet do not offer that same level of stability.

Table 1. Extrapolated nucleon mass (in GeV) evaluated at the physical pion mass. The two columns describe extrapolations based on the LNA and NLNA forms.

Regulator	$M_N^{(\mathrm{LNA})}$	$M_N^{(\mathrm{NLNA})}$
Dim. Reg.	0.784	0.884
Dim. Reg. (BP)	$\cdots$	0.923
Sharp cutoff	0.968	0.961
Monopole	0.964	0.960
Dipole	0.963	0.959
Gaussian	0.966	0.960

5. Conclusions

We have highlighted the need for some form of resummation of the chiral series by demonstrating the poor convergence of the expansion. Finite-range regularisation offers the best available technique to connect with lattice results. Only a small correction between leading and next-leading order extrapolations is observed, with minimal dependence on the choice of regulator.

The improved convergence properties of FRR mean that it provides the most reliable extrapolation form for modern lattice simulations. The robust form provided by FRR will become increasingly important as simulations explore excited state spectroscopy and three-point functions.

Acknowledgements

This work was supported by U.S. DOE Contract No. DE-AC05-06OR23177, under which Jefferson Science Associates, LLC operate Jefferson Lab.

References

1. R. D. Young, D. B. Leinweber and A. W. Thomas, Prog. Part. Nucl. Phys. **50**, 399 (2003) [arXiv:hep-lat/0212031].
2. V. Bernard, T. R. Hemmert and U. G. Meissner, Nucl. Phys. A **732**, 149 (2004) [arXiv:hep-ph/0307115].
3. D. Djukanovic, M. R. Schindler, J. Gegelia and S. Scherer, Phys. Rev. D **72**, 045002 (2005) [arXiv:hep-ph/0407170].
4. V. Pascalutsa and M. Vanderhaeghen, Phys. Lett. B **636**, 31 (2006) [arXiv:hep-ph/0511261].
5. A. Semke and M. F. M. Lutz, Nucl. Phys. A **778**, 153 (2006) [arXiv:nucl-th/0511061].
6. L. F. Li and H. Pagels, Phys. Rev. Lett. **26**, 1204 (1971).
7. J. Gasser, H. Leutwyler and M. E. Sainio, Phys. Lett. B **253**, 252 (1991).
8. D. B. Leinweber, A. W. Thomas and R. D. Young, Phys. Rev. Lett. **92**, 242002 (2004) [arXiv:hep-lat/0302020].
9. A. Ali Khan *et al.* [CP-PACS Collaboration], Phys. Rev. D **65**, 054505 (2002) [Erratum-ibid. D **67**, 059901 (2003)] [arXiv:hep-lat/0105015].
10. M. K. Banerjee and J. Milana, Phys. Rev. D **54**, 5804 (1996) [arXiv:hep-ph/9508340].

RECENT RESULTS FROM HAPPEX

ROBERT MICHAELS

Thomas Jefferson National Accelerator Facility
12000 Jefferson Ave
Newport News, VA, USA
E-mail: rom@jlab.org

New measurements of the parity-violating asymmetry $A_{\rm PV}$ in elastic scattering of 3 GeV electrons off hydrogen and ^{4}He targets $\langle\theta_{lab}\rangle \approx 6.0^\circ$ are reported. The ^{4}He result is $A_{\rm PV} = (+6.40 \pm 0.23$ (stat) $\pm$ 0.12 (syst)) $\times\ 10^{-6}$. The hydrogen result is $A_{\rm PV} = (-1.58 \pm 0.12$ (stat) $\pm$ 0.04 (syst)) $\times\ 10^{-6}$. The asymmetry for hydrogen is a function of a linear combination of G_E^s and G_M^s, the strange quark contributions to the electric and magnetic form factors of the nucleon respectively, and that for ^{4}He is a function solely of G_E^s. The combination of the two measurements separates G_E^s and G_M^s and provide new limits on the role of strange quarks in the nucleon charge and magnetization distributions. We extract $G_E^s = 0.002 \pm 0.014 \pm 0.007$ at $\langle Q^2\rangle = 0.077\,\mathrm{GeV}^2$, and $G_E^s + 0.09\,G_M^s = 0.007 \pm 0.011 \pm 0.006$ at $\langle Q^2\rangle = 0.109\,\mathrm{GeV}^2$.

Keywords: Elastic Electron Scattering, Parity Violation, Nucleon Structure, Strangeness in Nucleon

1. Introduction

Measurements of the contribution of strange quarks to nucleon structure provide a unique window on the quark-antiquark sea and make an important impact on our understanding of the non-perturbative QCD structure of nucleons. Since the mass of the strange quark m_s is comparable to the strong interaction scale $\Lambda_{\rm QCD}$ it is reasonable to think that strangeness $s\bar{s}$ pairs could make observable contributions to the properties of nucleons, i.e. the mass, spin, momentum, and the electromagnetic form factors. Data from the European Muon Collaboration (EMC)[1] showed that valence quarks contribute less than half of the proton spin and also suggested that significant spin may be carried by the strange quarks. Based on these results, Kaplan and Manohar[2] pointed out that strange quarks might also contribute to the magnetic moment and charge radius of the proton, *i.e.* to the vector matrix elements. These strange vector matrix elements can be accessed by measuring the electroweak asymmetry in polarized electron scattering.[3–5] See also the reviews by Kumar and Souder,[6] Beck and McKeown,[7] Beck and Holstein,[8] and Musolf *et al.*[9]

2. Strange Form Factors

Electron scattering at a given elastic kinematics can measure only two linear combinations of the Sachs form factors. In terms of u, d, and s quark form factors, the proton form factors are:

$$G^{\gamma p}_{E,M} = \frac{2}{3}G^{u}_{E,M} - \frac{1}{3}G^{d}_{E,M} - \frac{1}{3}G^{s}_{E,M} \tag{1}$$

And for the neutron :

$$G^{\gamma n}_{E,M} = \frac{2}{3}G^{d}_{E,M} - \frac{1}{3}G^{u}_{E,M} - \frac{1}{3}G^{s}_{E,M} \tag{2}$$

where $G^{f}_{E,M}$ is the electric (E) or magnetic (M) form factor for quark flavor f. Here it is assumed that the quark flavors u, d, and s contribute and the goal is to determine $G^{s}_{E,M}$, the strangeness form factors. Charge symmetry between proton p and neutron n is normally assumed, so that for the quark form factors

$$G^{u}_{p} = G^{d}_{n}\ ;\quad G^{d}_{p} = G^{u}_{n}\ ;\quad G^{s}_{p} = G^{s}_{n} \tag{3}$$

where now the subscripts p and n are for proton and neutron. We note that a recent theoretical analysis[10] suggests the effects of charge symmetry violation (CSV) may be as large as the HAPPEX experimental error.

Since there are too many quark form factors for the given number of nucleon electromagnetic form factors, one needs additional information to determine whether or not there is a contribution from the strangeness form factors $G^{s}_{E,M}$. This is provided by parity violation because the Z^0 of the weak interaction accesses the same flavor structure with different coupling constants given by the Standard Model. For a proton:

$$\begin{aligned} G^{Zp}_{E,M} = &\left(\frac{1}{4} - \frac{2}{3}\sin^2\theta_W\right) G^{u}_{E,M} + \\ &\left(-\frac{1}{4} + \frac{1}{3}\sin^2\theta_W\right) \times \left[G^{d}_{E,M} + G^{s}_{E,M}\right] \end{aligned} \tag{4}$$

The parity violating asymmetry is $A_{\rm PV} = (\sigma_R - \sigma_L)/(\sigma_R + \sigma_L)$ where $\sigma_{R(L)}$ is the differential cross section for elastic scattering of right(R) and left(L) handed longitudinally polarized electrons from protons. This asymmetry arises from an interference between the weak and electromagnetic amplitudes and is a function of the neutral weak form factors as well as the ordinary electromagnetic form factors. By measuring $A_{\rm PV}$ for a proton we can extract a linear combination of G_E^s and G_M^s. The asymmetry also contains a term with the neutral weak axial form factor G_A^{Zp} which is a small contribution at the HAPPEX kinematics.

Parity violation in elastic scattering from the ^{4}He nucleus is directly sensitive to the G_E^s. Neglecting well-known radiative corrections the asymmetry is

$$A_{\rm PV} = A_0 \tau \left\{ 4 \sin^2 \theta_W + \frac{2G_E^s}{G_E^p + G_E^n} \right\} \tag{5}$$

where

$$A_0 = \frac{G_F M_P^2}{\sqrt{2}\pi\alpha} = 316.7 \text{ ppm} \tag{6}$$

Thus by combining measurements from the proton and from helium, we can separate G_E^s and G_M^s. The interpretation of the experiments is rather clean, requiring only the assumptions of charge symmetry in the proton and that at our low Q^2 nuclear effects in ^{4}He are under sufficient control theoretically.

Numerous calculations of strange matrix elements have been computed in the context of various models. The theoretical approaches include dispersion relations,[23–26] vector dominance models with $\omega - \phi$ mixing,[27] meson-exchange models,[28] kaon loops,[29–31] light-cone diquark model,[32] chiral quark model,[33,34] Skyrme model,[35,36] the chiral bag model,[37] unquenched quark model,[38] perturbative chiral quark model,[39] Nambu-Jona-Lasinio soliton model,[40] an SU(3) chiral quark-soliton model,[41] heavy baryon chiral perturbation theory,[42,43] quenched chiral perturbation theory,[44] as well as lattice QCD calculations.[44–46] These calculations have elucidated the physics behind strange matrix elements and have provided numerical estimates of the size of possible effects to be observed by experiment.

The strategy of the most recent HAPPEX measurements is to run at very low Q^2 where one is sensitive mainly to the static moments ρ_s and μ_s, the strangeness radius and magnetic moment. Most models focus on predictions of these static moments at $Q^2 = 0$, while only a subset of the models attempt to predict the form factors at higher Q^2. In some cases the models are considered to be only an order of magnitude estimate, and in other cases only an upper bound to the strangeness effects. The large variety of models with very different physics assumptions is indicative of the difficulty in making solid predictions in the regime of nonperturbative QCD.

3. Parity Violation Measurements

The work presented here has been submitted for publication,[11] and the experimental technique was described in detail in.[12–14]

The experiment measured the helicity-dependent left-right asymmetry in the scattering of longitudinally polarized electrons from a hydrogen or a helium target. The kinematics were $E_{\text{beam}} \sim 3$ GeV, $\theta_{\text{lab}} \sim 6^\circ$ and $Q^2 \sim 0.1$ (GeV/c)^2. Results are present from our 2005 run in Hall A at the Thomas Jefferson National Accelerator Facility (Jefferson Lab). Two identical 3.7 msr spectrometer systems consisting of the Hall A septum magnets plus HRS spectrometers focus elastically scattered electrons onto total-absorption detectors in their focal planes. A 100μA, 85% polarized beam with a 30 Hz helicity reversal scattered from the targets. Čerenkov light from each detector was collected by a photomultiplier tube, integrated over the duration of each helicity window and digitized by analog to digital converters (ADCs). The HRS pair has sufficient momentum resolution to spatially separate the elastic electrons from inelastic electrons. The amount of background was measured in separate calibration runs in counting mode at low rate using drift chambers. Low rate counting measurements were also done to measure Q^2. Polarization was measured every few days by a Møller polarimeter and monitored continuously with a Compton polarimeter.

A major challenge for these measurements of tiny asymmetries of order 10^{-6} is to maintain systematic errors associated with helicity reversal at the $\approx 10^{-8}$ level. The two beams corresponding to the two helicity states must be as close to identical as possible in all their parameters, i.e. intensity, position, angle, and energy. There must be at least one, and preferably several, methods to reverse the helicity. Many reversals are needed during an experiment, and they should follow a rapid and random sequence to avoid any correlation with noise. The helicity reversals should be uncoupled to other parameters which affect the cross section. Experiments must measure the sensitivity of the cross section to these parameters, as well as the helicity correlated differences in them. Electronic pickup of the helicity correlated signals can cause a false asymmetry. In a count rate limited experiment in which the detected particles must be integrated in order to get the desired accuracy in a reasonable time, the linearity of the detection system and the susceptibility to backgrounds are important issues. The electrical environment around the data acquisition and control system were configured so that the observed fluctuations in the integrated scattered flux were dominated by counting statistics.

We have made tremendous progress in controlling helicity correlated systematics, see fig 1 which shows the helicity correlated position differences, which average to $\sim$1 nanometer during the run. This accomplishment is due to a long-term effort to improve the optics setup at the polarized source as well as the tune of the accelerator. A GaAs photocathode was optically pumped by circularly polarized laser light to produce polarized electrons, with the ability to rapidly and randomly flip the sign of the electron beam polarization. Most of the work to control systematic errors is done at the polarized source. The helicity-correlated intensity asymmetry was maintained to be less than 1 ppm by an active feedback loop. We have a well-developed model for reducing and controlling the laser systematics. For the accelerator, progress in understanding the betatron matching has helped achieve maximum dampening of position differences, while improved understanding of beam tuning provided good phase advances along our beamline that permit us to simultaneously have good measurements of the sensitivities of each of the beam parameters.

The data collection took place over 55 days (^{4}He) and 36 days (^{1}H) in 2005. A half-wave ($\lambda/2$) plate was periodically inserted into the laser optical path which passively reversed the sign of the electron beam polarization. There were 121 (^{4}He) and 41 (^{1}H) such reversals. The data set between two successive $\lambda/2$ reversals is referred to as a "slug."

Loose requirements were imposed on beam quality to remove periods of instability, leaving about 95% of the data sample for further analysis. No helicity-dependent cuts were applied. The final data sample consisted of 35.0×10^6 (^{4}He) and 26.4×10^6 (^{1}H) pairs. The right-left helicity asymmetry in the integrated detector response, normalized to the beam intensity, was computed for each pair to form the raw asymmetry $A_{\rm raw}$. The dependence of $A_{\rm raw}$ on fluctuations in the five correlated beam parameter differences Δx_i is quantified as $A_{\rm beam} = \sum c_i \Delta x_i$, where the coefficients c_i quantify the $A_{\rm raw}$ beam parameter sensitivity. The electroweak physics of the signal and backgrounds is contained in $A_{\rm corr} = A_{\rm raw} - A_{\rm beam}$.

The $A_{\rm corr}$ window-pair distributions for the two complete data samples were Gaussian over more than 4 orders of magnitude with an RMS width of 1130 ppm (^{4}He) and 540 ppm (^{1}H); the dominant source of noise in the PMT response was counting statistics. To further test that the data behaved statistically and the errors were being accurately calculated, $A_{\rm corr}$ averages and statistical errors for typical one hour runs, consisting of about 50k pairs each, were studied. Each set of roughly 400 average $A_{\rm corr}$ values, normalized by the corresponding statistical errors, populated a Gaussian distribution of unit variance as expected.

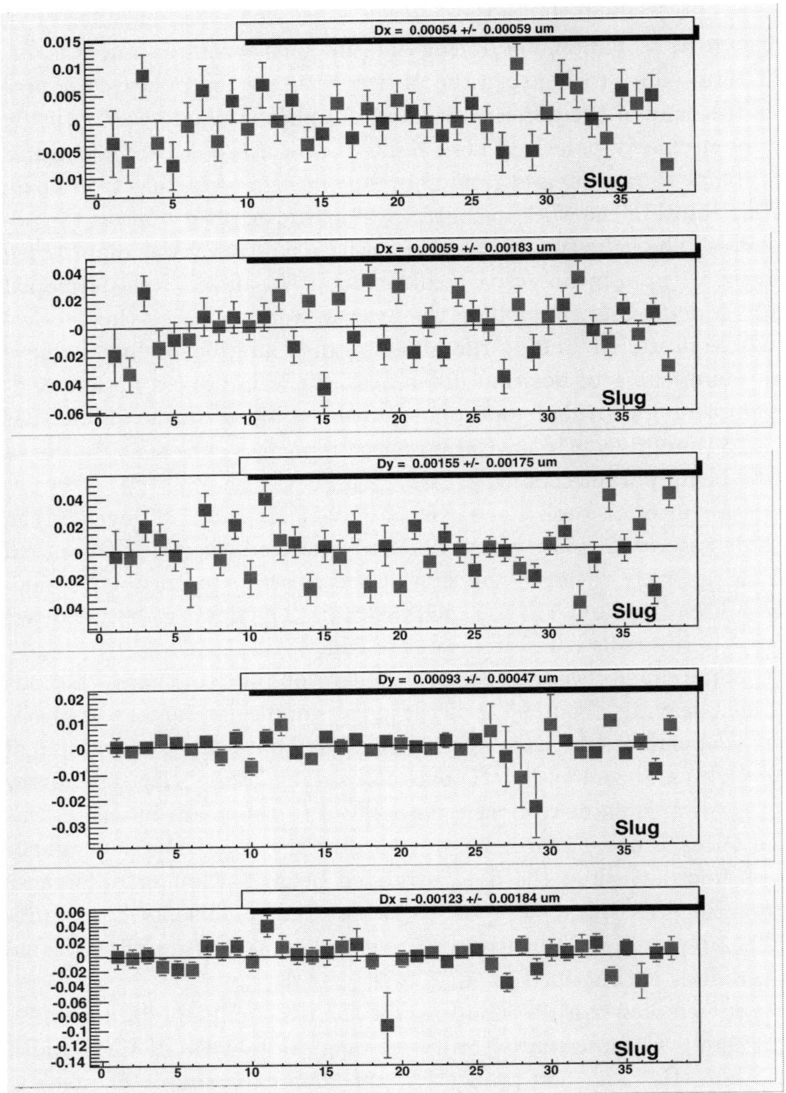

Fig. 1. Helicity correlated beam monitor differences (μm) versus slug number (1 "slug" $\sim$ 1 day of running) for HAPPEX. The top four plots are X and Y monitors near the target, and the bottom plot is a monitor in the dispersive section of the magnet ARC leading into the hall, providing a relative energy monitor. The averages over a month of running are $\sim$ 1 nanometer.

An important validation was to use two independent methods to calculate c_i. The first relied on linear regression of the observed response of the detector PMTs to intrinsic beam fluctuations. The other used calibration data in which the beam was modulated, by amounts large compared to intrinsic beam fluctuations, using steering magnets and an accelerating cavity. Differences in the two $A_{\rm beam}$ calculations were always much smaller than corresponding $A_{\rm corr}$ statistical errors.

Final $A_{\rm corr}$ results were calculated using the beam modulation technique and are summarized in table 1. Due to the excellent control of beam parameter differences Δx_i, $A_{\rm corr} - A_{\rm raw}$ values are of the order of, or much smaller than, the corresponding statistical errors. Under $\lambda/2$ reversal, the absolute values of $A_{\rm corr}$ are consistent within statistical errors. The reduced χ^2 for $A_{\rm corr}$ "slug" averages is close to one in every case, indicating that any residual beam-related systematic effects were small and randomized over the time period of $\lambda/2$ reversals (typically 5 to 10 hours). The final $A_{\rm corr}$ results are $A^{\rm He}_{\rm corr} = +5.25 \pm 0.19({\rm stat}) \pm 0.05({\rm syst})$ ppm and $A^{\rm H}_{\rm corr} = -1.42 \pm 0.11({\rm stat}) \pm 0.02({\rm syst})$ ppm.

Table 1. Raw and corrected asymmetries (in ppm) and reduced "slug" chi-squares (rχ^2), broken up by $\lambda/2$ reversals.

	$\lambda/2$ **OUT**		$\lambda/2$ **IN**		**BOTH**	
^{4}He	(DOF = 59)		(DOF = 60)		(DOF = 120)	
	Asym	rχ^2	Asym	rχ^2	Asym	rχ^2
$A_{\rm raw}$	4.80±0.27	0.75	-5.41±0.27	1.12	5.10±0.19	0.95
$A_{\rm corr}$	5.12±0.27	0.78	-5.38±0.27	1.07	5.25±0.19	0.92
^{1}H	(DOF = 20)		(DOF = 19)		(DOF = 40)	
$A_{\rm raw}$	-1.40±0.15	0.73	1.42±0.15	1.04	-1.41±0.11	0.86
$A_{\rm corr}$	-1.41±0.15	0.81	1.43±0.15	1.02	-1.42±0.11	0.89

The physics asymmetry $A_{\rm phys}$ is formed from $A_{\rm corr}$,

$$A_{\rm phys} = \frac{K}{P_{\rm b}} \frac{A_{\rm corr} - P_{\rm b} \sum_i A_i f_i}{1 - \sum_i f_i}, \tag{7}$$

with corrections for the beam polarization $P_{\rm b}$, background fractions f_i with asymmetries A_i and finite kinematic acceptance K. These corrections are summarized in Table 2. The first line lists the cumulative $A_{\rm beam}$ corrections discussed above, scaled by $K/P_{\rm b}$.

The high resolution spectrometers focus the elastically scattered electrons into a compact region. Thus, $\ll$1% of the flux intercepted by the detectors originated from inelastic scattering in the target. The main background was quasi-elastic scattering from target windows, measured using

Table 2. Corrections to $A_{\rm corr}$ and systematic errors.

Correction (ppb)	Helium			Hydrogen		
Beam Asyms.	183	±	59	−10	±	17
Target windows	113	±	32	7	±	19
Helium QE	12	±	20		-	
Rescatter	20	±	15	2	±	4
Nonlinearity	0	±	58	0	±	15
Scale Factor	**Helium**			**Hydrogen**		
Acceptance K	1.000	±	0.001	0.979	±	0.002
Q^2 Scale	1.000	±	0.009	1.000	±	0.017
Polarization $P_{\rm b}$	0.844	±	0.008	0.871	±	0.009

an equivalent aluminum target and computed to be $1.8 \pm 0.2\%$ (^{4}He) and $0.76 \pm 0.25\%$ (^{1}H).

An electron must give up more than 19 MeV to break up the ^{4}He nucleus and undergo quasi-elastic scattering off nucleons. The quasi-elastic threshold lies beyond the edge of the detector. A limit of $0.15{\pm}0.15\%$ on this background was placed by studies of the low-intensity data. For ^{1}H, the π^0 threshold is even further removed from the detector and therefore negligible.

Background from rescattering in the spectrometer was studied by varying the spectrometer momentum in counting mode to measure inelastic spectra and to obtain the detector response as a function of scattered electron energy. From these two distributions, the rescattering background was estimated to be $0.25 \pm 0.15\%$ (^{4}He) and $0.10 \pm 0.05\%$ (^{1}H).

For each source of background, a theoretical estimate for $A_{\rm PV}$ was used, with relative uncertainties taken to be 100% or more to account for kinematic variations and resonance contributions. The resulting corrections and the associated errors are shown in Table 2. Upper limits on rescattering contributions from exposed iron in the spectrometer led to an additional uncertainty of 5 ppb.

Nonlinearity in the PMT response was limited to 1% in bench-tests that mimicked running conditions. The relative nonlinearity between the PMT response and those of the beam intensity monitors was $< 2\%$. The acceptance correction K accounted for the non-linear dependence of the asymmetry with Q^2. A nuclear recoil technique using a water-cell target[13] was used to determine the scattering angle $\theta_{\rm lab}$, thus keeping the scale error on $\langle Q^2 \rangle$ due to $\theta_{\rm lab}$ to be $< 0.2\%$.

The beam polarization, $P_{\rm b}$, was continuously monitored by a Compton polarimeter; results, averaged over the duration of each run, are listed in

Tab. 2. Redundant cross-calibration of the recoil Compton electron spectrum restricted the relative systematic error to $\approx 1\%$. The results were consistent, within systematic uncertainties, with those obtained from recoil Compton photon asymmetries, and with dedicated measurements using Møller scattering in the experimental hall and Mott scattering at low energy. After all corrections:

$$A^{\rm He}_{\rm phys} = +6.40 \pm 0.23 \text{ (stat) } \pm 0.12 \text{ (syst) ppm},$$
$$A^{\rm H}_{\rm phys} = -1.58 \pm 0.12 \text{ (stat) } \pm 0.04 \text{ (syst) ppm}.$$

4. Results on G^s_E and G^s_M

The theoretical predictions $A^{\rm He}_{\rm NS}$ and $A^{\rm H}_{\rm NS}$ with $G^s = 0$ were estimated using the formalism in[9] and described in our publications.[13,14]

Assuming a pure isoscalar $0^+ \to 0^+$ transition, $A^{\rm He}_{\rm NS}$ is completely independent of nuclear structure and determined purely by electroweak parameters. Effects due to D-state and isospin admixtures and meson exchange currents are negligible at the level of the experimental fractional accuracy of $\sim 3\%$.[18] For our kinematics (E_b=2.75 GeV, $\langle Q^2 \rangle = 0.077\,{\rm GeV}^2$) we obtain $A^{\rm He}_{\rm NS} = +6.37$ ppm.

Electromagnetic form factors from a phenomenological fit to the world data at low Q^2[19] were used to calculate $A^{\rm H}_{\rm NS}$, with uncertainties governed by data near $Q^2 \sim 0.1\,{\rm GeV}^2$. The value used for $G^{\gamma n}_E = 0.037$, with a 10% relative uncertainty based on new data from the BLAST experiment.[20] For our kinematics (E_b=3.18 GeV, $\langle Q^2 \rangle = 0.109\,{\rm GeV}^2$) we obtain $A^{\rm H}_{\rm NS} = -1.66 \pm 0.05$ ppm. This includes a contribution from the axial form factor G^Z_A, and associated radiative corrections,[21] of -0.037 ± 0.018 ppm.

Comparing our results to the theoretical expectations, we extract $G^s_E = 0.002 \pm 0.014 \pm 0.007$ at $Q^2 = 0.077\,{\rm GeV}^2$ and $G^s_E + 0.09 G^s_M = 0.007 \pm 0.011 \pm 0.004 \pm 0.005$ (FF) at $Q^2 = 0.109\,{\rm GeV}^2$, where the uncertainties in the nucleon electromagnetic form factors govern the last error. Figure 2 displays the combined result for these and our previous measurements,[13,14] taken with $\langle Q^2 \rangle$ between 0.077-0.109 ${\rm GeV}^2$. The requisite small extrapolation to a common $Q^2 = 0.1\,{\rm GeV}^2$ was made assuming that $G^s_E \propto Q^2$ and that G^s_M is constant. The values $G^s_E = -0.005 \pm 0.019$ and $G^s_M = 0.18 \pm 0.27$ (correlation coefficient $=-0.87$) are obtained. The results are quite insensitive to variations in G^Z_A, as evidenced by the negligible change induced by an alternate fit similar to that in,[22] where G^Z_A is constrained by other $A_{\rm PV}$ data.

Figure 2 also displays predictions from selected theoretical mod-

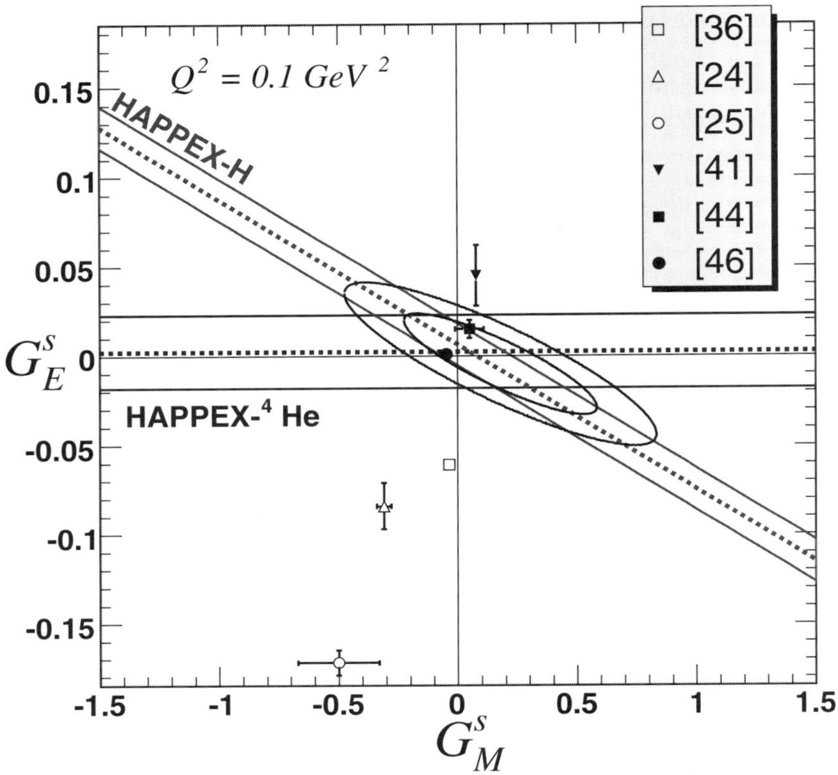

Fig. 2. 68 and 95% C.L. constraints in the $G_E^s - G_M^s$ plane from data from this apparatus ([13,14] and this paper). Various theoretical predictions are plotted. The 1-σ bands (a quadrature sum of statistical and systematic errors) and central values from the new results alone are also shown.

els.[24,25,36,41,44,46] Those that predict little strange quark dynamics in the vector form factors are favored.[44,46] Other measurements[15–17] that had suggested non-zero strangeness effects are consistent, within quoted uncertainties, with our results at $Q^2 = 0.1\,\mathrm{GeV}^2$. Due to the improved statistical precision and lower G_A^Z sensitivity of our result, adding these earlier published measurements does not alter our conclusions.

5. Summary and Outlook

In summary, we have reported the most precise constraints on the strange form factors at $Q^2 \sim 0.1\,\mathrm{GeV}^2$. The results are consistent within errors with other A_{PV} measurements and suggest a very small contribution of strange quarks at low Q^2. It will be important to complete the world-wide program of approved parity experiments to precisely quantify the strangeness effects over a range of kinematics and over various distance scales in the nucleon. These experiments should yield a detailed mapping of the spatial dependence of $s\bar{s}$ contributions to nucleon structure and will provide constraints on theoretical approaches to non-perturbative QCD dynamics.

References

1. J. Ashman *et al.*, Phys. Lett. B **206** (1988) 364; Nucl. Phys. B **328** (1988) 527. Nucl. Phys. B **328** (1989) 1.
2. D. B. Kaplan and A. Manohar, Nucl. Phys. B **310** (1988) 527.
3. R. D. McKeown, Phys. Lett. B **219** (1989) 140.
4. E. J. Beise and R. D. McKeown, Comments Nucl. Part. Phys. **20** (1991) 105.
5. D. H. Beck, Phys. Rev. D **39** (1989) 3248.
6. K.S. Kumar and P.A. Souder, Prog. Part. Nucl. Phys. 45 (2000) S333.
7. D.H. Beck and R.D. McKeown, Ann. Rev. Nucl. Part. Sci. 51 (2001) 189.
8. D.H. Beck and B.R. Holstein, Int. J. Mod. Phys. E10 (2000) 1.
9. M.J. Musolf *et al.*, Phys. Rep. 239 (1994) 1.
10. B. Kubis and R. Lewis, Phys. Rev. **74**, 015204 (2006).
11. A. Acha, *et al.*, nucl-ex/0609002.
12. K. A. Aniol *et al.*, Phys. Rev. C **69**, 065501 (2004).
13. K.A. Aniol *et al.*, Phys. Rev. Lett. **96**, 022003 (2006).
14. K.A. Aniol *et al.*, Phys. Lett. B **635**, 275 (2006).
15. D.T. Spayde *et al.*, Phys. Lett. B **583**, 79 (2004).
16. F.E. Maas *et al.*, Phys. Rev. Lett. **94**, 152001 (2005).
17. D.S. Armstrong *et al.*, Phys. Rev. Lett. **95**, 092001 (2005).
18. M.J. Musolf, R. Schiavilla and T.W. Donnelly, Phys. Rev. C **50**, 2173 (1994); S. Ramavataram, E. Hadjimichael and T.W. Donnelly, Phys. Rev. C **50**, 1175 (1994); M.J. Musolf and T.W. Donnelly, Phys. Lett. B **318**, 263 (1993).
19. J. Friedrich and Th. Walcher, Eur. Phys. J. A **17**, 607 (2003).
20. V. Ziskin, PhD. thesis, MIT, 2005.
21. S.-L. Zhu *et al.*, Phys. Rev. D **62**, 033008 (2000).
22. R. Young *et al.*, Phys. Rev. Lett. **97**, 102002 (2006).
23. R.L. Jaffe, Phys. Lett. B **229**, 275 (1989).
24. H.-W. Hammer, U.-G. Meissner, and D. Drechsel, Phys. Lett. B **367**, 323 (1996).
25. H.-W. Hammer and M.J. Ramsey-Musolf, Phys. Rev. C **60**, 045204 (1999); *ibid* **60**, 045205 (1999); *erratum ibid* **62**, 049902 (2000); 63, 049903 (2000).
26. H. Forkel, Phys. Rev. C **56**, 510 (1997)..
27. H. Forkel, M. Nielsen, X. Jin, and T. Cohen, Phys. Rev. C **50**, 3108 (1994).
28. U.-G. Meissner, V. Mull, J. Speth, J.W. Van Orden, Phys. Lett. B **408**, 381 (1997).
29. W. Koepf, E.M. Henley, and J.S. Pollock, Phys. Lett. B **288**, 11 (1992).

30. M.J. Musolf and M. Burkhardt, Z. Phys. C **61**, 433 (1994).
31. H. Ito, Phys. Rev. C **52**, R1750 (1995).
32. B.-Q. Ma, Phys. Lett. B **408**, 387 (1997).
33. D.O. Riska, Nucl. Phys. A **678**, 79 (2000).
34. L. Hannelius, D.O. Riska, and L.Ya. Glozman, Nucl. Phys. A **665**, 353 (2000).
35. N.W. Park, J. Schechter, and H. Weigel, Phys. Rev. D **43**, 869 (1991).
36. N.W. Park and H. Weigel, Nucl. Phys. A **541**, 453 (1992).
37. S.-T. Hong, B.-Y. Park, and D.-P. Min, Phys. Lett. B **414**, 229 (1997).
38. P. Geiger and N. Isgur, Phys. Rev. D **55**. 299 (1997).
39. V. E. Lyubovitskij, P. Wang, Th. Gutsche, and A. Faessler, Phys. Rev. C **66**, 055204 (2002).
40. H. Weigel *et al.*, Phys. Lett. B **353**, 20 (1995).
41. A. Silva, H.-C. Kim, and K. Goeke, Phys. Rev. D **65**, 014016 (2002), Erratum-ibid. D **66**, 039902 (2002).
42. T. Hemmert, U.-G. Meissner, and S. Steininger, Phys. Lett. B **437**, 184 (1998).
43. T. Hemmert, B. Kubis, and U.-G. Meissner, Phys. Rev. C **60**, 045501 (1999).
44. R. Lewis, W. Wilcox, and R.M. Woloshyn, Phys. Rev. D **67**, 013003 (2003).
45. S.J. Dong, K.F.Liu, and A.G. Williams, Phys. Rev. D **58**, 074504 (1998).
46. D.B. Leinweber *et al.*, Phys. Rev. Lett. **94**, 212001 (2005); D.B. Leinweber *et al.*, Phys. Rev. Lett. **97**, 022001 (2006).

CHIRAL SYMMETRIES AND LOW ENERGY SEARCHES FOR NEW PHYSICS

MICHAEL J. RAMSEY-MUSOLF*

Kellogg Radiation Laboratory, California Institute of Technology
Pasadena, CA 01125 U.S.A
and
Department of Physics, University of Wisconsin-Madison
Madison, WI 53706 U.S.A.
**E-mail: mjrm@caltech.edu*
www.its.caltech.edu/~mjrm/

I discuss low energy searches for new physics beyond the Standard Model, identifying the role played by chiral symmetries in these searches and in various new physics scenarios. I focus in particular on electric dipole moment searches; precision studies of weak decays and electron scattering; and neutrino properties and interactions.

Keywords: Standard Model, Electroweak Interactions, Chiral Symmetry

1. Introduction

The search for physics beyond the Standard Model (BSM) lies at the forefront of the intersection of nuclear physics with particle physics and cosmology. In this talk, I attempt to give an overview of low-energy studies that are being used in this search and try to describe ways in which they complement present and future high energy collider studies. As theme for this meeting is the broken chiral symmetry of QCD, I will endeavor to highlight the role played by chiral symmetries in both the low energy BSM searches and various BSM scenarios. In particular, I will address four questions:

i) What were the fundamental symmetries that governed the microphysics of the early universe?
ii) Were there additional (broken) chiral symmetries?
iii) What insights can precision low energy ($E << M_Z$) studies provide?
iv) How does the approximate chiral symmetry of QCD affect the low

energy search for new symmetries?

As I have described elsewhere,[2,3] in thinking about fundamental symmetries and the microphysics of the early universe, I like to break cosmic history into three periods: (A) the era of broken Standard Model symmetry, starting from the moment when electroweak symmetry-breaking (EWSB) occurred until the present; (B) the era from the Big Bang until EWSB; and (C) the brief period of EWSB itself. The broken symmetries of the SM – including the approximate $SU(3)_L \times SU(3)_R$ chiral symmetry of strong dynamics involving light quarks – provide a remarkably successful framework for explaining many phenomena of the present universe, such as the abundance of light elements, weak interactions in stars, and the chiral dynamics of pions and nucleons. Of course, there remain many important but poorly understood aspects of SM dynamics, such as the mechanism for confinement and the possible deconfinement-confinement and chiral symmetry breaking (χSB) phase transitions that the universe experienced subsequent to EWSB. By and large, however, the SM represents a triumph of 20th century physics with its simple, symmetry-based framework for explaining so much of what we observe today.

Where the SM starts to run into shaky ground begins with EWSB. We believe that the electroweak symmetry of the SM broke down to that of electromagnetism through the Higgs mechanism, whereby elementary particles received non-vanishing mass proportional to the vacuum expectation value (vev) of the neutral Higgs field: $m \propto \langle H^0 \rangle \equiv v/\sqrt{2}$. One should emphasize, however, that we have no direct experimental proof that this idea is correct, as no Higgs particle has yet been observed. Direct searches at LEP II place a lower bound of about 114 GeV on its mass, whereas global analyses of precision electroweak data suggest that the mass of the SM Higgs should be less than 166 GeV.[1] However, it may be that the SM picture of EWSB is too simple. There may be more than one Higgs field, as in supersymmetric models, for example, or EWSB may not even occur through the Higgs mechanism at all. One hopes that the searches for the Higgs boson at the Tevatron and LHC will either find the SM Higgs or tell us what the mechanism of EWSB truly is.

Looking back beyond the era of EWSB, all bets are off when it comes to the SM. Most strikingly, the origin of matter and energy in the cosmos cannot be explained within the SM at all. The most anthropically relevant component of this cosmic energy density – the visible, baryonic matter component (about 5%) – could have been explained by the SM but it turns out the SM interactions fall short of what is required for such an explanation.

The next biggest fraction corresponding to the cold dark matter (CDM) (about 25%) has no particle candidate within the SM. Finally, the dark energy responsible for cosmic acceleration and comprising about 70% of the cosmic energy density is the biggest mystery. While various BSM scenarios provide potentially feasible explanations for the baryonic and dark matter, our ideas about the dark energy remain the most speculative at present. For nuclear physicists, understanding the origin of the baryonic matter is quite important, as baryons[a] and their interactions with leptons and other hadrons are the bread and butter of the field.

The pre-EWSB era presents additional puzzles having to do with the unification of forces, stability of the electroweak scale, and neutrinos. If – as many believe – all forces were unified into a single interaction that included gravity at the end of the Big Bang, then the electroweak and strong couplings of the SM ought to meet at a common point when evolved to scales around 10^{16} GeV. The problem is that they don't quite do so; there is something of a "near miss" for grand unification. Consequently, one would need new physics to bring about unification, just as one does to explain the origin of matter. Generally speaking, the introduction of new physics tends to push the electroweak scale, given by the Higgs vev $v \approx 246$ GeV, up to higher scales. A larger value of v would be problematic for the present SM universe because the Fermi constant that characterizes the strength of low-energy weak interactions is given by $G_F = (\sqrt{2}v^2)^{-1}$. Thus, larger the values of v would lead to more feeble low-energy weak interactions, a sun that burns less brightly, and likely a different combination of light elements produced in Big Bang Nucleosynthesis than we observe in our universe. Clearly, new symmetries are required to preserve the relatively large value of G_F in the presence of new physics needed to explain the origin of matter and bring about unification[b].

Finally, neutrinos cause all kinds of consternation for the SM. The SM can be minimally extended to include Dirac neutrino mass terms with $SU(2)_L \times U(1)_Y$ right-handed (RH) neutrino fields, but the corresponding Yukawa couplings would have to be considerably smaller than for the other SM particles. This may, in fact, be what nature has given us, but many people believe a more natural explanation of the tiny neutrino masses arises if neutrinos are Majorana particles. In this case, the scale of neutrino mass is naturally given by $\sim m_e^2/M_{\rm new}$, where $M_{\rm new}$ is about 10^{12} GeV. At present,

---

[a] I include nuclei under this rubric

[b] One could get around the need for new symmetries if one allows for fine tuning.

however, we have no evidence that this scenario is correct[c]. Even more puzzling is the pattern of neutrino mixing. In contrast to the situation with quarks, where the mixing of weak eigenstates into mass eigenstates involves relatively small angles, just the opposite is true for neutrinos. Clearly, new physics is required to explain the differences in the patterns of mass generation or chiral symmetry-breaking for charged fermions and neutrinos.

In the search for new symmetries that address these issues, the next high energy frontier is obviously the Large Hadron Collider. A second frontier – the precision frontier – now lies in the domain of low-energy studies that are being carried out by nuclear and atomic physicists. Information from the latter complements the former, as I hope to describe below. In doing so, I will divide the discussion in three parts: the origin of matter and EDM searches; precision studies of SM -allowed processes; and neutrinos. In doing so, I will not provide a comprehensive list of references, as space considerations preclude this possibility. More extensive reviews with reference to the literature can be found in Refs.[6–10]

2. EDMs and the Origin of Matter

EDM searches are a particularly apt topic for this chiral dynamics meeting since the EDM operator is chiral odd. It is useful to write down the relevant $SU(2)_L \times U(1)_Y$ dimension six operators that involve the electric dipole couplings to the W and B gauge fields and include the relevant Higgs field insertions as needed for gauge invariance:

$$\begin{aligned}\mathcal{L}^{(6)}_{\rm CPV} &= \frac{i\, g_1 d_u^B}{\Lambda^2}\bar{Q}\sigma_{\mu\nu}\gamma_5 B^{\mu\nu}\tilde{H}U + \frac{i\, g_1 d_d^B}{\Lambda^2}\bar{Q}\sigma_{\mu\nu}\gamma_5 B^{\mu\nu}HD \\ &+ \frac{i\, g_2 d_u^W}{\Lambda^2}\bar{Q}\sigma_{\mu\nu}\gamma_5 \tau^A W^{\mu\nu\, A}\tilde{H}U + \frac{i\, g_2 d_d^W}{\Lambda^2}\bar{Q}\sigma_{\mu\nu}\gamma_5 \tau^A W^{\mu\nu\, A}HD + \cdots\end{aligned} \tag{1}$$

where Q is the LH quark doublet; U and D are RH quark singlets; $\tilde{H} = \epsilon_{ij}H_j^*$: Λ is a mass scale associated with the CP-violating (CPV) physics; and the $+\cdots$ indicate other CPV interactions, such as those involving gluons. The expression for charged leptons is similar. In Eq. (1). After EWSB, one obtains from these interactions the corresponding EDMs

$$\mathcal{L}_{EDM} = -\frac{i\, d_u^\gamma}{2\Lambda}\bar{U}_L\sigma_{\mu\nu}F^{\mu\nu}U_R - \frac{i\, d_d^\gamma}{2\Lambda}\bar{D}_L\sigma_{\mu\nu}F^{\mu\nu}D_R - \frac{i\, d_\ell^\gamma}{2\Lambda}\bar{\ell}_L\sigma_{\mu\nu}F^{\mu\nu}\ell_R \tag{2}$$

[c]In many string-inspired models studied to date, it appears to be quite difficult to obtain Majorana mass terms or the minimal see-saw mechanism.[4,5]

where, for example, the EDM of the u-quark is given by

$$d_u^\gamma = -\frac{\sqrt{2}\, v_u \left(c_W\, d_U^B + s_W\, d_U^W\right)}{\Lambda} \quad . \tag{3}$$

Note that the EDM operators are expressly chiral odd, as they involve one LH and one RH quark field.

The present experimental bounds on EDMs of the electron and systems built out of quarks are quite stringent. In the case of the electron and neutron, for example, one has (for references to the experimental literature, see, *e.g.*, Refs.[6–8])

$$\begin{aligned} |d_e^\gamma/\Lambda| &< 1.6 \times 10^{-27} \quad e - \text{cm} \\ |d_n^\gamma/\Lambda| &< 3.0 \times 10^{-26} \quad e - \text{cm} \end{aligned} \tag{4}$$

at 90% confidence. The expectations for these EDMs within the electroweak sector of the SM lie many orders of magnitude below these values, but different BSM scenarios can lead to significantly larger EDMs. In the minimal supersymmetric standard model (MSSM), for example, one has for the electron at one-loop order

$$\frac{d_e^\gamma}{\Lambda} \approx 5 \times 10^{-25} \left(\frac{100\,\text{GeV}}{\tilde{m}}\right)^2 [\tan\beta \sin\phi_\mu - 0.05 \sin\phi_A] \; e - \text{cm} \quad , \tag{5}$$

where ϕ_μ and ϕ_A are CPV phases, $\tilde{m}$ is the mass of the supersymmetric particles (assuming their masses are degenerate), and $\tan\beta$ is the ratio of the vevs of the two Higgs doublets that arise in the MSSM. For $\tan\beta$ of $\mathcal{O}(1)$ and $\tilde{m} = 100$ GeV, the bounds in Eq. (4) imply that $\phi_\mu \lesssim 3 \times 10^{-3}$.

From the standpoint of the cosmology, it turns out that this value of the CPV phase is too small to explain the matter-antimatter asymmetry within SUSY. The asymmetry itself is quite small and can be characterized by the baryon to photon entropy density

$$Y_B \equiv \frac{n_B}{s} = \begin{cases} (7.3 \pm 2.5) \times 10^{-11}, & \text{BBN} \\ (9.2 \pm 1.1) \times 10^{-11}, & \text{WMAP} \end{cases} \tag{6}$$

where "BBN" and "WMAP" indicate values derived from Big Bang Nucleosynthesis[11] and the cosmic microwave background,[12] respectively.

One can relax the EDM constraints while producing the value of Y_B given in Eq. (6) by allowing the scalar superpartners of leptons and quarks become heavier – of order a few TeV – while keeping the masses of the gauge boson superpartners relatively light. In this scenario, the EDMs of elementary fermions are dominated by two-loop graphs, and CPV phases of $\mathcal{O}(1)$ are consistent with the experimental EDM bounds. In order to obtain the observed value of Y_B one would need superpartner masses and CPV phases consistent with an electron EDM of roughly 10^{-28} e-cm or larger (and similarly for the neutron). What makes the next several years particularly interesting from this standpoint is that new experiments are poised to search for EDMs with precisely this magnitude or even smaller. Thus, one has some chance of either seeing the CPV that could explain the matter-antimatter asymmetry or ruling out conventional SUSY models as a mechanism for producing it.

Making a robust connection between the results of future EDM experiments and the value of Y_B requires careful analysis of the dynamics of baryon number generation during the electroweak phase transition (EWPT). Baryon number is produced by anomalous, topological transitions known as sphaleron processes. The effect of sphalerons live on the presence non-zero chiral charge density, and the latter is produced by CPV interactions of matter fields with the spacetime varying Higgs vevs at the boundary between regions of broken and unbroken electroweak symmetry. Thus, the CPV must be sufficiently effective to produce enough chiral charge to make the baryon number we observe today. Moreover, the phase transition must be strongly first order to ensure that as the region of broken electroweak symmetry expands, the sphalerons get sufficiently quenched that they cannot wash out the produced baryon number. In the SM, the lower bounds on the mass of the Higgs imply that a SM phase transition cannot be first order, but various BSM scenarios with extended Higgs sectors can produce such a strong first order phase transition.

Although considerable theoretical progress has been made in performing refined computations of these effects (for the recent literature, see, *e.g.*, Ref.[7]), there remains considerable room for future theoretical progress. The prospect of significantly more sensitive EDM searches provides powerful motivation for making such progress. A similar comment applies to computing EDMs within various BSM CPV scenarios. Clearly, considerations of chiral symmetry – both through the EDMs themselves and through the dynamics of the electroweak phase transition – are central to this problem of the origin of matter.

3. Precision Electroweak Probes of New Symmetries

Chiral symmetries can be a similarly interesting consideration when considering precision measurements of observables that are not suppressed in the SM. The presence or absence of tiny deviations from SM expectations can provide important clues for BSM physics. Perhaps the most widely-known recent example is the anomalous magnetic moment of the muon, where many argue one now sees a deviation of nearly three standard deviations from the SM prediction. As with the EDM, the magnetic moment is a chiral-odd operator. The pure QED contributions to the corresponding operator coefficient, $a_\mu = (g_\mu - 2)/2$, have been computed to high precision, as have one-loop electroweak contributions. The effects of strong interactions that enter the hadronic two-loop vacuum polarization and three-loop hadronic light-by-light contributions have proven more challenging theoretically. Assuming the relevant theoretical uncertainties are under control, the deviation from the SM expectation could be a signature of supersymmetric loop effects if $\tan\beta$ is large.

Chiral-odd BSM effects can also enter low-energy weak interaction observables at an observable level. In the case of weak decays of hadrons, such as neutron β-decay and pion-decay, the low-energy semileptonic interaction can be described by a dimension six four fermion Lagrangian[13]

$$\mathcal{L}^{\beta-\text{decay}} = -\frac{4G_\mu}{\sqrt{2}} \sum_{\gamma,\,\epsilon,\,\delta} a^\gamma_{\epsilon\delta}\ \bar{e}_\epsilon \Gamma^\gamma \nu_e\ \bar{u}\Gamma_\gamma d_\delta \tag{7}$$

where the $a^\gamma_{\epsilon\delta}$ coefficients are determined by the SM and its possible extensions. At tree-level in the SM, $a^V_{LL} = V_{ud}$ with all others being zero. There exist several equivalent representations of the low-energy effective semileptonic interaction,[9,10] but I prefer the form in Eq. (7) because of its similarity to the muon decay effective Lagrangian. Note that none of these forms is invariant under the SM gauge symmetries and must, therefore, be used only for interactions taking place at energies well below the weak scale.

As discussed elsewhere,[9,10,13] the study of β-decay correlations can probe for the existence of non $(V-A)\times(V-A)$ interactions appearing in Eq. (7). Bounds on these interactions obtained from existing measurements can be found, *e.g.*, in Ref.[9] Future, more precise probes may be obtained with studies of cold and ultracold neutrons at LANSCE, NIST, ILL, the SNS, and various other laboratories. From the standpoint of chiral symmetry, the scalar and tensor interactions in Eq. (7) are interesting since they are chiral odd. Such operators can be generated in various BSM scenarios. In SUSY, for example, mixing between the superpartners of LH and RH fermions in loop graphs can give rise to the operators proportional to a^S_{RR}, a^S_{RL}, and a^T_{RL}. These operators generate β energy-dependent contributions to the parity-violating correlation of the neutrino with the spin of the decaying nucleus (the "B-term") as well as the so-called Fierz interference term (the "b-coefficient") that also depends on the β energy. Current limits on b

are at the 10^{-3} level while future measurements of the energy-dependence of B at the 10^{-4} level may be achievable with cold or ultracold neutrons.

Effects of this size could be generated in SUSY if the LH-RH first generation superpartner mixing is nearly maximal – a situation that would be interesting since it would imply that the extra Higgs bosons in SUSY are too heavy to be seen at the LHC or that one must admit considerable fine-tuning to obtain appropriate electroweak symmetry-breaking. It is interesting to note that the mixing between LH and RH superpartners implies the explicit breaking of chiral symmetry in the scalar superpartner sector. The question is whether this breaking is small (*i.e.*, proportional to the fermion Yukawa couplings) for all but the third generation sfermions – as for the SM particles - or anomalously large. To date, I am not aware of explicit experimental tests of this possibility for the first generation quark and lepton superpartners.

Chiral symmetry considerations are clearly important for discussions of pion decay as well as for β-decay. In the present context, it is interesting to consider the purely leptonic decays of the pion, from which we obtain the value of the pion decay constant, F_π, that is so important for the chiral dynamics of QCD. If we include the effects of electroweak radiative corrections in the SM as well as possible effects of new physics, the decay rate is given by (for reference to the literature, see, *e.g.*, Ref.[7])

$$\Gamma[\pi^+ \to \ell^+ \bar{\nu}_\ell(\gamma)] = \frac{G_\mu^2 |V_{ud}|^2}{4\pi} F_\pi^2 m_\pi m_\ell^2 \left[1 - \frac{m_\ell^2}{m_\pi^2}\right] \quad (8)$$
$$\times \left\{1 + \left(2\left[\Delta\hat{r}_\pi^A - \Delta\hat{r}_\mu\right] + \text{brem}\right)_{\text{SM}} + 2\left(\Delta\hat{r}_\pi^A - \Delta\hat{r}_\mu\right)_{\text{new}}\right\}$$

where $\Delta\hat{r}_\pi^A$ and $\Delta\hat{r}_\mu$ are corrections that come from SM radiative corrections to the fundamental semileptonic and μ-decay amplitudes, respectively (the "SM " subscript) or from BSM physics ("new" subscript) and "brem" indicates the contributions from real photon radiation (required to keep the rate infrared finite). Taking just the SM contributions and the corresponding low-energy QCD related uncertainties in the SM radiative corrections as well as the experimental error in the rate, one obtains $F_\pi = 92.4 \pm 0.025(\text{expt}) \pm 0.25(\text{theory})$. Allowing for possible contributions from new physics can lead to further increases in the error. For example, if one allows for new tree-level SUSY interactions that violate lepton number conservation, present experimental constraints on these interactions leaves room for an additional $\sim 0.25\%$ uncertainty in the value of F_π – comparable to the present SM QCD uncertainty.

A well-known way to circumvent the largest QCD uncertainties and to actually use $\pi_{\ell 2}$ decays to probe BSM physics is to consider the ratio $R_{e/\mu}$ of the rates for decays to an electron-neutrino and muon-neutrino final state. The SM prediction for this quantity which tests "lepton universality" is $R_{e/\mu} = (1.2352 \pm 0.0005) \times 10^{-4}$ where the error is dominated by uncertainties in various low energy constants. This departure from unity follows simply from the different masses of the final state charged leptons. Experimentally, one finds

$$\frac{R^{\rm exp}_{e/\mu}}{R^{\rm SM}_{e/\mu}} = 0.9966 \pm 0.0030 \pm 0.0004, \tag{9}$$

where the first error is experimental and the second theoretical (experimental references may be found in Refs.[6,7]). A new generation of experiments are poised to measure $R_{e/\mu}$ with five to ten times smaller error bars, making the experimental uncertainty comparable to the theory error. Tests at this level could be quite interesting for SUSY, where lepton universality can be broken by differences in the lepton superpartner masses, leading to corrections as large as a few times 10^{-3}. One could imagine further improvements in this experimental probe of "slepton universality" by reducing the theoretical in the SM prediction – clearly a task for chiral dynamics in QCD.

4. Neutrinos and Chiral Symmetry

The fact that neutrinos have small, but non-vanishing masses can have significant implications for other properties of neutrinos that would be forbidden in the presence of exact chiral symmetry. For example, the magnetic moments of neutrinos, which correspond to chiral odd operators, are constrained by the scale of neutrino mass and "naturalness" considerations.[14–16] The basic idea is quite simple. Insertions of the magnetic moment operator in loop graphs generates contributions to the neutrino mass operator, as both are chiral odd[d]. In order to avoid requiring large, "un-natural" cancellations between these loop effects and tree-level contributions as needed to obtain the small scale of neutrino mass, the magnetic momenta operator coefficients cannot be too large.

Model-independent constraints implementing this idea can be obtained by considering an effective Lagrangian containing SM fields and RH neu-

[d]Here, I focus on the case of Dirac neutrinos for simplicity.

trino fields

$$\mathcal{L}_{\text{eff}} = \sum_{n,j} \frac{C_j^n(\mu)}{\Lambda^{n-4}} \mathcal{O}_j^{(n)}(\mu) \ + \text{h.c.} \tag{10}$$

where Λ is the mass scale associated with BSM physics and j labels all operators of dimension $n \geq 4$. The $n = 4$ operators as just those of the SM plus a neutrino Dirac mass term while Majorana mass terms appear at $n = 5$. Magnetic moment operators for Dirac neutrinos appear at $n = 6$ and have the gauge-invariant form

$$\mathcal{O}_B^{(6)} = g_1(\bar{L}\sigma^{\mu\nu}\widetilde{H})\nu_R B_{\mu\nu} \tag{11}$$

$$\mathcal{O}_W^{(6)} = g_2(\bar{L}\sigma^{\mu\nu}\tau^a\widetilde{H})\nu_R W_{\mu\nu}^a \quad . \tag{12}$$

After EWSB, these operators generate a neutrino magnetic moment $\mu_\nu/\mu_B = -4\sqrt{2}(m_e v/\Lambda^2)[C_B^6(v) + C_W^6(v)]$. Radiative contributions to the neutrino mass operators and naturalness considerations lead to bounds on the coefficients $C_{B,W}$. For $\Lambda >> v$ the most stringent expectations arise from considering contributions to the $n = 4$ neutrino mass operators and matching of the effective theory described by Eq. (10) onto the (unspecified) full theory. However, for Λ not too different from v mixing among the magnetic moment operators and the $n = 6$ mass leads to bounds of comparable magnitude.

From these arguments one expects the Dirac neutrino magnetic moments to be bounded above by

$$|\mu_\nu| / \mu_B \lesssim 10^{-14} \times (m_\nu / 1\,\text{eV}) \quad . \tag{13}$$

These bounds are two or more orders of magnitude more stringent than the present experimental bounds on μ_ν. For Majorana neutrinos, the situation is more subtle. For $\Lambda \sim v$, the bounds on the transition moments are weaker than present experimental limits, while for $\Lambda \gtrsim 100$ GeV, the expectations are that the transition moments would be smaller than present direct constraints. Given the expected sensitivity of future neutrino magnetic moment searches, the discovery of a non-zero moment would imply that the neutrino is a Majorana particle and that the mass scale Λ of the corresponding BSM physics is well below the standard see-saw scale of $\sim 10^{12}$ GeV. These conclusions could only be altered in specific models wherein these chiral symmetry-based expectations allow for considerably larger magnetic moments through the introduction of other mechanisms that protect m_ν from larger corrections or through the presence of non-SM mechanisms for generating charged lepton masses. Applications of these neutrino mass naturalness arguments to other processes can be found in Refs.[17,18]

5. Conclusions

I hope to have convinced the reader that low-energy searches for BSM physics are an appropriate topic for a meeting on chiral dynamics. Chiral symmetries can play a significant role in both the computation of SM observables as well as in BSM scenarios that generate corrections to SM expectations. Indeed, the presence of (broken) chiral symmetries are a key element in explaining the origin of matter and CPV in the early universe; the properties and interactions of neutrinos; and the weak decays of leptons and systems built from light quarks. The next several years promises to be an exciting time in the study of these phenomena, and we may expect reports of interesting experimental and theoretical developments at future chiral dynamics conferences.

6. Acknowledgments

I would like to thank N. Bell, V. Cirigliano, J. Erler, R. Erwin, B. Filippone, M. Gorshteyn, B. Holstein, J. Kile, P. Langacker, C. Lee, W. Marciano, R. McKeown, B. Nelson, S. Profumo, S. Su, S. Tulin, P. Vogel, P. Wang, and M. Wise for useful conversations. This work was supported under U.S. Department of Energy contract DE-FG02-05ER41361 and National Science Foundation Award PHY-05556741.

References

1. http://lepewwg.web.cern.ch/LEPEWWG
2. M. J. Ramsey-Musolf, AIP Conf. Proc. **842** (2006) 661 [arXiv:hep-ph/0603023].
3. M. J. Ramsey-Musolf, arXiv:nucl-th/0608035.
4. J. Giedt, G. L. Kane, P. Langacker and B. D. Nelson, Phys. Rev. D **71**, 115013 (2005) [arXiv:hep-th/0502032].
5. P. Langacker and B. D. Nelson, "String-inspired triplet see-saw from diagonal embedding of SU(2)L in Phys. Rev. D **72**, 053013 (2005) [arXiv:hep-ph/0507063].
6. J. Erler and M. J. Ramsey-Musolf, Prog. Part. Nucl. Phys. **54** (2005) 351 [arXiv:hep-ph/0404291].
7. M. J. Ramsey-Musolf and S. Su, arXiv:hep-ph/0612057.
8. M. Pospelov and A. Ritz, Annals Phys. **318**, 119 (2005) [arXiv:hep-ph/0504231].
9. N. Severijns, M. Beck and O. Naviliat-Cuncic, arXiv:nucl-ex/0605029.

10. P. Herczeg, Prog. Part. Nucl. Phys. **46** (2001) 413.
11. S. Eidelman *et al.* [Particle Data Group], Phys. Lett. B **592** (2004) 1.
12. D. N. Spergel *et al.* [WMAP Collaboration], Astrophys. J. Suppl. **148** (2003) 175 [arXiv:astro-ph/0302209].
13. S. Profumo, M. J. Ramsey-Musolf and S. Tulin, arXiv:hep-ph/0608064.
14. N. F. Bell, V. Cirigliano, M. J. Ramsey-Musolf, P. Vogel and M. B. Wise, Phys. Rev. Lett. **95** (2005) 151802 [arXiv:hep-ph/0504134].
15. S. Davidson, M. Gorbahn and A. Santamaria, Phys. Lett. B **626** (2005) 151 [arXiv:hep-ph/0506085].
16. N. F. Bell, M. Gorchtein, M. J. Ramsey-Musolf, P. Vogel and P. Wang, Phys. Lett. B **642** (2006) 377 [arXiv:hep-ph/0606248].
17. G. Prezeau and A. Kurylov, Phys. Rev. Lett. **95** (2005) 101802 [arXiv:hep-ph/0409193].
18. R. J. Erwin, J. Kile, M. J. Ramsey-Musolf and P. Wang, arXiv:hep-ph/0602240.

KAON PHYSICS: RECENT EXPERIMENTAL PROGRESS

MATTHEW MOULSON
Laboratori Nazionali di Frascati, 00044 Frascati RM, Italy
E-mail: moulson@lnf.infn.it

Numerous recent measurements of kaon decays are described, with an emphasis on results offering constraints on chiral perturbation theory calculations. An up-to-date estimate of $|V_{us}|f_+(0)$ based on semileptonic kaon decay rates is presented.

Keywords: Kaon decays, chiral perturbation theory, CKM matrix, V_{us}.

1. Introduction

The last three years have been marked by very rapid progress in experimental studies of kaon decays. This review is an attempt to summarize those results that offer direct comparison for the predictions of chiral perturbation theory (ChPT), including in particular the determination of the $I = 0, 2$ $\pi\pi$ scattering lengths from $K \to 3\pi$ and K_{e4} data, recent results on a selection of rare and radiative kaon decays, and the determination of $|V_{us}|f_+(0)$ from $K_{\ell 3}$ rates.

2. Kaon Decays and the $\pi\pi$ Scattering Lengths

2.1. $K^\pm \to \pi^\pm\pi^0\pi^0$ *decays*

In the $M^2_{\pi^0\pi^0}$ distribution for 23×10^6 $K^\pm \to \pi^\pm\pi^0\pi^0$ decays from 2003 data, NA48/2 observes a cusp at $M^2_{\pi^0\pi^0} = 4m^2_{\pi^+}$.[1] Cabibbo[2] has explained the cusp in terms of the interference between two amplitudes: the direct amplitude illustrated in Fig. 1a, and the rescattering amplitude illustrated in Fig. 1b. The latter amplitude is proportional to $a_0 - a_2$, the difference between the $I = 0, 2$ $\pi\pi$ scattering lengths; this quantity can thus be determined from fits to the $M^2_{\pi^0\pi^0}$ distribution. In the NA48/2 analysis, the fits are based on an extension of Cabibbo's original treatment, which includes contributions from the relevant two-loop diagrams.[3] The theoretical

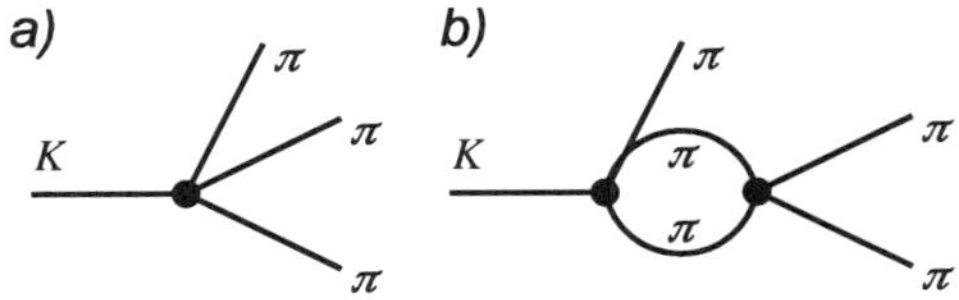

Fig. 1. $K \to 3\pi$ amplitudes used in the description of the cusp at $M^2_{\pi^0\pi^0} = 4m^2_{\pi^+}$ in the treatment of Ref. 2.

contribution to the uncertainty on the determination of $a_0 - a_2$ is estimated to be 5%. The NA48/2 fits account for a possible contribution from pionium formation in the vicinity of the cusp, but the seven bins nearest the position of the cusp are excluded from the fit to avoid possible bias arising from the lack of radiative corrections in the model used. NA48/2 obtains a fit to the $M^2_{\pi^0\pi^0}$ distribution with $\chi^2/\text{ndf} = 146/139$ (32.5%) giving $(a_0 - a_2)m_{\pi^+} = 0.268(10)_{\text{st}}(4)_{\text{sy}}(13)_{\text{th}}$. This result compares favorably to the prediction of Colangelo *et al.* ,[4] based on a matching procedure between representations of the $\pi\pi$ scattering amplitude from $\mathcal{O}(p^6)$ ChPT calculations and from the Roy equations as constrained by $\pi\pi$ scattering data at higher energies: $(a_0 - a_2)m_{\pi^+} = 0.265(4)$. NA48/2 statistics will increase by a factor of five when all 2003–2004 data are analyzed, which should allow the experimental uncertainty on $a_0 - a_2$ to be reduced to a level comparable to the uncertainty on the predicted value. A precise comparison will require a reduction of the uncertainty arising from the theoretical description of the cusp. Recent work on a nonrelativistic effective field theory treatment by Colangelo *et al.*[5,6] is promising in this regard. This scheme provides consistent power counting and allows electromagnetic corrections to be included in a standard way. NA48/2 is currently working on fits to the cusp using this treatment.[7]

2.2. K_{e4} *decays*

K_{e4} decays ($K \to \pi\pi e\nu$) also provide an opportunity to study the $\pi\pi$ interaction down to threshold. Fits to the kinematic distributions for K_{e4} decays provide sensitivity to the axial form factors F and G, and the vector form factor H. Recent work[8,9] makes use of the parameterization of Ref. 10. In this parameterization, the form factors are expanded in partial waves, e.g. , $F = F_s e^{i\delta_0^0} + F_p e^{i\delta_1^1} \cos\theta_\pi$, where θ_π is the angle between the π^+ momentum in the $\pi\pi$ system and the momentum of the $\pi\pi$ system in the kaon rest frame, and $\delta^{I=0}_{L=0}$ and $\delta^{I=1}_{L=1}$ are the $\pi\pi$ phase shifts. The coefficients

F_s, F_p, etc., are expanded in powers of $q^2 = (M_{\pi\pi}^2 - 4m_\pi^2)/m_\pi^2$, so that, e.g. , $F_s = f_s + f_s' q^2 + f_s'' q^4 + f_e(M_{e\nu}^2/4m_\pi^2) + \cdots$. Fits to the kinematic distributions thus allow determination of the parameters f_s, f_s', etc., which can be used to constrain ChPT couplings (F and G have been calculated to two loops in ChPT[11]), as well as values for $\delta_0^0 - \delta_1^1$ in bins of $M_{\pi\pi}$.

The Roy equations can be used to relate the $\pi\pi$ phase shifts to a_0 and a_2 (see, e.g. , Ref. 12), but measurements of $\delta_0^0 - \delta_1^1$ provide very little constraint on the value of a_2. However, consistency with the Roy equations and $\pi\pi$ scattering data above 0.8 GeV restricts possible values of a_0 and a_2 to a region in the a_0-a_2 plane often called the universal band (UB). The UB constraint can be used to eliminate a_2 in fits to K_{e4} data.

The most recent published data on $K^\pm \to \pi^+\pi^- e\nu$ decays are from BNL E865.[8] Using a sample of 388k $K^+ \to \pi^+\pi^- e^+\nu$ decays, E865 measures BR $= 4.11(11) \times 10^{-5}$. For the kinematic analysis, the data are divided into 28800 bins (six in $M_{\pi\pi}^2$). For each bin in $M_{\pi\pi}^2$, the expansion parameters for the form factors F, G, and H are obtained, as well as values for $\delta_0^0 - \delta_1^1$. A separate fit for a_0 is performed without binning the data in $M_{\pi\pi}^2$. From this latter fit, with the application of the UB constraint, E865 obtains $a_0 m_{\pi^+} = 0.228(12)_{\rm st}(4)_{\rm sy}(^{+12}_{-16})_{\rm th}$, which compares well with the prediction from Ref. 4, $a_0 m_{\pi^+} = 0.220(5)$.

NA48/2 has recently obtained preliminary results on $K^\pm \to \pi^+\pi^- e\nu$ decays as well.[7,9] The NA48/2 $K_{e4}^\pm$ sample includes 235k K^+ and 135k K^- events. The data are divided into 15000 bins (ten in $M_{\pi\pi}$). Like E865, NA48/2 measures the expansion parameters for the form factors F, G, and H, as well as $\delta_0^0 - \delta_1^1$, in bins of $M_{\pi\pi}$. With the UB constraint, NA48/2 obtains $a_0 = 0.256(8)_{\rm st}(7)_{\rm sy}(18)_{\rm th}$, which is in marginal ($\sim 1.7\sigma$) agreement with the prediction from Ref. 4. The data from the two experiments are compared in Fig. 2.

For either experiment, the experimental and theoretical contributions to the error on a_0 are 4–5% and 5–7%, respectively. In addition to the UB constraint, E865 uses the tighter constraint on a_0 vs. a_2 from Refs. 13 and 14 to obtain a value with a greatly reduced theoretical uncertainty: $a_0 m_{\pi^+} = 0.216(13)_{\rm st}(4)_{\rm sy}(2)_{\rm th}$. NA48/2 also expects to obtain values with smaller theoretical uncertainties using alternate treatments.

NA48/2 has also obtained preliminary results in the channel $K^\pm \to \pi^0\pi^0 e^\pm\nu$.[9] In this channel, Bose symmetry considerations imply that only the $L = 0$ partial wave contributes. With 37k events from 2003 and 2004 data, NA48/2 has obtained values for the form-factor parameters f_s'/f_s and f_s''/f_s that are consistent with the results obtained for $K^\pm \to \pi^+\pi^- e^\pm\nu$.

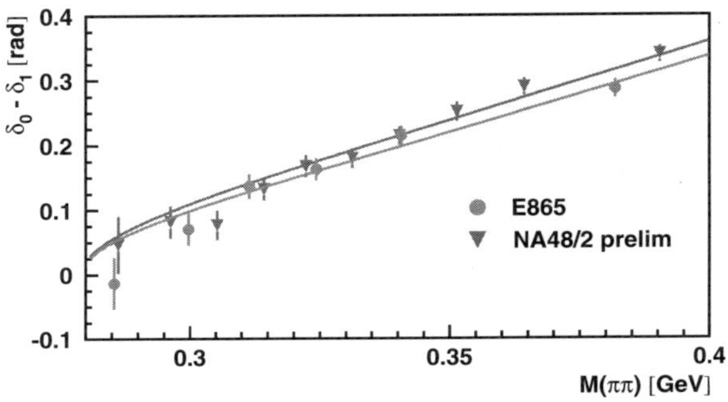

Fig. 2. Measurements of $\delta_0^0 - \delta_1^1$ from E865[8] and NA48/2[7,9] (NA48/2 data are preliminary). The solid curves illustrate the Roy equation solutions as parameterized in Ref. 12, with the values for a_0 quoted by each experiment. Adapted from a plot in Ref. 9.

3. Rare and Radiative Kaon Decays

3.1. *Radiative $K_{\ell 3}$ decays*

The amplitudes for radiative kaon decays can be divided into two components: an internal bremsstrahlung (IB) amplitude arising from radiation from external charged particles, and a direct-emission (DE), or structure-dependent (SD), amplitude arising from radiation from intermediate hadronic states. The components are isolated by analyzing the energy spectrum of the radiative photon; the study of the SD component can provide information about intermediate hadronic states in the decay.

For radiative $K_{\ell 3}$ decays, a convenient observable is

$$R_{\ell 3\gamma}(E_{\min}, \theta_{\min}) = \frac{\mathrm{BR}(K_{\ell 3\gamma}, E_\gamma > E_{\min}, \theta_{\ell\gamma} > \theta_{\min})}{\mathrm{BR}(K_{\ell 3(\gamma)})},$$

where $E_{\min}$ and $\theta_{\min}$ are cuts on the energy of the radiated photon and on its angle of emission (in the kaon CM frame) with respect to the momentum of the lepton, and $\mathrm{BR}(K_{\ell 3(\gamma)})$ signifies the inclusive branching ratio. These cuts are dictated by experimental necessity or by convention, bearing in mind that the IB amplitudes diverge for $E_\gamma \to 0$.

For $K_L \to \pi\mu\nu\gamma$, there is a recent measurement from KTeV:[15] $R^0_{\mu 3\gamma}(E_\gamma > 30\ \mathrm{MeV}) = 0.209(9)\%$. This is in good agreement with the earlier result from NA48:[16] $R^0_{\mu 3\gamma} = 0.208(26)\%$, as well as the predictions of Ref. 17. For $K_L \to \pi e\nu\gamma$, the situation is more complicated. Precise ChPT predictions are available—Gasser *et al.*[18] have performed an $\mathcal{O}(p^6)$ ChPT calculation to obtain $R^0_{e3\gamma}(E_\gamma > 30\ \mathrm{MeV}, \theta_{e\gamma} > 20°) = 0.96(1)\%$. On the experimental side, in 2001 KTeV published[19] the result $R^0_{e3\gamma} = 0.908(8)(^{+13}_{-12})\%$; the data have recently been reanalyzed using more restrictive cuts that provide better control over systematic effects, but which reduce the statistics by a factor of three. The new KTeV result[15] is $R^0_{e3\gamma} = 0.916(17)\%$. Both results are at variance with the value obtained in Ref. 18. On the other hand, NA48[20] obtains $R^0_{e3\gamma} = 0.964(8)(^{+11}_{-9})\%$, which is in good agreement with Ref. 18. KLOE has announced a preliminary value[21] based on 20% of its total data set: $R^0_{e3\gamma} = 0.92(2)(2)\%$. As the errors on this result are reduced, KLOE will be able to comment on the apparent discrepancy between the KTeV and NA48 results.

For $K^\pm \to \pi^0\mu^\pm\nu\gamma$ decays, experimental results have become available only recently. KEK E470 recently published[22] the measurement $\mathrm{BR}(K^\pm_{\mu3\gamma}, E_\gamma > 30\ \mathrm{MeV}, \theta_{\mu\gamma} > 20°) = 2.4(5)(6) \times 10^{-5}$, obtained using stopped K^+s. The ISTRA+ experiment at Protvino has used in-flight K^- decays to obtain measurements[23,24] of $\mathrm{BR}(K^\pm_{\mu3\gamma})$ for $5 < E_\gamma < 30$ MeV and $30 < E_\gamma < 60$ MeV. All of these results are in good agreement with $\mathcal{O}(p^4)$ ChPT estimates.[25,26] For $K^\pm \to \pi^0 e^\pm\nu\gamma$ decays, ISTRA+ has the preliminary result[24] $\mathrm{BR}(K^\pm_{e3\gamma}, E_\gamma > 30\ \mathrm{MeV}, \theta_{e\gamma} > 20°) = 3.05(2) \times 10^{-4}$, which is also in agreement with ChPT estimates.

3.2. *Radiative $K \to \pi\pi$ decays*

For the decay $K^\pm \to \pi^\pm\pi^0\gamma$, the IB component is suppressed by the $\Delta I = 1/2$ rule, leading to a relative enhancement of the DE component. The main contributions to the DE component are the M1 amplitude, arising from the chiral anomaly, and the E1 component. Both appear at $\mathcal{O}(p^4)$ in the chiral expansion. Interference between the M1 and E1 components vanishes in measurements inclusive with respect to the polarization of the radiative photon, but interference between the IB and E1 components is in principle observable. The decay is analyzed by looking at the variables T^*, the kinetic energy of the $\pi^\pm$ in the $K^\pm$ CM frame, and $W \equiv (p_K \cdot p_\gamma)(p_{\pi^+} \cdot p_\gamma)/(m_{K^+} m_{\pi^+})^2$. For a selected interval in T^*, chosen to reduce background from other $K^\pm$ decays with π^0s, the distribution in W is obtained. This can be written as the sum of a term from IB, a term from DE, and a term from IB/E1 interference. Fits to the W spectrum allow these components to be isolated.

Two K^+_{stop} experiments have recently measured the DE BR for $55 < T^* < 90$ MeV: the 2005 preliminary value from BNL E787[27] is $3.9(5)(^{+3}_{-4}) \times 10^{-6}$; the newly published value from KEK E470[28] is $3.8(8)(7) \times 10^{-6}$. Both measurements are based on samples of order 10k decays. No interference term is included in the fit in either case—both measurements are improvements of previous analyses in which no evidence for an interference term was found. On the other hand, NA48/2, has a preliminary result based on 124k in-flight decays from the 2003 data[7] obtained with an interference term included in the fit. The weights of the DE and interference terms are 3.35(35)(25)% and −2.67(81)(73)%. The errors on these two values are highly correlated, but the interference term is observed with 3σ significance. The NA48/2 measurement is for $0 < T^* < 80$ MeV because of trigger considerations. For comparison with the $K^+_{\rm stop}$ experiments, NA48/2 fits with no interference term and extrapolates to $55 < T^* < 90$ MeV, obtaining a DE fraction of 0.85(5)(2)%, which gives a DE BR of about 2.2×10^{-6}.

For the decay $K_L \to \pi^+\pi^-\gamma$, the IB and E1 amplitudes violate CP, leading to a relative enhancement of the M1 contribution. KTeV has recently published a new measurement in this channel[29] based on 112k events with $E_\gamma > 20$ MeV, updating their 2001 result with an increase in statistics of more than a factor of ten. In fits to the $(E_\gamma, \cos\theta_{\pi^+\gamma})$ distribution, the form factor describing the M1 amplitude is based on a pole model with VMD photon couplings.[30] KTeV obtains the best measurement to date of the M1 form-factor parameters, as well as the 90% CL limit $|g_{\rm E1}| < 0.21$. Setting $|g_{\rm E1}|$ to zero, KTeV obtains the fraction of the radiation spectrum from M1 DE: $\rm DE/(DE + IB) = 0.689(21)$ for $E_\gamma > 20$ MeV.

KTeV has also recently published a new measurement of the decay $K_L \to \pi^+\pi^-e^+e^-$.[31] In this decay, the polarization of the virtual photon is measured by the plane of the $\gamma^* \to e^+e^-$ conversion, so that the interference between the IB/E1 and M1 amplitudes can be observed. In addition, the process contributes in which the K_L emits a virtual photon and is transformed into a K_S, which then decays to $\pi^+\pi^-$; the amplitude for this process is proportional to $\langle r_K^2 \rangle$, the charge radius of the neutral kaon. With ~5200 events, KTeV obtains results for $\langle r_K^2 \rangle$ and the M1 form-factor parameters that agree with and improve upon the previous results from KTeV and NA48. The M1 form-factor parameter values also agree with those KTeV obtains for the $K_L \to \pi^+\pi^-\gamma$ channel, while the new KTeV limit on $|g_{\rm E1}|/|g_{\rm M1}|$ from $K_L \to \pi^+\pi^-e^+e^-$ is a stronger constraint on the size of the E1 amplitude than is obtained from $K_L \to \pi^+\pi^-\gamma$. The asymmetry A_ϕ about zero in the distribution of $\sin\phi\cos\phi$, where ϕ is the angle between the $\pi^+\pi^-$ and e^+e^- planes in the decay, parameterizes CP violation in the interference between IB and M1 amplitudes. The new KTeV measurement, $A_\phi = 0.136(14)(15)$, significantly improves on previous results from KTeV and NA48.

3.3. $K_S \to \gamma\gamma$

In ChPT calculations of the amplitude for $K_S \to \gamma\gamma$, since all particles involved are neutral, there are no tree-level contributions. Moreover, at $\mathcal{O}(p^4)$, only finite chiral-meson loops contribute. $BR(K_S \to \gamma\gamma)$ is predicted unambiguously at this level in terms of the couplings G_8 and G_{27}, giving 2.1×10^{-6}.[32] The most precise published measurement of this BR is from NA48: BR $= 2.78(6)(4) \times 10^{-6}$.[33] This result would suggest the need for a significant $\mathcal{O}(p^6)$ correction in the ChPT calculation of the BR. The NA48 result may soon be confirmed by KLOE.[21] While the number of $K_S \to \gamma\gamma$ events observed by KLOE is ~600, as compared to the ~ 7500 observed by NA48, KLOE profits from the use of a tagged K_S beam and does not have to contend with irreducible background from $K_L \to \gamma\gamma$. The KLOE measurement is in progress; a total error of 5% is expected, which is sufficient to confirm the NA48 result.

4. Determination of $|V_{us}|f_+(0)$ from $K_{\ell 3}$ Decays

A precise test of CKM unitarity can be obtained from the first-row constraint $|V_{ud}|^2 + |V_{us}|^2 + |V_{ub}|^2 = 1$ (with $|V_{ub}|^2$ negligible). At present, the most precise value for $|V_{us}|$ is obtained from $K_{\ell 3}$ decay rates, via

$$\Gamma(K_{l3(\gamma)}) = N_K\,|V_{us}|^2\,|f_+^{K^0\pi^-}(0)|^2\,I_{K\ell}\,(1 + 2\Delta_K^{SU(2)} + 2\Delta_{K\ell}^{\mathrm{EM}}), \tag{1}$$

where the subscripts K and ℓ indicate dependence on the kaon charge state ($K^\pm$, K^0) and lepton flavor, and N_K is a well-determined constant. The value of the hadronic matrix element at zero momentum transfer, $f_+(t=0)$, differs from unity because of $SU(2)$- and $SU(3)$-breaking effects; conventionally, the value for $K^0 \to \pi^-$ decays is used in Eq. (1) and $SU(2)$-breaking corrections are encoded in $\Delta_K^{SU(2)}$. $\Delta_{K\ell}^{\mathrm{EM}}$ is the correction to the form factor for the effects of long-distance electromagnetic interactions. These theory inputs to Eq. (1), and especially the status of $f_+^{K^0\pi^-}(0)$, are discussed in the contribution to these proceedings by V. Cirigliano (see also Ref. 34). The inputs from experiment are $\Gamma(K_{l3(\gamma)})$, the radiation-inclusive decay rates, or in practice, BR and lifetime measurements; and $I_{K\ell}$, the phase-space integrals of the form factors, which are calculated from measurements of the form-factor slopes λ as discussed in Sec. 4.3.

In the 2002 PDG evaluation, $|V_{ud}|^2 + |V_{us}|^2 = 0.9965(15)$, a 2.3σ hint of CKM unitarity violation. The 2003 result from BNL E865,[35] BR$(K_{e3}^\pm) =$ 5.13(2)(10)%, is 2.7σ higher than the 2002 PDG average for this BR and seemed to resolve the problem. With respect to the older measurements of $K_{\ell 3}$ decays, the E865 result made use of much higher statistics, and in addition was the first measurement with a well-defined treatment of the contribution from radiative decays. All newer measurements from KTeV, KLOE, ISTRA+, and NA48 share this feature.

4.1. K_L *and* K_S *branching ratios and lifetimes*

KTeV, NA48, and KLOE have recently published measurements of the BRs for the dominant K_L decay channels, including the $K_{\ell 3}$ decays.

KTeV has measured five ratios of the BRs for the principal K_L decays.[36] The six BRs involved account for 99.93% of the K_L width; the ratios are combined to determine the absolute BR values.

NA48 has measured the ratio of BR(K_{e3}) to the sum of the BRs for all decays to two tracks.[37] This is essentially BR(K_{e3})/[1 − BR($3\pi^0$)]. NA48 normalizes using the average of the KTeV and NA31 measurements of BR($3\pi^0$); they also have a preliminary measurement of BR($3\pi^0$) normalized to $K_S \to \pi^0\pi^0$ decays that confirms this average.[38]

KLOE has measured absolute BRs for the four dominant K_L modes using $\phi \to K_S K_L$ events in which a $K_S \to \pi^+\pi^-$ decay is used to tag the K_L decay, providing normalization.[39] The dominant contribution to the uncertainties on the absolute BRs comes from the uncertainty on τ_L, the K_L lifetime, which is needed to calculate the overall geometrical efficiency. Expressing the BRs as functions of τ_L and imposing the constraint $\sum \text{BR} = 1$ (with the 2004 PDG values used for the smaller K_L BRs), final values are obtained for the four BRs and for τ_L, with greatly reduced uncertainties. KLOE has also measured τ_L directly, using 10^7 $K_L \to 3\pi^0$ events,[40] for which the reconstruction efficiency is high and uniform inside the fiducial volume ($0.37\lambda_L$). The result, $\tau_L = 50.92(17)(25)$ ns, is consistent with the value obtained from the sum of the K_L BRs.

The results from the experiments are summarized in Table 1. Because of their interdependence for the purposes of normalization, they are best incorporated into a new evaluation of $|V_{us}|f_+(0)$ via a global fit akin to that performed by the PDG. The fit performed here uses the data in Table 1 in addition to four other measurements used in the 2006 PDG fit. The free parameters are the seven largest K_L BRs and τ_L; the BRs in the fit are constrained to sum to unity. The principal difference between the fit performed here and the 2006 PDG fit is that here, the intermediate KTeV and KLOE values (i.e., , before applying constraints) are the inputs, and the complete error matrix is used to handle the correlations between the measurements from each experiment. (In the 2006 PDG fit, the final KTeV

Table 1. Recent K_L BR and lifetime measurements used in fit

KTeV[a]	$\mathrm{BR}(K_{\mu 3}/K_{e3}) = 0.6640(26)$ $\mathrm{BR}(\pi^+\pi^-\pi^0/K_{e3}) = 0.3078(18)$ $\mathrm{BR}(\pi^0\pi^0/3\pi^0) = 4.446(25) \times 10^{-3}$	$\mathrm{BR}(3\pi^0/K_{e3}) = 0.4782(55)$ $\mathrm{BR}(\pi^+\pi^-/K_{e3}) = 4.856(29) \times 10^{-3}$
KLOE[a,b]	$\mathrm{BR}(K_{e3}) = 0.4049(21)$ $\mathrm{BR}(3\pi^0) = 0.2018(23)$ $\tau_L = 50.92(30)$ ns	$\mathrm{BR}(K_{\mu 3}) = 0.2726(16)$ $\mathrm{BR}(\pi^+\pi^-\pi^0) = 0.1276(15)$
NA48	$\mathrm{BR}(K_{e3}/2\text{ track}) = 0.4978(35)$	$\mathrm{BR}(3\pi^0) = 0.1966(34)$[c]

Note: [a] In the fit, errors on these BRs are parameterized by the complete covariance matrix. [b] In the fit, these BR values are expressed as functions of τ_L as described in Ref. 39. [c] Preliminary.

and KLOE BR results were used and one measurement involving $\mathrm{BR}(3\pi^0)$ was removed in each case.) Scale factors for the errors are calculated and used as per the PDG prescription. The fit has $\chi^2/\mathrm{ndf} = 13.2/9$ (15.4%) and gives $\mathrm{BR}(K_{e3}) = 0.4047(11)$ ($S = 1.4$), $\mathrm{BR}(K_{\mu 3}) = 0.2698(9)$ ($S = 1.3$), and $\tau_L = 51.11(21)$ ns ($S = 1.1$). These values are quite similar to those from the 2006 PDG fit.

KLOE also has recently published[41] a measurement of $\mathrm{BR}(K_S \to \pi e \nu)$ that is precise enough to contribute meaningfully to the evaluation of $|V_{us}|f_+(0)$. For this measurement, K_S decays are tagged by the observation of a K_L interaction in the KLOE calorimeter. The quantity directly measured is $\mathrm{BR}(\pi e \nu/\pi^+\pi^-)$. Together with the recently published KLOE value $\mathrm{BR}(\pi^+\pi^-/\pi^0\pi^0) = 2.2459(54)$, the constraint that the K_S BRs must sum to unity, and the assumption of universal lepton couplings, this completely determines the K_S BRs for $\pi^+\pi^-$, $\pi^0\pi^0$, K_{e3}, and $K_{\mu 3}$ decays.[42] In particular, $\mathrm{BR}(K_S \to \pi e \nu) = 7.046(91) \times 10^{-4}$. The KLOE measurement is performed separately for each lepton charge state, yielding the first result for the semileptonic charge asymmetry from K_S decays, $A_S = 15(96)(29) \times 10^{-4}$. This value has been used in tests of CPT symmetry and the $\Delta S = \Delta Q$ rule.[43]

4.2. *$K^\pm$ branching ratios and lifetime*

NA48/2, ISTRA+, and KLOE all have preliminary measurements of $K^\pm$ BRs. These new measurements have a significant impact on the evaluation of $|V_{us}|f_+(0)$, as demonstrated by the updated fit performed here.

NA48/2 measures $\mathrm{BR}(K_{e3}/\pi\pi^0)$ and uses the 2004 PDG value for $\mathrm{BR}(\pi\pi^0)$ to quote $\mathrm{BR}(K_{e3}) = 5.14(2)(6)\%$.[38] The fit performed here makes use of the value $\text{BR}(K_{e3}/\pi\pi^0) = 0.2433(25)$, as well as of the NA48/2 result $\text{BR}(K_{\mu3}/K_{e3}) = 0.6749(35)(24)$.

ISTRA+ also measures $\text{BR}(K_{e3}/\pi\pi^0)$; they use the 2006 PDG value for $\text{BR}(\pi\pi^0)$ to quote $\text{BR}(K_{e3}) = 5.170(11)(57)\%$.[24] The fit performed here makes use of the value $\text{BR}(K_{e3}/\pi\pi^0) = 0.2471(23)$.

KLOE measures the absolute K_{e3} and $K_{\mu3}$ BRs.[44] In $\phi \to K^+K^-$ decays, K^+ decays into $\mu\nu$ or $\pi\pi^0$ are used to tag a K^- beam, and vice versa. KLOE performs four separate measurements for each $K_{\ell3}$ BR, corresponding to the different combinations of kaon charge and tagging decay. The final averages are $\text{BR}(K_{e3}) = 5.047(46)(80)\%$ and $\text{BR}(K_{\mu3}) = 3.310(40)(70)\%$. The fit performed here takes into account the dependence of these BRs on the $K^\pm$ lifetime.

The world average value for $\tau_\pm$ is nominally quite precise; the 2006 PDG quotes $\tau_\pm = 12.385(25)$ ns. However, the error is scaled by 2.1; the confidence level for the average is 0.2%. It is important to confirm the value of $\tau_\pm$. KLOE has a preliminary measurement based on $K^\pm$ decays tagged by $K^\mp \to \mu\nu$ and observed in a fiducial volume of $\sim 1\lambda_\pm$.[45] The result, $\tau_\pm = 12.336(44)(65)$ ns, agrees with the PDG average, although at present the KLOE uncertainty is significantly larger.

The fit performed here makes use of all preliminary results cited above, plus the data used in the 2006 PDG fit, for a total of 30 measurements. The free parameters are the six main $K^\pm$ BRs and $\tau_\pm$; the BRs are constrained to sum to unity. The fit gives $\chi^2/\text{ndf} = 38/24$ (3.6%). The poor fit quality principally derives from the scatter in the five older measurements of $\tau_\pm$; when these are replaced with their PDG average with scaled error, $\tau_\pm = 12.385(25)$ ns, the fit gives $\chi^2/\text{ndf} = 20.5/20$ (42%), with no significant changes in the results. The results are $\text{BR}(K_{e3}) = 5.056(37)\%$ ($S = 1.3$), $\text{BR}(K_{\mu3}) = 3.399(29)\%$ ($S = 1.2$), and $\tau_\pm = 12.384(21)$ ns ($S = 1.8$). The significant evolution of the average values of the BRs for $K^\pm_{\ell3}$ decays and for the important normalization channels is evident in Fig. 3.

4.3. $K_{\ell3}$ *form-factor slopes*

Only the vector part of the weak current contributes to the hadronic matrix element for $K_{\ell3}$ decays:

$$\langle\pi|\, J_\alpha \,|K\rangle = f_+(t)(p_K + p_\pi)_\alpha + f_-(t)(p_K - p_\pi)_\alpha,$$

where $t = (p_K - p_\pi)^2$. When the squared matrix element is evaluated, a factor of m_ℓ^2/m_K^2 multiplies all terms containing $f_-(t)$. This form factor can be neglected for K_{e3} decays. For the description of $K_{\mu3}$ decays, it is

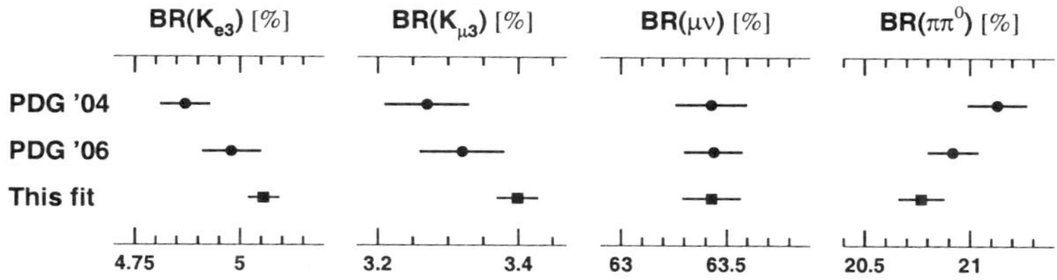

Fig. 3. Evolution of average values for main $K^{\pm}$ BRs.

customary to use $f_+(t)$ and the scalar form factor $f_0(t) \equiv f_+(t) + [t/(m_K^2 - m_\pi^2)]\, f_-(t)$. The form factors are written as $f_{+,0}(t) = f_+(0)\tilde{f}_{+,0}(t)$, with $\tilde{f}_{+,0}(0) = 1$, and often expanded in powers of t as

$$\begin{aligned} \tilde{f}(t) &= 1 + \lambda \frac{t}{m_{\pi^+}^2} && \text{(linear)}, \\ \tilde{f}(t) &= 1 + \lambda' \frac{t}{m_{\pi^+}^2} + \tfrac{1}{2}\lambda'' \left(\frac{t}{m_{\pi^+}^2}\right)^2 && \text{(quadratic)}. \end{aligned} \tag{2}$$

The slopes λ are obtained from fits to the measured t distributions, but sensitivity to the quadratic terms is poor, in large part because the kinematic density of the matrix element drops to zero at large t, where the form factor itself is maximal. The vector form factor f_+ is dominated by the vector $K\pi$ resonances, e.g. , $K^*(892)$; this fact suggests the pole parameterization, $\tilde{f}_+(t) = M_V^2/(M_V^2 - t)$. This one-parameter form generally fits experimental data better than does the linear parameterization; its expansion gives $\lambda' = (m_{\pi^+}/M_V)^2$; $\lambda'' = 2\lambda'^2$. A dispersive representation for $f_0(t)$ featuring a single experimental parameter has also been proposed.[46]

KTeV, KLOE, and NA48 have measured the form-factor slopes in K_L decays; ISTRA+ has measured the slopes in K^- decays. KTeV and ISTRA+ have reported fit results for λ'_+, λ''_+, and λ_0 for both K_{e3} and $K_{\mu 3}$ decays; KLOE and NA48 have values for λ'_+ and λ''_+ from K_{e3} decays. These data are collected in Table 2. The experiments use different conventions in reporting the form-factor slopes; the data in the table have been adjusted for use with Eq. (2). Most experiments also quote various other combinations of linear, quadratic, and pole fit results. NA48 has preliminary results for λ_+ and λ_0 (linear fit) from $K_{\mu 3}$ decays.[47]

To obtain reference values of the form-factor slopes for the phase-space integrals, the data in Table 2 have been averaged. Correlation coefficients are available for the KTeV and KLOE data; for NA48 and ISTRA+, their values have been inferred, in part by assuming the correlations to be intrinsic to the measurement and largely independent of the experimental

Table 2. Measurements of $K_{\ell 3}$ form-factor slopes

Experiment	$\lambda'_+ \times 10^3$	$\lambda''_+ \times 10^3$	$\lambda_0 \times 10^3$
KTeV K_L e3-μ3 avg. (Ref. 48)	20.6 ± 1.8	3.2 ± 0.7	13.7 ± 1.3
KLOE K_L e3 (Ref. 49)	25.5 ± 1.8	1.4 ± 0.8	
NA48 K_L e3 (Ref. 50)	28.0 ± 2.4	0.4 ± 0.9	
ISTRA+ K^- e3 (Ref. 51)	24.9 ± 1.7	1.9 ± 0.9	
ISTRA+ K^- μ3 (Ref. 52)[a]	23.0 ± 6.4	2.3 ± 2.3	17.1 ± 2.2

Note: [a] No systematic uncertainties are quoted for this fit.

details. In principle, this fact could be exploited to fix all correlations *a priori*.[53] The results are $\lambda'_+ = 24.72(84) \times 10^{-3}$, $\lambda''_+ = 1.67(36) \times 10^{-3}$, and $\lambda_0 = 15.72(97) \times 10^{-3}$, with $\rho(\lambda'_+, \lambda''_+) = -0.94$, $\rho(\lambda'_+, \lambda_0) = +0.30$, and $\rho(\lambda''_+, \lambda_0) = -0.40$. The fit gives $\chi^2/\text{ndf} = 11.6/9$ (23.9%).

KTeV, KLOE, and NA48 all quote values for M_V for K_{e3} decays. The average value is $M_V = 875.3 \pm 5.4$ MeV with $\chi^2/\text{ndf} = 1.83/2$ (40%). Using this or the avergage of the quadratic fit results stated above to calculate the phase-space integral for the $K_{L\,e3}$ form factor makes a 0.02% difference in the result. In the calculation of $|V_{us}|f_+(0)$, no additional error is assigned to account for differences obtained with quadratic and pole parameterizations for $\tilde{f}_+(t)$.

4.4. *Discussion*

Using the results of the fits discussed above for the BRs, lifetimes, and form-factor slopes, $|V_{us}|f_+(0)$ has been evaluated for each of the five decay modes using Eq. (1). The results are summarized in the left panel of Fig. 4. The most precise determination is from $K_{L\,e3}$ decays: $|V_{us}|f_+(0) = 0.2161(6)$. Indicatively, for K_L decays, the precison is limited by the uncertainties on the decay widths, and in particular, by the uncertainty on τ_L. For the $K_{L\,\mu3}$ mode, the uncertainty on Δ^{EM} is also a significant issue. For both $K^\pm$ decays, the uncertainties on the BR measurements are the limiting factor; for $K^\pm_{\mu3}$, the uncertainty on Δ^{EM} is nearly as important. Although not a limiting factor at the moment, for $K^\pm$ decays, the present ~10% uncertainty on $\Delta^{SU(2)}_{K^\pm}$ would ultimately prohibit obtaining the same level of precision as for K_L decays. The uncertainty on the phase-space integral is not currently a limiting factor for any mode.

A fit to all five values for $|V_{us}|f_+(0)$ taking all correlations into account has $\chi^2/\text{ndf} = 2.76/4$ (60%) and gives the average value $|V_{us}|f_+(0) = 0.2162(4)$. When $|V_{us}|f_+(0)$ is evaluated separately for K^0 and $K^\pm$ decays

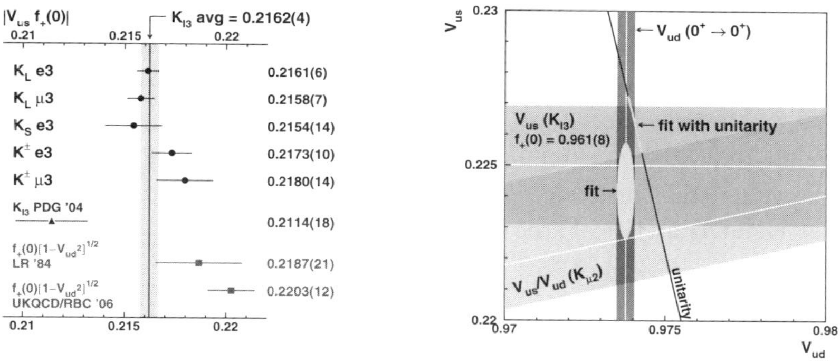

Fig. 4. Left: Determinations of $|V_{us}|f_+(0)$ for five $K_{\ell 3}$ decay channels, with average and comparison values. Right: Results of a fit to values for $|V_{ud}|$, $|V_{us}|$, and $|V_{us}|/|V_{ud}|$.

(making use of separate averages for the form-factor slopes for each case), the result for $K^{\pm}$ decays is 1σ higher than that for K^0 decays. At present, the results obtained from all modes are consistent, though it is worth noting that all of the new BR($K^{\pm}_{\ell 3}$) results are still preliminary.

To test CKM unitarity, a value for $f_+(0)$ is needed. Conventionally, the original estimate of Leutwyler and Roos,[54] $f_+(0) = 0.961(8)$, is used; this gives $|V_{us}| = 0.2250(19)$. Using the most recent evaluation of $|V_{ud}|$ from $0^+ \to 0^+$ nuclear beta decays,[55] $|V_{ud}| = 0.97377(27)$, one has $|V_{ud}|^2 + |V_{us}|^2 - 1 = -0.0012(10)$, a result perfectly compatible with unitarity. As is evident from Fig. 4, this represents a significant evolution of the experimental picture since the 2004 PDG evaluation. However, lattice evaluations of $f_+(0)$ are rapidly improving in precision. For example, the UKQCD/RBC Collaboration has announced a preliminary result[56] from a lattice calculation with $2+1$ flavors of dynamical domain-wall quarks: $f_+(0) = 0.9680(16)$. This value implies $|V_{us}| = 0.2234(6)$ and $|V_{ud}|^2 + |V_{us}|^2 - 1 = -0.0019(6)$, a 3.2σ discrepancy with CKM unitarity.

Marciano[57] has observed that $\Gamma(K_{\mu 2})/\Gamma(\pi_{\mu 2})$ can be precisely related to the product $(|V_{us}|/|V_{ud}|)^2(f_K/f_\pi)^2$. The recent measurement BR($K^+ \to \mu^+\nu$) $= 0.6366(9)(15)$ from KLOE,[58] together with the preliminary lattice result $f_K/f_\pi = 1.208(2)(^{+7}_{-14})$ from the MILC Collaboration,[59] gives $|V_{us}|/|V_{ud}| = 0.2286(^{+27}_{-15})$. This ratio can be used in a fit together with the values of $|V_{ud}|$ from Ref. 55 and $|V_{us}|$ from $K_{\ell 3}$ decays as above. Using the value for $|V_{us}|$ obtained with $f_+(0) = 0.961(8)$, the fit gives $|V_{ud}| = 0.97377(27)$ and $|V_{us}| = 0.2242(16)$, with $\chi^2/\text{ndf} = 0.52/1$ (47%). The unitarity constraint can also be included, in which case the fit gives $\chi^2/\text{ndf} = 3.43/2$ (18%). Both results are illustrated in Fig. 4, right. If instead the newer lattice result for $f_+(0)$ is used to obtain $|V_{us}|$, the fit gives $|V_{ud}| = 0.97377(27)$ and $|V_{us}| = 0.2233(6)$, with $\chi^2/\text{ndf} = 0.074/1$ (79%). When the unitarity constraint is imposed, the fit gives $\chi^2/\text{ndf} = 10.5/2$, corresponding to a probability of 0.53%. These results reinforce the conclusion that the result of the first-row test of CKM unitarity depends mainly on the value and uncertainty assumed for $f_+(0)$. Confirmation of the new lattice result is a critical step towards a clearer understanding of the situation.

Acknowledgments

I would like to congratulate the members of the E470, E787, E865, KTeV, ISTRA+, and NA48 collaborations, as well as my fellow KLOE collaborators, for their hard work. I apologize for omitting many results for reasons of space. Special thanks go to M. Antonelli, V. Cirigliano, and P. Franzini for many useful discussions; to B. Sciascia for help with the fits; and to G. Isidori and T. Spadaro for comments on the manuscript.

References

1. NA48/2 Collaboration, J. Batley *et al.*, *Phys. Lett. B* **633**, 173 (2006).
2. N. Cabibbo, *Phys. Rev. Lett.* **93**, 121801 (2004).
3. N. Cabibbo and G. Isidori, *JHEP* **0503**, 021 (2005).
4. G. Colangelo, J. Gasser and H. Leutwyler, *Phys. Lett. B* **488**, 261 (2000).
5. G. Colangelo *et al.*, *Phys. Lett. B* **638**, 187 (2006).
6. B. Kubis, these proceedings.
7. S. Goy López, for the NA48/2 Collaboration, these proceedings.
8. S. Pislak *et al.*, *Phys. Rev. D* **67**, 072004 (2003).
9. B. Bloch-Devaux, for the NA48/2 Collaboration, talk at QCD'06 conference (Montpellier, France, 2006).
10. G. Amorós and J. Bijnens, *J. Phys. G* **25**, 1607 (1999).
11. G. Amorós, J. Bijnens and P. Talavera, *Nucl. Phys. B* **585**, 293 (2000).
12. B. Ananthanarayan *et al.*, *Phys. Rep.* **353**, 207 (2001).
13. G. Colangelo, J. Gasser and H. Leutwyler, *Nucl. Phys. B* **603**, 125 (2001).
14. G. Colangelo, J. Gasser and H. Leutwyler, *Phys. Rev. Lett.* **86**, 5008 (2001).
15. KTeV Collaboration, T. Alexopoulos *et al.*, *Phys. Rev. D* **71**, 012001 (2005).
16. M. Bender *et al.*, *Phys. Lett. B* **418**, 411 (1998).
17. H. Fearing, E. Fischbach and J. Smith, *Phys. Rev. D* **2**, 542 (1970).
18. J. Gasser *et al.*, *Eur. Phys. J. C* **40**, 205 (2005).
19. KTeV Collaboration, A. Alavi-Harati *et al.*, *Phys. Rev. D* **64**, 112004 (2001).
20. NA48 Collaboration, A. Lai *et al.*, *Phys. Lett. B* **605**, 247 (2005).
21. M. Martini, for the KLOE Collaboration, these proceedings.
22. KEK-E470 Collaboration, S. Shimizu *et al.*, *Phys. Lett. B* **633**, 190 (2006).
23. O. Tchikilev *et al.*, hep-ex/0506023, (2005).

24. V. Duk, talk at ICHEP'06 conference (Moscow, 2006).
25. V. Braguta, A. Likhoded and A. Chalov, *Phys. Rev. D* **65**, 054038 (2002).
26. J. Bijnens, G. Ecker and J. Gasser, *Nucl. Phys. B* **396**, 81 (1993).
27. T. Tsunemi, talk at Kaon'05 conference (Evanston IL, USA, 2006).
28. KEK-E470 Collaboration, M. Aliev *et al.*, *Eur. Phys. J. C* **46**, 61 (2006).
29. KTeV Collaboration, E. Abouzaid *et al.*, *Phys. Rev. D* **74**, 032004 (2006).
30. Y. Lin and G. Valencia, *Phys. Rev. D* **37**, 143 (1988).
31. KTeV Collaboration, E. Abouzaid *et al.*, *Phys. Rev. Lett.* **96**, 101801 (2006).
32. G. D'Ambrosio *et al.*, in *Second DAΦNE Physics Handbook*, eds. L. Maiani, G. Pancheri and N. Paver (Laboratori Nazionali di Frascati, 1995) pp. 265–312.
33. NA48 Collaboration, A. Lai *et al.*, *Phys. Lett. B* **551**, 7 (2003).
34. V. Cirigliano, in *Proc. Flavor Physics and CP Violation Conf. (FPCP'06)*, (Vancouver, 2006). hep-ph/0606020.
35. A. Sher *et al.*, *Phys. Rev. Lett.* **91**, 261802 (2003).
36. T. Alexopoulos *et al.*, *Phys. Rev. D* **70**, 092006 (2004).
37. NA48 Collaboration, A. Lai *et al.*, *Phys. Lett. B* **602**, 41 (2004).
38. L. Litov, for the NA48 Collaboration, in *Proc. 32nd Int. Conf. on High-Energy Physics (ICHEP'04)*, (Beijing, 2004). hep-ex/0501048.
39. KLOE Collaboration, F. Ambrosino *et al.*, *Phys. Lett. B* **632**, 43 (2006).
40. KLOE Collaboration, F. Ambrosino *et al.*, *Phys. Lett. B* **626**, 15 (2005).
41. KLOE Collaboration, F. Ambrosino *et al.*, *Phys. Lett. B* **636**, 173 (2006).
42. KLOE Collaboration, F. Ambrosino *et al.*, hep-ex/0601025, (2006).
43. KLOE Collaboration (F. Ambrosino *et al.*), G. D'Ambrosio and G. Isidori, hep-ex/0610034, (2006).
44. B. Sciascia, for the KLOE Collaboration, in *Proc. EPS Int. Europhysics Conf. on High-Energy Physics (HEP-EPS'05)*, (Lisbon, 2005). hep-ex/0510028.
45. M. Palutan, for the KLOE Collaboration, in *Proc. 41st Rencontres de Moriond on Electroweak Interactions and Unified Theories*, (La Thuile, Italy, 2006). hep-ex/0605055.
46. J. Stern, these proceedings.
47. A. Winhart, for the NA48 Collaboration, in *Proc. EPS Int. Europhysics Conf. on High-Energy Physics (HEP-EPS'05)*, (Lisbon, 2005). PoS(HEP2005)289.
48. T. Alexopoulos *et al.*, *Phys. Rev. D* **70**, 092007 (2004).
49. KLOE Collaboration, F. Ambrosino *et al.*, *Phys. Lett. B* **636**, 166 (2006).
50. NA48 Collaboration, A. Lai *et al.*, *Phys. Lett. B* **604**, 1 (2004).
51. O. Yushchenko *et al.*, *Phys. Lett. B* **589**, 111 (2004).
52. O. Yushchenko *et al.*, *Phys. Lett. B* **581**, 31 (2004).
53. P. Franzini, unpublished note, (2006).
54. H. Leutwyler and M. Roos, *Z. Phys. C* **25**, 91 (1984).
55. W. Marciano and A. Sirlin, *Phys. Rev. Lett.* **96**, 032002 (2006).
56. UKQCD/RBC Collaboration, D. Antonio *et al.*, in *Proc. 24th Int. Symposium on Lattice Field Theory*, (Tucson, 2006). hep-lat/0610080.
57. W. Marciano, *Phys. Rev. Lett.* **93**, 231803 (2004).
58. KLOE Collaboration, F. Ambrosino *et al.*, *Phys. Lett. B* **632**, 76 (2006).
59. C. Bernard, these proceedings.

STATUS OF THE CABIBBO ANGLE

VINCENZO CIRIGLIANO*

Theoretical Division, T-16 Group, MS B283
Los Alamos National Laboratory
Los Alamos, NM 87545, USA
**E-mail: cirigliano@lanl.gov*

I review the current most precise determinations of the CKM elements V_{ud} and V_{us}, that constrain the Cabibbo angle at the 1% level.

Keywords: CKM unitarity; Weak Decays; Chiral Perturbation Theory

1. Introduction: weak universality and paths to V_{ud} and V_{us}

The CKM matrix[1,2] is a unitary matrix parameterizing quark mixing in charged current weak interactions within the Standard Model:

$$\mathcal{L}_{CC} = \frac{g}{2\sqrt{2}} W_\mu^+ \, \bar{u}_i \, \gamma^\mu \left(1 - \gamma^5\right) V_{ij} \, d_j + \text{h.c.} \; . \tag{1}$$

The most precise constraints on the size of the elements of the matrix V_{ij} are obtained from the low-energy $s \to u$ and $d \to u$ semileptonic transitions. Combining the precise determinations of $|V_{ud}|$ and $|V_{us}|$ extracted from these processes one can perform the most stringent test of CKM unitarity, namely one can probe the validity of the relation

$$|V_{ud}|^2 + |V_{us}|^2 + |V_{ub}|^2 = 1 \; . \tag{2}$$

at the 0.1% level. A deviation from unitarity at the 2σ level has been around for a while and has spurred recent experimental and theoretical investigations of this issue. The present accuracy on $|V_{ud}|$ and $|V_{us}|$ is such that the contribution of $|V_{ub}|^2 \approx 2 \times 10^{-5}$ in the relation (2) can be safely neglected, and the uncertainty of the first two terms is comparable. Provided unitarity is satisfied, one can then set

$$V_{us} = \sin\theta_c \; , \qquad V_{ud} = \cos\theta_c \; , \tag{3}$$

as in the original Cabibbo theory,[1] and $|V_{us}|$ and $|V_{ud}|$ provide two independent determinations of the Cabibbo angle both around the 1% level.

$\mathbf{V_{ud}}$	$0^+ \to 0^+$ $\pi^\pm \to \pi^0 e\nu$	$n \to pe\bar{\nu}$		
$\mathbf{V_{us}}$	$K \to \pi \ell \nu$	$\Lambda \to pe\bar{\nu},...$	$K \to \mu\nu$	$\tau \to h_s \nu_\tau$ *(inclusive)*

Fig. 1. Paths to V_{ud} and V_{us}.

Table 1 summarizes the semi-leptonic processes that are currently analyzed to determine V_{ud} and V_{us}. The various processes are grouped in several columns according to the nature of the weak transition.

In the first column I list the processes where only the vector current contributes at tree level (i.e. before considering radiative corrections). These are traditionally the gold-plated modes that provide the best determination of V_{ud} and V_{us}. The key observation which allows a precise extraction of the CKM factors is the conservation of the vector current at zero momentum transfer in the $SU(N)$ limit and the non-renormalization theorem. The latter implies that the relevant hadronic form factors are completely determined up to tiny second order isospin-breaking corrections in the $d \to u$ case[3] or SU(3)-breaking corrections in the $s \to u$ case.[4] As a result of this fortunate situation, the accuracy on $|V_{us}|$ is approaching the 1% level and the one on $|V_{ud}|$ is already below the 0.05% level.

In the second column I list processes that receive tree level contributions from both the vector and the axial current. While the vector current contribution enjoys the same advantage of the golden modes, the axial current contribution is not easily determined from theoretical arguments. The strategy here is to get information on the axial matrix element from experiment (such as the β asymmetry in neutron decay).

Less conventional processes and strategies are listed in the third and fourth column. The $K \to \mu\nu$ process is a purely axial current process. The viability of this method is due to recent progress in lattice QCD that allows one to calculate the decay constants F_π, F_K and in particular to pin down the ratio F_K/F_π to an accuracy of 1 %. Semi-leptonic τ decays into strange hadronic final states also provide information on V_{us}. Here the key ingredient is the presence of a hard scale in the problem (the τ mass), which allows one to use the OPE in the calculation of inclusive rates.

In the following sections I will review the determinations of $|V_{us}|$ and $|V_{ud}|$ mainly from the golden modes. A more detailed discussion of all approaches can be found in Ref.[5] and references therein.

2. V_{ud} from super-allowed nuclear beta decays

Currently the best determination of V_{ud} comes from the so-called super-allowed $0^+ \to 0^+$ Fermi transitions between nuclei. The expression for the inverse half-life reads:

$$\frac{1}{t} = \frac{G_\mu \, |V_{ud}|^2 m_e^5}{\pi^3 \log 2} \, f(Q) \; (1 + RC) \;\;, \tag{4}$$

where G_μ is the Fermi constant as extracted from muon decay, $f(Q)$ is a phase space factor (that depends on the experimentally determined Q-value for the given reaction) and RC collectively denotes the radiative corrections. This term is conventionally separated into three pieces:

$$1 + RC = (1 - \delta_C)\,(1 + \delta_R)\,(1 + \Delta_R) \;. \tag{5}$$

δ_C is a nuclear structure dependent correction that parameterizes the deviation of the weak current matrix element from the $SU(2)$ limit ($\langle f|\tau_+|i\rangle = \sqrt{2}\,(1 - \delta_C/2)$) due to Coulomb distortion of the nuclear wave-function. The typical size of these correction is $\delta_C \sim 0.5\%$.[6,7] δ_R collects the nucleus-dependent electromagnetic corrections, that depend on Z, E_e^{max} as well as on nuclear structure. The typical size is $\delta_R \sim 1.5\%$.[8,9] Finally, Δ_R denotes the nucleus-independent short distance corrections. These involve model-independent large logarithms of the ratio $M_Z/1\text{GeV}$, as well as model-dependent contributions induced by the axial current. A recent matching calculation by Marciano and Sirlin[10] has provided an improved determination of $\Delta_R \sim 2.4\%$ and its uncertainty.

The crucial test of nucleus-dependent radiative corrections is provided by checking that while the uncorrected ft values are nucleus-dependent, the corrected values $f\,t(1 + RC)$ are indeed nucleus-independent within uncertainties. This is shown in Fig. 2 where both uncorrected and corrected ft values are reported as a function of the Z of daughter nuclei for the most precisely known transitions. The overall consistency of the data is interpreted as a validation of the theoretical corrections, although the new Q-value measurement of ^{46}V[11,12] (obtained with a penning trap) implies a worse consistency of the data-set. Other Q-values have been measured with the new penning trap technique.[12] The results are consistent with previous measurements, suggesting that the ^{46}V constitutes an isolated anomaly. This, however, raises doubts on the reliability of theoretical corrections for this particular transition.

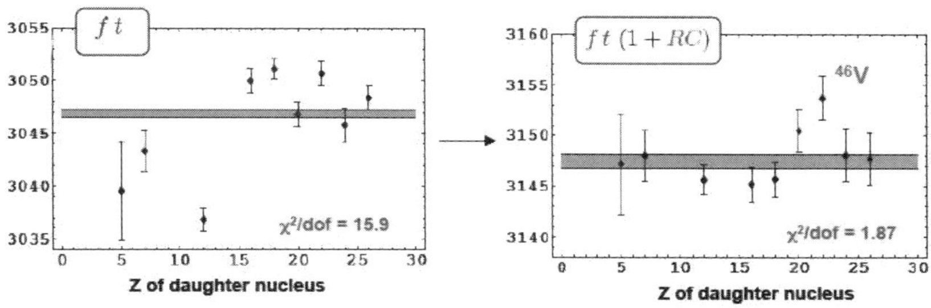

Fig. 2. Uncorrected (left) and corrected (right) ft values (in unit of s) as a function of Z of the daughter nucleus.

Putting all the ingredients together, the nuclear $0^+ \to 0^+$ transitions lead to[10]

$$V_{ud} = 0.97377\,(11)_{ft}\,(15)_{\delta_C}\,(19)_{\Delta_R}\ , \tag{6}$$

which is the current best determination of V_{ud} and sets the standard for V_{us} and unitarity tests.

3. V_{us} from $K_{\ell 3}$ decays

The decay rates for all four $K_{\ell 3}$ modes ($K = K^{\pm}, K^0, \ell = \mu, e$) can be written compactly as follows:

$$\begin{aligned}\Gamma(K_{\ell 3[\gamma]}) = & \frac{G_F^2\, S_{\rm ew}\, M_K^5}{128\pi^3}\, C^K I^{K\ell}(\lambda_i) \times |V_{us} \times f_+^{K^0\pi^-}(0)|^2 \\ & \times \left[1 + 2\,\Delta^K_{SU(2)} + 2\,\Delta^{K\ell}_{\rm EM}\right]\ .\end{aligned} \tag{7}$$

Here G_F is the Fermi constant as extracted from muon decay, $S_{\rm ew} = 1 + \frac{2\alpha}{\pi}\left(1 - \frac{\alpha_s}{4\pi}\right) \times \log\frac{M_Z}{M_\rho} + O(\frac{\alpha\alpha_s}{\pi^2})$ represents the short distance electroweak correction to semileptonic charged-current processes, C^K is a Clebsh-Gordan coefficient equal to 1 ($1/\sqrt{2}$) for neutral (charged) kaon decay, while $I^{K\ell}(\lambda_i)$ is a phase-space integral depending on slope and curvature of the form factors. The latter are defined by the QCD matrix elements

$$\langle \pi^j(p_\pi)|\bar{s}\gamma_\mu u|K^i(p_K)\rangle = f_+^{K^i\pi^j}(t)\,(p_K + p_\pi)_\mu + f_-^{K^i\pi^j}(t)\,(p_K - p_\pi)_\mu\ . \tag{8}$$

In the physical region these can be conveniently parameterised as

$$f_0^{K^i\pi^j}(t) \equiv f_+^{K^i\pi^j}(t) + \frac{t}{M_K^2 - M_\pi^2} f_-^{K^i\pi^j}(t) \ , \tag{9}$$

$$f_{+,0}^{K^i\pi^j}(t) = f_+^{K^i\pi^j}(0) \left(1 + \lambda'_{+,0} \frac{t}{M_\pi^2} + \lambda''_{+,0} \frac{t^2}{M_\pi^4} + \dots \right) \ , \tag{10}$$

where $t = (p_K - p_\pi)^2$.

As shown explicitly in Eq. (7), it is convenient to normalise the form factors of all channels to $f_+^{K^0\pi^-}(0)$, which in the following will simply be denoted by $f_+(0)$. The channel-dependent terms $\Delta_{SU(2)}^K$ and $\Delta_{\rm EM}^{K\ell}$ represent the isospin-breaking and long-distance electromagnetic corrections, respectively. A determination of V_{us} from $K_{\ell 3}$ decays at the 1% level requires

- Knowledge of the decay rates and the phase space integrals. This is mainly an experimental issue and I refer to the contribution of Matthew Moulson to these proceedings for a discussion of recent developments.
- Inclusion of $\Delta_{SU(2)}^K$ and $\Delta_{\rm EM}^{K\ell}$.
- Knowledge of $f_+(0)$ at the $\sim 1\%$ level.

In the next sections I discuss the last two items, involving theoretical input.

3.1. *SU(2) breaking and radiative corrections*

The natural framework to analyze these corrections is provided by chiral perturbation theory[13–15] (CHPT), the low energy effective theory of QCD. Physical amplitudes are systematically expanded in powers of external momenta of pseudo-Goldstone bosons (π, K, η) and quark masses. When including electromagnetic corrections, the power counting is in $(e^2)^m \, (p^2/\Lambda_\chi^2)^n$, with $\Lambda_\chi \sim 4\pi F_\pi$ and $p^2 \sim O(p_{\rm ext}^2, M_{K,\pi}^2) \sim O(m_q)$. To a given order in the above expansion, the effective theory contains a number of low energy couplings (LECs) unconstrained by symmetry alone. In lowest orders one retains model-independent predictive power, as these couplings can be determined by fitting a subset of available observables. Even in higher orders the effective theory framework proves very useful, as it allows one to bound unknown or neglected terms via power counting and dimensional analysis arguments.

Strong isospin breaking effects $O(m_u - m_d)$ were first studied to $O(p^4)$ in Ref.[16] Both loop and LECs contributions appear to this order. Using updated input on quark masses and the relevant LECs, the results quoted in Table 1 for $\Delta_{SU(2)}^K$ were obtained in Ref.[17]

Long distance electromagnetic corrections were studied within CHPT to order e^2p^2 in Refs.[17,18] To this order, both virtual and real photon corrections contribute to $\Delta_{\rm EM}^{K\ell}$. The virtual photon corrections involve (known) loops and tree level diagrams with insertion of $O(e^2p^2)$ LECs. The relevant LECs have been estimated in[19,20] using large-N_C techniques. The resulting uncertainty is reported in Table 1, and does not affect the extraction of V_{us} at an appreciable level.

Radiation of real photons is also an important ingredient in the calculation of $\Delta_{\rm EM}^{K\ell}$, because only the inclusive sum of $K_{\ell 3}$ and $K_{\ell 3\gamma}$ rates is infrared finite to any order in α. Moreover, the correction factor depends on the precise definition of inclusive rate. In Table 1 we collect results for the fully inclusive rate ("full") and for the "3-body" prescription, where only radiative events consistent with three-body kinematics are kept. CHPT power counting implies that to order e^2p^2 one has to treat K and π as point-like (and with constant weak form factors) in the calculation of the radiative rate, while structure dependent effects enter to higher order in the chiral expansion.[21]

Radiative corrections to $K_{\ell 3}$ decays have been recently calculated also outside the CHPT framework.[22,23] Within these schemes, the UV divergences of loops are regulated with a cutoff (chosen to be around 1 GeV). In addition, the treatment of radiative decays includes part of the structure dependent effects, introduced by the use of form factors in the weak vertices. Table 1 shows that numerically the "model" approach of Ref.[22] agrees rather well with the effective theory.

Finally, it is worth stressing that the consistency of the calculated SU(2) and electromagnetic corrections can be probed by experimental data by comparing the determination of $V_{us} \times f_+(0)$ from the various decay modes. This consistency check is shown graphically in Fig. 3 using a compilation

Table 1. Summary of SU(2) and radiative correction factors for various $K_{\ell 3}$ decay modes. Refs.[16–18] work within chiral perturbation theory to order p^4, e^2p^2, while Ref.[22] works within a hadronic model for Kaon electromagnetic interactions.

	$\Delta^K_{SU(2)}$ (%)	$\Delta_{\rm EM}^{K\ell}$(%)	
		3-body	full
K^+_{e3}	2.31 ± 0.22[16,17]	-0.35 ± 0.16[17]	-0.10 ± 0.16[17]
K^0_{e3}	0	+0.30 ± 0.10[18]	+0.55 ± 0.10[18]
			+0.65 ± 0.15[22]
$K^+_{\mu 3}$	2.31 ± 0.22[16,17]		
$K^0_{\mu 3}$	0		+0.95 ± 0.15[22]

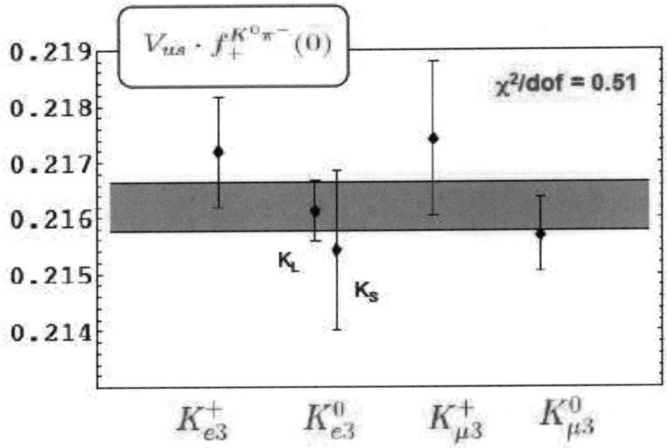

Fig. 3. The product $f_+(0) \times V_{us}$ as extracted by different $K_{\ell 3}$ decay modes.

of recent experimental data.[24] Averaging the different determinations leads to:

$$V_{us} \times f_+(0) = 0.2162 \pm 0.0004 \ . \tag{11}$$

3.2. *SU*(3) *breaking and* $f_+(0)$

Within CHPT we can break up the form factor according to its expansion in quark masses:

$$f_+(0) = 1 + f_{p^4} + f_{p^6} + \dots \ . \tag{12}$$

Deviations from unity (the octet symmetry limit) are of second order in SU(3) breaking.[4] The first correction arises to $O(p^4)$ in CHPT: a finite one-loop contribution[16,25] determines $f_{p^4} = -0.0227$ in terms of F_π, M_K and M_π, with essentially no uncertainty. The p^6 term receives contributions from pure two-loop diagrams, one-loop diagrams with insertion of one vertex from the p^4 effective Lagrangian, and pure tree-level diagrams with two insertions from the p^4 Lagrangian or one insertion from the p^6 Lagrangian:[26,27]

$$f_{p^6} = f_{p^6}^{2-\text{loops}}(\mu) + f_{p^6}^{L_i \times \text{loop}}(\mu) + f_{p^6}^{\text{tree}}(\mu) \ . \tag{13}$$

Individual components depend on the chiral renormalization scale μ, their sum being scale independent. Using as reference scale $\mu = M_\rho = 0.77$ GeV and the L_i from fit 10 in Ref.,[29] one has:[27]

$$f_{p^6}^{2-\text{loops}}(M_\rho) = 0.0113\ , \qquad f_{p^6}^{L_i\times\text{loop}}(M_\rho) = -0.0020 \pm 0.0005\ . \qquad (14)$$

The explicit form for the tree-level contribution in terms of the LECs L_5[15] and $C_{12,34}$[30] is then[27,28]

$$f_{p^6}^{\text{tree}}(M_\rho) = 8\frac{\left(M_K^2 - M_\pi^2\right)^2}{F_\pi^2}\left[\frac{(L_5^r(M_\rho))^2}{F_\pi^2} - C_{12}^r(M_\rho) - C_{34}^r(M_\rho)\right]\ . \qquad (15)$$

$L_5^r(M_\rho)$ is known from phenomenology to a level that induces less than 1% uncertainty in $f_{p^6}^{\text{tree}}(M_\rho)$. The p^6 constants $C_{12,34}$ can in principle be determined phenomenologically. It has been shown in Ref.[27] that combinations of C_{12} and C_{34} govern the slope λ_0 and curvature λ_0'' of the scalar form factor $f_0(t)$, accessible in $K_{\mu 3}$ decays. In order to extract C_{12} and C_{34} to a useful level (i.e. leading to 1% final uncertainty in V_{us}), one needs experimental errors at the level $\Delta\lambda_0 \sim 0.001$ (roughly a 5% measurement) and $\Delta\lambda_0'' \sim 0.0001$ (roughly a 20% measurement), as well as F_K/F_π at the 1% level from theory. At the moment there are no good prospects to measure the momentum-dependence of form factors to the required accuracy, and therefore further theoretical input on $f_+(0)$ is needed. I will discuss below two different approaches.

Large-N_C estimate of $f_{p^6}^{\text{tree}}$

In Ref.[31] a (truncated) large-N_C estimate of $f_{p^6}^{\text{tree}}$ was performed. It was based on matching a meromorphic approximation to the $\langle SPP\rangle$ Green function (with poles corresponding to the lowest-lying scalar and pseudoscalar resonances) onto QCD by imposing the correct large-momentum falloff, both off-shell and on one- and two-pion mass shells. In particular, C_{12} is uniquely determined by requiring the correct behavior of the pion scalar form factor $\langle\pi|S|\pi\rangle$, while C_{34} is fixed by the correct scaling of the one-pion form factors $\langle\pi|S|\mathcal{P}\rangle$ and $\langle\pi|P|\mathcal{S}\rangle$. The uncertainty of the large-N_C matching procedure was estimated by varying the chiral renormalization scale at which the matching is performed in the range $\mu \in [M_\eta, 1\text{GeV}]$, and is found to be $\delta f_{p^6}^{\text{tree}}|_{1/N_C} = \pm 0.008$. The final result is

$$f_{p^6}^{\text{tree}}(M_\rho) = -\frac{\left(M_K^2 - M_\pi^2\right)^2}{2\,M_S^4}\left(1 - \frac{M_S^2}{M_P^2}\right)^2 = -0.002 \pm 0.008_{\,1/N_C} \pm 0.002_{\,M_S}\ , \qquad (16)$$

and is much smaller than the ratio of mass scales $(M_K^2 - M_\pi^2)^2/M_S^4$ would suggest, due to interfering contributions. When combined with the p^6 loop corrections,[27] this estimate leads to $f_{p^6} = 0.007 \pm 0.012$. Variations of the hadronic ansatz lead to the conclusion that the smallness of the tree-level part compared to the loop contribution of $O(p^6)$ for $f_+(0)$ appears as generic feature of a few-resonance approximation for the set of large-momentum constraints considered. As a consitency check of this approach, it is worth mentioning that within the same framework one obtains a prediction for the slope of the scalar form factor, $\lambda_0 = 0.013 \pm 0.002_{\,1/N_C} \pm 0.001_{\,M_S}$, fully consistent with the value measured by KTeV in $K^L_{\mu 3}$ decays, $\lambda_0 = (13.72 \pm 1.31) \times 10^{-3}$.[32]

Combining this p^6 result with the well-known p^4 term, leads to the following global estimate of $f_+(0)$:

$$f_+(0)_{\mathrm{large}-N_C} = 0.984 \pm 0.012 \; . \tag{17}$$

This value is substantially higher –although compatible within the errors– with respect to the old estimate by Leutwyler Roos[25]

$$f_+(0)_{\mathrm{Leutwyler-Roos}} = 0.961 \pm 0.008 \; , \tag{18}$$

which for a long time has been the reference value of $f_+(0)$ (and it is still the value adopted by the PDG[33]) in the extraction of V_{us}. For another analytic approach to $f_+(0)$ based on a dispersive analysis of the scalar form factor see Ref.[34]

$f_+(0)$ *from lattice QCD*

Starting from Ref.[35] it has been realized that lattice QCD is a powerful tool to estimate $f_+(0)$ at a level of accuracy interesting for phenomenological purposes.[36–40] Both quenched and unquenched results are available. On general grounds, determining a form factor at the 1% level of accuracy seems very challenging –if not impossible– for present lattice-QCD calculations. However, the specific case of $f_+(0)$ is quite special: by taking appropriate ratios of correlation functions one can directly isolate the SU(3)-breaking quantity $[f_+(0)-1]$, or even better the quantity $[f_+(0)-1-f_{p^4}]$, which is the only irreducible source of uncertainty.[35] Estimating these SU(3)-breaking quantities with a relative error of about 30% is sufficient to predict $f_+(0)$ at the 1% level or below. Thus even with the present techniques there are good prospects to obtain lattice estimates of $f_+(0)$ of phenomenological interest.

Currently the dominant systematic uncertainty arises from the extrapolation of lattice results, obtained with unphysical quark masses, to physical light quark masses. The published results $f_+(0) = 0.960\,(5)_{\mathrm{stat}}\,(7)_{\mathrm{syst}}$[35]

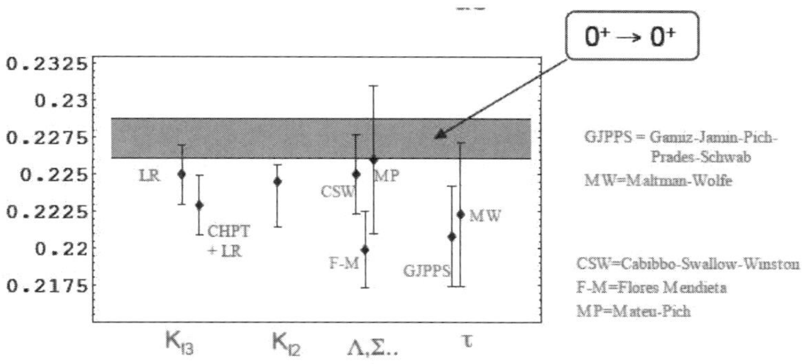

Fig. 4. Direct and indirect determinations of $f \times V_{us}$ (see text for more details).

(quenched) and $f_+(0) = 0.968\,(9)_{\rm stat}\,(6)_{\rm syst}$[40,41] (two dynamical fermions) are in reasonable agreement with the Leutwyler-Roos result, although the central value of the unquenched calculation lies on the upper end of the Leutwyler-Roos range.

Summary on $f_+(0)$

The dust does not seem to have settled yet. Recent analytic calculation including the $O(p^6)$ chiral logs[27] find values of $f_+(0)$ considerably higher than Leutwyler-Roos, while lattice QCD calculations find values only moderately above it. In the absence of a new "standard value", for current phenomenological analyses the Leutwyler-Roos result is still used as a reference value. Doing so and using the most recent experimental data one obtains:

$$V_{us}\Big|_{K_{\ell 3}} = 0.2250\,(4)_{\rm exp}\,(20)_{\rm th} \cdot \frac{0.961}{f_+(0)} \,. \tag{19}$$

4. Summary

In Fig. 4 I summarize the current status of the Cabibbo angle by plotting V_{us} as obtained by various independent determinations, many of which I have had no space to discuss in these proceedings. The solid band represents the indirect determination of V_{us} implied by V_{ud} imposing CKM unitarity. On the plot I also report the values of direct V_{us} determinations from $K_{\ell 3}$ decays (with two choices of the form factor: Leutwyler-Roos and Leutwyler-Roos plus chiral loops to $O(p^6)$), $K_{\ell 2}$ decays[42–45] , Hyperon decays,[46–48] and τ decays.[49,50]

The plot clearly shows that currently Kaon decays provide the best direct constraints on the size of V_{us}. Moreover, a meaningful unitarity test will be possible only once the dust on $SU(3)$ breaking settles. This remains one of the most interesting challenges for chiral dynamics and lattice QCD.

References

1. N. Cabibbo, Phys. Rev. Lett. 10 (1963) 531.
2. M. Kobayashi and T. Maskawa, Prog. Th. Phys **49** (1973) 652.
3. R. E. Behrends and A. Sirlin, Phys. Rev. Lett. **4** (1960) 186.
4. M. Ademollo and R. Gatto, Phys. Rev. Lett. **13** (1964) 264.
5. E. Blucher *et al.*, arXiv:hep-ph/0512039.
6. J. C. Hardy and I. S. Towner, "Superallowed 0+ -¿ 0+ nuclear beta decays: A critical survey with tests Phys. Rev. C **71**, 055501 (2005) [arXiv:nucl-th/0412056].
7. W. E. Ormand and B. A. Brown, Phys. Rev. Lett. **62** (1989) 866.
8. A. Sirlin and R. Zucchini, Phys. Rev. Lett. **57** (1986) 1994;
9. W. Jaus and G. Rasche, Phys. Rev. D **35**, 3420 (1987).
10. W. J. Marciano and A. Sirlin, Phys. Rev. Lett. **96**, 032002 (2006) [arXiv:hep-ph/0510099].
11. G. Savard *et al.*, Phys. Rev. Lett. **95**, 102501 (2005).
12. T. Eronen *et al.*, arXiv:nucl-ex/0606035.
13. S. Weinberg, Physica A **96** (1979) 327.
14. J. Gasser and H. Leutwyler, Ann. Phys. **158** (1984) 142.
15. J. Gasser and H. Leutwyler, Nucl. Phys. B **250** (1985) 465.
16. J. Gasser and H. Leutwyler, Nucl. Phys. B **250** (1985) 517.
17. V. Cirigliano, M. Knecht, H. Neufeld, H. Rupertsberger and P. Talavera, Eur. Phys. J. C **23** (2002) 121 [hep-ph/0110153].
18. V. Cirigliano, H. Neufeld and H. Pichl, Eur. Phys. J. C **35** (2004) 53 [hep-ph/0401173].
19. B. Moussallam, Nucl. Phys. B **504**, 381 (1997) [hep-ph/9701400].
20. S. Descotes-Genon and B. Moussallam, "Radiative corrections in weak semi-leptonic processes at low energy: A Eur. Phys. J. C **42**, 403 (2005) [arXiv:hep-ph/0505077].
21. J. Bijnens, G. Ecker and J. Gasser, Nucl. Phys. B **396**, 81 (1993) [hep-ph/9209261];
J. Gasser, B. Kubis, N. Paver and M. Verbeni, Eur. Phys. J. C **40**, 205 (2005) [hep-ph/0412130].
22. T. C. Andre, hep-ph/0406006.
23. V. Bytev, E. Kuraev, A. Baratt and J. Thompson, Eur. Phys. J. C **27** (2003) 57 [Erratum-ibid. C **34** (2004) 523] [hep-ph/0210049].
24. M. Moulson, contribution to these proceedings.
25. H. Leutwyler and M. Roos, Z. Phys. C **25** (1984) 91.
26. P. Post and K. Schilcher, Eur. Phys. J. C **25** (2002) 427 [hep-ph/0112352].
27. J. Bijnens and P. Talavera, Nucl. Phys. B **669** (2003) 341 [hep-ph/0303103]; see also http://www.thep.lu.se/∼bijnens/chpt.html.

28. J. Bijnens, G. Colangelo and G. Ecker, Phys. Lett. B **441** (1998) 437 [hep-ph/9808421].
29. G. Amorós, J. Bijnens and P. Talavera, Nucl. Phys. B **602** (2001) 87 [hep-ph/0101127].
30. J. Bijnens, G. Colangelo and G. Ecker, JHEP **9902** (1999) 020 [hep-ph/9902437].
31. V. Cirigliano, G. Ecker, M. Eidemueller, R. Kaiser, A. Pich and J. Portoles, hep-ph/0503108.
32. T. Alexopoulos *et al.* [KTeV Collaboration], Phys. Rev. D **70** (2004) 092007 [hep-ex/0406003].
33. S. Eidelman *et al.* [Particle Data Group Collaboration], Phys. Lett. B **592**, 1 (2004).
34. M. Jamin, J. A. Oller and A. Pich, JHEP **0402**, 047 (2004) [arXiv:hep-ph/0401080].
35. D. Becirevic *et al.*, Nucl. Phys. B **705**, 339 (2005) [hep-ph/0403217].
36. S. Hashimoto, A. X. El-Khadra, A. S. Kronfeld, P. B. Mackenzie, S. M. Ryan and J. N. Simone, Phys. Rev. D **61** (2000) 014502 [hep-ph/9906376].
37. M. Okamoto, hep-lat/0510113.
38. M. Okamoto [Fermilab Lattice Collaboration], hep-lat/0412044.
39. N. Tsutsui *et al.* [JLQCD Collaboration], Proc. Sci. **LAT2005** (2005) 357 [hep-lat/0510068].
40. C. Dawson, T. Izubuchi, T. Kaneko, S. Sasaki and A. Soni, arXiv:hep-ph/0607162.
41. D. J. Antonio *et al.*, arXiv:hep-lat/0610080.
42. W. J. Marciano, Phys. Rev. Lett. **93** (2004) 231803.
43. F. Ambrosino *et al.* [KLOE Collaboration], hep-ex/0509045 and submitted to Phys. Lett. B.
44. C. Aubin *et al.* [MILC Collaboration], Phys. Rev. D **70**, 114501 (2004).
45. C. Bernard *et al.* [MILC Collaboration], hep-lat/0509137.
46. N. Cabibbo, E. C. Swallow and R. Winston, Ann. Rev. Nucl. Part. Sci. **53**, 39 (2003).
47. R. Flores-Mendieta, J. J. Torres, M. Neri, A. Martinez and A. Garcia, and Phys. Rev. D **71**, 034023 (2005) and references therein.
48. V. Mateu and A. Pich, JHEP **0510**, 041 (2005) [arXiv:hep-ph/0509045].
49. E. Gámiz et al., Phys. Rev. Lett. **94** (2005) 011803.
50. K. Maltman and C. E. Wolfe, arXiv:hep-ph/0603215.

LATTICE QCD AND NUCLEON SPIN STRUCTURE

J. W. NEGELE

Center for Theoretical Physics, Massachusetts Institute of Technology,
Cambridge MA 02139, USA
** E-mail: negele@mitlns.mit.edu, avp@mit.edu*

Recent results providing insight into the spin structure of the nucleon are presented by the LHPC Collaboration based on lattice QCD calculations using a mixed action of domain wall valence quarks and asqtad staggered sea quarks for pion masses down to 359 MeV.

Keywords: Lattice QCD, Spin Structure of the Nucleon

1. Introduction

By solving nonperturbative QCD from first principles, lattice QCD provides a powerful tool for investigating the spin structure of the nucleon. In this work, we briefly describe two recent results[1] relevant to the nucleon spin. First, we report calculations of nucleon matrix elements of twist-two operators that specify the first three moments of the isovector helicity distribution, and show that a simple self-consistent improved one-loop chiral extrapolation is consistent with experiment. Second, by calculating the generalized form factors A_{20}^{u+d} and B_{20}^{u+d} and using the Ji sum rule[2] we determine the connected diagram contributions to the fraction of the nucleon spin arising separately from the spin and orbital angular momentum of the up and down quarks.

As explained in Ref. 3, we utilize a hybrid action combining domain wall valence fermions with improved staggered sea quarks. Improved staggered sea quarks offer the advantage that due to the relative economy of calculations with this action, lattices with large volumes, small pion masses, and several lattice spacings are publicly available from the MILC collaboration. Although the fourth root of the fermion determinant remains controversial, current evidence suggests it is manageable.[4,5] Renormalization group arguments indicate that the coefficient of the nonlocal term approaches zero in the continuum limit,[6] partially quenched staggered chiral perturbation theory accounts well for the artificial properties at finite lattice spacing,[7] and the action has the advantage of being improved to $\mathcal{O}(a^2)$. Domain wall valence quarks offer such compelling advantages that it is justified to invest resources in calculating hadron observables on staggered configurations that are roughly comparable to the resources required to generate the configurations themselves. Domain wall fermions prevent mixing of quark observables by chiral symmetry, are accurate to $\mathcal{O}(a^2)$, and possess a conserved five dimensional axial current that facilitates calculation of renormalization factors. In addition, hybrid action (often referred to as mixed action) chiral perturbation theory results are available for many observables, and by virtue of an exact lattice chiral symmetry, one loop results have the simple chiral behavior observed in the continuum. Details can be found in Refs. 1,3,8.

2. Moments of Parton Distributions

Parton distributions measure forward matrix elements of the gauge invariant light cone operators

$$\mathcal{O}_\Gamma(x) = \int \frac{d\lambda}{4\pi} e^{i\lambda x} \overline{q}(-\lambda n/2)\Gamma\, \mathcal{P} e^{-ig\int_{\lambda/2}^{-\lambda/2} d\alpha\, n\cdot A(\alpha n)} q(\lambda n/2), \tag{1}$$

where x is a momentum fraction, n is a light cone vector and $\Gamma = \not{n}$ or $\Gamma = \not{n}\gamma_5$. Using the operator product expansion, the operators in Eq. (1) yield towers of symmetrized, traceless local operators that can be evaluated on a Euclidean lattice

$$\mathcal{O}_{[\gamma_5]}^{\{\mu_1\ldots\mu_n\}} = \overline{q}\gamma^{\{\mu_1}[\gamma_5] i \overleftrightarrow{D}^{\mu_2} \cdots i \overleftrightarrow{D}^{\mu_n\}} q\,, \tag{2}$$

where $[\gamma_5]$ denotes the possible inclusion of γ_5, the curly brackets represent symmetrization over the indices μ_i and subtraction of traces, and $\overleftrightarrow{D} = 1/2(\overrightarrow{D} - \overleftarrow{D})$. A related operator for transversity distributions is

$$\mathcal{O}_{\sigma}^{\mu\{\mu_1\ldots\mu_n\}} = \overline{q}\gamma_5\sigma^{\mu\{\mu_1} i \overleftrightarrow{D}^{\mu_2} \cdots i \overleftrightarrow{D}^{\mu_n\}} q\,. \tag{3}$$

Using the notation and normalization of Ref. 9, the forward matrix elements $\langle P,S|\mathcal{O}^{\{\mu_1\ldots\mu_{n+1}\}}|P,S\rangle$ yield moments of the unpolarized quark distribution:

$$\langle x^n\rangle_q = \int_0^1 dx\, x^n [q(x) + (-1)^{n+1}\overline{q}(x)], \tag{4}$$

the forward matrix elements $\langle P,S|\mathcal{O}_{\gamma_5}^{\{\mu_1\ldots\mu_{n+1}\}}|P,S\rangle$ yield moments of the helicity distribution:

$$\langle x^n\rangle_{\Delta q} = \int_0^1 dx\, x^n[\Delta q(x) + (-1)^n \Delta\overline{q}(x)], \tag{5}$$

and the forward matrix elements $\langle P,S|\mathcal{O}_{\sigma}^{\mu\{\mu_1\ldots\mu_{n+1}\}}|P,S\rangle$ yield moments of the transversity distribution:

$$\langle x^n\rangle_{\delta q} = \int_0^1 dx\, x^n[\delta q(x) + (-1)^{n+1} \delta\overline{q}(x)]. \tag{6}$$

In this work, we calculate only connected diagrams, and hence concentrate as much as possible on isovector quantities. All quark bilinear operators in Eqs. (2) and (3) are renormalized as described in References 1 and 10. Ideally, we would like to perform high statistics calculations at pion masses below 350 MeV and extrapolate them in pion mass and volume using a chiral perturbation theory expansion of sufficiently high order to provide a quantitatively controlled approximation. In practice, our most convincing chiral extrapolation has been for g_A using the finite volume results including Δ intermediate states of Ref. 11, where the fit involving 6 low energy parameters yielded an excellent fit up to the order of a 700 MeV pion mass and agreed with experiment with 6.8% errors.[12] Similar extrapolations of g_A have been performed by other groups.[13] This success for g_A is particularly relevant to the subsequent discussion of nucleon spin, because it involves the same operators $\langle 1\rangle_{\Delta q}$ as $\Delta\Sigma$. We note that because the nucleon and Δ should be included together at large N_c[14] and indeed show large cancellations in the axial charge, we prefer to include the Δ as an explicit degree of freedom in the analysis.

An unresolved puzzle in calculating moments of structure functions is the relatively flat behavior of the momentum fraction $\langle x\rangle$ at a constant value substantially higher than experiment.[15,16] Hence, it is particularly interesting to ask whether a chiral perturbation theory fit determined without knowledge of the experimental result is in fact statistically consistent with experiment. Since there is presently insufficient data to perform a full analysis including the Δ, here we present a simple self-consistently improved one-loop analysis using only nucleon degrees of freedom that appears to work very well in our regime.

The details will be presented in a future publication,[17] but the basic

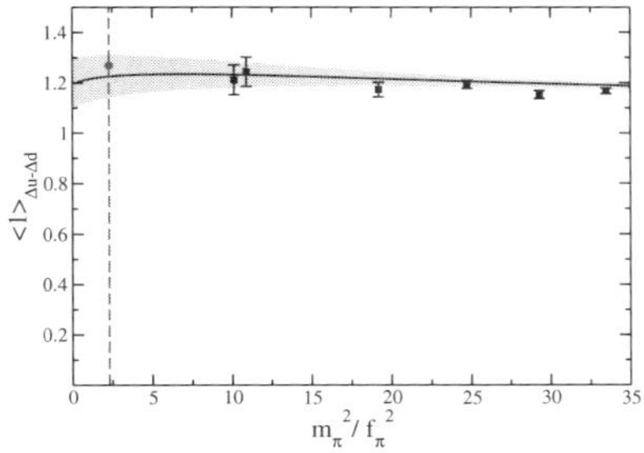

Fig. 1. Zeroth moment of helicity distribution.

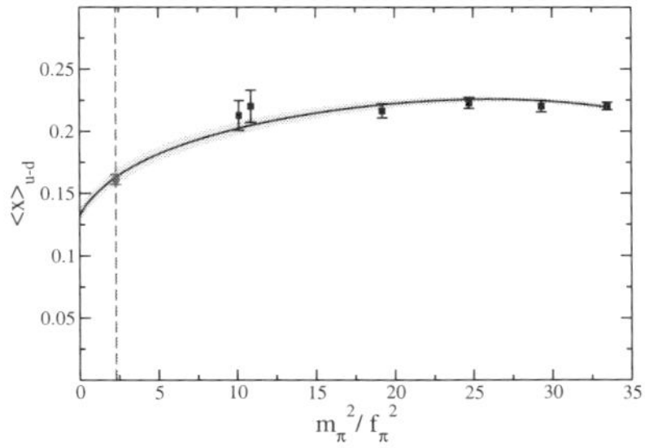

Fig. 2. First moment of unpolarized distribution.

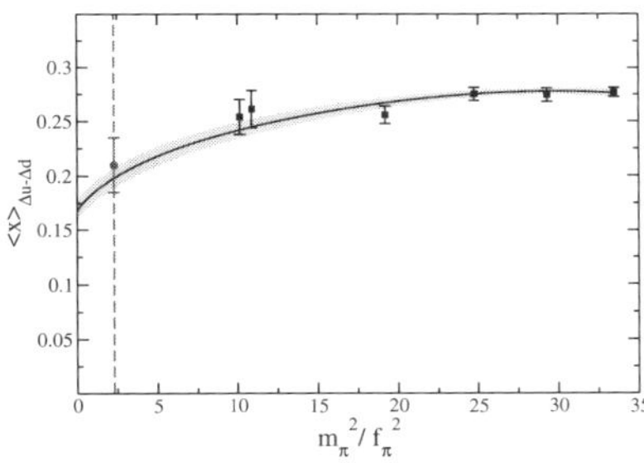

Fig. 3. First moment of helicity distribution.

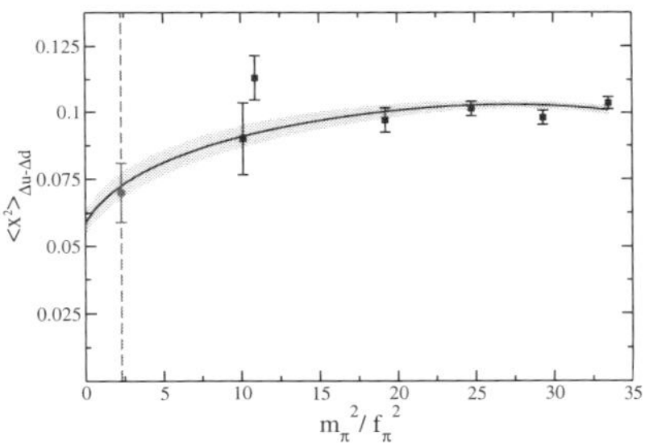

Fig. 4. Second moment of helicity distribution.

idea is as follows. We begin with the one loop expression at scale μ[18,19]

$$\langle x^n \rangle_{u-d} = a_n \left(1 - \frac{(3g_{A,0}^2 + 1)}{(4\pi f_{\pi,0})^2} m_\pi^2 \ln \left(\frac{m_\pi^2}{\mu^2} \right) \right) + b_n'(\mu) m_\pi^2 \tag{7}$$

in which we explicitly note that $g_{A,0}$ and $f_{\pi,0}$ are g_A and f_π in the chiral limit. We are free to choose the scale μ to be f_π. Additionally we replace $g_{A,0}$ and $f_{\pi,0}$ with their values at the given pion mass g_{A,m_π} and f_{π,m_π}, so that the result may be rewritten as

$$\langle x^n \rangle_{u-d} = a_n \left(1 - \frac{(3g_{A,m_\pi}^2 + 1)}{(4\pi)^2} \frac{m_\pi^2}{f_{\pi,m_\pi}^2} \ln \left(\frac{m_\pi^2}{f_{\pi,m_\pi}^2} \right) \right) + b_n \frac{m_\pi^2}{f_{\pi,m_\pi}^2}. \tag{8}$$

If we view this as an expansion in the ratio $r = \frac{m_\pi^2}{(4\pi)^2 f_{\pi,m_\pi}^2}$, one can show that shifting to an expansion around g_A and f_π defined at another mass

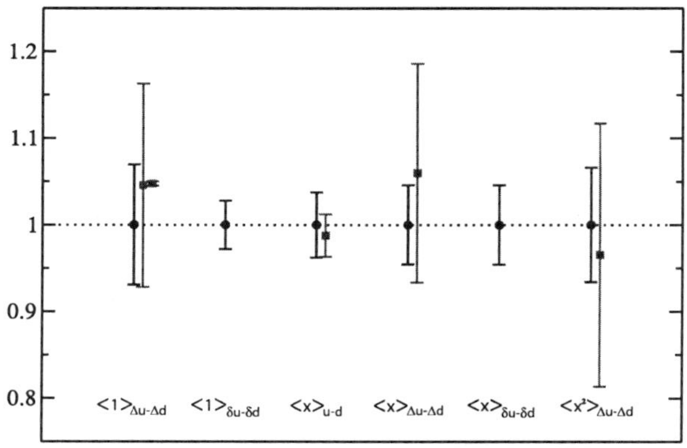

Fig. 5. The six moments considered in this work. Lattice results are shown as blue circles and experimental measurements as red squares, and each is normalized to the corresponding lattice result.

only introduces changes of $\mathcal{O}(r^2)$. Hence, to leading order, we may write an expression in which we use the values $g_{A,lat}$, $f_{\pi,lat}$, and $m_{\pi,lat}$ calculated on the lattice at specific values of the quark mass. Then, the expressions for the moments of the unpolarized, helicity, and transversity distributions are the following:

$$\begin{aligned}
\langle x^n\rangle_{u-d} &= a_n\left(1-\frac{(3g_{A,\mathrm{lat}}^2+1)}{(4\pi)^2}\frac{m_{\pi,\mathrm{lat}}^2}{f_{\pi,\mathrm{lat}}^2}\ln\left(\frac{m_{\pi,\mathrm{lat}}^2}{f_{\pi,\mathrm{lat}}^2}\right)\right)+b_n\frac{m_{\pi,\mathrm{lat}}^2}{f_{\pi,\mathrm{lat}}^2}\\
\langle x^n\rangle_{\Delta u-\Delta d} &= \Delta a_n\left(1-\frac{(2g_{A,\mathrm{lat}}^2+1)}{(4\pi)^2}\frac{m_{\pi,\mathrm{lat}}^2}{f_{\pi,\mathrm{lat}}^2}\ln\left(\frac{m_{\pi,\mathrm{lat}}^2}{f_{\pi,\mathrm{lat}}^2}\right)\right)+\Delta b_n\frac{m_{\pi,\mathrm{lat}}^2}{f_{\pi,\mathrm{lat}}^2}\\
\langle x^n\rangle_{\delta u-\delta d} &= \delta a_n\left(1-\frac{(4g_{A,\mathrm{lat}}^2+1)}{2(4\pi)^2}\frac{m_{\pi,\mathrm{lat}}^2}{f_{\pi,\mathrm{lat}}^2}\ln\left(\frac{m_{\pi,\mathrm{lat}}^2}{f_{\pi,\mathrm{lat}}^2}\right)\right)+\delta b_n\frac{m_{\pi,\mathrm{lat}}^2}{f_{\pi,\mathrm{lat}}^2}\,. \qquad (9)
\end{aligned}$$

These results allow a least-squares two-parameter fit to the lattice data for moments and provides an extrapolation to the physical pion mass with a corresponding error band. Note that the series is substantially rearranged, by virtue that the calculated values of g_A, f_π, and m_π are used at each value of the bare quark mass. Although we cannot prove that this self-consistent improved one-loop result should be accurate throughout the range of our data, to the extent to which it is successful, we believe its success arises from this self-consistent rearrangement. Additionally, the use of lattice rather than chiral limit values for f_π was first tried in Ref. 20 and has since been studied in chiral perturbation theory[21,22] and applied to a variety of lattice calculations.[23–27]

The results of this one-loop analysis are shown in Figures (1-5). Figure (1) shows the result for g_A, which is nearly as good as the complete analysis of Ref. 12, and yields a comparable extrapolation and error bar. Reassured by this result, we show analogous results for $\langle x \rangle_{u-d}$, $\langle x \rangle_{\Delta u - \Delta d}$, and $\langle x^2 \rangle_{\Delta u - \Delta d}$ in Figs. (2)-(4). Note that in every case for which there is experimental data, this analysis, which in no way includes the experimental result in the fit, yields an extrapolation consistent with experiment. The results are collected together in Fig. (5), where because experimental results are not available for all cases, we have normalized all results to the corresponding lattice result.

3. Generalized Parton Distributions

Off diagonal matrix elements of the tower of operators in Eq. (2) yield the generalized form factors

$$\langle P'|\mathcal{O}^{\mu_1}|P\rangle = \langle\!\langle \gamma^{\mu_1} \rangle\!\rangle A_{10}(t) + \frac{i}{2m} \langle\!\langle \sigma^{\mu_1 \alpha} \rangle\!\rangle \Delta_\alpha B_{10}(t) \,, \tag{10}$$

$$\begin{aligned} \langle P'|\mathcal{O}^{\{\mu_1 \mu_2\}}|P\rangle = {} & \overline{P}^{\{\mu_1} \langle\!\langle \gamma^{\mu_2\}} \rangle\!\rangle A_{20}(t) + \frac{i}{2m} \overline{P}^{\{\mu_1} \langle\!\langle \sigma^{\mu_2\}\alpha} \rangle\!\rangle \Delta_\alpha B_{20}(t) \\ & + \frac{1}{m} \Delta^{\{\mu_1} \Delta^{\mu_2\}} C_{20}(t) \,, \end{aligned} \tag{11}$$

$$\begin{aligned} \langle P'|\mathcal{O}^{\{\mu_1 \mu_2 \mu_3\}}|P\rangle = {} & \overline{P}^{\{\mu_1} \overline{P}^{\mu_2} \langle\!\langle \gamma^{\mu_3\}} \rangle\!\rangle A_{30}(t) \\ & + \frac{i}{2m} \overline{P}^{\{\mu_1} \overline{P}^{\mu_2} \langle\!\langle \sigma^{\mu_3\}\alpha} \rangle\!\rangle \Delta_\alpha B_{30}(t) \\ & + \Delta^{\{\mu_1} \Delta^{\mu_2} \langle\!\langle \gamma^{\mu_3\}} \rangle\!\rangle A_{32}(t) \\ & + \frac{i}{2m} \Delta^{\{\mu_1} \Delta^{\mu_2} \langle\!\langle \sigma^{\mu_3\}\alpha} \rangle\!\rangle \Delta_\alpha B_{32}(t), \end{aligned} \tag{12}$$

where we use the short-hand notation $\langle\!\langle \Gamma \rangle\!\rangle = \overline{U}(P', \Lambda') \Gamma U(P, \Lambda)$ for matrix elements of Dirac spinors U and where $\Delta = P' - P$ and $t = \Delta^2$.

Of particular interest in this work is the relationship between the generalized form factors and the origin of the nucleon spin. The contribution of the spin of the up and down quarks to the total spin of the nucleon is given by the zeroth moment of the spin dependent structure function $\langle 1 \rangle_{\Delta q}$ as

$$\frac{1}{2} \Delta\Sigma = \frac{1}{2} \langle 1 \rangle_{\Delta u + \Delta d} \,. \tag{13}$$

Note both that our previous calculation of g_A confirms our ability to calculate the connected contributions to $\langle 1 \rangle_{\Delta q}$ accurately on the lattice and that

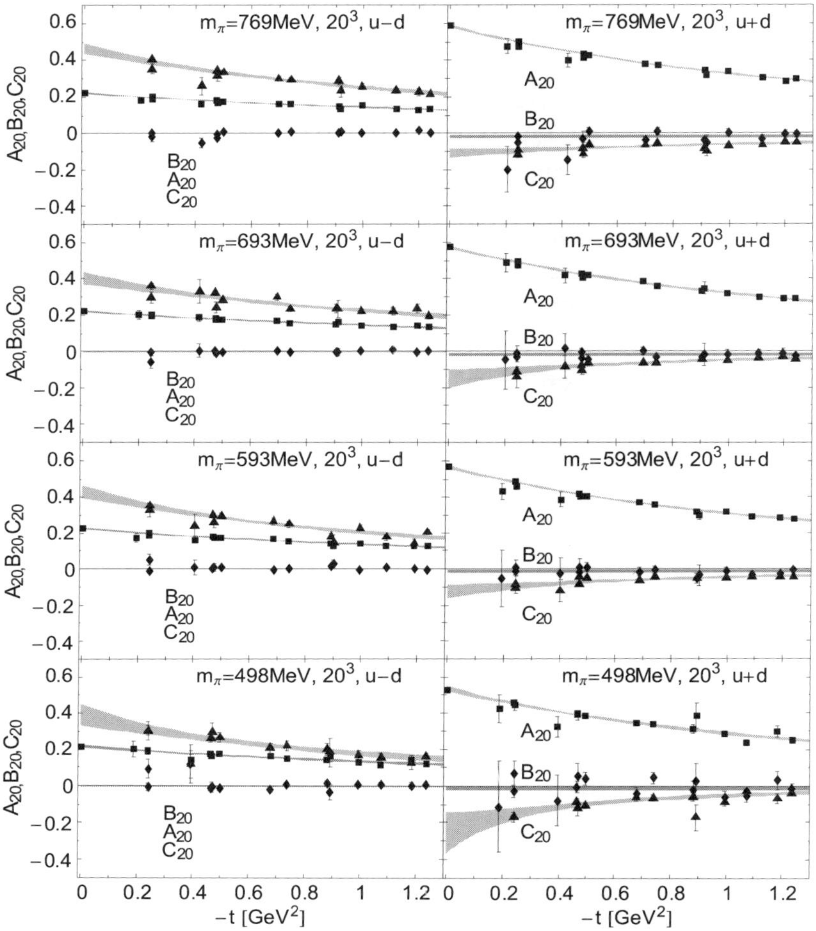

Fig. 6. Generalized form factors A_{20}, B_{20} and C_{20} for $u - d$ and $u + d$ at four masses.

$\Delta\Sigma$ requires the calculation of disconnected as well as connected contributions. Throughout this section, we will only discuss the results of connected diagrams, so all results will eventually need to be corrected for the effect of disconnected diagrams as well.

The total contribution to the nucleon spin from both the spin and orbital

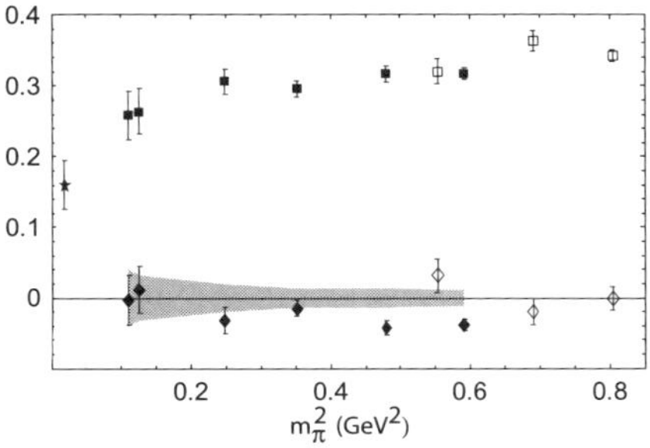

Fig. 7. Nucleon spin decomposition. Squares denote $\Delta\Sigma^{u+d}/2$, the star indicates the experimental quark spin contribution, and diamonds denote L^{u+d}.

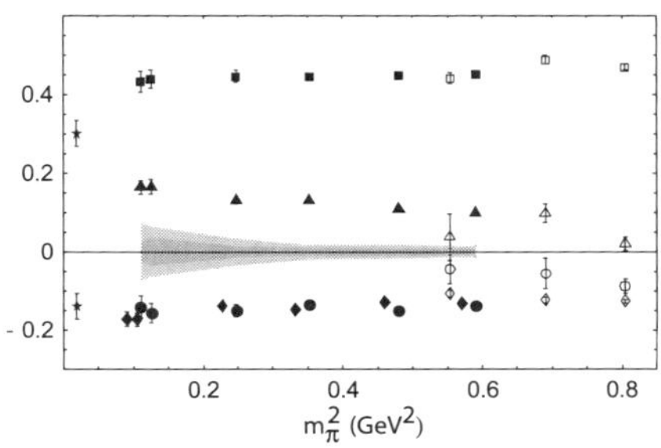

Fig. 8. Nucleon spin decomposition by flavor. Squares denote $\Delta\Sigma^u/2$, diamonds denote $\Delta\Sigma^d/2$, triangles denote L^u, and circles denote L^d.

angular momentum of quarks is given by the Ji sum rule:[2]

$$J_q = \frac{1}{2}\left(A_{20}^{u+d}(0) + B_{20}^{u+d}(0)\right), \tag{14}$$

so that the contribution of the orbital angular momentum is given by $L_q = J_q - \frac{1}{2}\Delta\Sigma$. Earlier calculations in the heavy quark regime showed that for 700 MeV pions, roughly 68% of the spin of the nucleon arises from the spin of quarks, 0% arises from orbital angular momentum, and hence the remaining 32% must come from gluons.[28–30] Figure (6) shows the recent lattice data for A_{20}, B_{20} and C_{20} for lighter pion masses, and the results for $\Delta\Sigma$ and L_q are shown in Fig. (7). The full decomposition showing the contribution of spin and orbital angular momentum from up and down quarks is shown in Fig. (8). Here, one observes that the angular momentum contributions of both up and down quark are separately substantial, and it is only the sum that is extremely small.

4. Conclusions

In summary, the hybrid combination of valence domain wall quarks on an improved staggered sea has enabled us to begin to quantitatively solve full lattice QCD in the chiral regime and to explore the spin structure of the nucleon. The axial charge, g_A, represents a successful, "gold-plated" test. The chiral extrapolation of moments of quark distributions using our self-consistently improved one-loop analysis is encouraging, but of course we must eventually directly calculate the turn over in the approach to the chiral regime. The calculation of the connected diagram contributions to the nucleon spin and orbital angular momentum of the up and down quarks is an important development, and future work will focus on the calculation of disconnected contributions.

Acknowledgments

This work was supported by the DOE Office of Nuclear Physics under contracts DE-FC02-94ER40818, DE-AC05-06OR23177 and DE-AC05-84150, the EU Integrated Infrastructure Initiative Hadron Physics (I3HP) under contract RII3-CT-2004-506078, the DFG under contract FOR 465 (Forschergruppe Gitter-Hadronen-Phänomenologie) and the DFG Emmy-Noether program. Computations were performed on clusters at Jefferson Laboratory and at ORNL using time awarded under the SciDAC initiative. We are indebted to the members of the MILC collaboration for providing the dynamical quark configurations which made our full QCD calculations possible.

References

1. R. G. Edwards *et al.* (2006).
2. X. D. Ji, *Phys. Rev. Lett.* **78**, 610 (1997).
3. D. B. Renner *et al.*, *Nucl. Phys. Proc. Suppl.* **140**, 255 (2005).
4. C. Bernard, M. Golterman and Y. Shamir, *Phys. Rev.* **D73**, p. 114511 (2006).
5. S. R. Sharpe, *hep-lat/0610094* (2006).
6. Y. Shamir, *hep-lat/0607007* (2006).
7. C. Bernard, *Phys. Rev.* **D73**, p. 114503 (2006).
8. R. G. Edwards *et al.*, *PoS* **LAT2005**, p. 056 (2006).
9. D. Dolgov *et al.*, *Phys. Rev.* **D66**, p. 034506 (2002).
10. B. Bistrovic *et al.*, *In preparation* (2006).
11. S. R. Beane and M. J. Savage, *Phys. Rev.* **D70**, p. 074029 (2004).
12. R. G. Edwards *et al.*, *Phys. Rev. Lett.* **96**, p. 052001 (2006).
13. A. Ali Khan *et al.*, *hep-lat/0603028* (2006).
14. R. F. Dashen, E. Jenkins and A. V. Manohar, *Phys. Rev.* **D49**, 4713 (1994).
15. W. Detmold *et al.*, *Phys. Rev. Lett.* **87**, p. 172001 (2001).
16. K. Orginos, T. Blum and S. Ohta, *Phys. Rev.* **D73**, p. 094503 (2006).
17. D. B. Renner *et al.*, *In preparation* (2006).
18. D. Arndt and M. J. Savage, *Nucl. Phys.* **A697**, 429 (2002).
19. J. W. Chen and X. D. Ji, *Phys. Lett.* **B523**, 107 (2001).
20. S. R. Beane, P. F. Bedaque, K. Orginos and M. J. Savage, *Phys. Rev.* **D73**, p. 054503 (2006).
21. J. W. Chen, D. O'Connell, R. S. Van de Water and A. Walker-Loud, *Phys. Rev.* **D73**, p. 074510 (2006).
22. D. O'Connell, *hep-lat/0609046* (2006).
23. S. R. Beane *et al.*, *hep-lat/0607036* (2006).
24. S. R. Beane, P. F. Bedaque, K. Orginos and M. J. Savage, *hep-lat/0606023* (2006).

25. S. R. Beane, K. Orginos and M. J. Savage, *hep-lat/0605014* (2006).
26. S. R. Beane, K. Orginos and M. J. Savage, *hep-lat/0604013* (2006).
27. S. R. Beane, P. F. Bedaque, K. Orginos and M. J. Savage, *Phys. Rev. Lett.* **97**, p. 012001 (2006).
28. P. Hägler *et al.*, *Phys. Rev.* **D68**, p. 034505 (2003).
29. M. Göckeler *et al.*, *Phys. Rev. Lett.* **92**, p. 042002 (2004).
30. N. Mathur *et al.*, *Phys. Rev.* **D62**, p. 114504 (2000).

SPIN SUM RULES AND POLARIZABILITIES: RESULTS FROM JEFFERSON LAB

Jian-ping Chen

Thomas Jefferson National Accelerator Facility
12000 Jefferson Ave., Newport News, Virginia 23606, USA
** E-mail: jpchen@jlab.org*

The nucleon spin structure has been an active, exciting and intriguing subject of interest for the last three decades. Recent experimental data on nucleon spin structure at low to intermediate momentum transfers provide new information in the *confinement* regime and the transition region from the *confinement* regime to the *asymptotic freedom* regime. New insight is gained by exploring moments of spin structure functions and their corresponding sum rules (i.e. the generalized Gerasimov-Drell-Hearn, Burkhardt-Cottingham and Bjorken). The Burkhardt-Cottingham sum rule is verified to good accuracy. The spin structure moments data are compared with Chiral Perturbation Theory calculations at low momentum transfers. It is found that chiral perturbation calculations agree reasonably well with the first moment of the spin structure function g_1 at momentum transfer of 0.05 to 0.1 GeV^2 but fail to reproduce the neutron data in the case of the generalized polarizability δ_{LT} (the δ_{LT} puzzle). New data have been taken on the neutron (^{3}He), the proton and the deuteron at very low Q^2 down to 0.02 GeV^2. They will provide benchmark tests of Chiral dynamics in the kinematic region where the Chiral Perturbation theory is expected to work.

Keywords: Nucleon; Spin; Sum Rule; Polarizability; Moment.

1. Introduction

In the last twenty-five years the study of the spin structure of the nucleon led to a very productive experimental and theoretical activity with exciting results and new challenges.[1] This investigation has included a variety of aspects, such as testing QCD in its perturbative regime *via* spin sum rules (like the Bjorken sum rule[2]) and understanding how the spin of the nucleon is built from the intrinsic degrees of freedom of the theory, quarks and gluons. Recently, results from a new generation of experiments performed at Jefferson Lab seeking to probe the theory in its non-perturbative and transition regimes have reached a mature state. The low momentum-transfer results offer insight in a region known for the collective behavior of the nucleon constituents and their interactions. Furthermore, distinct features seen in the nucleon response to the electromagnetic probe, depending on the resolution of the probe, point clearly to different regimes of description, i.e. a scaling regime where quark-gluon correlations are suppressed versus a coherent regime where long-range interactions give rise to the static properties of the nucleon.

In this talk we describe an investigation[4–9] of the spin structure of the nucleon through the measurement of the helicity-dependent photoabsorption cross sections or asymmetries using virtual photons across a wide resolution spectrum. These observables are used to extract the spin structure functions g_1 and g_2 and to evaluate their moments. These moments are powerful tools to test QCD sum rules and Chiral Perturbation Theory calculations.

2. Sum rules and Moments

Sum rules involving the spin structure of the nucleon offer an important opportunity to study QCD. In recent years the Bjorken sum rule at large Q^2 (4-momentum transfer squared) and the Gerasimov, Drell and Hearn (GDH) sum rule[10] at $Q^2 = 0$ have attracted large experimental and theoretical efforts[3] that have provided us with rich information. Another type of sum rules, such as the generalized GDH sum rule[11] or the polarizability sum rules,[12] relate the moments of the spin structure functions to real or virtual Compton amplitudes, which can be calculated theoretically. These sum rules are based on "unsubtracted" dispersion relations and the optical theorem.

Considering the forward spin-flip doubly-virtual Compton scattering (VVCS) amplitude g_{TT} and assuming it has an appropriate convergence behavior at high energy, an unsubtracted dispersion relation leads to the following equation for[9,12] g_{TT}:

$$\mathrm{Re}[g_{TT}(\nu,Q^2)-g_{TT}^{pole}(\nu,Q^2)] = (\frac{\nu}{2\pi^2})\mathcal{P}\int_{\nu_0}^{\infty}\frac{K(\nu',Q^2)\sigma_{TT}(\nu',Q^2)}{\nu'^2-\nu^2}d\nu', \quad (1)$$

where g_{TT}^{pole} is the nucleon pole (elastic) contribution, $\mathcal{P}$ denotes the principal value integral and K is the virtual photon flux factor. The lower limit of the integration ν_0 is the pion-production threshold on the nucleon. A low-energy expansion gives:

$$\mathrm{Re}[g_{TT}(\nu,Q^2) - g_{TT}^{pole}(\nu,Q^2)] = (\frac{2\alpha}{M^2})I_{TT}(Q^2)\nu + \gamma_0(Q^2)\nu^3 + O(\nu^5). \quad (2)$$

Combining Eqs. (1) and (2), the $O(\nu)$ term yields a sum rule for the generalized GDH integral:[3,11]

$$\begin{aligned} I_{TT}(Q^2) &= \frac{M^2}{4\pi^2\alpha}\int_{\nu_0}^{\infty}\frac{K(\nu,Q^2)}{\nu}\frac{\sigma_{TT}}{\nu}d\nu \\ &= \frac{2M^2}{Q^2}\int_0^{x_0}\Big[g_1(x,Q^2) - \frac{4M^2}{Q^2}x^2 g_2(x,Q^2)\Big]dx. \end{aligned} \tag{3}$$

The low-energy theorem relates I(0) to the anomalous magnetic moment of the nucleon, κ, and Eq. (3) becomes the original GDH sum rule:[10]

$$I(0) = \int_{\nu_0}^{\infty}\frac{\sigma_{1/2}(\nu) - \sigma_{3/2}(\nu)}{\nu}d\nu = -\frac{2\pi^2\alpha\kappa^2}{M^2}, \tag{4}$$

where $2\sigma_{TT} \equiv \sigma_{1/2} - \sigma_{3/2}$. The $O(\nu^3)$ term yields a sum rule for the generalized forward spin polarizability:[12]

$$\begin{aligned} \gamma_{TT}(Q^2) &= (\frac{1}{2\pi^2})\int_{\nu_0}^{\infty}\frac{K(\nu,Q^2)}{\nu}\frac{\sigma_{TT}(\nu,Q^2)}{\nu^3}d\nu \\ &= \frac{16\alpha M^2}{Q^6}\int_0^{x_0}x^2\Big[g_1(x,Q^2) - \frac{4M^2}{Q^2}x^2 g_2(x,Q^2)\Big]dx. \end{aligned} \tag{5}$$

Considering the longitudinal-transverse interference amplitude g_{LT}, the $O(\nu^2)$ term leads to the generalized longitudinal-transverse polarizability:[12]

$$\begin{aligned} \delta_{LT}(Q^2) &= (\frac{1}{2\pi^2})\int_{\nu_0}^{\infty}\frac{K(\nu,Q^2)}{\nu}\frac{\sigma_{LT}(\nu,Q^2)}{Q\nu^2}d\nu \\ &= \frac{16\alpha M^2}{Q^6}\int_0^{x_0}x^2\Big[g_1(x,Q^2) + g_2(x,Q^2)\Big]dx. \end{aligned} \tag{6}$$

Alternatively, we can consider the covariant spin-dependent VVCS amplitudes S_1 and S_2, which are related to the spin-flip amplitudes g_{TT} and g_{LT}. The unsubtracted dispersion relations for S_2 and νS_2 lead to a "superconvergence relation" that is valid for any value of Q^2,

$$\int_0^1 g_2(x,Q^2)dx = 0, \tag{7}$$

which is the Burkhardt-Cottingham (BC) sum rule.[13]

At high Q^2, the OPE[14] for the VVCS amplitude leads to the twist expansion. The leading-twist (twist-2) component can be decomposed into flavor triplet (g_A), octet (a_8) and singlet ($\Delta\Sigma$) axial charges. The difference between the proton and the neutron gives the flavor non-singlet term:

$$\Gamma_1^p(Q^2) - \Gamma_1^n(Q^2) = \frac{1}{6}g_A + O(\alpha_s) + O(1/Q^2), \tag{8}$$

which becomes the Bjorken sum rule at the $Q^2 \to \infty$ limit.

The leading-twist part provides information on the polarized parton distributions. The higher-twist parts are related to quark-gluon interations or correlations. Of particular interest is the twist-3 component, d_2, which is related to the second moment of the twist-3 part of g_1 and g_2:

$$d_2(Q^2) = \int_0^1 dx\ x^2 \Big(2g_1(x,Q^2) + 3g_2(x,Q^2)\Big)$$
$$= 3\int_0^1 dx\ x^2 \Big(g_2(x,Q^2) - g_2^{WW}(x,Q^2)\Big), \tag{9}$$

where g_2^{WW} is the twist-2 part of g_2 as derived by Wandzura and Wilczek[15]

$$g_2^{WW}(x,Q^2) = g_1(x,Q^2) + \int_x^1 dy \frac{g_1(y,Q^2)}{y}\ . \tag{10}$$

d_2 is related to the color electric and magnetic polarizabilities, which describe the response of the collective color electric and magnetic fields to the spin of the nucleon.[14]

3. Description of the JLab experiments

The inclusive experiments described here took place in JLab Halls A[18] and B.[19] The accelerator produces a polarized electron beam of energy up to 6 GeV.

A polarized high-pressure (~12 atm.) gaseous ^{3}He target was used as an effective polarized neutron target in the experiments performed in Hall A. The average target polarization, monitored by NMR and EPR techniques, was 0.4±0.014 and its direction could be oriented longitudinal or transverse to the beam direction. The measurement of cross sections in the two orthogonal directions allowed a direct extraction of g_1^{3He} and g_2^{3He}, or equivalently σ_{TT} and σ_{LT}.

The scattered electrons were detected by two High Resolution Spectrometers (HRS) with the associated detector package. The high luminosity of 10^{36} cm^{-2}s^{-1} allowed for statistically accurate data.

The spin structure functions g_1^n and g_2^n are extracted using polarized cross-section differences. Electromagnetic radiative corrections were performed. Nuclear corrections are applied via a PWIA-based model.[20] To form the moments, the integrands (e.g. σ_{TT} or g_1) were determined from the measured points by interpolation. To complete the moments for the unmeasured high-energy region, the Bianchi and Thomas parameterization[22] was used for $4 < W^2 < 1000$ GeV^2 and a Regge-type parameterization was used for $W^2 > 1000$ GeV^2. Polarized solid $^{15}NH_3$ and $^{15}ND_3$ targets using dynamic nuclear polarization were used in Hall B. The CEBAF Large Acceptance Spectrometer (CLAS) in Hall B, which has a large angular (2.5π sr) and momentum acceptance, was used to detect scattered electrons. The spin structure functions were extracted using asymmetry measurements together with the world unpolarized structure function fits.[21] Radiative corrections were applied.

4. Recent results from Jefferson Lab

4.1. *Results of the generalized GDH sum and BC sum for ^{3}He and the neutron*

Fig. 1 shows the extended GDH integrals $I(Q^2)$ (open circles) for ^{3}He (preliminary) (upper-left) and for the neutron (upper-right), which were extracted from Hall A experiment E94-010[4] , from break-up threshold for ^{3}He (from pion threshold for the neutron) to $W = 2$ GeV. The uncertainties, when visible, represent statistics only; the systematics are shown by the grey band. The solid squares include an estimate of the unmeasured high-energy part. The corresponding uncertainty is included in the systematic uncertainty band. The preliminary ^{3}He results rise with decreasing Q^2. Since the GDH sum rule at $Q^2 = 0$ predicts a large negative value, a drastic turn around should happen at Q^2 lower than 0.1 GeV^2. A simple model using MAID[3] plus quasielastic contributions indeed shows the expected turn around. The data at low Q^2 should be a good test ground for few-body Chiral Pertubation Theory Calculations.

The neutron results indicate a smooth variation of $I(Q^2)$ to increasingly negative values as Q^2 varies from $0.9\,\text{GeV}^2$ towards zero. The data are more negative than the MAID model calculation.[3] Since the calculation only includes contributions to $I(Q^2)$ for $W \leq 2\,\text{GeV}$, it should be compared with the open circles. The GDH sum rule prediction, $I(0) = -232.8\,\mu\text{b}$, is indicated in Fig. 1, along with extensions to $Q^2 > 0$ using two next-to-leading order χPT calculations, one using the Heavy Baryon approximation (HBχPT)[23] (dotted line) and the other Relativistic Baryon χPT (RBχPT)[24] (dot-dashed line). Shown with a grey band is RBχPT including resonance effects,[24] which have an associated large uncertainty due to the resonance parameters used.

The capability of transverse polarization of the Hall A ^{3}He target allows precise measuremetns of g_2. The integral of $\Gamma_2^{^3He}$ (preliminary) and Γ_2^n is plotted in the lower-left and lower-right panels of Fig. 1 in the measured

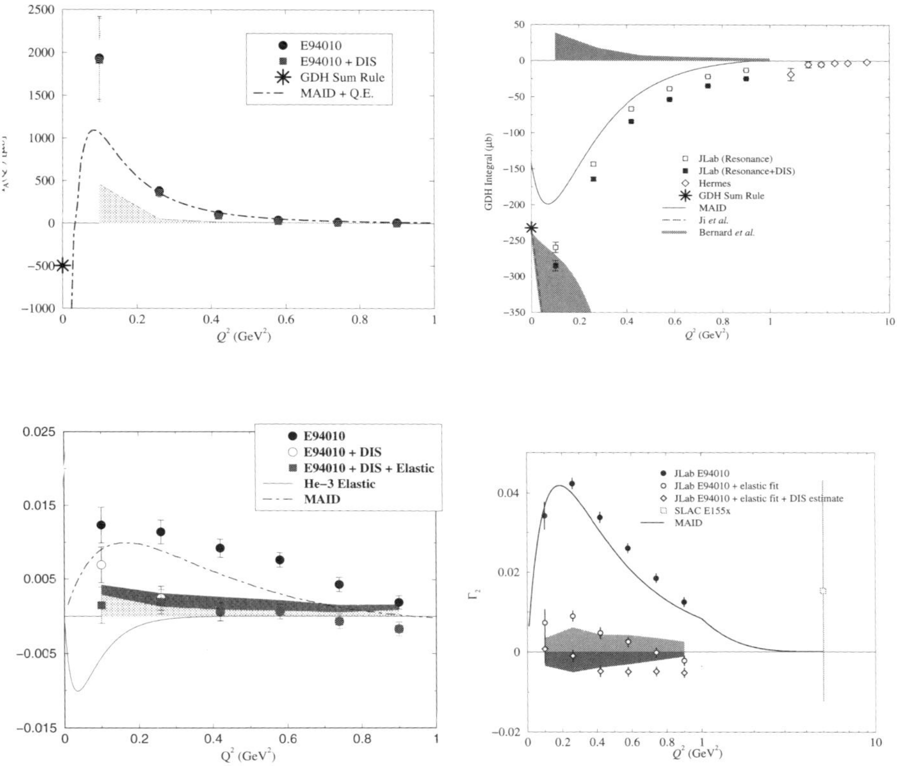

Fig. 1. Results of GDH sum $I(Q^2)$ and BC sum $\Gamma_2(Q^2)$ for ^{3}He (preliminary) and the neutron. The ^{3}He GDH results are compared with the MAID model plus quasielastic contribution. The neutron GDH results are compared with χPT calculations of ref.[23] (dotted line) and ref.[24] (dot-dashed line). The MAID model calculation of ref.,[3] is represented by a solid line. Data from HERMES[16] are also shown. The BC sum results (resonance only) are compared with MAID model calculations.

region (solid circles) and open circles show the results after adding an estimated DIS contribution for ^{3}He (elastic contribution for the neutron). The solid squares (open diamonds) correspond to the results obtained after adding the elastic contributions for ^{3}He, (adding an estimated DIS contribution assuming $g_2 = g_2^{WW}$ for the neutron). The MAID estimate agrees with the general trend but slightly lower than the resonance data. The two bands correspond to the experimental systematic errors and the estimate of the systematic error for the low-x extrapolation. The total results are consistent with the BC sum rule. The SLAC E155x collaboration[17] previously reported a neutron result at high Q^2 (open square), which is consistent with zero but with a rather large error bar. On the other hand, the SLAC proton result was reported to deviate from the BC sum rule by 3 standard deviations.

4.2. *First moments of g_1 and the Bjorken sum*

The preliminary results from Hall B EG1b[8] experiment on $\bar{\Gamma}_1(Q^2)$ at low to moderate Q^2 are shown together with published results from Hall A[4] and Hall B eg1a[5,6] in Fig. 2 along with the data from SLAC[17] and HERMES.[16] The new results are in good agreement with the published data. The inner uncertainty indicates the statistical uncertainty while the outer one is the quadratic sum of the statistical and systematic uncertainties.

At Q^2=0, the GDH sum rule predicts the slopes of moments (dotted lines). The deviation from the slopes at low Q^2 can be calculated with χPT. We show results of calculations by Ji *et al.*[23] using HBχPT and by Bernard *et al.* with and without[24] the inclusion of vector mesons and Δ degrees of freedom. The calculations are in reasonable agreements with the data at the lowest Q^2 settings of 0.05 - 0.1 GeV^2. At moderate and large Q^2 data are compared with two model calculations[25,26] . Both models agree well with the data. The leading-twist pQCD evolution is shown by the grey band. It tracks the data down to surprisingly low Q^2, which indicates an overall suppression of higher-twist effects.

4.3. *Spin Polarizabilities: γ_0, δ_{LT} and d_2 for the neutron*

The generalized spin polarizabilities provide benchmark tests of χPT calculations at low Q^2. Since the generalized polarizabilities have an extra $1/\nu^2$ weighting compared to the first moments (GDH sum or I_{LT}), these integrals have less contributions from the large-ν region and converge much faster, which minimizes the uncertainty due to the unmeasured region at large ν.

At low Q^2, the generalized polarizabilities have been evaluated with next-to-leading order χPT calculations.[24,32] One issue in the χPT calculations is how to properly include the nucleon resonance contributions, especially the Δ resonance. As was pointed out in Refs.[24,32] , while γ_0 is sensitive to resonances, δ_{LT} is insensitive to the Δ resonance. Measurements of the generalized spin polarizabilities are an important step in understanding the dynamics of QCD in the chiral perturbation region.

The first results for the neutron generalized forward spin polarizabilities

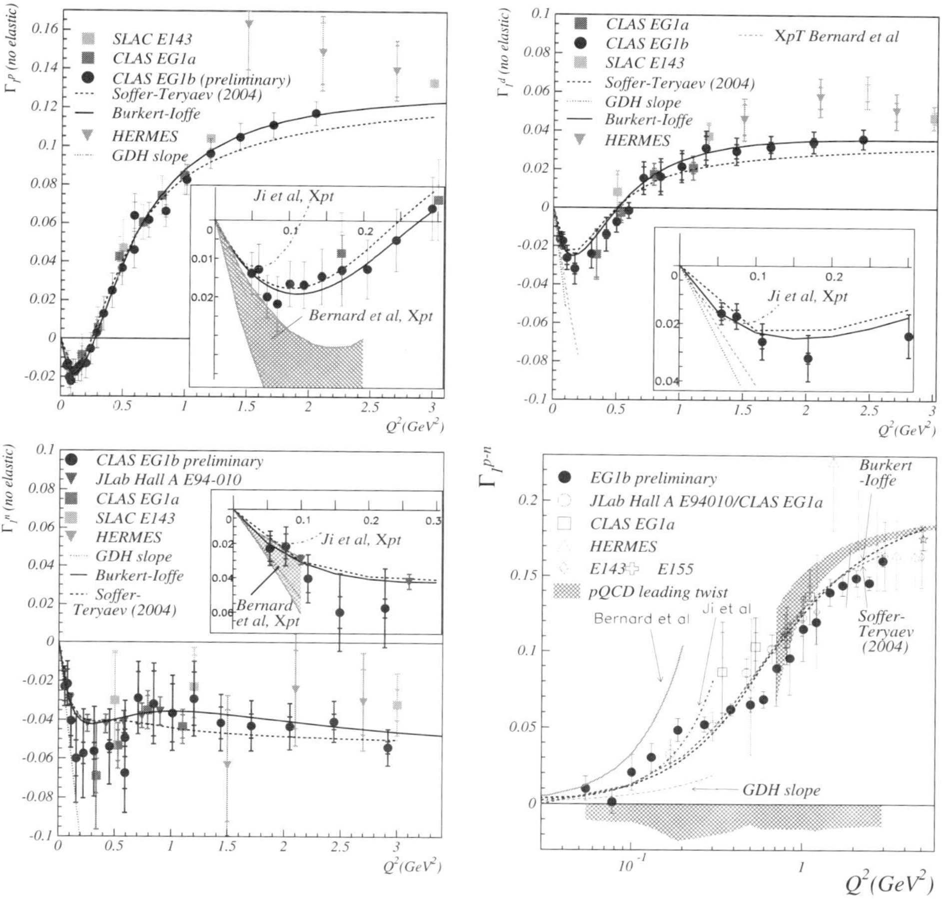

Fig. 2. Preliminary results of $\Gamma_1(Q^2)$ from CLAS eg1b[8] for p, d, n and p-n, together with the published results from Hall A[4] and CLAS eg1a[5,6]. The slopes at Q^2=0 predicted by the GDH sum rule are given by the dotted lines. The MAID model predictions that include only resonance contributions are shown by the full lines while the dashed (dot-dashed) lines are the predictions from the Soffer-Teryaev (Burkert-Ioffe) model. The leading twist Q^2-evolution of the moments is given by the grey band. The insets show comparisons with χPT calculatiosns. The full lines (bands) at low Q^2 are the next-to-leading order χPT predictions by Ji *et al.* (Bernard *et al.*).

$\gamma_0(Q^2)$ and $\delta_{LT}(Q^2)$ were obtained at Jefferson Lab Hall A.[4] The results for $\gamma_0(Q^2)$ are shown in the top-left panel of Fig. 3. The statistical uncertainties are smaller than the size of the symbols. The data are compared with a next-to-leading order ($O(p^4)$) HBχPT calculation,[32] a next-to-leading order RBχPT calculation and the same calculation explicitly including both the Δ resonance and vector meson contributions.[24] Predictions from the

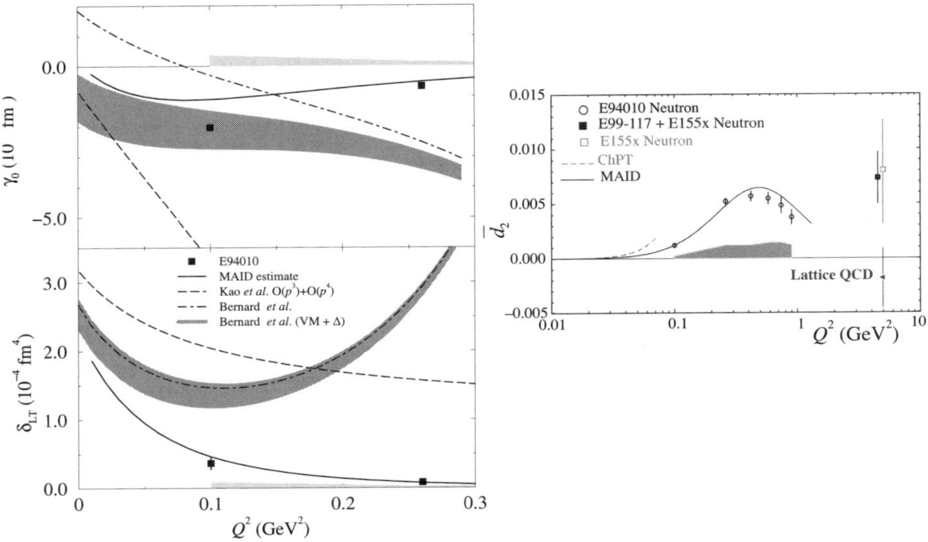

Fig. 3. Results for the neutron spin polarizabilities γ_0 (top-left panel) and δ_{LT} (bottom-left panel). Solid squares represent the results with statistical uncertainties. The light bands represent the systematic uncertainties. The dashed curves represent the HBχPT calculation.[32] The dot-dashed curves and the dark bands represent the RBχPT calculation without and with[24] the Δ and vector meson contributions, respectively. Solid curves represent the MAID model.[3] The right panel shows the $\bar{d}_2^n$ results from JLab[4,34] and SLAC,[17] together with the Lattice QCD calculations.[33]

MAID model[3] are also shown. At the lowest Q^2 point, the RBχPT calculation including the resonance contributions is in good agreement with the experimental result. For the HBχPT calculation without explicit resonance contributions, discrepancies are large even at $Q^2 = 0.1$ GeV^2. This might indicate the significance of the resonance contributions or a problem with the heavy baryon approximation at this Q^2. The higher Q^2 data point is in good agreement with the MAID prediction, but the lowest data point at $Q^2 = 0.1$ GeV^2 is significantly lower. Since δ_{LT} is insensitive to the Δ resonance contribution, it was believed that δ_{LT} should be more suitable than γ_0 to serve as a testing ground for the chiral dynamics of QCD.[24,32] The bottom-left panel of Fig. 3 shows δ_{LT} compared to χPT calculations and the MAID predictions. While the MAID predictions are in good agreement with the results, it is surprising to see that the data are in significant disagreement with the χPT calculations even at the lowest Q^2, 0.1 GeV^2. This disagreement presents a significant challenge to the present Chiral Pertubation Theory.

New experimental data have been taken at very low Q^2, down to 0.02

GeV^2 for the neutron (^3He)[27] and the proton and deuteron.[28] Analyses are underway. Preliminary asymmetry results just became available for the neutron. These results will shed light and provide benchmark tests to the χPT calculations at the kinematics where they are expected to work.

Another combination of the second moments, $d_2(Q^2)$, provides an efficient way to study the high Q^2 behavior of the nucleon spin structure, since it is a matrix element, related to the color polarizabilities and can be calculated from Lattice QCD. It also provides a means to study the transition from high to low Q^2. In Fig. 3, $\bar{d}_2(Q^2)$ is shown. The experimental results are the solid circles. The grey band represents the systematic uncertainty. The world neutron results from SLAC[17](open square) and from JLab E99-117[34] (solid square) are also shown. The solid line is the MAID calculation containing only the resonance contribution. At low Q^2 the HBχPT calculation[32] (dashed line) is shown. The RBχPT with or without the vector mesons and the Δ contributions[24] are very close to the HBχPT curve at this scale, and are not shown on the figure for clarity. The Lattice QCD prediction[33] at $Q^2 = 5$ GeV^2 is negative but close to zero. There is a 2σ deviation from the experimental result. We note that all models (not shown at this scale) predict a negative or zero value at large Q^2. At moderate Q^2, our data show that $\bar{d}_2^n$ is positive and decreases with Q^2.

Preliminary results at a Q^2 range of 1-4 GeV^2 for the neutron[29] are available now. New experiments are planned with 6 GeV beam[30] at average Q^2 of 3 GeV^2 and with future 12 GeV upgraded JLab[31] at constant Q^2 values of 3, 4 and 5 GeV^2. They will provide a benchmark test of the lattice QCD calculations.

5. Conclusion

A large body of nucleon spin-dependent cross-section and asymmetry data have been collected at low to moderate Q^2 in the resonance region. These data have been used to evaluate the Q^2 evolution of moments of the nucleon spin structure functions g_1 and g_2, including the GDH integral, the Bjorken sum, the BC sum and the spin polarizabilities.

At Q^2 close to zero, available next-to-leading order χPT calculations were tested against the data and found to be in reasonable agreement for Q^2 of 0.05 to 0.1 GeV^2 for the GDH integral $I(Q^2)$, $\Gamma_1(Q^2)$ and the forward spin polarizability $\gamma_0(Q^2)$. Above Q^2 of 0.1 GeV^2 a significant difference between the calculation and the data is observed, pointing to the limit of applicability of χPT as Q^2 becomes larger. Although it was expected that the χPT calculation of δ_{LT} would offer a faster convergence because of the absence of the Δ contribution, the experimental data show otherwise. None of the available χPT calculations can reproduce δ_{LT} at Q^2 of 0.1 GeV^2. This discrepancy presents a significant challenge to our theoretical understanding at its present level of approximations.

Overall, the trend of the data is well described by phenomenological models. The dramatic Q^2 evolution of I_{GDH} from high to low Q^2 was observed as predicted by these models for both the proton and the neutron. This behavior is mainly determined by the relative strength and sign of the Δ resonance compared to that of higher-energy resonances and deep inelastic processes. This also shows that the current level of phenomenological understanding of the resonance spin structure using these moments as observables is reasonable.

The BC sum rule for both the neutron and ^{3}He is observed to be satisfied within uncertainties due to a cancellation between the resonance and the elastic contributions. The BC sum rule is expected to be valid at all Q^2. This test validates the assumptions going into the BC sum rule, which provides confidence in sum rules with similar assumptions.

Overall, the recent JLab data have provided valuable information on the transition between the non-perturbative to the perturbative regime of QCD. They form a precise data set for a check of χPT calculations. New results at very low Q^2 for the neutron,[27] proton and deuteron[28] will be available soon. They will provide benchmark tests of the Chiral Pertubation Theory calculations in the kinematical region where they are expected to work.

Future precision measurements[30,31] of d_2^n at $Q^2 = 3 - 5$ GeV^2 will provide a benchmark test of Lattice QCD.

Acknowledgments

This work was supported by the U.S. Department of Energy (DOE). The Southeastern Universities Research Association operates the Thomas Jefferson National Accelerator Facility for the DOE under contract DE-AC05-84ER40150, Modification No. 175.

References

1. see, *e.g.*, B. W. Filippone and X. Ji, Adv. Nucl. Phys. **26**, 1 (2001).
2. J. D. Bjorken, Phys. Rev. **148**, 1467 (1966).
3. D. Drechsel, S. S. Kamalov and L. Tiator, Phys. Rev. **D 63**, 114010 (2001).
4. M. Amarian *et al.*, Phys. Rev. Lett. **89**, 242301 (2002);**92**, 022301 (2004);**93**, 152301 (2004); K. Slifer, *et al.*, to be submitted to Phys. Rev. Lett.
5. R. Fatemi *et al.*, Phys. Rev. Lett. **91**, 222002 (2003).

6. J. Yun *et al.*, Phys. Rev. **C 67**, 055204 (2003).
7. A. Deur *et al.*, Phys. Rev. Lett. **93**, 212001 (2004).
8. K.V. Dharmawardane *et al.*, Phys. Lett. **B 641** 11 (2006); Y. Prok *et al.*, to be published.
9. J.-P. Chen, A. Deur and Z.-E. Meziani, Mod. Phys. Lett. **A 20**, 2745 (2005).
10. S. B. Gerasimov, Sov. J. Nucl. Phys. **2**, 598 (1965); S. D. Drell and A. C. Hearn, Phys. Rev. Lett. **16**, 908 (1966).
11. X. Ji and J. Osborne, J. of Phys. G **27**, 127 (2001).
12. D. Drechsel, B. Pasquini and M. Vanderhaeghen, Phys. Rep. **378**, 99 (2003); D. Drechsel and L. Tiator, Ann. Rev. Nucl. Part. Sci. **54**, 69 (2004).
13. H. Burkhardt and W. N. Cottingham, Ann. Phys. (N.Y.) **56**, 453 (1970).
14. X. Ji and P. Unrau, Phys. Lett. **B 333**, 228 (1994).
15. S. Wandzura and F. Wilczek, Phys. Lett. **B 72**, 195 (1977).
16. K. Ackerstaff *et al.*, Phys. Lett. **B 404**, 383 (1997); **B 444**, 531 (1998).
17. K. Abe *et al.*, Phys. Rev. **D 58**,112003 (1998); P. L. Anthony, *et al.*, Phys. Lett. **B 493**, 19 (2000), **B 553**, 18 (2003).
18. J. Alcorn *et al.*, Nucl. Inst. Meth. **A522**, 294 (2004).
19. B. A. Mecking *et al.*, Nucl. Inst. Meth. **A503**, 513 (2003).
20. C. Ciofi degli Atti and S. Scopetta, Phys. Lett. **B 404**, 223 (1997).
21. Y. Liang et al., nucl-ex/0410027.
22. N. Bianchi and E. Thomas, Nucl. Phys. **B 82** (Proc. Suppl.), 256 (2000).
23. X. Ji, C. Kao, and J. Osborne, Phys. Lett. B **472**, 1 (2000).
24. V. Bernard, T. Hemmert and Ulf-G. Meissner, Phys. Lett. **B 545**, 105 (2002); Phys. Rev. **D 67**, 076008 (2003).
25. J. Soffer and O. V. Teryaev, Phys. Rev. **D 70**, 116004 (2004).
26. V. D. Burkert and B. L. Ioffe, Phys. Lett. **B 296**, 223 (1992).
27. JLab experiment E97-110, J. P. Chen, A. Deur, F. Garibaldi, spokespersons.
28. JLab run group eg4, M. Battaglieri, R. De Vita, A. Deur, M. Ripani spokespersons.
29. JLab E01-012, J. P. Chen, S. Choi and N. Liyanage, spokespersons.
30. JLab experiment E06-014, S. Choi, X. Jiang, Z.E. Meziani and B. Sawatzky spokespersons.
31. JLab experiment E12-06-121, T. Averett, W. Korsch, Z.E. Meziani and B. Sawatzky, spokespersons.
32. C. W. Kao, T. Spitzenberg and M. Vanderhaeghen, Phys. Rev. **D 67**, 016001 (2003).
33. M. Gockeler *et al.*, Phys. Rev. **D 63**, 074506, (2001).
34. X. Zheng *et al.*, Phys. Rev. **C 70**, 065207 (2004).

COMPTON SCATTERING AND NUCLEON POLARISABILITIES

JUDITH A. McGOVERN*

Theoretical Physics Group, Department of Physics and Astronomy
University of Manchester, Manchester, M13 9PL, U.K.
**judith.mcgovern@manchester.ac.uk*

We review the use of chiral EFT to extract values of the electric and magnetic polarisabilities of the proton and neutron from Compton scattering data, and consider the potential of planned experiments at MAXlab and HIγS for better pinning down these and also the spin polarisabilities.

Keywords: Chiral Perturbation Theory; Nucleon Polarisabilities; Compton Scattering.

1. Introduction to polarisabilities

Chiral dynamics in the baryonic sector is typically thought of as the study of the interactions of pions and nucleons. However the constraints of electromagnetic gauge invariance means that chiral symmetry also strongly constrains the interactions of both with photons, and so Compton scattering from the nucleon is as fundamental a probe of chiral dynamics as pion-nucleon or nucleon-nucleon scattering. The lowest order term in the scattering amplitude (the long-wavelength limit) is the Thomson term which is reproduced by χPT but which, depending as it does only on the nucleon charge and mass, is independent of chiral dynamics. However at shorter wavelengths the probing proton starts to be sensitive to the structure of the target. At NLO in HBχPT the dominant new contribution comes from a single pion loop with photons coupling to the pion or to the πN vertex (see Fig. 1), and hence a prediction can be made for these structure effects. This includes, but is not limited to, the numbers known as the polarisabilities of the nucleon.

If we expand the effective Hamiltonian for a nucleon in an electromagnetic field in numbers of derivatives, the leading and next to leading terms are[1]

$$H_{eff} = \frac{(\mathbf{p} - Q\mathbf{A})^2}{2m} + Q\phi - \frac{1}{2}4\pi\left(\alpha\mathbf{E}^2 + \beta\mathbf{H}^2 + \gamma_{E1}\sigma\cdot\mathbf{E}\times\cdot\mathbf{E}\right.$$
$$\left. + \gamma_{M1}\sigma\cdot\mathbf{H}\times\cdot\mathbf{H} - 2\gamma_{E2}E_{ij}\sigma_i H_j + 2\gamma_{M2}H_{ij}\sigma_i E_j\right) + \dots$$

where $E_{ij} = \frac{1}{2}(\nabla_i E_j + \nabla_j E_i)$ and $H_{ij} = \frac{1}{2}(\nabla_i H_j + \nabla_j H_i)$. This introduces the electric and magnetic polarisabilities α and β, and the four polarisabilities γ_{E1} etc which characterise the spin-dependent response. These numbers are not only fundamental properties of the proton and neutron which are of interest in themselves, they are also a test of chiral dynamics.

While this talk is concerned with the use of Compton scattering as a tool to extract the nucleon polarisabilities, a certain amount of information can be obtained from other sources.[2] One is the sum rules which relate the forward scattering amplitude to the total photoproduction cross section; the Baldin sum rule

$$\alpha + \beta = \frac{1}{2\pi^2}\int_{\omega_{\rm th}}^{\infty}\frac{\sigma_{\rm tot}(\omega)}{\omega^2}\mathrm{d}\omega$$

and the related one for polarised targets

$$\gamma_0 = \frac{1}{4\pi^2}\int_{\omega_{\rm th}}^{\infty}\frac{\sigma_{1/2}(\omega) - \sigma_{3/2}(\omega)}{\omega^3}\mathrm{d}\omega.$$

Here γ_0 is the combination of spin polarisabilities which contribute to forward scattering; it is also useful to define γ_π which governs backward scattering. They correspond to $-\gamma_{E1} - \gamma_{M2} \mp (\gamma_{M1} + \gamma_{E2})$ respectively. The Baldin sum rule provides strong constraints on $\alpha + \beta$ for the proton and rather more model-dependent ones for the neutron. Polarised photoproduction cross sections on the proton have been measured at Mainz and Bonn as part of the GDH collaboration but only above 200 MeV, so the extraction of $\gamma_0^{\rm p}$ requires some input from pion photoproduction amplitudes such as MAID.

These sum rules are special cases of a more general approach to estimating polarisabilities, namely through dispersion relations, which also make use of MAID and related analyses. For more details see Barbara Pasquini's talk at this conference. Table 1 summarises what is known about the polarisabilities from these approaches. It uses an alternative basis for the spin polarisabilities: $\gamma_1 \equiv -\gamma_{E1} - \gamma_{M2}$, $\gamma_2 \equiv \gamma_{E2} - \gamma_{M1}$; $\gamma_3 \equiv \gamma_{M2}$ and $\gamma_4 \equiv \gamma_{M1}$.

Table 1. Collection of non-compton predictions for polarisabilities. The units are as follows: α, β: 10^{-4}fm^3; γ_i: 10^{-4}fm^4. Q^3 and Q^4 represent the HBχPT predictions.[3,4] "DR" and "SR" refer to dispersion relation and sum rule respectively.[2] The asterisk indicates model input (sigma exchange).

Proton	$\alpha+\beta$	$\alpha-\beta$	γ_0	γ_π	γ_1	γ_2	γ_3	γ_4
Q^3	13.6	10.3	4.5	[−48.6]+4.5	4.5	2.3	1.1	1.1
Q^4	-	-	−3.9	[−48.6]+6.1	1.1	−1.5	0.2	3.3
DR	$12-14$	12 ± 3*	−(0.7-1.5)	[−48.6]+(8-9)	3-5	−1	0	3
SR	13.8 ± 0.4	-	-0.86 ± 0.13	-	-	-	-	-

Neutron	$\alpha+\beta$	$\alpha-\beta$	γ_0	γ_π	γ_1	γ_2	γ_3	γ_4
Q^3	13.6	10.3	4.5	[48.6]+4.5	4.5	2.3	1.1	1.1
Q^4	-	-	−0.9	[48.6]+8.3	4.7	−0.1	0.3	2.3
DR	13-15	12 ± 3*	−(0.1-0.5)	[48.6]+(9-14)	6-7	−1	−(0-1)	3-4
SR	15.2 ± 0.5	-	-	-	-	-	-	-

2. Polarisabilities and Compton scattering in chiral perturbation theory

Within Chiral Perturbation Theory, one can calculate the complete Compton scattering amplitude to any given order. To date this has been done to $\mathcal{O}(Q^4)$ (NNLO);[5] the relevant diagrams are shown in Fig. 1. The low-energy terms such as the Thomson term are automatically recovered, and pion loops give rise to the non-Born pieces of the amplitude. The polarisabilities are the leading terms in an expansion of these in powers of ω. However it is important to be clear that the full amplitudes, not just the polarisabilities, are predicted.

At $\mathcal{O}(Q^3)$ no unknown counter-terms enter, and we get the well-known – and astonishingly good – predictions of $\alpha = 12.5$ and $\beta = 1.2 \times 10^{-4}\text{fm}^3$. The spin polarisabilities are also predicted, and as can be seen from Table 1 they do not seem to be so good though the order of magnitude is correct. At $\mathcal{O}(Q^4)$ four counterterms with unknown coefficients enter, one each for α and β of the proton and neutron, and hence there is no further predictive power. These counterterms encode information about degrees of freedom not explicitly included in χPT, including – but not restricted to – the Δ. The equivalent terms for the spin polarisabilities do not enter till one order higher however, so these are still predictions at this order. (In general there is improvement in going from third to fourth order, but the change is often large and one cannot help questioning the convergence).

At first sight the lack of predictive power for α and β beyond NLO might seem like a drawback. However as we shall see, this can be turned to an advantage, and we can use HBχPT as a tool to extract experimental and model-independent values of these polarisabilities from Compton scattering

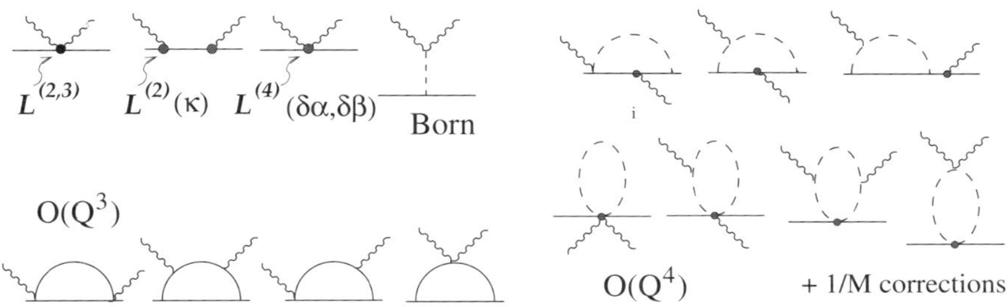

Fig. 1. Contributions to Compton Scattering in HBχPT to fourth order

data.

3. Compton Scattering on the Proton

Modern experiments measure the differential cross section for tagged photons over a range of energies and angles. Between 1956 and 1995, twelve Compton scattering experiments took data at energies below 200 MeV. In a recent paper Baranov *et al*[6] examined the world data on proton Compton scattering, and demonstrated that the data form the 50s and 60s was compatible with the more modern data from 1974 onwards, and was useful in reducing errors. In addition, the most recent and extensive experiment used the TAPS detector and MAMI tagged photon facility at Mainz.[7]

The traditional way to extract polarisabilities from Compton scattering was to expand the cross section in powers of the photon energy ω, discarding terms of higher order than those at which α and β enter. These could then be fit to the data. The advantage of this approach is that it is entirely model-independent; the disadvantage is that there is a very narrow window in which the deviation from the Born cross section is significant but the expansion is still valid. A desire to use the full range of data up to the region of the pion-production threshold meant that later authors turned to dispersion relations to obtain a prediction for the variation of the cross section with energy, leaving only certain combinations of the polarisabilities as free parameters to be fit. Olmos de León *et al.*[7] for instance fit α, β and γ_π. Unfortunately the dispersion relation approach is not entirely free of assumptions; for instance the asymptotic form of the backward spin-independent amplitude is assumed to be dominated by σ exchange. In Ref. 8, therefore, Beane *et al.* used the $\mathcal{O}(Q^4)$ HBχPT amplitudes, in which α and β are free parameters, to fit Compton scattering data below an appropriate cut-off value on energy and momentum transfer, and extracted

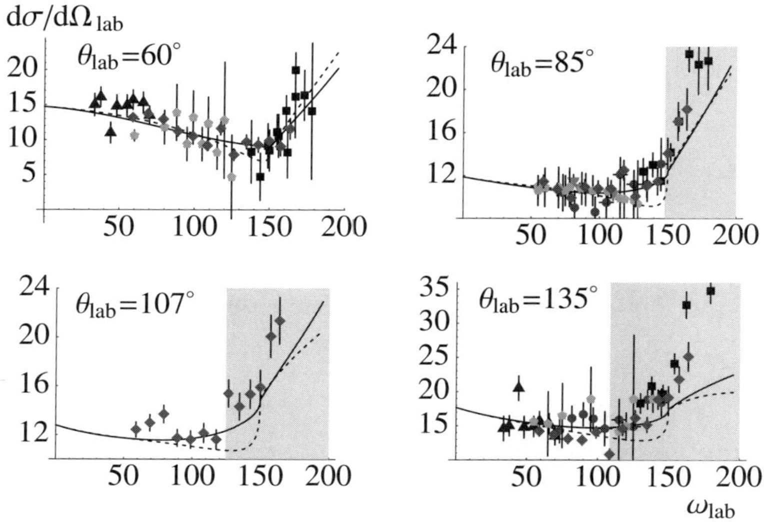

Fig. 2. Third- (dotted) and fourth- (solid) order Compton scattering cross sections from HBχPT. In the fourth-order curve, α and β are fitted. The grey shaded regions are excluded from the fit. Figure from Ref. 8.

the following values:

$$\alpha_p = (12.1 \pm 1.1 \text{ (stat.)})^{+0.5}_{-0.5} \text{ (theory)} \times 10^{-4} \text{ fm}^3$$
$$\beta_p = (3.4 \pm 1.1 \text{ (stat.)})^{+0.1}_{-0.1} \text{ (theory)} \times 10^{-4} \text{ fm}^3.$$

A sample of the fit is shown in Fig. 2.

It is noticeable that the HBχPT calculation fails to reproduce the steep rise of the cross section at backwards angles beyond the pion-production threshold. The reason is clear of course; the contribution of the Δ is only encoded in HBχPT through contact terms, which can only work well below the Δ-peak. The work of Beane *et al.* has been repeated by Hildebrand *et al.*[8] using an explicit Δ in the "small-scale expansion" to $\mathcal{O}(Q^3)$. The fit is shown in Fig. 3, and the values extracted are very similar: $\alpha_p = 11.52 \pm 2.43$, $\beta_p = 3.42 \mp 1.70 \times 10^{-4}$ fm^3 where only statistical errors are quoted.

4. Compton Scattering on the Deuteron

In the absence of free nucleon targets, all attempts to extract neutron polarisabilities from Compton scattering so far have used deuteron targets. Only three low-energy experiments have been done, at Illinois with photon energies of 49 and 69 MeV,[10] Saskatoon (95 MeV)[11] and Lund (60 MeV).[12]

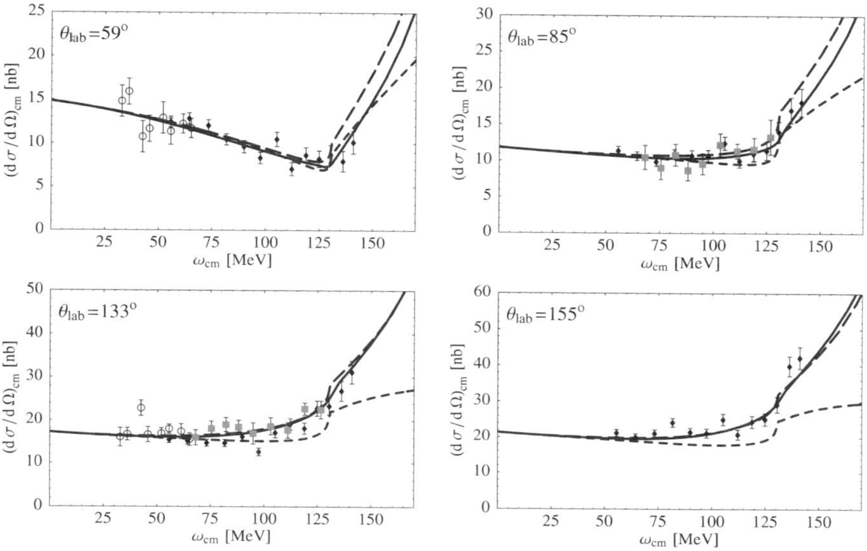

Fig. 3. Compton scattering cross sections from SSE calculations with an explicit Δ (long dash), 3rd order HBχPT (short dash) and dispersion relations (solid). In the SSE and DR curves, α and β are fitted. Figure from Ref. 9

(At higher energies, extractions have been done using quasi-free scattering from the neutron, but high-precision results are heavily dependent on the use of dispersion-relation amplitudes.[13])

However Compton scattering from the deuteron is not simply the sum of scattering from the constituent proton and neutron. A consistent extraction of neutron properties requires the binding of the nucleons in the nucleus also to be described in a framework compatible with χPT. Over the past ten years much progress has been made in developing such a chiral EFT for nuclear forces.[14] In Fig. 4 we show the diagrams which contribute to $\mathcal{O}(Q^4)$.

As the deuteron is isoscalar, Compton scattering is sensitive only to the isoscalar combinations $\alpha_N \equiv (\alpha_p + \alpha_n)/2$ and $\beta_N \equiv (\beta_p + \beta_n)/2$. Fig. 5 shows the best fit values from Ref. 8:

$$\alpha_N = (13.0 \pm 1.9)^{+3.9}_{-1.5} \times 10^{-4}\,\text{fm}^3,$$
$$\beta_N = (-1.8 \pm 1.9)^{+2.1}_{-0.9} \times 10^{-4}\,\text{fm}^3.$$

The theory error is large, for a number of reasons (which turn out to be related): chiefly, the power-counting used is not valid at very low energy, which means that the Thomson limit for the deuteron is not recovered. As

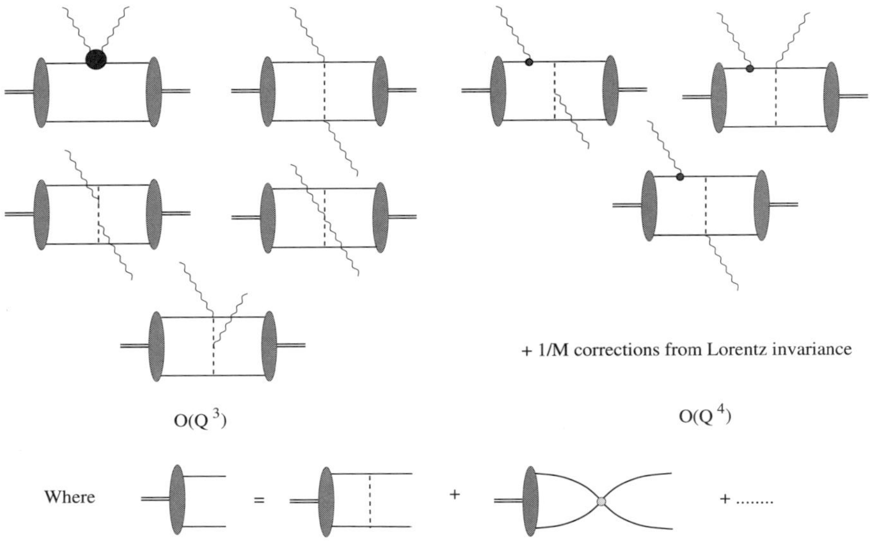

Fig. 4. Compton scattering on the deuteron in Chiral EFT. The large blob in the first diagram comprises all the single-nucleon graphs of Fig. 1.

a result, there is considerable sensitivity to the cut-off used in the wave function.

This problem has been solved by Hildebrandt *et al.* , who showed how to re-sum the NN rescattering diagrams in the kernel.[15] They have applied their method to the theory with an explicit Δ, and shown that the wave-function-dependence is dramatically reduced. Their results are shown in Fig. 6, and their best-fit results are

$$\alpha_N = 11.5 \pm 1.4_{\text{stat}} \times 10^{-4}\,\text{fm}^3,$$
$$\beta_N = 3.4 \pm 1.6_{\text{stat}} \times 10^{-4}\,\text{fm}^3.$$

These results strongly suggest that α and β for the proton and neutron are very similar to one another. Together with the results for the proton quoted above, they also suggest that the PDG value[16] of $\beta^{\text{p}} = 1.9 \times 10^{-4}$ fm^3 is probably too low.

It seems that statistical errors now dominate the extraction of the electromagnetic polarisabilities from both the proton and the neutron. In the case of the proton this is because the data are not all consistent (even when the normalisations are allowed to float), and it might be instructive to attempt to construct a reduced database of consistent experiments. For the deuteron on the other hand it is clear that more data at a wider range of en-

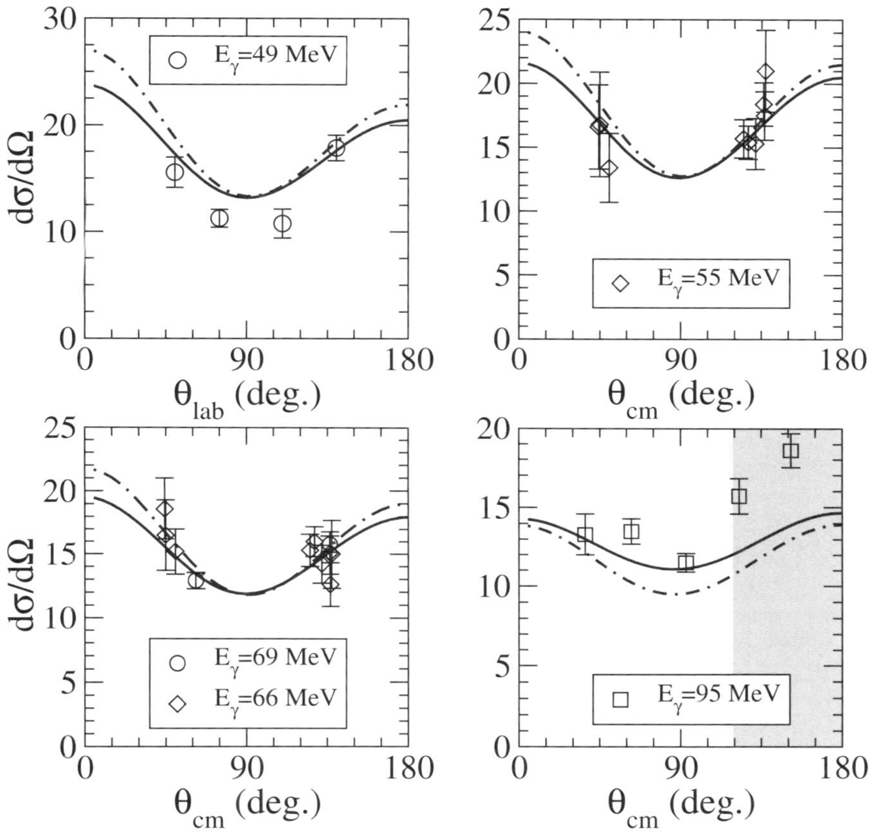

Fig. 5. EFT Compton scattering amplitudes for the deuteron. The solid line is the $O(Q^4)$ calculation with the best-fit values of α_N and β_N The gray area is the region excluded from the fit (ω, $\sqrt{|t|} > 160$ MeV). The dot-dashed line is the (parameter-free) $O(Q^3)$ calculation. The gray area is the region excluded from the fit. Figure from Ref. 8

ergies and, particularly, angles, is needed. Fortunately such an experiment is planned: work is underway by the COMPTON@MAXlab collaboration at the facility in Lund. They will use three NaI detectors (from Mainz, Boston and Kentucky), a liquid deuterium target, and cover an angular range of 30°-150° and an energy range of 40-110 MeV The aim is better than 5% precision (at least as good as the best to date). To make the best of such data, though, further theoretical work is required (for instance, to incorporate the Thomson limit and, ideally, include the Δ in the $\mathcal{O}(Q^4)$ calculation).

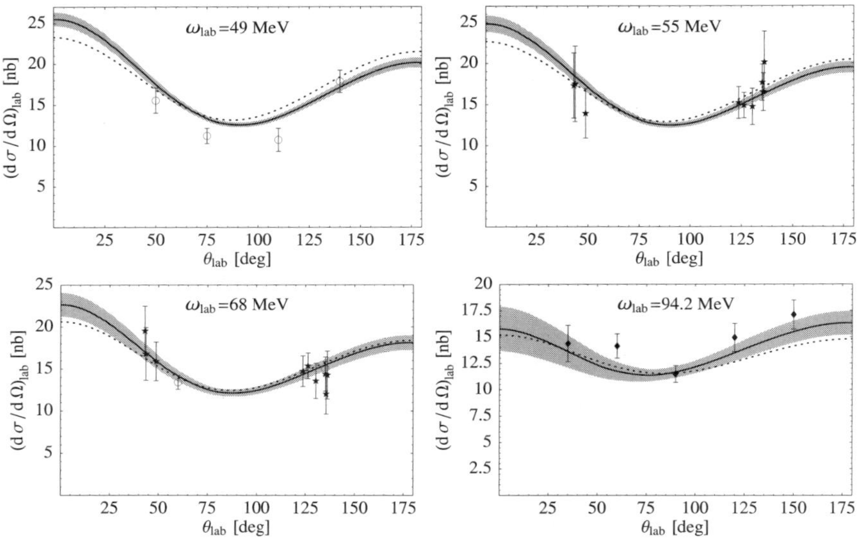

Fig. 6. Compton scattering amplitudes for the deuteron. The solid line with the grey spread indicating the statistical error is the best fit for the $\mathcal{O}(Q^3)$ EFT with explicit Δ; the dotted line is a $\mathcal{O}(Q^4)$ Δ-less EFT fit (very similar to that in Fig. 5). Figure from Ref. 15

5. Spin Polarisabilities

In contrast to the electromagnetic polarisabilities, the spin polarisabilities are not well known. Their contribution to low-energy unpolarised Compton scattering is negligible and it is unlikely they will be extracted from that source. However polarised Compton scattering is a different matter. When the current upgrade is complete, HIγS will deliver a high-intensity circularly-polarised photon beam at energies up to around 120 MeV, which will be ideal for the purpose. Plans are underway to used a polarised scintillating target for the proton and a ^{3}He target for the neutron; and a detector with an array of 16 NaI elements ("HINDA") is under construction. The talks of Barbara Pasquini and Deepshikha Choudhury at this conference indicate the kind of sensitivity to the spin polarisabilities which can be expected, and suggest that when we meet for CD2009 in Berne, experimental information on the spin polarisabilities is a real possibility.

Acknowledgments

I would like to thank the organisers for their invitation and support, both for the main CD2006 conference and for the workshop on planning experiments at HIγS which preceded it. Both were stimulating and enjoyable.

References

1. B. R. Holstein, D. Drechsel, B. Pasquini and M. Vanderhaeghen, Phys. Rev. C **61** 034316 (2000).
2. D. Drechsel, B Pasquini and M Vanderhaeghen, Phys. Rept. 378 99 (2003); F. Wissmann, Springer Tracts in Modern Physics 200 (2004); M. Schumacher, Prog. Part. Nucl. Phys. **55** 567 (2005).
3. V. Bernard, N. Kaiser, J. Kambor and U.-G. Meißner, Nucl. Phys. **B 388** 315 (1992).
4. K. B. V. Kumar, J. A. McGovern, M. C. Birse, Phys. Lett. **B479** 167 (2000).
5. J.A. McGovern, Phys. Rev. C **63**, 064608 (2001).
6. P.S. Baranov, A.I. L'vov, V.A. Petrun'kin, and L.N. Shtarkov, Phys. Part. Nucl. **32**, 376 (2001).
7. V. Olmos de León *et al.*, Eur. Phys. J. A **10**, 207 (2001).
8. S. R. Beane, M. Malheiro, J. A. McGovern, D. R. Phillips and U. van Kolck, Phys. Lett. **B567** 200 (2003); Erratum-ibid. **B607** 320 (2005); Nucl. Phys. **A747** 311 (2005).
9. R. P. Hildebrandt, H. W. Griesshammer, T. R. Hemmert and B. Pasquini, Eur. Phys. J. A **20** 293 (2004).
10. M. Lucas, PhD thesis, University of Illinois, unpublished (1994).
11. D. L. Hornidge *et al.* ,Phys. Rev. Lett. **84** 2334 (2000).
12. M. Lundin *et al.* , Phys. Rev. Lett. **90** 192501 (2003).
13. K.W. Rose *et al.*, Nucl. Phys. **A514**, 621 (1990); N.R. Kolb *et al.*, Phys. Rev. Lett. **85**, 1388 (2000); K. Kossert *et al.*, Phys. Rev. Lett. **88**, 162301 (2002) and Eur. Phys. J. **16**, 259 (2003).
14. P. F. Bedaque and U. van Kolck, Ann. Rev. Part. Nucl. Sci. **52** 339 (2002); S. R. Beane, P. F. Bedaque, W. C. Haxton, D. R. Phillips and M. J. Savage, in "At the Frontier of Particle Physics", M. Shifman ed. (World Scientific, Singapore, 2001); U. van Kolck, Prog. Part. Nucl. Phys. **43** 337 (1999).
15. R. P. Hildebrandt, H. W. Griesshammer and T. R. Hemmert, nucl-th/0512063.
16. W.-M. Yao et al., J. Phys. G **33** 1 (2006)

VIRTUAL COMPTON SCATTERING AT MIT-BATES

RORY MISKIMEN

Department of Physics, University of MassachusettsAmherst, MA 01003, USA

Recent results from a virtual Compton scattering experiment performed at the MIT-Bates out-of-plane scattering facility are presented. In this experiment the mean square electric and magnetic polarizability radii of the proton have been measured for the first time. The response functions, P_{LL}-P_{TT}/ϵ and P_{LT}, and the generalized polarizabilities, $\alpha(Q^2)$ and $\beta(Q^2)$, are obtained from a dispersion analysis of the data at Q^2=0.057 GeV^2/c^2 . The results are in good agreement with $O(p^3)$ heavy baryon chiral perturbation theory. The data support the dominance of mesonic effects in the polarizabilities.

1. Introduction

Hadron polarizabilities are of compelling experimental and theoretical interest,[1] providing a vital testing ground for theories of low-energy QCD and nucleon structure. In the case of atomic polarizabilities the electric polarizability is approximately equal to the atomic volume. By contrast, the electric polarizability of the nucleon is approximately 10^4 times smaller than the nucleon volume, demonstrating the extreme stiffness of the nucleon relative to the atom. Although the electric and magnetic polarizabilities of the proton, α and β are known with reasonable accuracy[2] from real Compton scattering (RCS), much less is known about the polarizability distributions inside the nucleon. To measure the polarizability distributions it is necessary to use the virtual Compton scattering (VCS) reaction,[3] where the incident photon is virtual.

At low Q^2 it is expected[4] that $\alpha(Q^2)$ should decrease with increasing Q^2 with a characteristic length scale given by the pion range. The first VCS experiments at Mainz[5] at Q^2=0.33 GeV^2/c^2 and later at JLab[6] at Q^2=0.92 and 1.76 GeV^2/c^2 established that $\alpha(Q^2)$ is falling off, but with a form inconsistent with a simple dipole shape.[6] By contrast, because of the destructive interference between paramagnetism and diamagnetism in the proton, necessary to generate a small β, $\beta(Q^2)$ is predicted[4] to be relatively flat as a function of Q^2 with a 20% peaking near Q^2=0.1 GeV^2/c^2. The rise in β results from a paramagnetic pion-loop contribution that increases with Q^2.

Recently a VCS experiment on the proton was performed at the out-of-plane scattering facility at the MIT-Bates linear accelerator at Q^2=0.057 GeV^2/c^2. Data at this low Q^2 can provide a test of chiral perturbation theory, and are sensitive to the mean square electric and magnetic polarizability radii. ChPT also predicts that at this low value of Q^2 there is increased sensitivity to the polarizabilities relative to the Mainz VCS kinematics.[4] The data from this experiment will be presented and discussed in this conference proceeding.

2. The VCS reaction

The relationship between VCS cross sections and the polarizabilities is most easily seen in the low energy expansion (LEX) of the unpolarized VCS cross section.[3]

$$d^5\sigma^{VCS} = d^5\sigma^{BH+Born} + q'\Phi\Psi_0(q,\epsilon,\theta,\phi) + O(q'^2) \qquad (1)$$

where q(q′) is the incident (final) photon 3-momenta in the photon-nucleon C.M. frame, ϵ is the photon polarization, $\theta(\phi)$ is the C.M. polar (azimuthal) angle for the outgoing photon, and Φ is a phase space factor. $d\sigma^{BH+Born}$ is the cross section for the Bethe-Heitler + Born amplitudes only, i.e. no nucleon structure, and is exactly calculable from QED and the nucleon form factors. The polarizabilites enter the cross section expansion at order $O(q')$ through the term Ψ_0 given as,[7]

$$\Psi_0 = V_1[P_{LL}(q) - \frac{P_{TT}(q)}{\epsilon}] + V_2 P_{LT}(q) \qquad (2)$$

where P_{LL}, P_{TT} and P_{LT} are VCS response functions, with $P_{LL} \approx \alpha(Q^2)$, $P_{LT} \approx \beta(Q^2)$ + spin-polarizabilities, and $P_{TT} \approx$ spin-polarizabilities. V_1 and V_2 are kinematic functions. The Bates VCS experiment was designed to make an azimuthal separation of P_{LL}-P_{TT}/ϵ and P_{LT} by taking data simultaneously at ϕ angles of 90°, 180° and 270°, at fixed θ=90°. For the out-of-plane cross sections at ϕ=90° and 270°, the cross sections are equal and the polarizability effect is proportional to P_{LL}-P_{TT}/ϵ. At ϕ=180°, the cross section is proportional to the sum of P_{LL}-P_{TT}/ϵ and P_{LT}. Data were taken at five different C.M. final photon energies q′ ranging from 43 MeV, where the polarizability effect is negligible, up to 115 MeV where the polarizability effect is approximately 20%. The data were taken at q=240 MeV, and ϵ=0.9, corresponding to $Q^2 \approx 0.06$ GeV^2/c^2. For these kinematics the sensitivities to $\alpha(Q^2)$ in P_{LL}-P_{TT}/ϵ and $\beta(Q^2)$ in P_{LT} are estimated[4] at 92% and 69%, respectively.

3. The MIT-Bates VCS experiment

This experiment was the first to use extracted CW beam from the MIT-Bates South Hall Ring. The extracted beams had duty factors of approximately 50%, currents of up to 7 mA. The five electron beam energies ranged from 570 to 670 MeV. The target was 1.6 cm of liquid hydrogen. This experiment marked the first use of the full Out-of-Plane Spectrometer (OOPS) system with gantry for proton detection,[8] and a new OHIPS electron spectrometer focal plane[9] that increased the momentum acceptance of the spectrometer from 9% to 13%, giving increased acceptance in q$'$. Optics studies were performed to measure OHIPS transport matrix elements over the extended focal plane instrumentation. A new OOPS optics tune using a 2.5 m drift distance was developed for the running at q$'$=43 and 65 MeV because of the close packing of the OOPS's at those energies. Data taken at higher q$'$ used the standard 1.4 m drift for the OOPS. The lowest proton kinetic energy in the experiment was 30 MeV. Because of concerns about protons ranging out in the three OOPS trigger scintillators, the OOPS trigger was modified to a two-fold trigger of the first two scintillators in the focal plane. A GEANT simulation of the OOPS trigger predicts a trigger efficiency of ≈99%. The acceptance montecarlo was based on the program Turtle,[10] and the measured spectrometer matrix elements were used for calculating focal-plane coordinates from target coordinates. The multiple scattering model[11] from GEANT4 was implemented in the acceptance montecarlo. Good agreement was achieved between measured and calculated angular and momentum distributions.

The final state photon was identified through missing mass and time-of-flight techniques. Photon yields were obtained by fitting the missing mass squared (MM^2) distributions using the radiated line shape calculated with the montecarlo and an empirical background to account for A(e,e$'$ p)X events on the havar target cell wall. Polynomial and skewed gaussian shapes for the MM^2 backgrounds gave identical yields within errors to fits that used the accidental MM^2 distributions for the background shape, and the latter distribution was utilized for peak fitting. Radiative corrections were applied to the data,[12] approximately 22% in these kinematics.

The VCS cross sections are shown in Fig. 1 with the statistical and

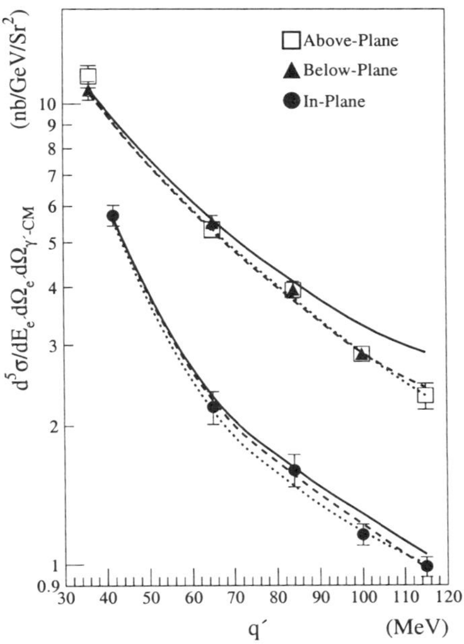

Fig. 1. VCS cross sections as a function of $< q' >$. The solid curves are Bethe-Heitler + Born, the dashed and dotted curves are fits with LEX and dispersion analyses, respectively.

systematic errors combined in quadrature. The dominant error is statistical, with the largest systematic uncertainty the OOPS tracking efficiency, at $\approx$1.6%.

4. LEX Analysis of the data

The solid lines in Fig. 1 are the Bethe-Heitler+Born (BH+Born) calculations, i.e. no polarizability effect, using Hoehler form factors.[13] The agreement between data and the BH+Born calculation is good at low q′, while at higher q′ the out-of-plane data falls significantly below the calculation because of destructive interference between the BH+Born and polarizability amplitudes. The in-plane cross sections show a much smaller deviation from the BH+Born cross sections at high q′ because the kinematic multipliers V_1 and V_2 in Eq. 2 have the same sign, and therefore much of the polarizability effect is canceled at O(q′) (note: P_{LL}-P_{TT}/ϵ is positive, and

P_{LT} is negative).

The dashed lines in Fig. 1 are fits to the data using the LEX, which give P_{LL}-P_{TT}/ϵ=54.5±4.8±2.0 GeV^2, and P_{LT}=-20.4±2.9±0.8 GeV^2, where the first error is statistical and the second is systematic. The largest systematic error results from the ±0.1% uncertainty in the beam energies, which introduces an error in the response functions through the energy dependence of $d^5\sigma^{BH+Born}$. A LEX analysis using the Friedrich-Walcher form factors[14] gives identical results, within errors, to the analysis presented here using the Hoehler form factors.

The LEX result for P_{LL}-P_{TT}/ϵ is shown in Fig. 2, where the statistical and systematic errors have been combined in quadrature. Also shown in the figure is the parameter free $O(p^3)$ calculation in heavy baryon chiral perturbation theory (HBChPT),[4] which is in good agreement with experiment for P_{LL}-P_{TT}/ϵ. However, the LEX result for P_{LT} is much larger than the the RCS result and the HBChPT prediction.

5. Dispersion Analysis of the Data

A dispersion analysis of the data was performed using the VCS dispersion model.[15] In this analysis the VCS amplitudes are obtained from the MAID $\gamma^*p\to N\pi$ multipoles,[16] and the unconstrained asymptotic contributions to 2 out of the 12 VCS amplitudes are varied to fit the VCS data. The dotted curves in Fig. 1 show the best dispersion fits to the VCS cross sections. The polarizabilities and response functions are found by summing the fitted asymptotic terms with calculated πN dispersive contributions. The best fit response functions from the dispersion analysis are P_{LL}-P_{TT}/ϵ =46.7±4.9±2.0 GeV^2 and P_{LT}=-8.9±4.2±0.8 GeV^2. The dispersion results are shown in Fig. 2 with the statistical and systematic errors combined in quadrature. The figure shows that the dispersion result for P_{LL}-P_{TT}/ϵ is in near agreement with the LEX analysis and the HBChPT predictions.

The dispersion result for P_{LT} is in good agreement with the HBChPT prediction, and is much smaller than the LEX result. The disagreement with LEX results from the near cancellation of the electric and magnetic polarizability responses at $O(q')$ for the in-plane kinematics, causing the polarizability effect to be predominantly quadratic in q$'$. The LEX analysis is only valid in kinematics where the polarizability effect is linear in q$'$ (see Eq. 1), while the dispersion analysis is valid to all orders in q$'$.[17]

The dispersion model fits give α=7.85±0.87±0.3610^{-4} fm^3, and β=2.69±1.48±0.2810^{-4} fm^3. These results are shown in Fig. 3 with the statistical and systematic errors combined in quadrature, along with pre-

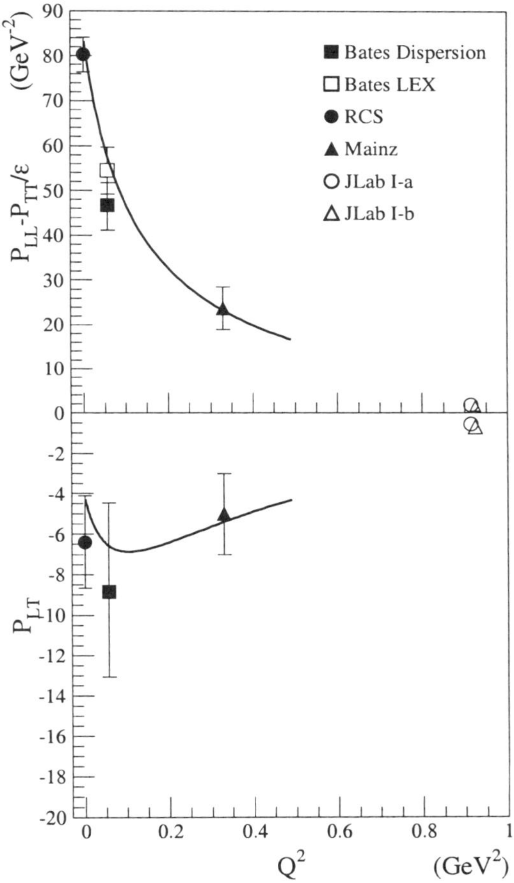

Fig. 2. VCS response functions from this experiment, RCS,[2] Mainz[5] and JLab.[6] The solid curves are $O(p^3)$ HBChPT[4].

vious results from RCS,[2] Mainz[18] and JLab.[6] The Bates results for α and β are in near agreement with the HBChPT prediction, shown as the solid curves in Fig. 3. Taken as a group, the $\beta(Q^2)$ data below Q^2=0.4 GeV^2/c^2 suggest a flat or nearly flat magnetic response as a function of Q^2, in agreement with the HBChPT prediction. The theoretical errors[19] for $\alpha(Q^2)$ and $\beta(Q^2)$ at Q^2=0.06 GeV^2/c^2 are estimated to be comparable to the errors for an $O(p^4)$ calculation[20] of α and β, approximately $\pm$2.0 and $\pm$3.6 in

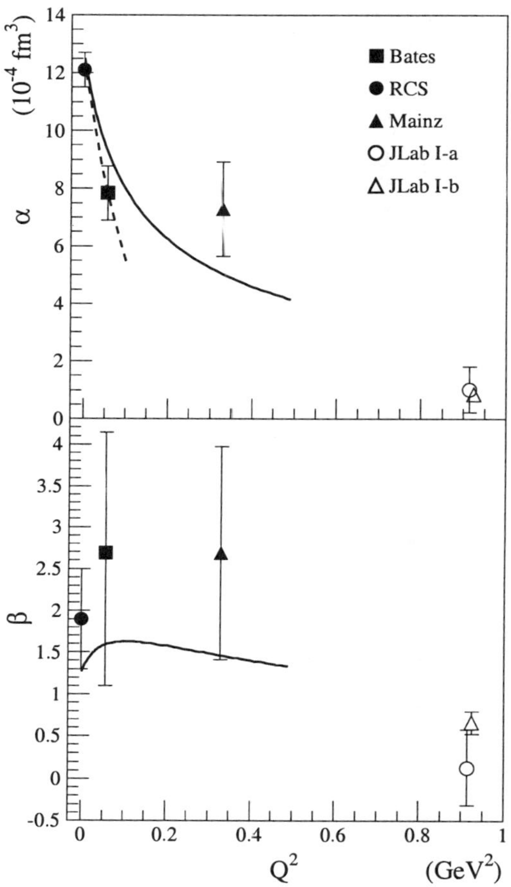

Fig. 3. Dispersion analysis results for $\alpha(Q^2)$ and $\beta(Q^2)$. The references are the same as in Fig. 2 except for Mainz.[18] The solid curves are $O(p^3)$ HBChPT,[4] the dashed curve is a low Q^2 fit to $\alpha(Q^2)$.

units of 10^{-4} fm^3, respectively.

The mean square electric polarizability radius $< r_\alpha^2 >$ was determined from a HBChPT fit to the RCS and Bates $\alpha(Q^2)$ data points, where the $O(Q^2)$ term in the momentum expansion of the HBChPT prediction was varied to fit the data. The fit gives $< r_\alpha^2 > =2.16\pm0.31$ fm^2, in near agreement with the HBChPT prediction[21] of 1.7 fm^2. The experimental

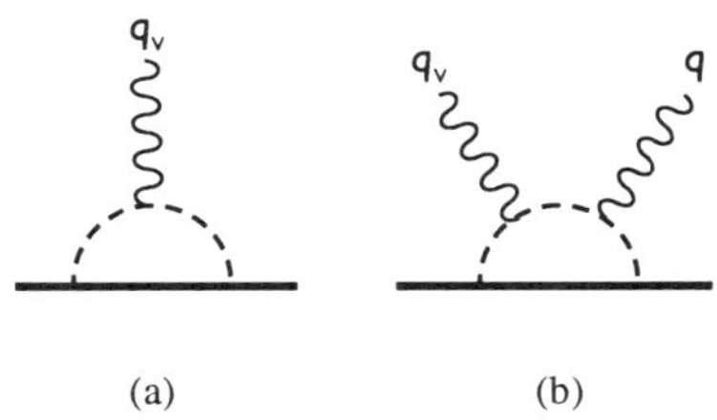

Fig. 4. The dominant pion loop diagrams for (a) the proton charge form factor and (b) the proton polarizability.

value is significantly larger than the proton mean square charge radius[22] of $0.757\pm.014$ fm^2, and is evidence for the dominance of mesonic effects in the electric polarizability. Figure 4 shows the dominant pion loop diagrams for the proton form factor and the proton polarizability. The additional electromagnetic vertex in the polarizability diagram increases the range of the interaction by 70% as compared to the form factor range. It is interesting to note that the experimental result for $< r_\alpha^2 >$ is close to the uncertainty principle estimate of 2.0 fm^2 for the size of the pion cloud.

An $O(\mathrm{p}^4)$ calculation by L'vov for $\beta(\mathrm{Q}^2)$ shows a nearly linear increase with Q^2 in the low Q^2 region, and therefore a straight line anstatz for fitting the RCS and Bates $\beta(\mathrm{Q}^2)$ data points was used to make an estimate for $< r_\beta^2 >$. The value obtained from this fit, $< r_\beta^2 > = -1.91\pm2.12$ fm^2, agrees with the HBChPT prediction[20] of -2.4 fm^2. Given the sign, size and uncertainty in $< r_\beta^2 >$, the confidence level that $\beta(\mathrm{Q}^2)$ is increasing from the real photon point is 82%.

6. Summary and conclusion

The experimental results are summarized in table 1. The experiment supports two long accepted, although arguably not fully tested, tenets of the proton polarizability. The first is that the electric polarizability is dominated by mesonic effects, and this is confirmed by the large size of $< r_\alpha^2 >$ relative to the proton charge radius. The second is the cancellation of positive paramagnetism by negative diamagnetism, a critical element in explanations for the small size of β relative to α since in HBChPT the size of the polarizabilities are predicted to be of the same order. Because paramagnetism from the $\Delta(1232)$ is predicted to be nearly independent of Q^2 in this low Q^2 range, the paramagnetic-diamagnetic interference will also have a relatively flat dependence on Q^2. The data suggest that $\beta(Q^2)$ is relatively flat as a function of Q^2, in agreement with the HBChPT prediction.

Future developments in this field can be expected from ongoing VCS experiments at Mainz,[23] where measurements with polarized beams and recoil polarimetry will determine the generalized spin-polarizabilities of the proton. RCS experiments at TUNL/HIGS[24] will utilize polarized beam and targets to measure the four spin-polarizabilities of the nucleon. Two of the four spin-polarizabilities of RCS, γ_{E1M2} and γ_{M1E2}, are related to VCS generalized polarizabilities[4] in the limit q→0.

Table 1. Response function units are GeV−2, the polarizabilities 10^{-4} fm^3, and the mean square radii fm^2. The errors are statistical and systematic, respectively.

Observable	LEX analysis	Dispersion analysis	HBChPT[4]
P_{LL}-P_{TT}/ϵ	54.5±4.8±2.0	46.7±4.9±2.0	56.9
P_{LT}		-8.9±4.2 ±0.8	-6.5
$\alpha(Q^2=.06)$		7.85±0.87±0.36	9.27
$\beta(Q^2=.06)$		2.69±1.48±0.28	1.59
$< r_a^2 >$		2.16±0.31	1.7
$< r_b^2 >$		-1.91±2.12	-2.4

Acknowledgements

The author thanks P. Bourgeois, T. Hemmert, B. Holstein, I. L'vov, B. Pasquini, and M. Vanderhaeghen for their comments and for communicating the results of their calculations. The author also thanks the OOPS collaboration and the staff of the MIT-Bates linear accelerator facility for their efforts on this experiment. This work was supported in part by D.O.E. grant DE-FG02-88ER40415.

References

1. B. Holstein, Comm. Nuc. Part. Phys. **19**, 221.
2. M. Schumacher, Prog. Part. and Nucl. Phys. **55**, 567.
3. P.A.M. Guichon *et al.*, Nucl. Phys. A **591**, 606.
4. T. R. Hemmert *et al.*, Phys. Rev. Lett. **79**, 22, and T.R. Hemmert *et al.*, Phys. Rev. D **62**, 014013.
5. J. Roche *et al.*, Phys. Rev. Lett. **85**, 708.

6. G. Laveissiere *et al.*, Phys. Rev. Lett. **93**, 122001.
7. P.A.M. Guichon, and M. Vanderhaeghen, Prog. Part. Nucl. Phys. **41**, 125.
8. S. Dolfini *et al.*, Nucl. Instum. Methods Phys. Res., Sect. A **344**, 571; J. Mandeville *et al.*, Nucl. Instrum. Methods Phys. Res., Sect. A **344**, 583; Z. Zhou *et al.*, Nucl. Instrum. Methods Phys. Res., Sect. A **487**, 365.
9. X. Jiang, Ph.D. thesis, University of Massachusetts, 1998, unpublished.
10. Fermi National Accelerator Labororatory report, NAL-64.
11. H.W. Lewis, Phys. Rev. **78**, 526.
12. M. Vanderhaeghen, *et al.*, Phys. Rev. C **62**, 025501.
13. G. Hoehler, E. Pietarinen, and I.Sabba-Stefanescu, Nuc. Phys. B **114**, 505.
14. J. Friedrich and Th. Walcher, Eur. Phys. J. **A** 17, 607 (2003).
15. B. Pasquini *et. al.*, Eur. Phys. J. A **11**, 185, and D. Drechsel, B. Pasquini, and M. Vanderhaeghen, Phys. Rep. 378, 99.
16. D. Drechsel, O. Hanstein, S. S. Kamalov, and L. Tiator, Nucl. Phys **A** 645, 145.
17. Private communication, B. Pasquini and M. Vanderhaeghen.
18. Private communication, H. Fonvieille, and also Fig. 2 of Ref. 6.
19. 19..Private communication, B. Holstein.
20. 20..V. Bernard, N. Kaiser, A. Schmidt, and Ulf-G. Meibner, Phys. Lett. **B** 319, 269, and V. Bernard, N. Kaiser, Ulf-G. Meibner, and A. Schmidt, Z. Phys. **A** 348, 317.
21. T. R. Hemmert, B. R. Holstein, G. Knochlein, and S. Scherer, Phys. Rev. D **55**, 2630.
22. S. Eidelman *et al.*, Phys. Lett. B **592**, 1.
23. Mainz experiment A1/1-00, N. D'Hose and H. Merkel spokespersons.
24. "Compton Scattering on Nucleon and Nuclear Targets", M. Ahmed *et al.*, unpublished TUNL report.

PHYSICS RESULTS FROM THE BLAST DETECTOR AT THE BATES ACCELERATOR

R.P. REDWINE

Department of Physics and Bates Linear Accelerator Center
Massachusetts Institute of Technology
77 Massachusetts Avenue
Cambridge, MA 02139, USA

We report preliminary physics results from recent runs of the BLAST detector on the South Hall Ring at the Bates Linear Accelerator Center at MIT. A highly polarized electron beam of 850 MeV was incident on internal polarized targets of hydrogen and deuterium, and the BLAST detector was used to provide data on spin observables in elastic and quasi-elastic scattering. These data were used to extract information on nucleon form factors, deuteron structure, and the nucleon-nucleon interaction.

1. Introduction

The study of the electromagnetic structure of nucleons and light nuclei has been an area of much interest, both experimentally and theoretically, in recent years. Such studies address issues in traditional models of nuclear structure as well as in QCD-inspired theories. The use of polarization observables has proven to be a very powerful tool in separating different aspects of nucleon and nuclear structure.

The BLAST experimental program reported here focused on the study of the structures of the proton, neutron, and deuteron using a polarized electron beam and polarized internal targets. In this report we present preliminary results for proton and neutron form factors and for elastic and quasi-elastic scattering from the deuteron.

2. Description of the Experiment

The BLAST experiment was designed to measure spin-dependent electron scattering from nucleons and light nuclei at intermediate energies in an environment essentially background-free. This required excellent performance by the accelerator, a highly-polarized internal target, and a large-acceptance detector capable of characterizing a variety of scattering events.

2.1. *The Bates Accelerator*

The Bates Linear Accelerator Center has the ability to provide high-quality polarized electron beams up to about 1 GeV in energy. For the BLAST experiment an 850 MeV electron beam from the Bates linac/recirculator was stored in the South Hall Ring to provide large current incident on the H/D internal target. Stored currents up to 225 mA and electron polarizations of about 65% were used. Longitudinal polarization at the position of the internal target was maintained using a Siberian Snake on the opposite side of the Ring. The lifetime of a fill in the South Hall Ring was typically 20 minutes. When the beam current dropped below a certain level, the beam was dumped and the Ring refilled from the linac/recirculator, a process which took about 2 minutes.

2.2. *The Polarized H/D Target*

The polarized Hydrogen/Deuterium target consisted of an Atomic Beam Source (ABS) providing a flow of 2.2×10^{16} atoms/s to a target cell 1.5 cm in diameter and 60 cm in length.[1] This resulted in a typical luminosity of 6×10^{31} cm^{-2}s^{-1}. The ABS provided proton and deuteron polarizations in the range of 70–80%. The target polarization state was changed about every 5 minutes to reduce systematic effects.

Because the target cell had no entrance and exit windows, large pumping capacity just upstream and downstream of the target was necessary to keep the overall South Hall Ring gas pressure sufficiently low. Such a windowless target assured that the events detected were extremely clean in terms of background from sources other than the target.

2.3. *The BLAST Detector*

A schematic diagram of the BLAST detector is shown in Figure 1. BLAST was designed to be a symmetric, large-acceptance, general purpose detector for electrons, pions, protons, neutrons, and deuterons.[2] The heart of the spectrometer was an 8-sector toroidal magnet with maximum magnetic field of 0.4 T. As shown in the figure, the two sectors to the left and right of the beam were instrumented with detectors designed to identify emerging particles and to measure their momenta and/or energies. Moving radially from

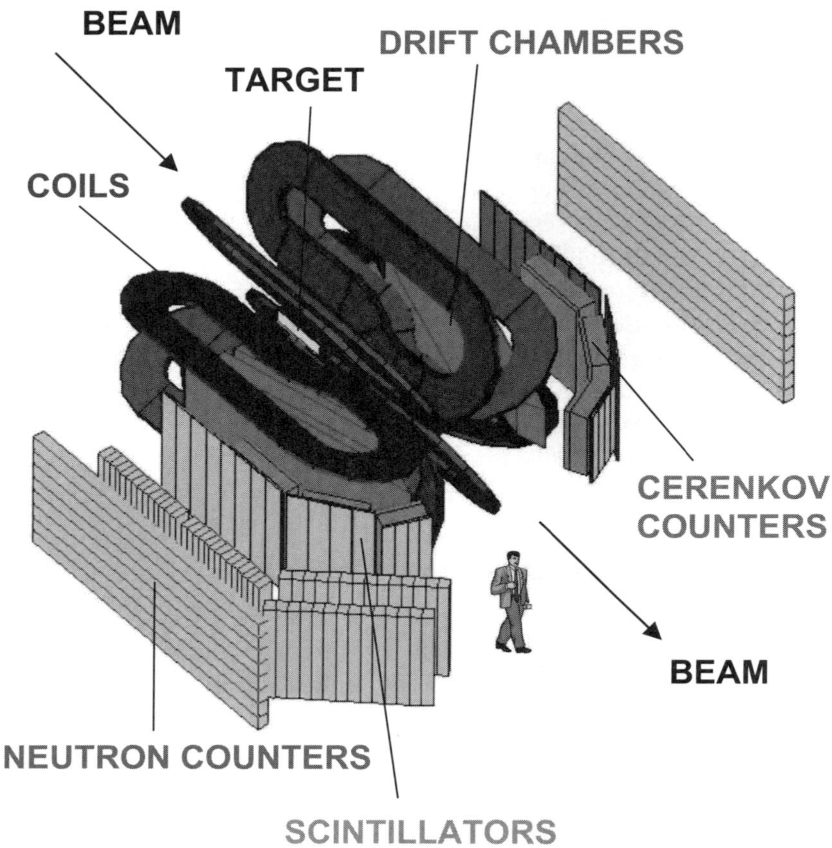

Fig. 1. Schematic view of the BLAST detector, showing the main components.

the beam axis, the BLAST components consisted of drift chambers (for charged particle trajectory measurement), Čerenkov counters (for distinguishing electrons from pions), thin plastic scintillators (for event triggers and time-of-flight measurement), and thick plastic scintillators (for neutron detection).

Overall, the relatively large acceptance of BLAST allowed detection of electron scattering events with Q^2 between 0.1 and 0.8 $(\mathrm{GeV/c})^2$, polar angles between 20° and 80°, and azimuthal angles of $\pm 15\,°$.

A critical feature of BLAST and its internal target was the ability to orient the target spin angle so as to optimize the physics addressed by BLAST. In particular, the in-plane spin angle was generally adjusted so that the left sector detected events in which the direction of the virtual photon was roughly perpendicular to the target spin angle and the right sector detected events in which the direction of the virtual photon was roughly parallel to the target spin angle.

2.4. *Data Collection and Analysis*

In year 2004 a total of 90 pb^{-1} integrated luminosity was delivered to a polarized hydrogen target and 169 pb^{-1} delivered to a polarized deuterium target. In year 2005 a total of 236 pb^{-1} integrated luminosity was delivered to a polarized deuterium target. In all, more than 10^6 Coulombs of electron beam passed through the internal target during the experiment.

Events were classified according to the charge and identities of emerging particles. For elastic and quasi-elastic events, requiring co-planarity and appropriate times-of-flight yielded $(e, e'p)$, $(e, e'd)$, and $(e, e'n)$ events which were essentially background-free.

3. Nucleon Elastic Form Factors

As indicated in the introduction, a major focus of the BLAST physics program was to provide precise information on nucleon elastic form factors. We present here preliminary results from BLAST.

3.1. *Elastic Hydrogen Data*

Our $^1\mathbf{H}(\mathbf{e}, e'p)$ data are an example of a double polarization experiment in elastic e–p scattering; in our case a polarized electron beam scattered from a polarized proton target and the resulting polarized cross section was

$$\sigma = \sigma_0(1 + P_e P_t A), \tag{1}$$

where A is the double spin asymmetry.

Because BLAST could simultaneously measure the double spin asymmetry with the target polarization perpendicular and parallel to the direction of q, we can form the asymmetry ratio $A_\perp/A_\parallel$, which is proportional to G_E/G_M and which is independent of polarization or analyzing power. The measured asymmetries are shown in Figure 2 as a function of Q^2.

Also shown are results for $\mu_n G_E^p/G_M^p$ along with previous measurements of this quantity and predictions of several models. The BLAST ratios can

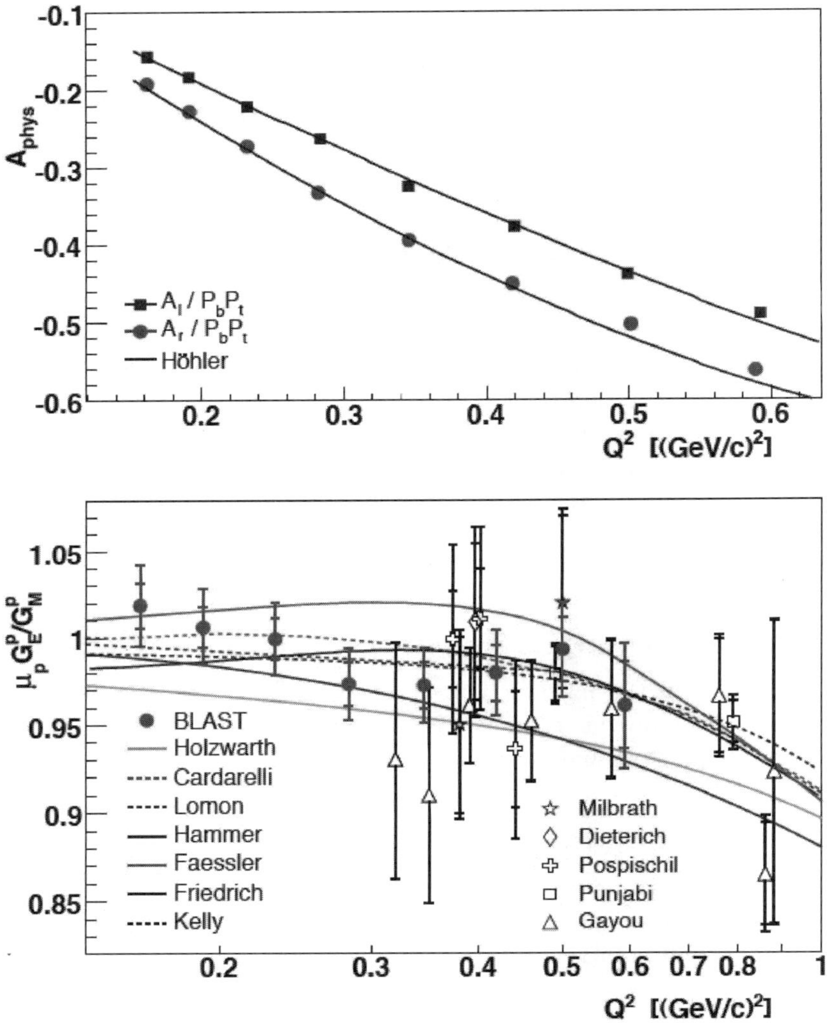

Fig. 2. Upper panel: Measured spin-dependent asymmetries[3] compared to expected asymmetries based on a parameterization[4] for the nucleon form factors. Lower panel: Results[3] for the ratio of electric to magnetic form factors shown with data from other polarization experiments[4–9] and several models.[10–16]

be used, along with previous differential cross section data on e–p elastic scattering, to yield results for the electric and magnetic form factors separately. These data are shown in Figure 3. It is clear that inclusion of

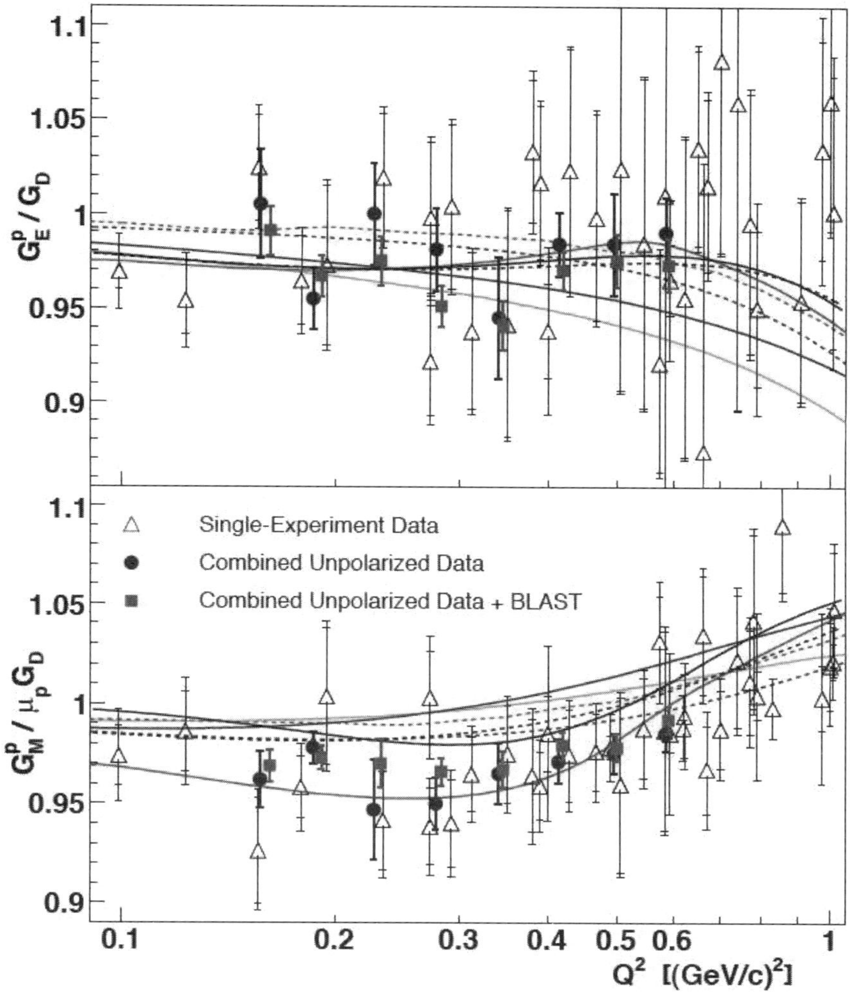

Fig. 3. Compilation of data on the proton electric and magnetic form factors at BLAST kinematics with (red) and without (blue) BLAST input, shown with total uncertainties. The results from single experiment L-T separations[17–24] are shown with open symbols. The curves have the same meaning as in Figure 2.

the BLAST results significantly reduces the uncertainties in extracting the proton electromagnetic form factors.

The extracted form factor ratio in Figure 2 is consistent with unity. However, the separated form factors suggest a deviation from the dipole form factor around $Q^2 = 0.3$ to 0.4 $(\mathrm{GeV/c})^2$, which may be an indication of a manifestation of the pion cloud at low momentum transfer.[10,11]

3.2. *Neutron Electric Form Factor*

The neutron electric form factor can be extracted from the quasi-elastic ${}^2\mathbf{H}(\mathbf{e}, e'n)$ reaction on the deuteron. The spin-perpendicular beam-target vector asymmetry A_{ed}^V shows high sensitivity to G_E^n when the target spin direction is perpendicular to the direction of q. Using this technique we extracted the preliminary BLAST results shown in Figure 4; this Figure

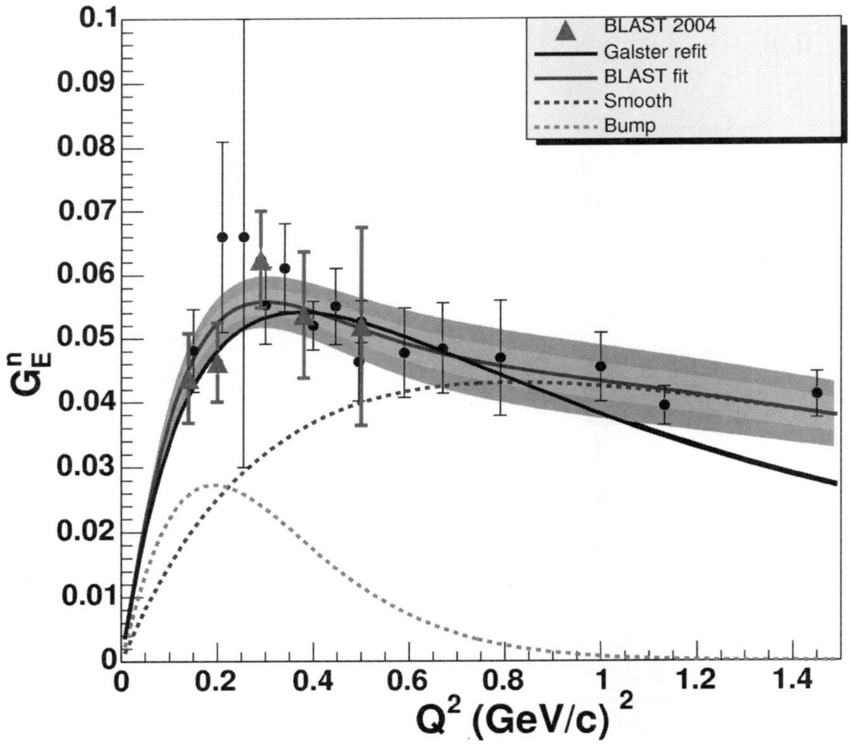

Fig. 4. Preliminary BLAST results[25] for the charge form factor of the neutron, along with previous results[26] and some fits[25,27] to the data.

also contains previous data as well as fits to the data. The BLAST data are consistent with other recent results in this Q^2 region. The BLAST data presented in Figure 4 come from only a subset of all collected data, so the final data analysis should reduce the uncertainties considerably.

3.3. *Neutron Magnetic Form Factor*

The sensitivity of the BLAST experiment to the neutron magnetic form factor comes from measurement of the ratio

$$\frac{A_\perp}{A_\parallel} \approx \frac{\kappa\frac{G_E^p}{G_M^p}}{1+\left(\frac{G_M^n}{G_M^p}\right)^2}, \tag{2}$$

where $A_\perp$ ($A_\parallel$) is the beam-target vector asymmetry for inclusive quasi-elastic scattering from deuteron and the target polarization is roughly perpendicular (parallel) to the direction of q. This ratio is independent of the exact polarization of the beam and target. There is some model dependence in the extraction of the neutron magnetic form factor, but this model dependence is small.

Figure 5 shows the preliminary BLAST results for the neutron magnetic form factor, along with previously reported data and several calculations.

4. Elastic Electron-Deuteron Scattering

As described in section 2, with BLAST we were able to measure elastic e–d events from the polarized deuterium target essentially background free. This allowed the study of the elastic form factors of the deuteron in our Q^2 region with unprecedented precision. Measurement of the unpolarized elastic electron-deuteron scattering cross section yields two form factors which are related to the charge form factor G_C^d, the magnetic form factor G_M^d, and the quadrupole form factor G_Q^d.

$$\sigma_0 = \sigma_{Mott}\left(A + B\tan^2(\theta_e/2)\right) \tag{3}$$

$$A(Q^2) = {G_C^d}^2 + \frac{8}{9}\eta^2 {G_Q^d}^2 + \frac{2}{3}\eta {G_M^d}^2 \tag{4}$$

$$B(Q^2) = \frac{4}{3}\eta(1+\eta){G_M^d}^2; \quad \eta = Q^2/(4M_D^2) \tag{5}$$

At least one other variable must be measured to allow a complete separation of the three elastic form factors. As described by R. Fatemi in another session of this conference, the BLAST experiment yielded measurements of T_{20}, T_{21}, and T_{22}, with the T_{20} results being most precise. Figure 6 shows the preliminary results for T_{20}, with previous data and relevant calculations. The precise BLAST data greatly constrain models for T_{20} through

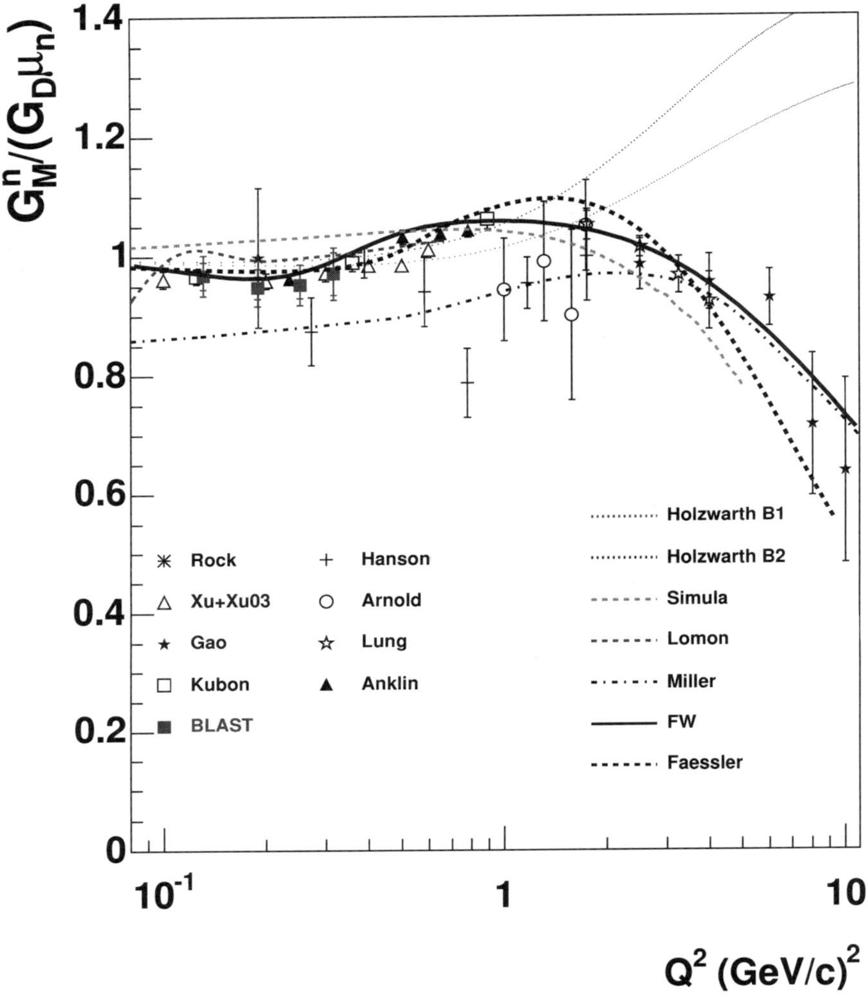

Fig. 5. Preliminary BLAST results[28] for the magnetic form factor of the neutron, along with previous data[29–34] and representative calculations.[10–12,35]

the region of the minimum at $Q = 3.3$ fm^{-1}. With these data the minimum in the charge form factor is determined to occur near $Q = 4.2$ fm^{-1}.

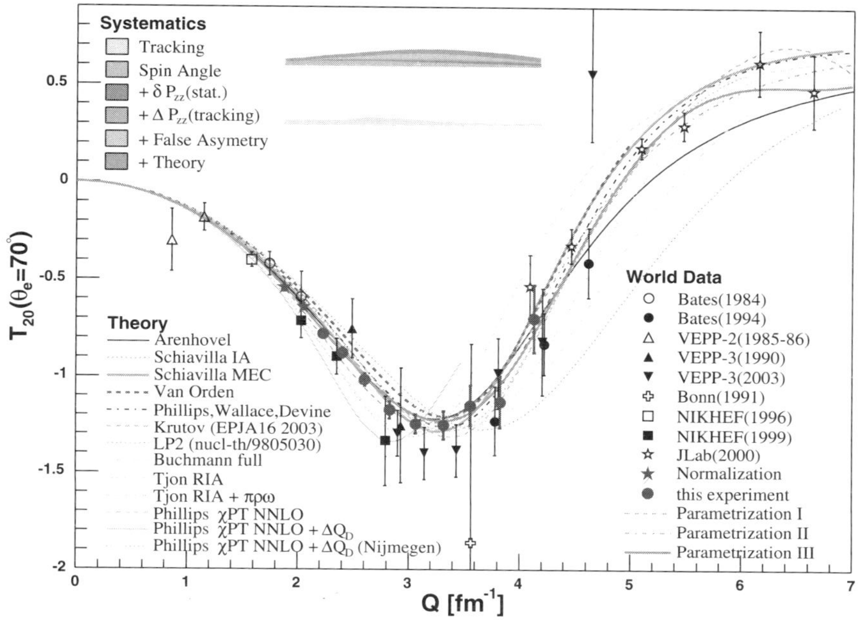

Fig. 6. Preliminary BLAST results[36] for the tensor polarization T20 in elastic electron-deuteron scattering. Previous data[37] are also shown, along with results of several calculations.[37]

5. Quasi-Elastic Electron Scattering from the Deuteron

The BLAST experiment also has taken extensive data[38] on quasi-elastic scattering from the deuteron. Such measurements from vector-polarized deuterium provide additional, and complementary, information about nucleon-nucleon interactions, especially at high missing momentum. These data are sensitive to several important effects, such as contributions of the D-state, final state interactions, meson exchange currents, isobar contributions, and relativistic effects. In general the agreement between our data and predictions from a Plane Wave Born Approximation model[39] which incorporates these effects is excellent. The talk by R. Fatemi in another session of this conferences describes these data in more detail.

6. Conclusions

In summary, the BLAST experiment is yielding a variety of proton, neutron, and deuteron spin observables in the Q^2 range 0.1–0.8 $(\mathrm{GeV/c})^2$. These observables are measured with high precision and with small systematic uncertainties.

We are providing a consistent and precise determination of elastic nucleon form factors at low momentum transfer. We see possible evidence for structure in the form factors at low Q^2 .

Precise measurement of the tensor analyzing powers of the deuteron allows us to separate the charge and quadrupole deuteron form factors, yielding a very accurate determination of the minimum in the charge form factor. We have also made a first measurement of T_{11}^e.

Measurement of polarization observables in quasi-elastic electron scattering from the deuteron has yielded results which allow direct comparison with models which include important exchange effects in the nucleon-nucleon potential. The comparisons show in general good agreement with the models.

All of the BLAST data are in the process of publication or are in final analysis.

Acknowledgments

The author thanks his colleagues in the BLAST collaboration for their hard work and their effective and enjoyable collaboration. The collaboration thanks the staff at the MIT-Bates Linear Accelerator Center for the delivery of high-quality electron beam and for their technical support. This work was supported in part by the U. S. Department of Energy and by the U. S. National Science Foundation.

References

1. D. Cheever *et al.*, *Nucl. Instr. Meth.* **A556**, 410 (2006); L. D. van Buuren *et al.*, *Nucl. Instr. Meth.* **A474**, 209 (2001).
2. D. Hasell *et al.*, The BLAST Experiment, in preparation.
3. C.B. Crawford, Ph.D. thesis, Massachusetts Institute of Technology (2005); C.B. Crawford *et al.*, *Phys. Rev. Lett.* 98, 052301 (2007).
4. G. Hoehler *et al.*, *Nucl. Phys.* **B114**, 505 (1976).
5. M. Jones *et al.*, *Phys. Rev. Lett.* **84**, 1398 (2000); V. Punjabi *et al.*, *Phys. Rev.* **C71**, 055202 (2005); Erratum-ibid. *Phys Rev.* **C71** 069902(E) (2005).
6. B. Milbrath *et al.*, *Phys Rev. Lett.* **80**, 452 (1998); Erratum-ibid. *Phys. Rev. Lett.* **82**, 2221(E) (1999).
7. S. Dieterich *et al.*, *Phys. Lett.* **B500**, 47 (2001).
8. T. Pospischil *et al.*, *Eur. Phys. J.* **A12**, 125 (2001).
9. O. Gayou *et al.*, *Phys. Rev.* **C64**, 038202 (2001).
10. A. Faessler, T. Gutsche, V.E. Lyubovitskij, and K. Pumsaard, *Phys. Rev.* **D73**, 114021 (2006).

11. J. Friedrich and T. Walcher, *Eur. Phys. J.* **A17**, 607 (2003).
12. G. Holzwarth, *Z. Phys.* **A356**, 339 (1996).
13. F. Cardarelli and S. Simula, *Phys. Rev.* **C62**, 065201 (2000); S. Simula, Proc. Of Workshop on the Physics of Excited Nucleons, World Scientific, 2001, p. 135.
14. E. L. Lomon, *Phys. Rev.* **C66**, 045501 (2002).
15. H.-W. Hammer and Ulf-G. Meissner, *Eur. Phys. J.* **A20**, 469 (2004).
16. J.J. Kelly, *Phys. Rev.* **C70**, 068202 (2004).
17. R. C. Walker *et al.*, *Phys. Rev.* **D49**, 5671 (1994).
18. M. E. Christy *et al.*, *Phys. Rev.* **C70**, 015206 (2004).
19. T. Janssens *et al.*, *Phys. Rev.* **142**, 922 (1996).
20. M. Goitein *et al.*, *Phys. Rev.* **D1**, 2449 (1970); L.E. Price *et al.*, *Phys. Rev.* **D4**, 45 (1971).
21. C. Berger *et al.*, *Phys. Lett.* **B35**, 87 (1971).
22. W. Bartel *et al.*, *Nucl. Phys.* **B58**, 429 (1973).
23. F. Borkowski *et al.*, *Nucl. Phys.* **A222**, 269 (1974); *Nucl. Phys.* **B93**, 461 (1975).
24. P.E. Bosted *et al.*, *Phys. Rev.* **C42**, 38 (1990).
25. V. Ziskin, Ph.D. thesis, Massachusetts Institute of Technology (2005).
26. D.I. Glazier *et al.*, *Eur. Phys. J.* **A24**, 101 (2005), and references therein.
27. S. Galster *et al.*, *Nucl. Phys.* **B32**, 221 (1971).
28. N. Meitanis, Ph.D. thesis, Massachusetts Institute of Technology (2006).
29. S. Rock *et al.*, *Phys. Rev. Lett.* **49,** 1139 (1982).
30. R. G. Arnold *et al.*, *Phys. Rev. Lett.* **61**, 806 (1988).
31. A. Lung *et al.*, *Phys. Lett.* **B336**, 313 (1994).
32. G. Kubon *et al.*, *Phys. Lett.* **B524**, 26 (2002).
33. H. Gao *et al.*, *Phys. Rev.* **C50** R546 (1994).
34. W. Xu *et al.*, *Phys. Rev. Lett.* **85**, 2900 (2000); W. Xu *et al.*, *Phys. Rev.* **C67**, R012201 (2003).
35. G.A. Miller, *Phys. Rev.* **C66**, 032201 (2002).
36. C. Zhang, Ph.D. thesis, Massachusetts Institute of Technology (2006).
37. See ref. 36 and references contained therein.
38. A. Maschinot, Ph.D. thesis, Massachusetts Institute of Technology (2005).
39. H. Arenhoevel, W. Leidemann, and E.L. Tomusiak, *Phys. Rev.* **C46**, 455 (1992).

THE πNN SYSTEM — RECENT PROGRESS

C. HANHART

IKP (Theorie), Forschungszentrum Jülich,
Jülich, D–52425, Germany
E-mail: c.hanhart@fz-juelich.de
www.fz-juelich.de/ikp/theorie/ikpth_en.shtml

Recent progress towards an understanding of the πNN system within chiral perturbation theory is reported. The focus lies on an effective field theory calculation and its comparison to phenomenological calculations for the reaction $NN \to d\pi$. In addition, the resulting absorptive and dispersive corrections to the πd scattering length are discussed briefly.

Keywords: πd scattering, pion production

1. Introduction

Pion reactions on few–nucleon systems provide access to various physics phenomena: deuterons can be used as effective neutron targets, null–experiments for isospin violation can be designed, and they are an important test of our understanding of the nuclear structure. What is therefore necessary is a controlled theoretical framework — a proper effective field theory needs to be constructed.

A first step in this direction was taken by Weinberg already in 1992.[1] He suggested that all that needs to be done is to convolute transition operators, calculated perturbatively in standard chiral perturbation theory (ChPT), with proper nuclear wave functions to account for the non–perturbative character of the few–nucleon systems. This procedure combines the distorted wave born approximation, used routinely in phenomenological calculations, with a systematic power counting for the production operators. Within ChPT this idea was already applied to a large number of reactions like $\pi d \to \pi d$,[2] $\gamma d \to \pi^0 d$,[3,4] $\pi^3\mathrm{He} \to \pi^3\mathrm{He}$,[5] $\pi^- d \to \gamma nn$,[6] and $\gamma d \to \pi^+ nn$,[7] where only the most recent references are given.

The central concept to be used in the construction of the transition operators is that of reducibility, for it allows one to disentangle effects of the wave functions and those from the transition operators. As long as the operators are energy independent, the scheme can be applied straight forwardly,[8] however, as we will see below, for energy dependent interactions more care is necessary.

Using standard ChPT especially means to treat the nucleon as a heavy field. Corrections due to the finite nucleon mass, M_N, appear as contact interactions on the lagrangian level that are necessarily analytic in M_N. However, some pion–few-nucleon diagrams employ few–body singularities that lead to contributions non–analytic in m_π/M_N, with m_π for the pion mass. In Ref.[9] it is explained how to deal with those.

A problem was observed when the original scheme by Weinberg was applied to the reactions $NN \to NN\pi$:[10,11] Potentially higher order corrections turned out to be large and lead to even larger disagreement between theory and experiment. For the reaction $pp \to pp\pi^0$ one loop diagrams that in the Weinberg counting appear only at NNLO where evaluated[12,13] and they turned out to give even larger corrections putting into question the convergence of the whole series. However, already quite early the authors of Refs.[14,15] stressed that an additional new scale enters, when looking at reactions of the type $NN \to NN\pi$, that needs to be accounted for in the power counting. Since the two nucleons in the initial state need to have sufficiently high kinetic energy to put the pion in the final state on–shell, the initial momentum needs to be larger than

$$p_{thr} = \sqrt{M_N m_\pi} \ .$$

The proper way to include this scale was presented in Ref.[16] and implemented in Ref.[17] — for a recent review see Ref.[18] As a result, pion p-waves are given by tree level diagrams up to NNLO in the modified power counting and the corresponding calculations showed satisfying agreement with the data.[16] However, for pions s–waves loops appear already at NLO. In the next section we will discuss their effect on the reaction $NN \to d\pi$ near threshold. In some detail we will compare the effective field theory result to that of phenomenological calculations.

Since the Delta–nucleon mass difference, Δ, is numerically of the order of p_{thr}, also the Delta–isobar should be taken into account explicitly as a dynamical degree of freedom.[14] We will use a scheme where

$$\Delta \sim p_{thr} \ .$$

Once the reaction $NN \to d\pi$ is understood within effective field theory one is in the position to also calculate the so–called dispersive and

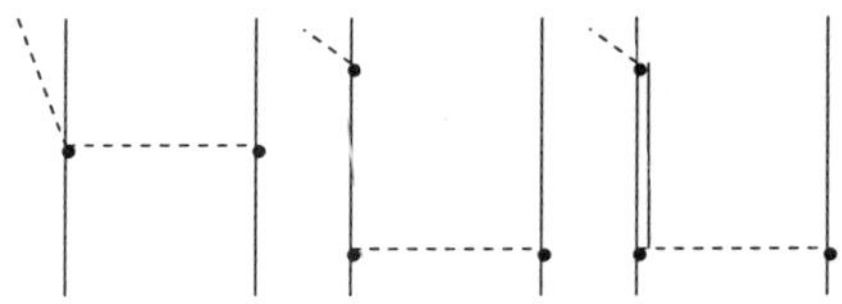

Fig. 1. Tree level diagrams that contribute to $pp \to d\pi^+$ s–waves up to NLO. Solid lines denote nucleons, dashed ones pions and the double line the propagation of a Delta–isobar.

absorptive corrections to the πd scattering length. This calculation will be presented in section 3.

We close with a brief summary and outlook.

2. $NN \to d\pi$

The tree level amplitudes that contribute to $pp \to d\pi^+$ are shown in Fig. 1. In Ref.[17] all NLO contributions of loops that start to contribute[a] to $NN \to NN\pi$ at NLO were calculated in threshold kinematics — that is neglecting the distortions from the NN final– and initial state interaction and putting all final states at rest. At threshold only two amplitudes contribute, namely the one with the nucleon pair in the final and initial state in isospin 1 (measured, e.g., in $pp \to pp\pi^0$) and the one where the total NN isospin is changed from 1 to 0 (measured, e.g., in $pp \to d\pi^+$)[b]. It was found that the sum of all loops that contain Delta–excitations vanish in both channels. This was understood, since the loops were divergent and at NLO no counter term is allowed by chiral symmetry. On the other hand the nucleonic loops were individually finite. It was found that the sum of all nucleonic loops that contribute to $pp \to pp\pi^0$ vanish, whereas the sum of those that contribute to $pp \to d\pi^+$ gives a finite answer. The resulting amplitude grows linear with the initial momentum.

At that time it appeared as a puzzle why the loops vanished for the reaction $pp \to pp\pi^0$ — no obvious symmetry reason could be identified. However, in Ref.[19] it was pointed out that the linear growth of the ampli-

[a]In a scheme with two expansion parameters — here m_π and p_{thr} — loops no longer contribute at a single order but, in addition, to all orders higher than where they start to contribute.

[b]The third independent amplitude, where the NN isospin is changed from 0 to 1 in the production process and that can be extracted from $pn \to pp\pi^-$, vanishes at threshold as a consequence of selection rules.

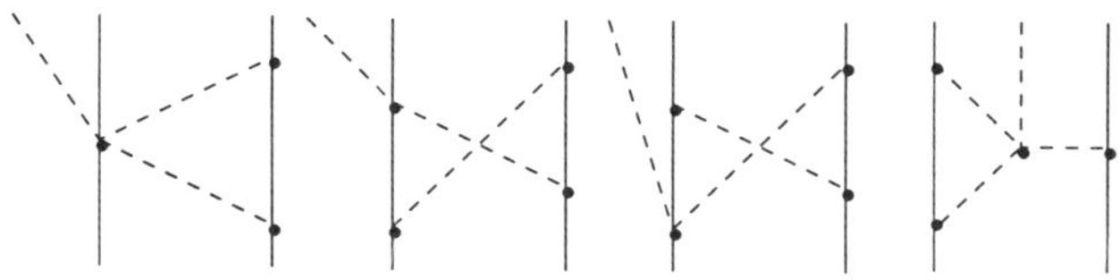

Fig. 2. Irreducible pion loops with nucleons only that start to contribute to $NN \to NN\pi$ at NLO that were considered in Ref.[17]

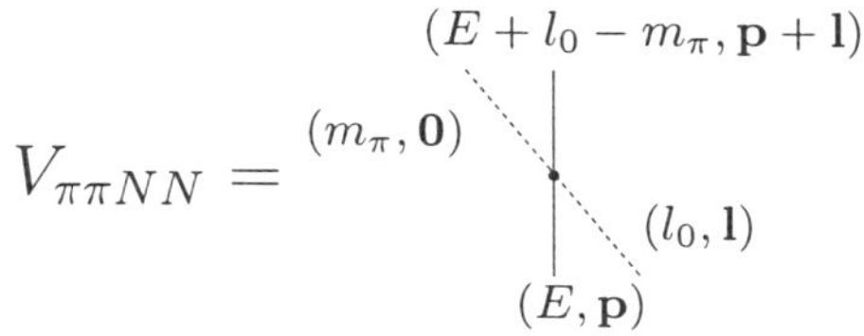

Fig. 3. The $\pi N \to \pi N$ transition vertex: definition of kinematic variables as used in the text.

tude for the charged pion production is the problematic one: when evaluated for finite outgoing NN momenta, the transition amplitudes turned out to scale as the momentum transfer. The amplitudes then grow linearly with the external NN momenta. As a consequence, once convoluted with the NN wavefunctions, a large sensitivity was found, in conflict with general requirements from field theory. In light of these insights it was acknowledged that the loops for $pp \to d\pi^+$ where the ones not understood. The solution to this puzzle was presented in Ref.[20] and will be reported now.

The observation central to the analysis is that the leading $\pi N \to \pi N$ transition vertex, as it appears in Fig. 1a, is energy dependent. Using the notation of Fig. 3 its momentum and energy dependent part may be written as[c]

$$\begin{aligned} V_{\pi\pi NN} &= l_0 + m_\pi - \frac{\mathbf{l}\cdot(2\mathbf{p}+\mathbf{l})}{2M_N} \\ &= \underbrace{2m_\pi}_{\text{on-shell}} + \underbrace{\left(l_0 - m_\pi + E - \frac{(\mathbf{l}+\mathbf{p})^2}{2M_N}\right)}_{(E'-H_0)=(S')^{-1}} - \underbrace{\left(E - \frac{\mathbf{p}^2}{2M_N}\right)}_{(E-H_0)=S^{-1}} . \end{aligned} \quad (1)$$

For simplicity we skipped the isospin part of the amplitude. The first term

[c]The expressions for the vertices can be found in Ref.[21] Note that the $\pi N \to \pi N$ vertex from $\mathcal{L}^{(1)}_{\pi N}$ as well as its recoil correction from $\mathcal{L}^{(2)}_{\pi N}$ are to be used already at leading order as a consequence of the modified power counting.

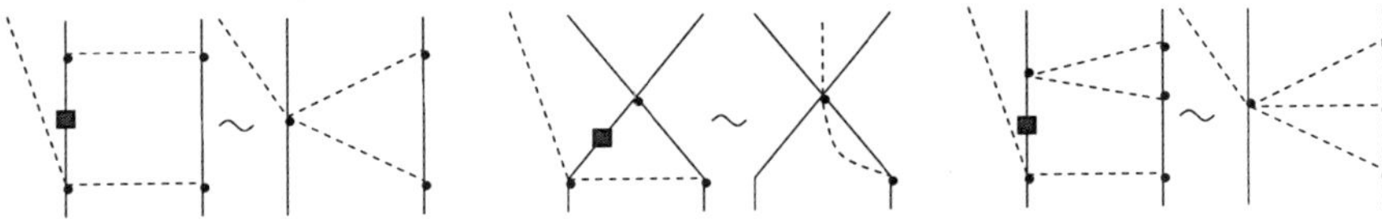

a) b) c)

Fig. 4. Induced irreducible topologies, when the off–shell terms of Eq. (1) hit the NN potential in the final state. The filled box on the nucleon line denotes the propagator canceled by the off–shell part of the vertex.

in the last line denotes the transition in on–shell kinematics, the second the inverse of the outgoing nucleon propagator and third the inverse of the incoming nucleon propagator. First of all we observe that for on–shell incoming and outgoing nucleons, the $\pi N \to \pi N$ transition vertex takes its on–shell value $2m_\pi$ — even if the incoming pion is off–shell, as it is for diagram *a* of Fig. 1. This is in contrast to standard phenomenological treatments,[22] where l_0 was identified with $m_\pi/2$ — the energy transfer in on–shell kinematics — and the recoil terms were not considered. Note, since $p_{thr}^2/M_N = m_\pi$ the recoil terms are to be kept.

The second consequence of Eq. (1) is even more interesting: when the $\pi N \to \pi N$ vertex gets convoluted with NN wave functions, only the first term leads to a reducible diagram. The second and third term, however, lead to irreducible contributions, since one of the nucleon propagators gets canceled. This is illustrated in Fig. 4, where those induced topologies are shown that appear when one of the nucleon propagators is canceled (marked by the filled box) in the convolution of typical diagrams of the NN potential with the $NN \to NN\pi$ transition operator. Power counting gives that diagrams *b* and *c* appear only at order N^{4}LO and N^{3}LO, respectively. However, diagram *a* starts to contribute at NLO and it was found in Ref.[20] that those induced irreducible contributions cancel the finite remainder of the NLO loops in the $pp \to d\pi^+$ channel. Thus, up to NLO only the diagrams of Fig. 1 contribute to $pp \to d\pi^+$, with the rule that the $\pi N \to \pi N$ vertex is put on–shell.

The result found in Ref.[20] is shown in Fig. 5, where the total cross section (divided by the energy dependence of phase space) is plotted against the normalized pion momentum. The dashed line is the result of the model by Koltun and Reitan,[22] as described above, whereas the solid line shows the result of the ChPT calculation of Ref.[20] .

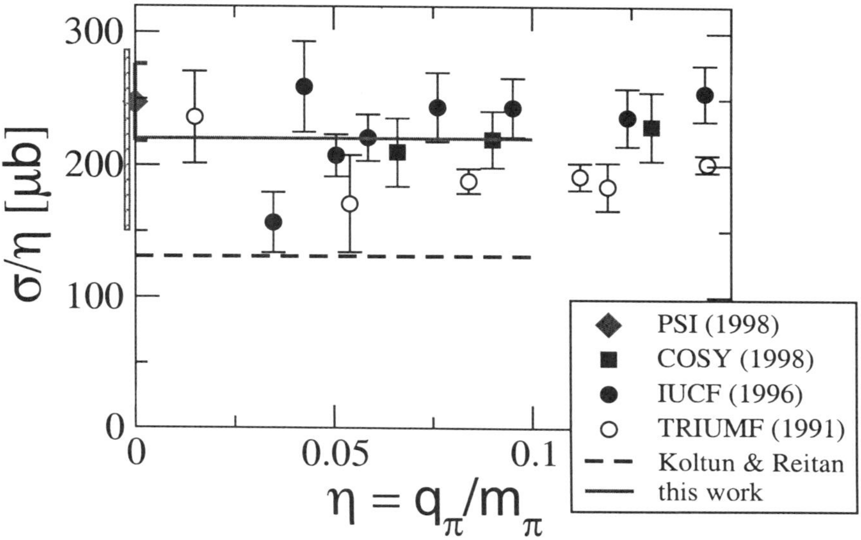

Fig. 5. Comparison of our results to experimental data for $NN \to d\pi$. The dashed line corresponds to the model of Koltun and Reitan,[22] whereas the solid line is the result of the ChPT calculation of Ref.[20] The estimated theoretical uncertainty (see text) is illustrated by the narrow box. The data are from Refs.[25] (open circles),[26] (filled circles) and[27] (filled squares). The first data set shows twice the cross section for $pn \to d\pi^0$ and the other two the cross section for $pp \to d\pi^+$.

3. Comparison to phenomenological works

Phenomenological calculations for the reaction $pp \to d\pi^+$ in near threshold kinematics are given, e.g., in Ref.[23] and Ref.[24] . In both works in addition to the diagrams of Ref.[22] some Delta–loops as well as short range contributions are included — heavy meson exchanges for the former and off–shell πN scattering[d] for the latter. Based on this the cross section for $pp \to d\pi^+$ is now even overestimated near threshold. How can we interpret this discrepancy in light of the discussion above?

First of all, the NLO parts of the Delta–loops cancel, as was shown already in Ref.[17] . However, in both Refs.[23,24] only one of these diagrams was included and, especially for Ref.,[23] gave a significant contribution. The only diagram of those NLO loops shown in Fig. 2 that is effectively included in Ref.[24] is the fourth, since the pion loop there can be regarded as part of the $\pi N \to \pi N$ transition T–matrix. However, as described, the contri-

[d]That those are also short range contributions is discussed in Ref.[18]

bution of this diagram gets canceled by the others shown in Fig. 2 and the induced irreducible pieces described above. Therefore, the physics that enhances the cross section compared to the work of Ref.[22] in Refs.[23,24] is completely different to that of Ref.[20] — the phenomenological calculations miss the essential contribution and are in conflict with both field theoretic consistency and chiral symmetry.

What are the observable consequences of the difference between the ChPT calculation and the phenomenological ones? As explained, in the former the near threshold cross section for $pp \to d\pi^+$ is basically given by a long–ranged pion exchange diagram, whereas the latter rely on short ranged operators with respect to the NN system. Obviously those observables are sensitive to this difference that get prominent contributions from higher partial waves in the final NN system. We therefore need to look at the reaction $pp \to pn\pi^+$. Unfortunately, the total cross section for this reaction is largely saturated by NN S–waves in the final state (see, e.g., Fig. 17 in Ref.[18]). On the other hand, linear combinations of double polarization observables allow one to remove the prominent components and the sub-leading amplitudes should be visible. We therefore expect from the above considerations that the phenomenological calculations give good results for polarization observables for $pp \to d\pi^+$, whereas there should be deviations for some of those for $pp \to pn\pi^+$. Predictions for these observables were presented in Ref.[28] and indeed the π^+ observables with the deuteron in the final state are described well whereas there are discrepancies for the pn final state (see Fig. 24 of Ref.[18]).

It remains to be seen how well the same data can be described in the effective field theory framework. Up to NNLO the number of counter terms is quite low: there are two counter terms for pion s–waves, that can be arranged to contribute to $pp \to pp\pi^0$ and $pp \to d\pi^+$ individually, and then there is one counter term for pion p–waves, that contributes only to a small amplitude in charged pion production.[16] On the other hand there is a huge amount of even double polarized data available[29–31] — and there is more to come especially for $pn \to pp\pi^-$.[32]

4. Corrections to $a_{\pi d}$

The πd scattering length is known to a high accuracy from measurements on pionic deuterium[33]

$$a_{\pi d}^{\text{exp}} = (-26.1 \pm 0.5 + i(6.3 \pm 0.7)) \times 10^{-3}\, m_\pi^{-1} \,, \tag{2}$$

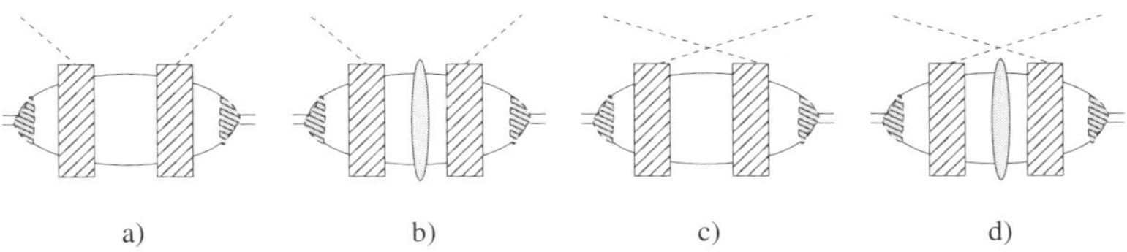

Fig. 6. Dispersive corrections to the πd scattering length.

where m_π denotes the mass of the charged pion. In the near future a new measurement with a projected total uncertainty of 0.5% for the real part and 4% for the imaginary part of the scattering length will be performed at PSI.[34] What is striking with this result is the quite large imaginary part that may be written as

$$4\pi \mathrm{Im}(a_{\pi d}) = \lim_{q\to 0} q \left\{\sigma(\pi d \to NN) + \sigma(\pi d \to \gamma NN)\right\} \ , \tag{3}$$

where q denotes the relative momentum of the initial πd pair. The ratio $R = \lim_{q\to 0} \left(\sigma(\pi d \to NN)/\sigma(\pi d \to \gamma NN)\right)$ was measured to be 2.83 ± 0.04.[35] At low energies diagrams that lead to a sizable imaginary part of some amplitude are expected to also contribute significantly to its real part. Those contributions are called dispersive corrections. As a first estimate Brückner speculated that the real and imaginary parts of these contributions should be of the same order of magnitude.[36] This expectation was confirmed within Faddeev calculations in Refs.[37] Given the high accuracy of the measurement and the size of the imaginary part of the scattering length, another critical look at this result is called for as already stressed in Refs.[38,39] . A consistent calculation is only possible within a well defined effective field theory — the first calculation of this kind was presented in Ref.[40] and is briefly sketched here.

To identify the diagrams that are to contribute we first need to specify what we mean by a dispersive correction. We define dispersive corrections as contributions from diagrams with an intermediate state that contains only nucleons, photons and at most real pions. Therefore, all the diagrams shown in Fig. 6 are included in our work. On the other hand, all diagrams that, e.g., have Delta excitations in the intermediate state do not qualify as dispersive corrections, although they might give significant contributions.[41]

The hatched blocks in the diagrams of Fig. 6 refer to the relevant transition operators for the reaction $NN \to NN\pi$ depicted in Fig. 1. Also in the kinematics of relevance here the $\pi N \to \pi N$ transitions are to be taken with their on–shell value $2m_\pi$. Using the CD–Bonn potential[42] for the NN

distortions we found for the dispersive correction from the purely hadronic transition

$$a_{\pi d}^{disp} = (-6.3 + 2 + 3.1 - 0.4) \times 10^{-3}\, m_\pi^{-1} = -1.6 \times 10^{-3}\, m_\pi^{-1} \; , \qquad (4)$$

where the numbers in the first bracket are the individual results for the diagrams shown in Fig. 6, in order. There are two points important to stress, first of all the inclusion of the intermediate NN interaction is necessary (and required based on power–counting) and the crossed diagrams (diagram c and d) give a numerically significant contribution. The latter finding might come as a surprise at first glance, however, please recall that in the chiral limit all four diagrams of Fig. 6 are kinematically identical and chiral perturbation theory is a systematic expansion around exactly this point. Thus, as a result we find that the dispersive corrections to the πd scattering length are of the order of 6 % of the real part of the scattering length. Note that the same calculation gave very nice agreement for the corresponding imaginary part.[40]

In Ref.[40] also the electro–magnetic contribution to the dispersive correction was calculated. It turned out that the contribution to the real part was tiny — $-0.1 \times 10^{-3}\, m_\pi^{-1}$ — while the sizable experimental value for the imaginary part (c.f. Eqs. (2) and 3) was described well.

To get a reliable estimate of the uncertainty of the calculation just presented a NNLO calculation is necessary. At that order two counter terms appear for pions at rest that can be fixed from $NN \to NN\pi$, as indicated above. For now we need to do a conservative estimate for the uncertainty by using the uncertainty of order $2\,m_\pi/M_N$ that one has for, e.g., the sum of all direct diagrams to derive a $\Delta a_{\pi d}^{disp}$ of around $1.4 \times 10^{-3}\, m_\pi^{-1}$, which corresponds to about 6% of $\mathrm{Re}\left(a_{\pi d}^{\mathrm{exp}}\right)$. However, given that the operators that contribute to both direct and crossed diagrams are almost the same and that part of the mentioned cancellations is a direct consequence of kinematics, this number for $\Delta a_{\pi d}^{disp}$ is probably too large.

In Ref.[40] a detailed comparison to previous works is given. Differences in the values found for the dispersive corrections were traced to the incomplete sets of diagrams included.

5. Summary and Outlook

The process $NN \to NN\pi$ has been a puzzle for more than a decade. Given the progress presented above we now have reason to believe that this puzzle will be solved soon. These results could only be obtained because a consistent effective field theory was used. For example, the potential problem with the transition operators of Ref.,[17] pointed at in Ref.,[19] would always be hidden in phenomenological calculations, since the form factors routinely used there always lead to finite, well behaved amplitudes. The very large number of observables available for the reactions $NN \to NN\pi$ will provide a non–trivial test to the approach described.

Once the scheme is established, the same field theory can be used to analyze the isospin violating observables measured in $pn \to d\pi^0$[43] and $dd \to \alpha\pi^0$.[44] First steps in this direction were already obtained in Ref.[45] for the former and in Refs.[46,47] for the latter.

Based on the calculation for $pp \to d\pi^+$ we also performed a calculation for the dispersive and absorptive corrections to the πd scattering length that were calculated for the first time within ChPT. The final answer turned out to be relatively small as a consequence of cancellations amongst various terms. This work is an important step forward towards a high accuracy calculation for the πd scattering length that will eventually allow for a reliable extraction of the isoscalar scattering length. However, before this can be done, isospin violating corrections[48] as well as the contributions from the Delta–isobar need to be evaluated.

Acknowledgments

I thank the organizers for a very well organized and educating conference and V. Lensky, V. Baru, J. Haidenbauer, A. Kudryavtsev, and U.-G. Meißner for a very fruitful collaboration that lead to the results presented. This research is part of the EU Integrated Infrastructure Initiative Hadron Physics Project under contract number RII3-CT-2004-506078, and was supported also by the DFG-RFBR grant no. 05-02-04012 (436 RUS 113/820/0-1(R)) and the DFG SFB/TR 16 "Subnuclear Structure of Matter". A. K. and V. B. acknowledge the support of the Federal Program of the Russian Ministry of Industry, Science, and Technology No 02.434.11.7091.

References

1. S. Weinberg, Phys. Lett. B **295** (1992) 114.
2. S. R. Beane, V. Bernard, E. Epelbaum, U.-G. Meißner and D. R. Phillips, Nucl. Phys. A **720** (2003) 399. [arXiv:hep-ph/0206219].
3. S. R. Beane, V. Bernard, T. S. H. Lee, U.-G. Meißner and U. van Kolck, Nucl. Phys. A **618**, 381 (1997) [arXiv:hep-ph/9702226].
4. H. Krebs, V. Bernard and U.-G. Meißner, Eur. Phys. J. A **22** (2004) 503 [arXiv:nucl-th/0405006].
5. V. Baru, J. Haidenbauer, C. Hanhart and J. A. Niskanen, Eur. Phys. J. A **16**, 437 (2003) [arXiv:nucl-th/0207040].

6. A. Gardestig and D. R. Phillips, Phys. Rev. C **73** (2006) 014002 [arXiv:nucl-th/0501049].
7. V. Lensky, V. Baru, J. Haidenbauer, C. Hanhart, A. E. Kudryavtsev and U.-G. Meißner, Eur. Phys. J. A **26**, 107 (2005) [arXiv:nucl-th/0505039].
8. D. R. Phillips, S. J. Wallace and N. K. Devine, Phys. Rev. C **72** (2005) 0140061. [arXiv:nucl-th/0411092].
9. V. Baru, C. Hanhart, A. E. Kudryavtsev and U.-G. Meißner, Phys. Lett. B **589**, 118 (2004) [arXiv:nucl-th/0402027].
10. B.Y. Park et al., Phys. Rev. C **53** (1996) 1519 [arXiv:nucl-th/9512023].
11. C. Hanhart, J. Haidenbauer, M. Hoffmann, U.-G. Meißner and J. Speth, Phys. Lett. B **424** (1998) 8 [arXiv:nucl-th/9707029].
12. V. Dmitrašinović, K. Kubodera, F. Myhrer and T. Sato, Phys. Lett. B **465** (1999) 43 [arXiv:nucl-th/9902048].
13. S. I. Ando, T. S. Park and D. P. Min, Phys. Lett. B **509** (2001) 253 [arXiv:nucl-th/0003004].
14. T.D. Cohen, J.L. Friar, G.A. Miller and U. van Kolck, Phys. Rev. C **53** (1996) 2661 [arXiv:nucl-th/9512036].
15. C. da Rocha, G. Miller and U. van Kolck, Phys. Rev. C **61** (2000) 034613 [arXiv:nucl-th/9904031].
16. C. Hanhart, U. van Kolck, and G.A. Miller, Phys. Rev. Lett. **85** (2000) 2905 [arXiv:nucl-th/0004033].
17. C. Hanhart and N. Kaiser, Phys. Rev. C **66** (2002) 054005 [arXiv:nucl-th/0208050].
18. C. Hanhart, Phys. Rep. **397** (2004) 155 [arXiv:hep-ph/0311341].
19. A. Gårdestig, talk presented at ECT* workshop 'Charge Symmetry Breaking and Other Isospin Violations', Trento, June 2005; A. Gardestig, D. R. Phillips and C. Elster, Phys. Rev. C **73** (2006) 024002 [arXiv:nucl-th/0511042].
20. V. Lensky, V. Baru, J. Haidenbauer, C. Hanhart, A. E. Kudryavtsev and U.-G. Meißner, Eur. Phys. J. A **27**, 37 (2006) [arXiv:nucl-th/0511054].
21. V. Bernard, N. Kaiser and U.-G. Meißner, Int. J. Mod. Phys. E **4** (1995) 193.
22. D. Koltun and A. Reitan, Phys. Rev. **141** (1966) 1413.
23. J. A. Niskanen, Phys. Rev. C **53**, 526 (1996) [arXiv:nucl-th/9502015].
24. C. Hanhart, J. Haidenbauer, O. Krehl and J. Speth, Phys. Lett. B **444** (1998) 25 [arXiv:nucl-th/9808020].
25. D. A. Hutcheon *et al.*, Nucl. Phys. A **535** (1991) 618.
26. P. Heimberg *et al.*, Phys. Rev. Lett. **77** (1996) 1012.
27. M. Drochner *et al.*, Nucl. Phys. A **643** (1998) 55.
28. C. Hanhart, J. Haidenbauer, O. Krehl and J. Speth, Phys. Rev. C **61** (2000) 064008 [arXiv:nucl-th/0002025].
29. H. O. Meyer et al., Phys. Rev. C **63** (2001) 064002.
30. B. von Przewoski et al., Phys. Rev. C **61** (2000) 064604.
31. W. W. Daehnick et al., Phys. Rev. C **65** (2002) 024003.
32. A. Kacharava *et al.*, arXiv:nucl-ex/0511028.
33. P. Hauser *et al.*, Phys. Rev. C **58** (1998) 1869; D. Chatellard *et al.*, Nucl. Phys. A **625** (1997) 855.

34. D. Gotta et al., PSI experiment R-06.03; D. Gotta, private communication.
35. V. C. Highland et al., Nucl. Phys. A **365** (1981) 333.
36. K. Brückner, Phys. Rev. **98** (1955) 769.
37. I.R. Afnan and A.W. Thomas, Phys. Rev. C**10** (1974) 109; D.S. Koltun and T. Mizutani, Ann. Phys. (N.Y.) **109** (1978) 1.
38. T. E. O. Ericson, B. Loiseau, A. W. Thomas, Phys. Rev. C **66** (2002) 014005 [arXiv:hep-ph/0009312].
39. V.Baru, A. Kudryavtsev, Phys. Atom. Nucl., **60** (1997) 1476.
40. V. Lensky, V. Baru, J. Haidenbauer, C. Hanhart, A. E. Kudryavtsev and U.-G. Meißner, arXiv:nucl-th/0608042.
41. M. Döring, E. Oset and M. J. Vicente Vacas, Phys. Rev. C **70** (2004) 045203 [arXiv:nucl-th/0402086].
42. R. Machleidt, Phys. Rev. C **63** (2001) 024001 [arXiv:nucl-th/0006014].
43. A. K. Opper *et al.*, Phys. Rev. Lett. **91** (2003) 212302 [arXiv:nucl-ex/0306027].
44. E. J. Stephenson *et al.*, Phys. Rev. Lett. **91** (2003) 142302 [arXiv:nucl-ex/0305032].
45. U. van Kolck, J. A. Niskanen, and G. A. Miller, Phys. Lett. B **493** (2000) 65 [arXiv:nucl-th/0006042].
46. A. Gårdestig *et al.*, Phys. Rev. C **69** (2004) 044606[arXiv:nucl-th/0402021].
47. A. Nogga *et al.*, Phys. Lett. B **639**, 465 (2006) [arXiv:nucl-th/0602003].
48. U.-G. Meißner, U. Raha and A. Rusetsky, Phys. Lett.B **639** (2006) 478 [arXiv:nucl-th/0512035].

APPLICATION OF CHIRAL NUCLEAR FORCES TO LIGHT NUCLEI

A. NOGGA*

Institut für Kernphysik, Forschungszentrum Jülich, 52425 Jülich, Germany
**E-mail: a.nogga@fz-juelich.de*

In these proceedings, we discuss the current status of nuclear bound state predictions based on chiral nuclear interactions. Results of ordinary s- and p-shell nuclei and light hypernuclei are shown.

Keywords: chiral nuclear interaction, nuclear binding energies, hypernuclei

1. Introduction

One main goal of nuclear physics is the understanding of the binding energies of nuclei. In the past, it was the aim to relate the binding energies to nuclear forces that describe the two-nucleon (NN) scattering data. This, however, can only be part of a complete understanding. Finally, it is necessary to related the binding energies to QCD.

The most promising approach to establish this relation is chiral effective field theory. It enables us to build into the nuclear interaction the symmetries of QCD and allows the determination of unknown parameters by adjustment of lattice calculations even at unphysically large quark masses. In this way, a direct connection of nuclear binding energies and QCD will be established in the future.[1]

At this time, chiral effective theory is an important guideline to identify the most important contributions to nuclear interactions and to pin down relations of nuclear interactions to other strong interaction processes, e.g. πN scattering and π production. In this context, it is of utmost importance to pin down the structure of three-nucleon forces (3NF's). Traditional calculations[2–4] clearly show that 3NF's are significant for a quantitative description of binding energies. Current models can provide correct binding for s-shell,[2] but fail for p-shell nuclei[5] and scattering observables[6–8] (see also contributions of Johan Messchendorp and Kimiko Sekiguchi to this conference[9,10]). Extensions of these models can improve some of these failures[5] in the regime of light nuclei, but again some deviations show up for the more complex systems.[11] A reliable extension, however, is the key to obtaining theoretical insight into the structure of, e.g., exotic nuclei.

Therefore, the application of chiral perturbation theory to the nuclear bound state problem is of interest to make nuclear structure calculations more reliable. We will argue below that such calculations are also important to confirm that we correctly extend chiral perturbation theory to the non-perturbative nuclear systems.

In these proceedings, we discuss the current status of bound state calculations for light nuclei and hypernuclei. In Section 2, we define the chiral interactions used for the calculations. We then look in detail at the dependence of the results on the cutoff necessary for the regularization of the problem in Section 3. Section 4 is devoted to predictions for p-shell nuclei. Then, we turn to first results for hypernuclei in Section 5 and conclude in Section 6.

2. Chiral nuclear forces

For a complete overview, we refer to the recent reviews on chiral nuclear interactions[12,13] (see also Evgeny Epelbaum's contribution to this conference[14]). Here we will only give a summary of the main results important for further discussion.

A direct application of the power counting of chiral perturbation theory to nuclear systems is not possible. The existence of nuclear bound states excludes any non-perturbative approach. Weinberg realized that this non-perturbative behavior is caused by an enhancement of diagrams with purely nucleonic intermediate states.[15] He classified such diagrams as "reducible" and conjectured that the power counting of chiral perturbation theory can be applied to the "irreducible" diagrams. These diagrams then need to be summed to all orders using a numerical technique, e.g. solving the Lippmann-Schwinger equation. In this way, one obtains in a systematic way a nuclear interaction based on a chiral Lagrangian. The interaction kernel is expanded in powers of a typical momentum in nuclei or the π mass (which is a generic small scale Q) over the chiral symmetry breaking scale $\Lambda_\chi \approx 1$ GeV. The power counting justifies straightforwardly the common assumption that NN interactions are much more important than 3NF's. Higher order forces are even further suppressed. Quantitative results[16–19] confirmed the approach for the NN system.

The more complex few-nucleon systems, however, promise further challenges for this approach. Since 3NF's become quantitatively important, the few-nucleon observables are sensitive to subleading parts of the nuclear interactions. Especially, the binding energies are sensitive to details of the interaction, since they are the result of a rather large cancelation of kinetic energy and potential energy.

3NF's first appear in order Q^3 (Ref[20]). Three topologies exist. Chiral symmetry relates the strength of the the 2π exchange 3NF to corresponding diagrams of the NN interaction and also to πN scattering. A quantitative confirmation that consistent values of the corresponding low energy constants (LEC's) (usually referred to as c_i) can be found for all of these processes is an important confirmation that chiral symmetry is realized in nuclear interactions in the way we assume now. At this point fairly consistent values have been extracted from NN and πN data.[21–25] For a more conclusive comparison, the extractions have to be more accurate or additional insight from few-nucleon systems is required.

Except for the 2π exchange 3NF's, chiral effective theory predicts two more leading 3NF structures. Here two a priori unknown LEC's appear. The first one determines the strength of the 1π exchange diagram and can in principle be related to π production in NN scattering[26] or weak processes in few-nucleon systems.[27] In practice, such extractions cannot be used at this time, since they were performed in frame works that are not consistent with the one used here. The second one enters via the 6N contact vertex. It can only be fixed by matching to few-nucleon observables.

3. Cutoff dependence of nuclear binding energies

In any order, the chiral potentials are singular interactions. If the singularities are attractive, the Hamiltonian becomes unbounded from below. Therefore, regularization is required before solving the Lippmann-Schwinger or Schrödinger equation based on chiral interactions. Usually the regularization is performed by means of a momentum cutoff Λ. The available realizations of chiral interactions mostly use $\Lambda \approx$ 500-600 MeV. Here Λ is chosen to be below a typical hadron mass, e.g. the ρ mass. A quantitative confirmation for this choice is desirable. To this aim, it is instructive to study the cutoff dependence in a much larger range of Λ's. For a few-body problem this was first done in[28] for the leading order nuclear interaction only. It turned out that in leading order, additional contact interactions beyond those required by naive dimensional analysis are required to describe the NN data cutoff independently.[28–30] These results triggered an ongoing discussion on the proper power counting for chiral nuclear interactions (see

Table 1. Expected variations of the results with the cutoff for different orders of the interaction. n is the order of the missing contact interactions, Λ the momentum cutoff used and $\Delta V/V$ ($\Delta E/E$) are the estimated relative variations of the potential (binding) energy.

order	n	Λ [MeV]	$\Delta V/V$	$\Delta E/E$
Q^0	2	500	7 %	70 %
Q^2	4	500	0.5 %	5 %
Q^3	4	500	0.5 %	5 %
Q^4	6	500	0.03 %	0.3 %
Q^0	4	700	0.1 %	1 %

e.g. the panel discussion in the few-body working group[31–33]). Whereas this issue will be of importance to extend the nuclear interaction to momenta close to the π production threshold,[34] where the currently used cutoffs become similar to typical momenta, we will argue below that it does not strongly affect progress for nuclear structure calculations. The differences discussed in the power counting become quantitatively negligible for high order interactions and for cutoffs of the order of 500 MeV.

It is useful to discuss first our expectations for the cutoff dependence of chiral interactions. To get a rough estimate, one can assume that the typical small scale (typical momentum or pion mass) is of the order $Q \approx 130$ MeV. One can expect that variations of the result are absorbed by contact interactions that are not considered at a given order. Assuming natural size, these contact interactions should give contributions that scale with $(Q/\Lambda)^n$. Given that NN contact interactions contribute only in even orders, one finds that $n = 2$ in leading order (Q^0), $n = 4$ in next-leading order (NLO,Q^2) and next-to-next-to-leading order (N2LO, Q^3) and $n = 6$ for order Q^4 (N3LO) interactions. $\Lambda \approx 500$ MeV for typical realizations of chiral interactions. Table 1 shows the expected variation of the potential energy for $\Lambda = 500$ and 700 MeV. Assuming that these missing contributions can be treated in perturbation theory, one finds that this is the actual change expected for the binding energy. This, however, implies that the relative variation of the binding energy becomes quite large. In the table, we simple assumed that it is one order of magnitude larger than the relative variation of the potential. The table also shows that the actual size of the cutoff variation is similar for NLO and N2LO (at least if only NN interactions are considered) and that the estimate strongly depends on the cutoff, especially for higher order interactions. Only in order Q^4 is the variation of the binding energy with the cutoff expected to be very small. It is now interesting to confront these expectations with actual calculations.

Fig. 1 shows the dependence of the ^{3}H binding energy on Λ for a wide

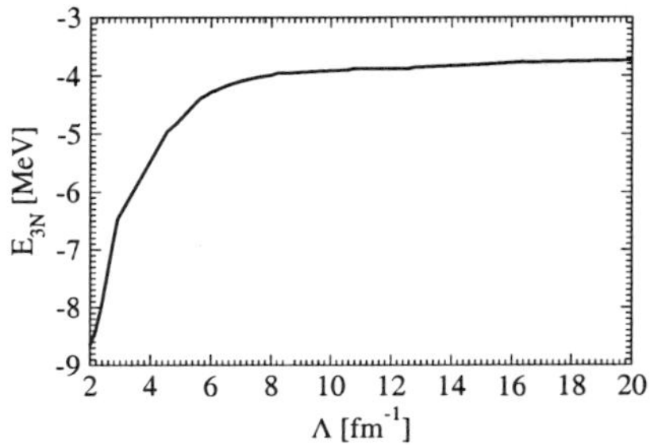

Fig. 1. Dependence of the ^{3}H binding energy on the cutoff Λ for the leading order interaction. Thereby additional contact interactions were required.[28]

Table 2. Calculated dependence of the ^{3}H binding energy for different chiral interactions. The Q^0, Q^2, Q^3 and Q^4 interactions are from Refs.[18,19,28] Λ is the momentum cutoff imposed on the Lippmann-Schwinger equation, $\tilde{\Lambda}$ are cutoffs imposed on internal loops (see Ref.[19]). DR notes that loops are dimensionally regulated.

order	Λ / $\tilde{\Lambda}$ [MeV]	E [MeV]	V [MeV]	ΔE [keV]	$\Delta E/V$
Q^0	500 / no loops	-7.50	-51.8	1430	3.0 %
	600 / no loops	-6.07			
Q^2	400 / 700	-8.46		650	1.6 %
	550 / 700	-7.81	-41.1		
Q^3	450 / 700	-8.42	-38.3	530	1.3 %
	600 / 700	-7.89			
Q^4	500 / DR	-7.84	-42.3	40	0.1 %
	600 / DR	-7.80			

range of cutoffs between Λ = 2-20 fm^{-1} ($\approx$ 400-5000 MeV). One observes that one can obtain a cutoff independent binding energy for large Λ, once the cutoff dependence of NN predictions is removed by promoting counter terms from naively higher orders. Contrary to the effective theory without pions,[35] one does not need to promote a 3NF to leading order to get cutoff independent results. In the range of cutoffs considered, one obtains a variation of the result of approximately 4 MeV consistent with the expected variation of a leading order calculation. Note also that in the low cutoff range between 500 and 600 MeV, one finds that the binding energy varies quite rapidly by approximately 1.5 MeV.

In view of the fact that so far no higher order realization of chiral interactions has been developed that covers a similarly large range of cutoffs, it is not possible to study the binding energies in a similar manner for the higher orders. An order of magnitude estimate, however, is possible by com-

Table 3. Strength constants of the 1π exchange and 6N contact 3NF's (see definition in Ref.[36]). Q^3 interactions are from Ref.[19] Q^4 interaction is from Ref.[18] 3NF-A and 3NF-B label two sets of parameters that describe the 3N and 4N binding energies equally well.

interaction	Λ / $\tilde{\Lambda}$ [MeV]	c_D	c_E
Q^3	450 / 700	1.20	-0.082
Q^3	600 / 700	-4.27	-1.25
Q^4-3NF-A	500 / DR	-1.11	-0.66
Q^4-3NF-B	500 / DR	8.14	-2.02

paring the variation within a small range of $\Lambda \approx$ 500-600 MeV. Neglecting the contribution of 3NF's, for which we assume a rather cutoff independent contribution to the binding energies, this is shown in Table 2 again for ^{3}H for several orders of the chiral expansion. Though the variations within the range of cutoffs are somewhat large compared to the estimate in Table 1, the results still confirm the power counting expectation. Quantitatively, the cutoff dependence becomes negligible in order Q^4 (N3LO).

For a quantitative calculation in higher orders, we need to fix the LEC's of the 3NF. As discussed above, there are two LEC's unrelated to the NN interaction in the leading 3NF's. Therefore, one needs two few-body data. Most naturally, we use the ^{3}H binding energy in all of our determinations of these LEC's. Both, the ^{4}He binding energy and the doublet neutron-deuteron scattering length are suitable to constrain the second LEC. The details of the determinations can be found in Refs.[36,37] We have performed the fits for combinations of the leading 3NF with the Q^3 chiral interactions of Ref.[19] and the Q^4 interaction of Ref.[18] The latter combination is not strictly consistent with the power counting, since we neglected Q^4 contributions to the 3NF and the 4NF.[38] The values of the parameters c_D and c_E in the notation of Ref.[36] are given in Table 3 for completeness. Note that we find two solutions for the LEC's in conjunction with the NN force of Ref.,[18] which are labeled 3NF-A and 3NF-B in the following.

The binding energies of ^{3}H and ^{4}He are well described by the chiral interactions (by construction). To confirm that the application of chiral interactions to s-shell nuclei gives consistent results, it is interesting to compare the contribution of 3NF's to the ^{3}H and ^{4}He potential energies to a power counting estimate. The 3NF is formally in order Q^3. An estimate similar to the one shown in Table 1 leads to an expected contribution of the 3NF of $\approx$ 2 % to the total potential energy. Our calculations show that the 3NF (even the various parts of it) do not contribute more than 7.5 % to the potential energy of ^{4}He, which is still in line with the power counting estimate. Therefore, we note that the bound state calculations for

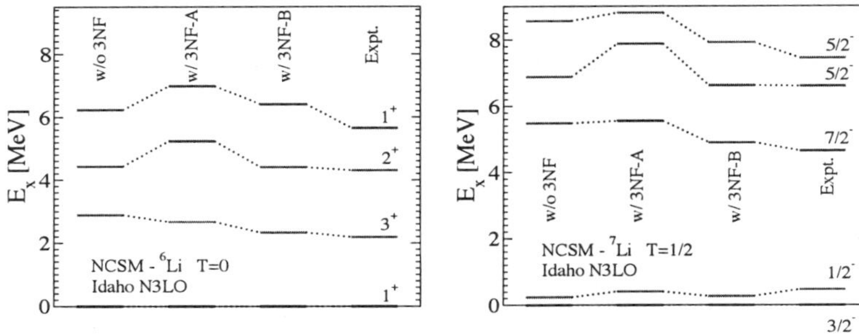

Fig. 2. Spectra of ^{6}Li and ^{7}Li nuclei based on the chiral interaction of Ref.[18] without 3NF or in combination with 3NF-A and 3NF-B. See text for definitions.

the s-shell nuclei confirm our expectations from the power counting.

4. Predictions for ^{6}Li and ^{7}Li

With the 3NF's completely fixed, we are now in the position to make predictions for p-shell nuclei. All results for the p-shell nuclei have been obtained for the Q^4 interaction of Ref.[18] Since the cutoff dependence for a Q^4 chiral interaction is negligibly small, we will restrict ourselves only to one cutoff $\Lambda = 500$ MeV.

To predict binding energies and spectra of p-shell nuclei, we need to use a technique for solving the Schrödinger equation based on non-local interactions. Here, we will show results based on the "no-core shell model" approach (NCSM). Details of the technique and the results for ^{7}Li are discussed in.[39] Here, it is sufficient to emphasize that excitation energies can be accurately obtained by the NCSM. The accuracy of the binding energy can be estimate to be approximately 1 MeV.

Since the NCSM results for the spectra are more accurate than the binding energies, they are especially important to study the chiral interactions. As can be seen in Fig. 2, the excitation energies are changed by the addition of the 3NF. Note that the two parameter sets 3NF-A and 3NF-B, that describe the s-shell nuclei equally well, result in different predictions for ^{6}Li and ^{7}Li. The expected sensitivity to the structure of the 3NF is confirmed by these calculations. For ^{6}Li and ^{7}Li, we find a consistently better description of the spectra for parameter set 3NF-B compared to the predictions of 3NF-A and without 3NF.

For the binding energies, the situation is somewhat different. Our results are compiled in Table 4. Again 3NF-B improves the description of the radii,

Table 4. Binding energies E and point proton radii r_p for ^{6}Li and ^{7}Li. Results for chiral interactions are compared to results based on phenomenological interactions[5] and to experimental values that a corrected for the finite size of the protons.

interaction	^{6}Li		^{6}Li	
	E [MeV]	r_p [fm]	E [MeV]	r_p [fm]
Q^4 – no 3NF	-30.0	2.20	-34.6	2.15
Q^4 – 3NF-A	-32.3	2.16	-38.0	2.11
Q^4 – 3NF-B	-31.1	2.25	-36.7	2.23
AV18 – IL2	-32.3	2.39	-38.9	2.25
AV18 – Urb-IX	-31.1	2.57	-37.5	2.33
Expt.	-32.0	2.43	-39.2	2.27

but both p-shell nuclei appear to be underbound.

This apparent inconsistency of the results deserves some further consideration. Here, it is important to note that we have fixed the strength of the 2π exchange part of the 3NF using the choice of c_i of the NN potential of Ref.[18] The description of the NN data is not very sensitive to the choice of these parameters. Nuclear matter calculations for low momentum interactions including the same 3NF's, however, indicate that a change of the c_i's by only 10 %, correcting the c_D and c_E so that the s-shell nuclei do not change their binding energy, may change the binding energy per particle in symmetric nuclear matter by 1 MeV.[40,41] It is therefore conceivable that a consistent description of binding energies and spectra can be obtained by a variation of the c_i's. This needs to be explicitly checked in the future.

Finally, we note that the addition of 3NF-B, though the relatively low cutoffs remove any strong short range repulsion, increases both the binding energy and the radii.

5. Hypernuclei

Now we turn to hypernuclear binding energies. Recently, Polinder and collaborators have developed a first realization of the chiral hyperon-nucleon (YN) interaction.[42] A systematic approach to the problem of the YN interaction is badly needed. It will enable us to understand the way flavor SU(3) symmetry is broken in nuclear systems. Also the impact of hyperons on the nuclear equation of state is possibly significant also for astrophysical applications,[43] but the poor knowledge of the interactions of hyperons hinders more insight. Most of these issues are due to the very scarce set of data in the YN sector. Moreover, most of the data are considerably above the ΛN threshold. Therefore, even the scattering lengths for ΛN scattering are essentially unknown (for a discussion on the current status see e.g.[44,45]).

Table 5. Λ separation energies of the 0^+ ($E_{sep}(0^+)$) and 1^+ ($E_{sep}(1^+)$) states and their difference ΔE_{sep} for ${}^4_\Lambda$H and the difference of the separation energies for the mirror hypernuclei ${}^4_\Lambda$He and ${}^4_\Lambda$H (CSB-0^+ and CSB-1^+). Results for the chiral YN interaction for various cutoffs Λ are compared to results for two phenomenological models[46,47] and the experimental values.

Λ [MeV]	500	550	650	700	Jülich 05	Nijm SC97f	Expt.
$E_{sep}(0^+)$ [MeV]	2.63	2.46	2.36	2.38	1.87	1.60	2.04
$E_{sep}(1^+)$ [MeV]	1.85	1.51	1.23	1.04	2.34	0.54	1.00
ΔE_{sep} [MeV]	0.78	0.95	1.13	1.34	-0.48	0.99	1.04
CSB-0^+ [MeV]	0.01	0.02	0.02	0.03	-0.01	0.10	0.35
CSB-1^+ [MeV]	-0.01	-0.01	-0.01	-0.01	—	-0.01	0.24

Current models of the YN interaction[46–48] all describe the available YN data, but predictions for non-measured observables vary very strongly. They also fail to describe the measured binding energies of the light hypernuclei[49] and, therefore, a more systematic insight into the YN interaction is even more badly needed.

Also the leading chiral interaction (one Goldstone-boson exchange and five non-derivative contact interactions) has been fitted to the scarce data base for the YN system. Additionally, the scattering lengths have been constrained, so that the ${}^3_\Lambda$H binding energy is in agreement with the experimentally know value of $E = -2.35$ MeV. Thereby, the cutoff was varied between 550 and 700 MeV (for details see Ref.[42]). Contrary to common expectation, the resulting ΛN cross section at low energies was much smaller than traditional models predict. In view of this surprisingly weak interaction, it is astonishing to find ${}^3_\Lambda$H binding energies in agreement with experiment.

It is now important to confront predictions based on the leading order chiral interaction for the more complex hypernuclei ${}^4_\Lambda$H and ${}^4_\Lambda$He with the data. Two states, the $J^\pi = 0^+$ ground and a $J^\pi = 1^+$ exited state, of these mirror hypernuclei are experimentally known. Though the YN interactions are not very strongly constrained by the YN data, it has proven to be difficult to obtain a consistent description of both of the states and also of the well known charge dependence of the Λ separation energies. It turned out that the splittings between the 0^+ and 1^+ state and the charge dependence of the separation energies are correlated with the strong Λ-Σ conversion process. Table 5 compiles the new results based on the chiral interaction together with model predictions and the experimental values. The dependence on the NN interaction is mild.[49] Here, we used the one of Ref.[18] The leading order YN interaction results in very reasonable Λ-separation energies. The separation energy of the 0^+ state appears to be very cutoff independent, whereas the one of the 1^+ state is more strongly dependent on the cutoff in leading order. Given that the results are leading order only, they are very encouraging. A NLO calculation will be important to confirm that the agreement improves and the cutoff dependence shrinks as we have seen for the ordinary nuclei. The leading order calculation does not include any charge symmetry breaking contributions in the interaction. As a result, the charge dependence of the separation energies is very small. (The small non-zero contributions is due to the Σ^+-Σ^- mass difference and the Coulomb interaction. Both have been included in the calculations of Table 5.) The NLO interaction will explicitly include first order charge symmetry breaking contributions. Therefore, an improvement of the leading order results in this respect is also conceivable.

6. Conclusions and Outlook

We discussed the results for nuclear binding energies of ordinary s-shell and p-shell nuclei and the lightest hypernuclei based on chiral interactions. A special emphasis was on the confirmation of the power counting underlying nuclear interactions.

For the ordinary nuclei, the results of LO, NLO and N2LO chiral interactions confirm our power counting expectations. We find that N3LO NN interactions give predictions that are only insignificantly cutoff dependent. In this order, chiral interactions will become quantitatively useful for predictions of nuclear binding energies. In this context, the outcome of an on-going discussion on possible promotions of contact interactions for formerly higher order will be important, since it might enable us to extend the current realizations of chiral interactions to slightly larger values of the cutoffs closer to Λ_χ and, thereby, possibly improve the convergence of the chiral expansion.

The predictions for p-shell nuclei confirm the expected sensitivity of the 3NF's. Spectra and binding energies are improved by their addition. It will be interesting to allow for small variations of the pertinent LEC's c_i to further improve the description, especially of the binding energies. In this way, a more accurate determination of the c_i might become possible.

We also discussed the first results for binding energies of the lightest hypernuclei and found that the separation energies of ${}^{3}_{\Lambda}$H and ${}^{4}_{\Lambda}$H can be consistently described. Now the extension to NLO will be very interesting. First, it should shrink the dependence on the cutoff and enable combined fits of the NN and YN interaction. Also first order charge-symmetry breaking terms contribute to the YN interaction in this order. Since phenomenological models do not describe the charge-symmetry breaking of the ${}^{4}_{\Lambda}$H-${}^{4}_{\Lambda}$He properly, the results of the chiral interaction for this observable will be especially interesting.

Acknowledgments

I very much thank B. Barrett, E. Epelbaum, W. Glöckle, J. Golak and U. Meißner, P. Navrátil, H. Polinder, R. Skibiński, R. Timmermans, U. van Kolck, J. Vary, and H. Witała for collaborating on the work presented here. The numerical calculations have in part been performed on the JUMP and JUBL computers of the NIC, Jülich, Germany.

References

1. M. Savage, Few-body lattice calculations, contribution to this conference.
2. A. Nogga, H. Kamada, W. Glöckle and B. R. Barrett, *Phys. Rev. C* **65**, p. 054003(May 2002).
3. S. C. Pieper and R. B. Wiringa, *Ann. Rev. Nuc. Part. Sci.* **51**, 53 (2001).
4. P. Navrátil and B. R. Barrett, *Phys. Rev. C* **57**, 3119(June 1998).
5. S. C. Pieper, V. R. Pandharipande, R. B. Wiringa and J. Carlson, *Phys. Rev. C* **64**, p. 014001(July 2001).
6. K. Sekiguchi *et al.*, *Phys. Rev. C* **65**, p. 034003(March 2002).
7. S. Kistryn *et al.*, *Phys. Rev. C* **72**, p. 044006(October 2005).
8. K. Ermisch *et al.*, *Phys. Rev. C* **71**, p. 064004(June 2005).
9. J. Messchendorp, Results from kvi: Nd elastic scattering, contribution to this conference.
10. K. Sekiguchi, Results on three-nucleon experiments from riken, contribution to this conference.
11. S. C. Pieper, Recent results from quantum monte carlo calculations of light nuclei, in *Workshop on "Electron-Nucleus Scattering IX"*, (Isola d'Elba, Italy, 2006). http://conferences.jlab.org/elba/talks/pieper.pdf.
12. E. Epelbaum, *Prog. Part. Nucl. Phys.* **57**, 654 (2006).
13. P. F. Bedaque and U. van Kolck, *Ann. Rev. Nuc. Part. Sci.* **52**, 339 (2002).
14. E. Epelbaum, Towards a systematic theory of nuclear forces and nuclear currents, contribution to this conference.
15. S. Weinberg, *Nucl. Phys. B* **363**, 3(September 1991).
16. C. Ordóñez, L. Ray and U. van Kolck, *Phys. Rev. C* **53**, 2086(May 1996).
17. E. Epelbaum, W. Glöckle and U.-G. Meißner, *Nucl. Phys. A* **671**, 295(May 2000).
18. D. R. Entem and R. Machleidt, *Phys. Rev. C* **68**, p. 041001(October 2003).
19. E. Epelbaum, W. Glöckle and U.-G. Meißner, *Nucl. Phys. A* **747**, 362(January 2005).
20. U. van Kolck, *Phys. Rev. C* **49**, 2932(June 1994).
21. U.-G. Meißner, *PoS* **LAT2005**, p. 009 (2006).
22. M. C. M. Rentmeester, R. G. E. Timmermans and J. J. de Swart, *Phys. Rev. C* **67**, p. 044001(April 2003).

23. D. R. Entem and R. Machleidt, *Phys. Rev. C* **66**, p. 014002(July 2002).
24. P. Büttiker and U.-G. Meißner, *Nucl. Phys. A* **668**, 97(March 2000).
25. N. Fettes, U.-G. Meißner and S. Steininger, *Nucl. Phys. A* **640**, 199(September 1998).
26. C. Hanhart, U. van Kolck and G. A. Miller, *Phys. Rev. Lett.* **85**, 2905(October 2000).
27. A. Gårdestig and D. R. Phillips, *Phys. Rev. Lett.* **96**, p. 232301(June 2006).
28. A. Nogga, R. G. E. Timmermans and U. van Kolck, *Phys. Rev. C* **72**, p. 054006(November 2005).
29. M. Pavón Valderrama and E. Ruiz Arriola (2005), nucl-th/0507075.
30. M. C. Birse, *Phys. Rev. C* **74**, p. 014003(July 2006).
31. U. van Kolck, Panel discussion on power counting, contribution to this conference.
32. U.-G. Meißner, On the consistency of weinberg's power counting, contribution to this conference.
33. M. Pavón Valderrama, Renormalization of singular potentials and power counting, contribution to this conference.
34. A. Nogga, A. C. Fonseca, A. Gårdestig, C. Hanhart, C. J. Horowitz, G. A. Miller, J. A. Niskanen and U. van Kolck, *Phys. Lett. B* **639**, 465(2006).
35. E. Braaten and H. W. Hammer, *Phys. Rep.* **428**, 259(June 2006).
36. E. Epelbaum, A. Nogga, W. Glöckle, H. Kamada, U.-G. Meißner and H. Witała, *Phys. Rev. C* **66**, p. 064001(December 2002).
37. A. Nogga, E. Epelbaum, P. Navrátil, W. Glöckle, H. Kamada, U.-G. Meißner, H. Witała, B. R. Barrett and J. P. Vary, *Nucl. Phys. A* **737**, 236(June 2004).
38. E. Epelbaum, *Phys. Lett. B* **639**, 456(August 2006).
39. A. Nogga, P. Navrátil, B. R. Barrett and J. P. Vary, *Phys. Rev. C* **73**, p. 064002(June 2006).
40. A. Nogga, S. K. Bogner and A. Schwenk, *Phys. Rev. C* **70**, p. 016002(December 2004).
41. S. K. Bogner, A. Schwenk, R. J. Furnstahl and A. Nogga, *Nucl. Phys. A* **763**, 59(December 2005).
42. H. Polinder, J. Haidenbauer and U.-G. Meißner, *Nucl. Phys.* **A779**, 244 (2006).
43. B. D. Lackey, M. Nayyar and B. J. Owen, *Phys. Rev. D* **73**, p. 024021(January 2006).
44. A. Gasparyan, J. Haidenbauer, C. Hanhart and J. Speth, *Phys. Rev. C* **69**, p. 034006(March 2004).
45. A. Gasparyan, J. Haidenbauer and C. Hanhart, *Phys. Rev. C* **72**, p. 034006(September 2005).
46. J. Haidenbauer and U.-G. Meißner, *Phys. Rev. C* **72**, p. 044005(October 2005).
47. T. A. Rijken, V. G. J. Stoks and Y. Yamamoto, *Phys. Rev. C* **59**, 21(January 1999).
48. T. A. Rijken and Y. Yamamoto, *Phys. Rev. C* **73**, p. 044008(April 2006).
49. A. Nogga, H. Kamada and W. Glöckle, *Phys. Rev. Lett.* **88**, p. 172501(April 2002).

NEW RESULTS ON FEW BODY EXPERIMENTS AT LOW ENERGY *

Y. NAGAI

Research Center for Nuclear Physics, Osaka University, Ibaraki, Osaka 567-0047, Japan

The simultaneous measurement of the two-body and three-body photodisintegration reaction cross sections of ^{3}He and ^{4}He were for the first time carried out in the energy range from 10 to 30 MeV using a quasi mono-energetic pulsed photon beam and a 4π time projection chamber containing He gas as an active target. A charged fragment from the reactions was detected with 100 % efficiency with a good signal to noise ratio. The obtained two-body cross sections for ^{4}He increase with energy up to 30 MeV, and the two-body cross section ratio, $\sigma(\gamma,p)/\sigma(\gamma,n)$, is consistent with the value expected from charge symmetry of the strong interaction. The first measurement of the keV neutron capture reaction cross section by deuteron was made successfully by detecting a discrete γ -ray from the reaction by means of an anti-Compton NaI(Tl) spectrometer. These results were compared to recent theoretical calculations.

1. Introduction

The photodisintegration reaction and its inreverse reaction on few body systems provides important information on nucleon-nucleon (NN) interactions, meson exchange currents, and the primordial and rapid neutron capture processes involved in nucleosynthesis. The proton and/or neutron capture reactions and/or their reverse reactions on light nuclei at a temperature relevant to the primordial nucleosynthesis are key reactions in estimating the primordial light-element abundance in the early universe. The comparison of the calculated light element abundance up to ^{7}Li to observed ones provides a critical test of the standard big bang cosmology.[1] The remaining uncertainties of these reaction cross sections give rise to the uncertainties in the estimated abundance.[2] Here it is noteworthy that

*Work supported in part by Grant-in-Aid for Specially Promoted Research of the Japan Ministry of Education, Science, Sports and Culture and in part by Grant-in-Aid for Scientific Research of the Japan Society for the Promotion of Science (JSPS)

several radiative and/or inverse reactions relevant to the primordial nucleosynthesis such as p(n, γ)^{2}H and ^{2}H(n, γ)^{3}H reactions are characterized as being hindered and/or forbidden transitions in an impulse approximation. A typical example of the hindrance can be seen in the thermal neutron capture reaction cross section by deuterium, which is very small, 1/660 of the thermal neutron capture by proton.[3] This small cross section is due to the nuclear structure of ^{2}H and ^{3}H, and the property of the scattering state in ^{3}H.[4] Hence, the electromagnetic transition proceeds via a small component of the wave function, and thus one could learn the role of sub-nucleonic degrees of freedom in these reaction processes.[5]

In this paper we present experimental studies of the cross section measurements of the ^{3}He and ^{4}He photodisintegrations and of the keV neutron capture reaction by deuteron. The photodisintegration study of ^{4}He has been a quite interesting subject in nuclear physics, since the study could provide a testing ground for the theory of NN, three-body forces and collective nuclear motion.[6] Here, the fact that ^{4}He is the lightest self-conjugate nucleus with closed shell structure and the photodisintegration at low energy proceeds mainly by an electric dipole process will play crucial roles in the discussions mentioned.[6] The quenching of the observed $E1$ strength compared to theory remains as a long-standing problem.[6] In addition, the cross section ratio of the ^{4}He(γ ,p) to ^{4}He(γ ,n) in the giant dipole region has been used to test the validity of the charge symmetry of the strong interaction in nuclei.[7] The ^{4}He photodisintegration study could provide useful information on the scenario of the rapid process nucleosynthesis induced by neutrino driven wind from a nascent neutron star,[8] and of the delayed supernovae explosion, where the neutrino heating by its interaction with ^{4}He would influence the explosion process.[9] Here, it should be noted that the neutrino-nucleus interaction is analogous to the electromagnetic interaction of electric dipole with a nucleus.

Experimental and theoretical situations of the photodisintegration cross section of ^{4}He have not yet been settled, although significant progress has been achieved.[6] The experimental data of the cross sections for the ^{4}He(γ ,p)^{3}H and ^{4}He(γ ,n)^{3}He channels, whose measurements were mostly carried out separately, show a large discrepancy between different data sets.[10] Here it is worthwhile to note that the systematic uncertainties of the old data are much larger than the statistical uncertainties. Theoretical calculations predict different cross sections in the region of the electric dipole resonance ($25 \sim 26$ MeV).[6,11]

Under such situations both in experiment and theory, we performed a

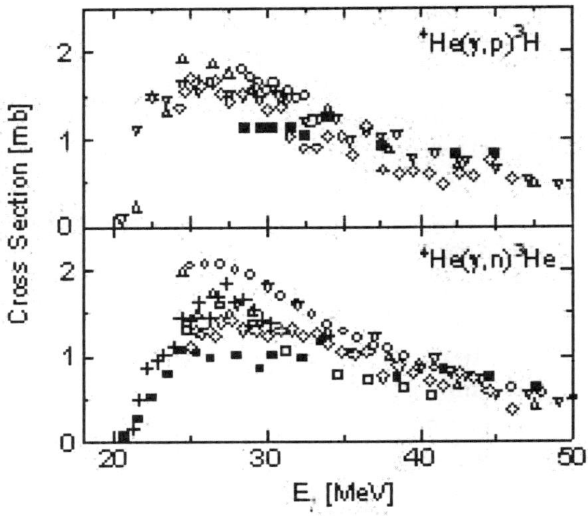

Fig. 1. Upper panel; (γ p) cross sections. Filled squares; Bernabei et al. (1988), open upward triangles; Gorbunov (1968), open downward triangles; Arkatov et al. (1978), open diamonds; Balestra et al. (1979), open circles; Hoorebeke et al. (1993). Lower panel; (γ,n) cross sections. Filled squares; Berman et al. (1980), open upward triangles; Gorbunov (1968), open circles; Malcom et al. (1973), crosses; Irish et al. (1973), open downward triangles; Arkatov et al. (1978), open diamonds; Balestra et al. (1979), open squares; Nilsson et al. (2005). (reference given here; see references in Ref.[10]).

new experiment of the ^{4}He photodisintegration in the energy range from 21.8 to 29.8 MeV by developing a new measurement system, which allowed us to determine the reaction cross section with small systematic uncertainty. In designing the new system we first compared the cross sections obtained by different photon probes as shown in Fig. 1, where the cross sections obtained by Bremsstrahlung photons are significantly larger than the data taken by monochromatic γ -rays below 35 MeV.[10] There might be a systematic uncertainty related to the background subtraction in the former measurements. Next, we reconsidered important points in the study of the reaction at low energy. Since photon flux is low, 10^4/s, and the cross section is small, around a few mb, it is necessary to detect a charged fragment from the reaction with high efficiency and a large solid angle. In addition, the energy of the emitted charged fragments is quite low, less than a few MeV, and therefore it is preferable to use an active target to detect a fragment with 100% efficiency. We have developed a new experimental method to meet these important requirements.[12]

The calculated cross-sections for the ^{3}He(γ,p) and ^{3}He(γ,pp) reactions based on the Faddeev approach[13] are shown to be quite sensitive to the *NN* and three-body forces at the respective peak energies of $\sim$ 10 MeV and $\sim$ 16 MeV. The calculated total cross section based on the LIT method[14] has a sensitivity similar to that given by the Faddeev approach. Experimental data of the (γ,p) and (γ,pp) cross sections of ^{3}He, which have been measured separately using real, virtual photons, and/or electrons,. are controversial in the sensitive γ-ray energy region of $E_\gamma < 20$ MeV, and therefore hardly compared to theoretical calculations.[15] Note that in the vicinity of the (γ,p) cross section peak the old data are divided into two discrete groups with the cross section of 0.75 mbarn and of about 1.0 mbarn. The uncertainty mentioned above seems to remain in the normalization of these cross sections.[15] Hence, we measured the ^{3}He photodisintegration cross section at $E_\gamma = 10.2$ and 16 MeV to accurately determine the normalization.[15]

Recently, a theoretical calculation of the neutron radiative capture reaction by deuteron with pionless effective field theory has been made in the big bang energy regime.[16] The experimental study of the keV neutron capture cross section, however, has not yet been carried out. It is therefore interesting to measure the cross section to compare to the calculated cross sections. Since the photodisintegration study of tritium is not easy due to radioactive target of tritium, we carried out the cross section measurement using the direct neutron capture reaction by deuteron by detecting a discrete γ-ray from the reaction to the ground state in ^{3}H by means of an anti-Compton shielded NaI(Tl) spectrometer.

2. Experimental procedures and results

2.1. *The photodisintegrations of ^{4}He and ^{3}He*

We first discuss the measurement of the cross sections of the ^{4}He and ^{3}He photodisintegrations. The experimental setup is shown in Fig. 2. We used pulsed laser Compton backscattering (LCS) photons, and a time projection chamber (TPC) with an active He target, which allowed us to simultaneously measure the (γ,p) and (γ,n) reaction channels and to clearly distinguish events due to the former channel from those of the latter one. The measurement was carried out at the National Institute of Advanced Science and Technology at Tsukuba. There are several key ingredients necessary to obtain real events with a large signal to noise ratio in the present method. First, real events are only produced along the LCS photon axis with a diameter of 2 mm, when the pulsed photon beam entered the TPC. Second,

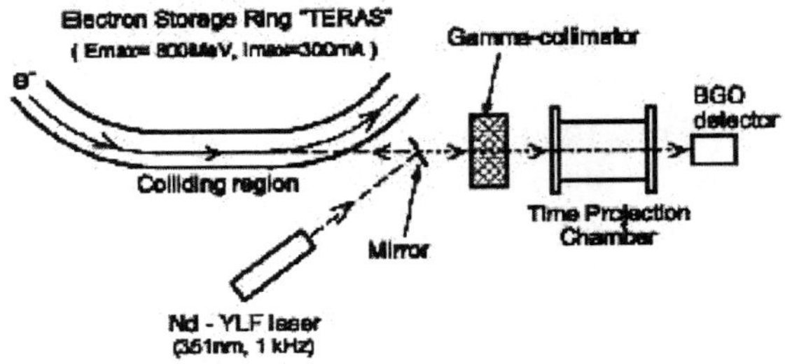

Fig. 2. Experimental setup for the photodisintegration reaction cross section of ^{4}He at AIST.

with the use of the TPC we could obtain information on the track shape of a charged fragment, energy loss deposited by the fragment, and the reaction point. Such information is also necessary to determine the angular distribution of a fragment. Third, the solid angle is large, nearly 4 π , and the detection efficiency is as high as 100%. Note that the energy loss rate of the fragment in the TPC increases with decreasing energy of the fragment, which makes it easier to detect a low energy fragment. These features of the present method allow for an accurate determination of the photodisintegration cross section of a nucleus near the reaction threshold region.

The photodisintegration cross section of ^{4}He is given as the product of the reaction yield, the number of ^{4}He target nuclei, the incident LCS γ-ray flux, and the detection efficiency of the TPC. The target number was obtained by measuring the gas pressure and temperature in the TPC, and the efficiency was determined by using the ^{241}Am α-ray source. The reaction yield and the γ-ray flux were obtained as follows: events taken by the TPC consisted of electron and beam uncorrelated background events, and real events due to the photodisintegration of He and C. We observed C events, since we used methane gas as a quenching gas in the TPC.[12] Most events were electron background events produced by the scattering of the LCS γ-rays with atomic electrons of He and C. These events could be discriminated, since their pulse height was quite low compared to that of the photodisintegration of ^{4}He. Using the thus obtained background free events, the identification of the reaction channel for each event was made by referring to the track shape of the charged fragment and its pulse height.[12] A typical track shape and pulse height of charged fragment for the ^{4}He(γ,p)^{3}H reaction is shown in Fig. 3. Here two tracks cross the LCS γ-ray beam axis, and are in the same straight line. The measured pulse height spectrum

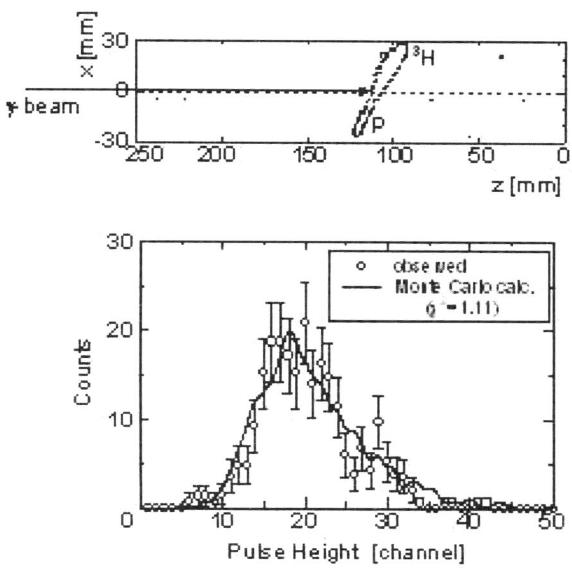

Fig. 3. Typical track shape and pulse height of charged fragment for the ^{4}He(γ,p)^{3}H reaction.

agrees nicely with the calculated one for the channel.[10] Note that we only see one track shape crossing the beam axis for the (γ,n) channel. Similar to the case mentioned above, we could clearly distinguish the events due to the photodisintegration of ^{4}He from those of ^{12}C.

The incident γ-ray flux was measured using a BGO detector as shown in Fig. 4, where multiple peaks were observed due to pile-up effects of LCS γ-rays.[10] A detailed description to derive the incident flux from the observed multiple peaks is given in Ref.[10]

The photodisintegration yield of ^{4}He is proportional to an averaged number M of multiple LCS γ-rays per laser pulse. The number M was obtained by comparing a measured BGO spectrum to one calculated by a Monte Carlo method.[10] Using the resultant number M the LCS γ-ray total flux is obtained as the product of M, the frequency of the laser pulse, and the live time of the measurement system of the TPC with an error of about 2%.

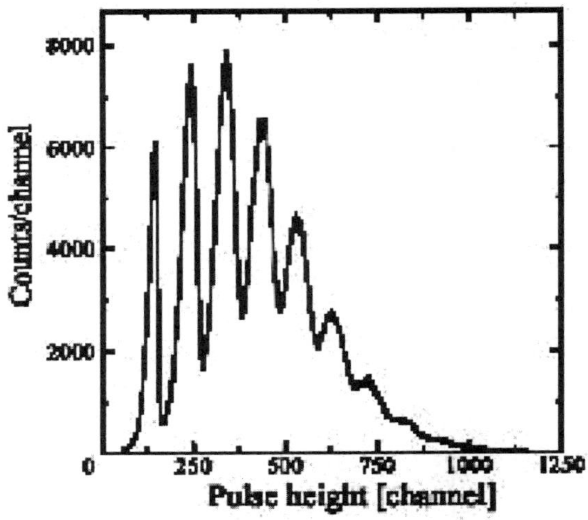

Fig. 4. Typical γ-ray pulse height spectrum measured with a BGO detector.

2.2. *keV neutron capture reaction by deuteron*

The measurement of the keV neutron capture reaction by deuterium was carried out using pulsed keV neutrons, which were produced by the ^{7}Li(p,n)^{7}Be reaction. The proton beam was provided from the 3.2 MV Pelletron accelerator at Tokyo Institute of Technology. An experimental setup is shown in Fig. 5. A discrete γ-ray promptly emitted from the neutron capture reaction by the deuteron was detected by means of an anti-Compton NaI(Tl) spectrometer,[17] which was heavily shielded against neutrons scattered by a sample (deuteron) and external γ-rays produced by thermalized neutron capture reactions by various materials in the measurement room. Gold was used to normalize the neutron capture cross section of a sample, since the cross section for Au is well known within an uncertainty of 3 %.[18]

The background subtracted (net) γ-ray spectrum from the D(n,γ)^{3}H reaction was for the first time measured in the keV region as shown in Fig. 6 (see Ref.[19]), where one sees clearly the γ-ray from the D(n,γ)^{3}H reaction with a good signal to noise ratio. Since we used a D_2O sample, we also observed several discrete γ-rays from the ^{16}O(n,γ)^{17}O reaction to the low-lying states in ^{17}O.

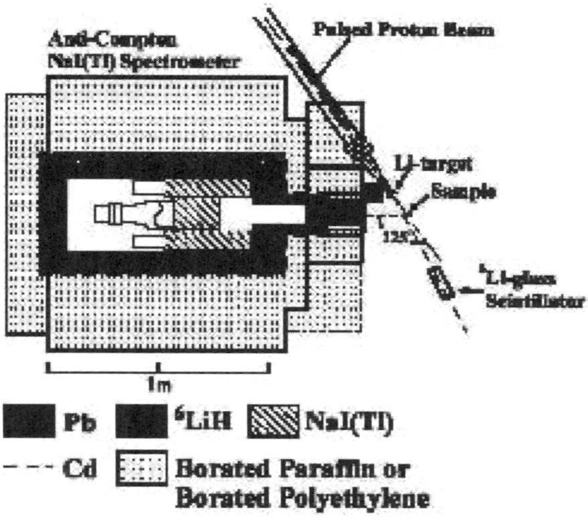

Fig. 5. Experimental setup for the measurement of the $^{2}H(n,\gamma)^{3}H$ reaction cross section.

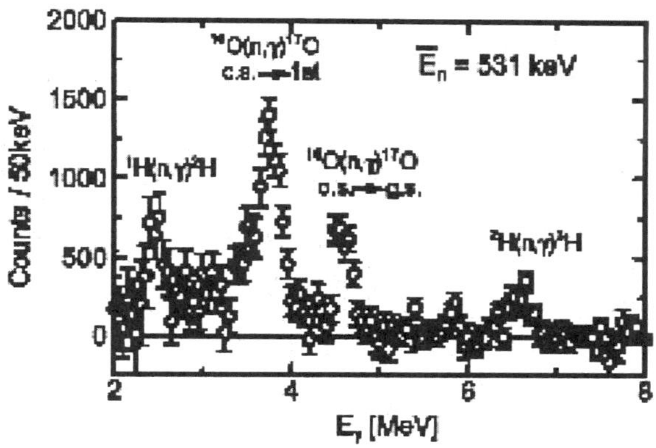

Fig. 6. A typical γ-ray spectrum from the $^{2}H(n,\gamma)^{3}H$ reaction at En= 531 keV.

3. Analysis

3.1. *Cross sections of the ^{4}He and ^{3}He photodisintegration reactions*

We obtained the reaction yield, the photon flux, the target number of ^{4}He nuclei, and the detection efficiency of the TPC as described above. Hence, we could obtain the photodisintegration cross section of ^{4}He. Here, it should be mentioned that in order to learn about any possible systematic uncertainty of the present experimental method, we measured the photodisintegration cross section of deuteron using CD_4 gas at E_γ=22.3 MeV. Note the reaction cross section has been well known with good accuracy. The result is in good agreement with old data and with theoretical values,[10] confirming the validity of the new method including its analysis.

The thus obtained cross sections for the (γ,p) and (γ,n) reactions on ^{4}He are shown together with previous data and theoretical calculations in Fig. 7. They increase monotonically with increasing γ-ray energy up to 30 MeV, and do not show a prominent peak in the region of 25 $\sim$ 26 MeV, contrary to several old data and a recent theoretical calculation.[6] (Note that the measured cross section at 30 MeV increases significantly compared to that at 26 MeV. The origin of the drastic change of the cross section should be clarified in the near future). The present (γ,p) data differ significantly from the old data. The present (γ,n) data agree marginally with the data by Berman et al.,[10] and with a theoretical calculation based on Faddeev type equations.[11] The cross section ratio of the (γ,p) to the (γ,n) reactions derived from the present measurement agrees with the expected value assuming the isospin symmetry of the strong interaction in nuclei. Note that the present simultaneous measurement with a 4π geometry provides the most stringent limit for the ratio. The obtained cross section for the three body ^{4}He(γ,pn)^{2}H channel, 0.05 mb, is in good agreement with previous data. Theoretical calculation is required to compare to the present result.

3.2. *Cross section of the photodisintegration of ^{3}He*

The ^{3}He photodisintegration cross sections for the ^{3}He(γ,p)^{2}H and ^{3}He(γ,pp)n reactions measured at E_γ=10.2 and 16.0 MeV are shown in Fig. 8 together with the theoretical values based on the Faddeev approach with the Urbana IX $3N$ force.[15] Both cross sections at 16 MeV agree marginally with the calculations. However, the present (γ,p) and (γ,pp) cross sections at 10.2 MeV differ by 20% and by a factor 3 compared to the calculated ones, respectively. In all cases the $3N$ force effects are small, 5 $\sim$ 10%.[15] Hence, the situation for the (γ,pp) channel at 10.2 MeV remains an open problem.

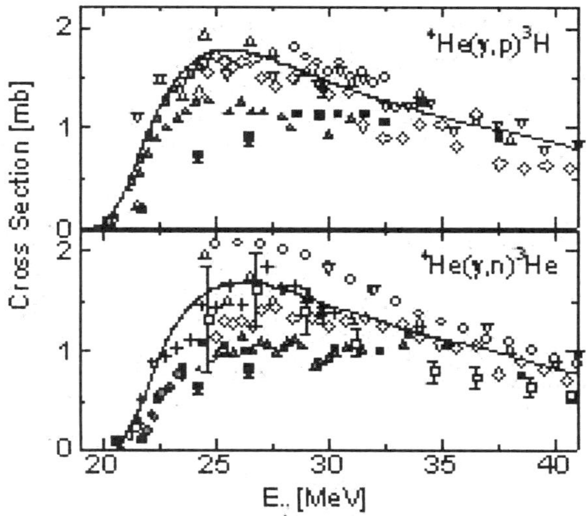

Fig. 7. Upper panel; (γ,p) cross sections. Filled circles; present, filled squares; Bernabei et al. (1988), open upward triangles; Gorbunov (1968), open downward triangles; Arkatov et al. (1978), open diamonds; Balestra et al. (1979), open circles; Hoorebeke et al. (1993), gray triangles; Feldman et al. (1980), gray squares; Hahn et al. (1995). The solid curve represents the calculated excitation function using the LIT method by Quaglioni et al. (2004). Lower panel; (γ,n) cross sections. Filled circles; present, filled squares; Berman et al. (1980), open upward triangles; Gorbunov (1968), open circles; Malcom et al. (1973), crosses; Irish et al. (1973), open downward triangles; Arkatov et al. (1978), open diamonds; Balestra et al. (1979), open squares; Nilsson et al. (2005) (here the data includes a reported systematic uncertainty), gray triangles; Ward et al. (1981), gray circles; Komar et al. (1993). The solid curve and the dashed curve represent the calculated excitation functions using the LIT method by Quaglioni et al. (2004) and the AGS formalism (Ellerkmann et al. (1996), Sandhas et al. (1998). (Reference given here; see references in Ref.[10]).

3.3. *Cross section for neutron capture by the deuteron*

The present result for neutron capture by deuterium is compared to theoretical calculations based on the Faddeev approach and the pionless effective field theory. The former calculations were made with single-nucleon current and meson-exchange current, and in addition with the three-nucleon force, and they marginally agree with the new data.[19] The present data differ from the latter calculations. It would be quite interesting to compare the present results of the ^{2}H(n,γ)^{3}H and ^{3}He(γ,p)^{2}H reactions, which were performed at quite low energy, to calculated ones based on the effective field theory.

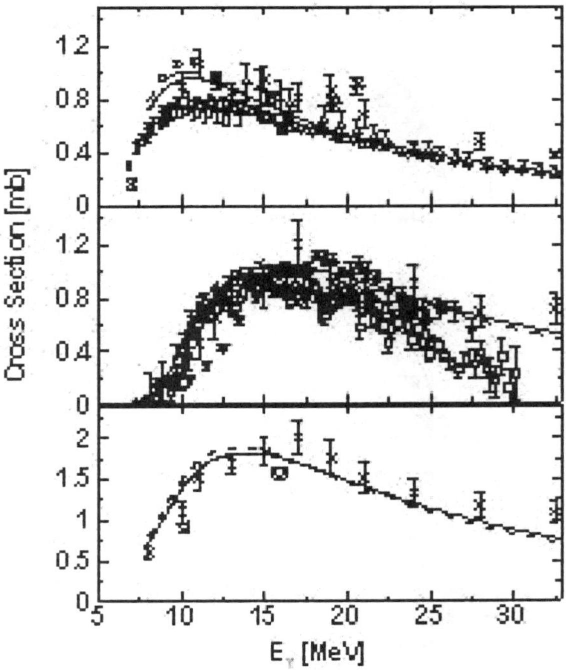

Fig. 8. The ^{3}He(γ,p)^{2}H (top), ^{3}He(γ,pp)n (middle), and total cross sections (bottom) of the ^{3}He photodisintegration.[15]

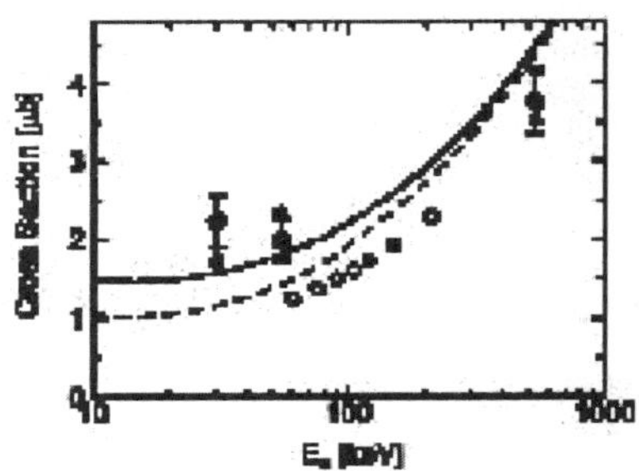

Fig. 9. The measured ^{2}H(n,γ)^{3}H cross section vs. neutron energy (lab.). • : present result, △ : Faddeev., ○ : pionless effective field theory, solid line : JENDL, and dotted line: ENDF.

4. Summary

We have successfully established a new measurement system to determine the photodisintegration cross section of a nucleus accurately with small systematic uncertainty. The present studies for the direct simultaneous measurements of the ^{4}He and ^{3}He two-body and three-body photodisintegration cross sections using the thus developed method have solved a longstanding problem of the discrepancy of the old data for the ^{4}He two-body photodisintegration cross sections. Further theoretical developments are required to get a deeper insight into the excitation function in case of ^{4}He photodisintegration.

Acknowledgements

We are deeply grateful to Drs. Baba, Naito, and Tomyo for their great contributions to the present studies. We also thank to Dr. Skibinski, Profs. Glockle and Witala for useful discussions.

References

1. P. J. E. Peebles, *Phys. Rev. Lett.* 16, 410 (1966), H. Sato, *Prog. Theor. Phys.*, 38, 1083 (1967), R. V. Wagoner, W. A. Fowler, and F. Hoyle *Astrophys. J.*, 148, 3, (1967)
2. R.H. Cyburt, Phys. Rev. D 70, 023505.
3. L. Kaplan, G. R. Ringo, and K. E. Wilzbach, *Phys. Rev.* 87, 785, (1952), V. P. Alfimenkov et al., *Sov. J. Nucl.Phys.* 32, 771, (1980), E. T. Jurney, P. T. Bendt, & J. C. Browne, *Phys. Rev.* C25, 2810, (1982)
4. L. I. Schiff, *Phys. Rev.* 52, 242, (1937), A. C. Philips, *Nucl. Phys.* A184, 337, (1972)
5. J. L. Friar, B. F. Gibson, & G. L. Payne, *Phys. Lett.* B251, 11, (1990), and J. Carlson, et al., *Phys. Rev.* C42, 830, (1990)
6. D. Gazit et al., *Phys. Rev. Lett.* 96,112301, (2006) and references therein.
7. F.C. Barker and A.K. Mann, *Philos. Mag.* 25, 5, (1957)
8. S.E. Woosley, D.H. Hartmann, R.D. Hoffman, and W.C. Haxton, *Astrophys. J.* 356, 272, (1990)
9. H.A. Bethe, and J.R. Wilson, *Ap.J*, 295, 14, (1895)
10. T. Shima et al., *Phys. Rev.* C72, 044004, (2005)
11. G. Ellerkmann, W. Sandhas, S.A. Sofianos, and H. Fiedeldey, *Phys. Rev.* C 53, 2638, (1996)
12. T. Kii, T. Shima, T. Baba and Y. Nagai, *Nucl. Instr. and Meth.* A 552, 329, (2005)
13. R. Skibiski et al. *Phys. Rev.* C67, 054001, (2003), R. Skibinski et al. *Phys. Rev.* C67, 054002, (2003)
14. V.D. Efros, W. Leidemann, G. Orlandini, and E.L. Tomusiak, *Phys. Lett.* B484, 223, (2000)
15. S. Naito et al., *Phys. Rev.* C 73, 034003, (2006)
16. H. Sadeghi and S. Bayegan, *Nucl. Phys.* A 753, 291, (2005)
17. T. Ohsaki et al., *Nucl. Instr. and Meth.* A 425, 302, (1999), M. Igashira,

K. Tanaka, and K. Masuda, *Proc. Conf. of the 8th. Int. Symp. on Capture Gamma-Ray and Related Topics*, World Scientific, Singapore, 992, (1993)
18. ENDF/B-VI data file for 197Au (MAT=7925). 1993, evaluated by P.G. Young and E.D. Arthur.
19. Y. Nagai et al., *Phys. Rev.* C 74. 025804, (2006)

FEW-BODY LATTICE CALCULATIONS

M. J. SAVAGE

Department of Physics, University of Washington, Seattle, WA 98195-1560.

I will discuss the recent progress toward computing few-body observables using numerical lattice techniques. The focus is overwhelmingly on the latest results from lattice QCD calculations. I present preliminary results from a lattice calculation of the central and tensor potentials between B-mesons in the heavy-quark limit.

Keywords: Lattice QCD, effective field theory.

1. Introduction

Perhaps the greatest challenge facing those of us working in the area of strong interaction physics is to be able to rigorously compute the properties and interactions of nuclei. The many decades of theoretical and experimental investigations in nuclear physics have, in many instances, provided a very precise phenomenology of the strong interactions in the non-perturbative regime. However, at this point in time we have little understanding of much of this phenomenology in terms of the underlying theory of the strong interactions, Quantum Chromo Dynamics (QCD).

The ultimate goal is to be able to rigorously compute the properties and interactions of nuclei from QCD. This includes determining how the structure of nuclei depends upon the fundamental constants of nature. Any nuclear observable is essentially a function of only five constants, the length scale of the strong interactions, Λ_{QCD} , the quark masses, m_u, m_d, m_s, and the electromagnetic coupling, α_e (at low energies the dependence upon the top, bottom and charm quarks masses is encapsulated in Λ_{QCD}). Perhaps as important, we would then be in the position to reliably compute quantities that cannot be accessed, either directly or indirectly, by experiment.

The only way to rigorously compute strong-interaction quantities in the nonperturbative regime is with lattice QCD. One starts with the QCD Lagrange density and performs a Monte-Carlo evaluation of Euclidean space

Green functions directly from the path integral. To perform such an evaluation, space-time is latticized and computations are performed in a finite volume, at finite lattice spacing, and at this point in time, with quark masses that are larger than the physical quark masses. To compute any given quantity, contractions are performed in which the valence quarks that propagate on any given gauge-field configuration are "tied together". For simple processes such as nucleon-nucleon scattering, such contractions do not require significant computer time compared with lattice or propagator generation. However, as one explores processes involving more hadrons, the number of contractions grows rapidly (for a nucleus with atomic number A and charge Z, the number of contractions is $(A+Z)!(2A-Z)!$), and a direct lattice QCD calculation of the properties of a large nucleus is quite impractical simply due to the computational time required. The way to proceed is to establish a small number of effective theories, each of which have well-defined expansion parameters and can be shown to be the most general form consistent with the symmetries of QCD. Each theory must provide a complete description of nuclei over some range of atomic number. Calculations in two "adjacent" theories are performed for a range of atomic numbers for which both theories converge. One then matches coefficients in one EFT to the calculations in the other EFT or to the lattice, and thereby one can make an indirect, but rigorous connection between QCD and nuclei. It appears that four different matchings are required:

(1) **Lattice QCD**. Lattice QCD calculations of the properties of the very lightest nuclei will be possible at some point in the not so distant future.[1] Calculations for $A \leq 4$ as a function of the light-quark masses, would uniquely define the interactions between nucleons up to and including the four-body operators. Depending on the desired precision, one could possibly imagine calculations up to $A \sim 8$.
The chiral potentials and interactions have been determined out to the order where four-body interactions contribute.[2] As with the EFT constructions in the meson-sector and single-nucleon sector, the number of counterterms proliferates with increasing order in the expansion, and at some order one looses rigorous predictive power without external input. Lattice QCD will make model independent determinations of these counterterms.

(2) **Exact Many-Body Methods**. During the past decade one has seen remarkable progress in the calculation of nuclear properties using Green Function Monte-Carlo (GFMC) with the AV_{18}-potential (e.g. Ref.[3]) and also the No-Core Shell Model (NCSM) (e.g. Ref.[4]) using chiral

potentials. Starting with the chiral potentials, which are the most general interactions between nucleons consistent with QCD, one would calculate the properties of nuclei as a function of all the parameters in the chiral potentials out to some given order in the chiral expansion. A comparison between such calculations and lattice QCD calculations will determine these parameters to some level of precision. These parameters can then be used in the calculation of nuclear properties up to atomic numbers $A \sim 20 - 30$. The computer time for these many-body theories suffers from the same $\sim (A!)^2$ blow-up that lattice QCD does, and for a sufficiently large nucleus, such calculations become impractical.

Another recent development that shows exceptional promise is the latticization of the chiral effective field theories.[5–9] This should provide a model-independent calculation of nuclear processes once matched to lattice QCD calculations.

(3) **Coupled Cluster Calculations**. In order to move to larger nuclei, $A \lesssim 100$, a technique that has shown promise is to implement a coupled-clusters expansion (e.g. Ref.[10]). One uses the same chiral potential that will have been matched to lattice QCD calculations, and then performs a diagonalization of the nuclear Hamiltonian, after truncating the cluster expansion, which itself contains arbitrary coefficients. The results of these calculations will be matched to those of the NCSM or GFMC for $A \sim 20 - 30$ to determine the arbitrary coefficients. This method is unlikely to be practical for very large atomic numbers.

(4) **Density Functional Theory and Very Large Nuclei** To complete the periodic table one needs to have an effective theory that is valid for very large nuclei and nuclear matter. A candidate that has received recent attention is Density Function Theory (DFT) (e.g. Refs.[11,12]). It remains to be seen if this is in fact a viable candidate. There is reason to hope that this will be useful because there is clearly a density expansion in large nuclei with a power-counting that is consistent with the Naive Dimensional Analysis (NDA) of Georgi and Manohar.[13] The application of DFT to large nuclei is presently the least rigorously developed component of this program.

The latticized chiral theory mentioned previously can also be applied to the infinite nuclear matter problem. This work is still in the very earliest stages of exploration, but this looks promising.[7]

2. Lattice QCD Calculations of Single Nucleon Properties

While the lattice QCD calculations have historically focused on the meson sector, both the properties of single mesons and the interactions between two mesons, primarily due to limitations in computational power, the last few years has seen an increasing number of precise calculations of the properties of nucleons at the available pion masses. I wish to show two that will be of interest to the participants of this meeting.

2.1. *The Matrix Element of the Axial Current*

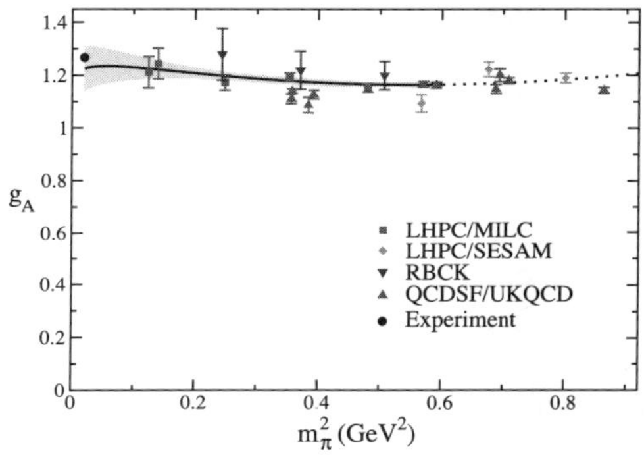

Fig. 1. g_A as a function of m_π^2. This figure is reproduced with the permission of LHPC.

The matrix element of the axial current in the nucleon, g_A, is a fundamental quantity in nuclear physics as it is related to the strength of the long-range part of the nucleon-nucleon interaction via PCAC. During the last year LHPC[14] has for the first time calculated g_A at small enough pion masses where Heavy Baryon Chiral Perturbation Theory (HBχPT) should converge. The results of this calculation, previous calculations and the physical value, are shown in Fig. 1. Also shown is the chiral extrapolation along with its uncertainty (the shaded region). The physical value lies within the range predicted by chiral extrapolation of the lattice calculations.

2.2. *The Neutron-Proton Mass Difference*

During the last year it was realized that one could use isospin-symmetric lattices to compute isospin-breaking quantities to NLO in the chiral expansion.[15] This is achieved by performing partially-quenched (unphysical) calculations of the nucleon mass[16] in which the valence quark masses differ from the sea-quark masses (those of the configurations) and determining counterterms in the partially-quenched chiral Lagrangian. The NPLQCD collaboration found that, in the absence of electromagnetic interactions, the neutron-proton mass difference at the physical value of the quark masses is

$$M_n - M_p\Big|^{m_d - m_u} = +2.26 \pm 0.57 \pm 0.42 \pm 0.10 \text{ MeV} \quad , \tag{1}$$

which is to be compared with estimates derived from the Cottingham sum-rule of $M_n - M_p\Big|^{m_d - m_u}_{\text{physical}} = +2.05 \pm 0.3$. We see that the lattice determination is consistent with what is found in nature. The lattice calculation is expected to become significantly more precise during the next year.

3. Hadron Scattering from Lattice QCD

To circumvent the Maiani-Testa theorem,[17] which states that one cannot compute Green functions at infinite volume on the lattice and recover S-matrix elements except at kinematic thresholds, one computes the energy-eigenstates of the two particle system at finite volume to extract the scattering amplitude.[18] The scattering amplitude of two pions in the $I = 2$-channel has been the process of choice to explore this technique, and to determine the reliability and systematics of lattice calculations.

3.1. $\pi\pi$ *Scattering*

NPLQCD has calculated $\pi\pi$ scattering in the $I = 2$ channel at relatively low pion masses using the mixed-action technique of LHPC. Domain-wall valence propagators are calculated on MILC configurations containing staggered sea-quarks. The lattice calculations were performed with the *Chroma* software suite[19,20] on the high-performance computing systems at the Jefferson Laboratory (JLab).

This calculation demonstrates the predictive capabilities of lattice QCD combined with low-energy EFT's. By writing the expansion of $m_\pi a_2$ as a function of m_π^2/f_π^2, it has been shown that, when inserting the values of m_π and f_π as calculated on the lattice, the lattice spacing effects in this mixed action calculation, which naively appear at $\mathcal{O}(b^2)$, are further suppressed, appearing only at higher orders.[22,23] The first non-trivial contributions from the partial-quenching in this calculation are shown to be numerically very small, and therefore a clean extraction of the counterterm in the χPT Lagrangian is possible. The chiral extrapolation, as shown in Fig. 2, is found to have a smaller uncertainty at the physical point than the experimental value.

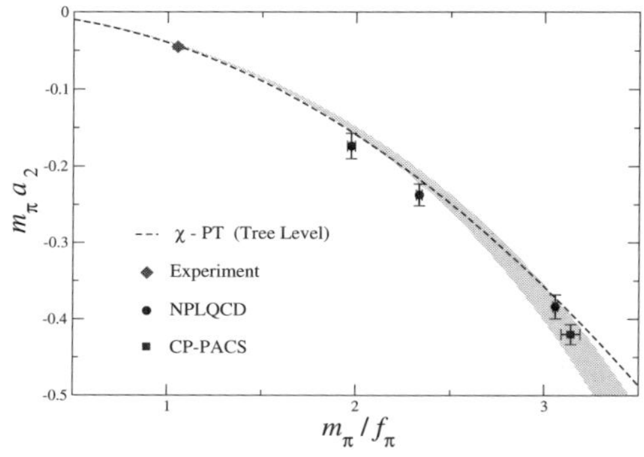

Fig. 2. The $\pi\pi$ scattering length in the $I = 2$ channel as a function of m_π/f_π. The dashed curve corresponds to the unique prediction of tree-level χPT, while the shaded region is the fit to the results of the NPLQCD calculations.[21] The NPLQCD calculations are performed at a single lattice spacing of $b \sim 0.125$ fm.

3.2. *$K\pi$ Scattering*

Studying the low-energy interactions between kaons and pions with $K^+\pi^-$ bound-states allows for an explicit exploration of the three-flavor structure of low-energy hadronic interactions, an aspect that is not directly probed in $\pi\pi$ scattering. Experiments have been proposed by the DIRAC collaboration[24] to study $K\pi$ atoms at CERN, J-PARC and GSI, the results of which would provide direct measurements or constraints on combinations of the scattering lengths. In the isospin limit, there are two isospin channels available to the $K\pi$ system, $I = \frac{1}{2}$ or $I = \frac{3}{2}$. The width of a $K^+\pi^-$ atom depends upon the difference between scattering lengths in the two channels, $\Gamma \sim (a_{1/2} - a_{3/2})^2$, (where $a_{1/2}$ and $a_{3/2}$ are the $I = \frac{1}{2}$ and $I = \frac{3}{2}$ scattering lengths, respectively) while the shift of the ground-state depends upon a different combination, $\Delta E_0 \sim 2a_{1/2} + a_{3/2}$.

The NPLQCD collaboration calculated the $K^+\pi^+$ scattering length, $a_{K^+\pi^+}$, in the same way as we computed $\pi\pi$ scattering, requiring only the additional generation of strange quark valence propagators.[25] As $a_{K^+\pi^+}$ was calculated (again at a single lattice spacing of $b \sim 0.125$ fm) at three different pion masses, a chiral extrapolation was performed. This extrapolation depends upon two counterterms, one from the crossing-even and one from the crossing-odd amplitudes, and the important point is that their coefficients have different dependence upon the meson masses. Therefore, both can be determined from the results of the lattice calculation, as shown in Fig. 3, and therefore, the scattering lengths in **both** isospin channels can

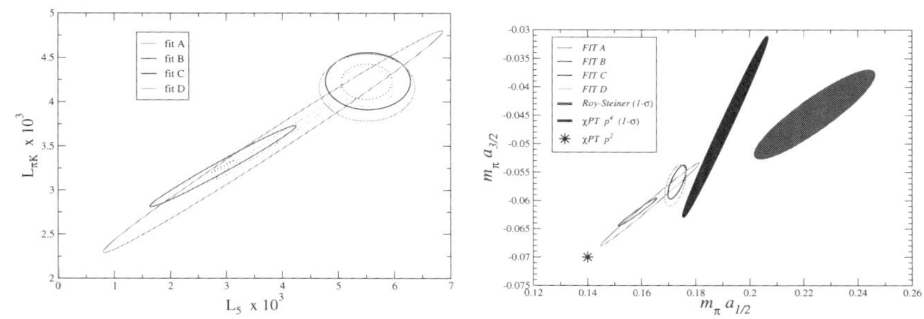

Fig. 3. The left panel shows the 95% confidence ellipses for the counterterms L_5 and $L_{K\pi}$ that contribute to $K\pi$ scattering at NLO in χPT as extracted from the mixed-action lattice QCD calculation of NPLQCD at a lattice spacing of $b \sim 0.125$ fm.[25] The right panel shows the 95% confidence ellipses for the $K\pi$ scattering lengths from a combination of a lattice QCD calculation and χPT. Also shown are the 38% confidence ellipses from a Roy-Steiner analysis and from a $\mathcal{O}(p^4)$ χPT analysis.

be predicted, as shown in Fig. 3. This is a another demonstration of the combined power of lattice QCD and χPT.

3.3. *Nucleon-Nucleon Scattering*

A few years ago we realized[26,27] that even though the scattering lengths in the nucleon-nucleon system are unnaturally large, and much larger than the spatial dimensions of currently available lattices, rigorous calculations in the NN-sector could be performed today. There are two aspects to this. First, it is unlikely that the scattering lengths in the NN sector are unnaturally large when computed on lattices with the lightest pion masses that are presently available, $m_\pi^{\rm sea} \gtrsim 250$ MeV. Second, it is not the scattering length that dictates the lattice volumes that can be used in a Luscher-type analysis, but it is the range of the interaction, which is set by m_π for the NN interaction. While the Luscher asymptotic formulae are not applicable when the scattering length becomes comparable to the spatial dimensions, the complete relation is still applicable,

$$p \cot \delta(p) \ = \ \frac{1}{\pi L}\, \mathbf{S}\left(\frac{pL}{2\pi}\right) \quad , \quad \mathbf{S}\,(\eta) \ \equiv \ \sum_{\mathbf{j}}^{|\mathbf{j}|<\Lambda} \frac{1}{|\mathbf{j}|^2 - \eta^2} \ - \ 4\pi\Lambda \ . \qquad (2)$$

where $\delta(p)$ is the elastic-scattering phase shift, and the sum in eq. (2) is over all triplets of integers $\mathbf{j}$ such that $|\mathbf{j}| < \Lambda$ and the limit $\Lambda \to \infty$ is implicit.[27]

The same set of configurations that was used to produce the scattering lengths for $\pi\pi$ and $K\pi$ scattering, discussed above, were used to construct the NN correlation functions, and hence to extract the energy shifts between two nucleons at finite volume and twice the single nucleon mass at finite volume. At the pion masses accessible, the NN scattering lengths are found

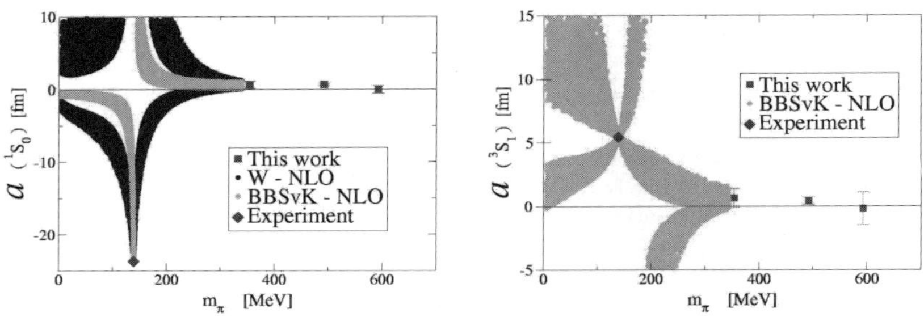

Fig. 4. Scattering lengths in the $^1S_0$ and $^3S_1 - ^3D_1$ NN channels a sa function of pion mass. The experimental value of the scattering length and NDA have been used to constrain the extrapolation in both BBSvK[28–30] and W[31–33] power-countings at NLO.

to be of natural size, set by the inverse pion mass. Only one of the pion masses is within the region described by the low-energy EFT's, and as such a prediction of the scattering length from lattice QCD alone cannot be made. However, when combined with the physical values of the scattering lengths, an allowed region for the scattering lengths as a function of pion mass can be made, as shown in Fig. 4.

4. Hadronic Potentials

Lattice QCD cannot, in general, isolate the potential between hadrons, as the potential by itself is not an observable quantity. An exception to this statement is when both hadrons are infinitely massive, and consequently their spatial separation can be fixed (this of course generalizes to N-hadrons). Currently, Silas Beane, William Detmold, Kostas Orginos and I are performing a quenched calculation of the potential between two B-mesons in the heavy quark limit. In this limit the light degrees of freedom (dof) of each B-meson decouple from the heavy-quark dynamics, and their spin, s_l, becomes a good quantum number. The potential between the two B-mesons when constructed from χPT has the same form as that for NN, but with different values of the counterterms that enter. Therefore, calculating the potential between two B-mesons may provide qualitative insight into the NN-system. Our calculation is not the first attempt to extract the potential between heavy hadrons, but is performed with a lattice spacing approximately half of that of the previous calculations.[34,35] The quark mass was chosen to give $m_\pi \sim 420$ MeV and $m_\rho \sim 700$ MeV. Fig. 5

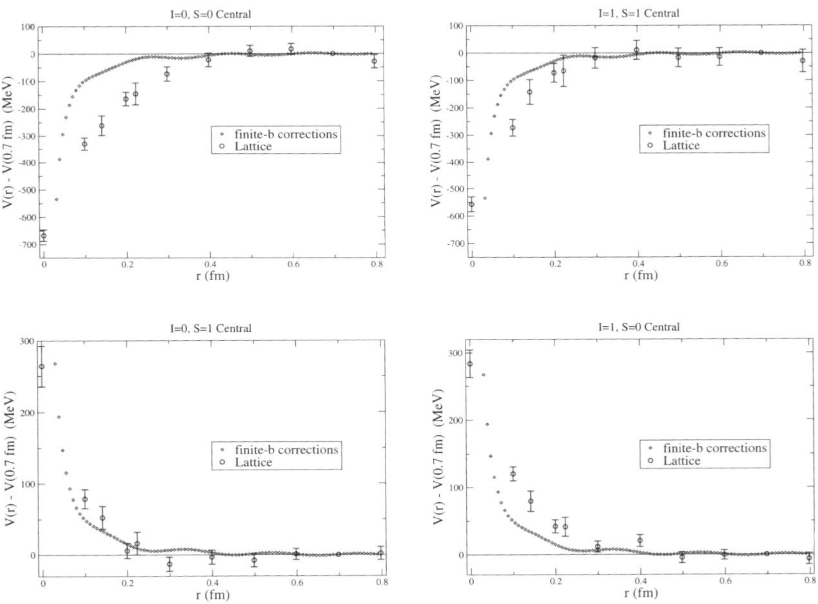

Fig. 5. Preliminary calculations of the potential between two B-mesons in the heavy quark limit. The vertical axis is the difference between the central potential at a separation r and that at $r = 7b$. The fine (red) dots correspond to the leading finite-lattice spacing correction that must be added to the potential obtained in the lattice calculation.

shows our preliminary results for the central potentials as a function of B-meson separation (the tensor potential is found to be very small, and so we have also shown the potentials at $r = \sqrt{2}$ and $\sqrt{5}$ which are not separated into tensor and central). One clearly sees short-distance repulsion in the channel with the quantum numbers of s-wave NN interactions, while one finds large attraction in the other channels. There is a correction due to the finite lattice spacing that is to be added to the results of the lattice calculation, and the leading correction is shown in Fig. 5. Adding this factor gives the continuum potential between two infinitely massive B-mesons in the lattice volume, while the extrapolation to infinite volume potential has not yet been performed.

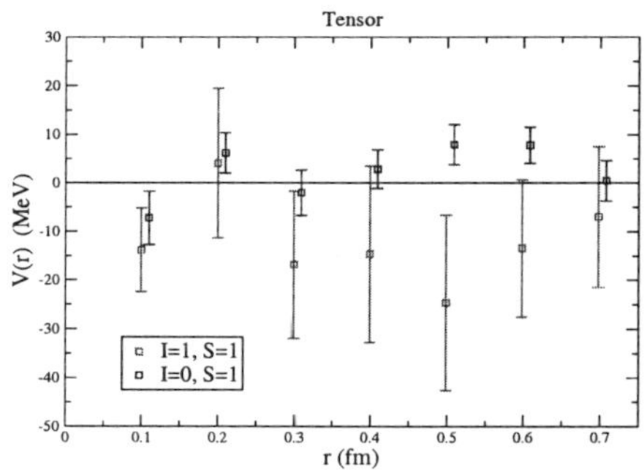

Fig. 6. Preliminary calculations of the tensor potential between two B-mesons in the heavy quark limit.

The tensor interaction between hadrons has not been directly isolated in previous works. In Fig. 6 we show our preliminary calculation of the tensor potential between two infinitely massive B-mesons. We find a potential that is consistent with zero within the uncertainty of the calculation, indicating that the tensor potential in this system is no larger than $V_T \lesssim 40$ MeV.

5. The Three-Body Sector

A discussion of the three-body sector would require multiple presentations all by itself. I want to say just a few words about it. In order for lattice QCD to make connections with the three-body sector in nuclear physics there must be both theoretical and numerical developments. So far essentially none of the formalism has been developed with which to analyze lattice results. The analogous relations to the Luscher relation in the NN sector must be developed, and numerically explored to determine parameters that can be used in lattice calculations. On the numerical side, it is clear from the poor quality of the NN correlation functions obtained by NPLQCD, that extracting any signal from the NNN sector will be very difficult, and likely not possible on the coarse MILC lattices. We are currently discussing performing the contractions for the triton with the propagators we currently have to quantify the situation. Performing the contractions rapidly becomes computationally expensive. For the proton there are just

2 contractions, for pp there are 48 contractions while for the triton there are 2880 contractions, this is simply $u!d!$. At this stage we believe that the extraction of scattering parameters from lattice QCD in the three-body sector (in the near future) will require the asymmetric Clover lattices that JLab has applied for computer resources to produce during the next couple of years.

6. Nuclear Effective Field Theory Lattice Calculations

The last subject I wish to mention, but again I have no time to actually discuss it, is the recent development of performing EFT calculations on the lattice. I think that this is an area that has a bright future and I refer those that are interested to recent papers such as Refs.[8,36]

7. Conclusions

It is clear that lattice QCD is an important part of the future of nuclear physics. Lattice QCD, when combined with effective field theory, is just starting to make rigorous predictions of few-body observables, and we can expect significantly more progress as the computational resources dedicated to these calculations is increased. The formal tools are in place to explore the two-body sector, all that is required is computational power. However, a coherent effort involving both numerical calculations and formal developments is presently required to move beyond the two-body sector.

I would like to thank my collaborators in this work, Silas Beane, Paulo Bedaque, Kostas Orginos, William Detmold, Tom Luu, Elisabetta Pallante and Assumpta Parreno.

References

1. M. J. Savage, PoS **LAT2005**, 020 (2005) [arXiv:hep-lat/0509048].
2. E. Epelbaum, Phys. Lett. B **639**, 456 (2006).
3. S. C. Pieper, R. B. Wiringa and J. Carlson, Phys. Rev. C **70**, 054325 (2004) [arXiv:nucl-th/0409012].
4. C. Forssen, P. Navratil, W. E. Ormand and E. Caurier, Phys. Rev. C **71**, 044312 (2005) [arXiv:nucl-th/0412049].
5. H. M. Muller and R. Seki, *Given at Caltech / INT Mini Workshop on Nuclear Physics with Effective Field Theories, Pasadena, CA, 26-27 Feb 1998*
6. H. M. Muller, S. E. Koonin, R. Seki and U. van Kolck, Phys. Rev. C **61**, 044320 (2000) [arXiv:nucl-th/9910038].
7. D. Lee, B. Borasoy and T. Schafer, Phys. Rev. C **70**, 014007 (2004) [arXiv:nucl-th/0402072].

8. B. Borasoy, H. Krebs, D. Lee and U. G. Meissner, arXiv:nucl-th/0510047.
9. R. Seki and U. van Kolck, arXiv:nucl-th/0509094.
10. M. Wloch, J. R. Gour, P. Piecuch, D. J. Dean, M. Hjorth-Jensen and T. Papenbrock, J. Phys. G **31**, S1291 (2005).
11. R. J. Furnstahl, J. Phys. G **31**, S1357 (2005) [arXiv:nucl-th/0412093].
12. A. Schwenk and J. Polonyi, arXiv:nucl-th/0403011.
13. A. Manohar and H. Georgi, Nucl. Phys. B **234**, 189 (1984).
14. R. G. Edwards *et al.* [LHPC Collaboration], Phys. Rev. Lett. **96**, 052001 (2006) [arXiv:hep-lat/0510062].
15. S. R. Beane, K. Orginos and M. J. Savage, arXiv:hep-lat/0605014.
16. S. R. Beane and M. J. Savage, Nucl. Phys. A **709**, 319 (2002) [arXiv:hep-lat/0203003].
17. L. Maiani and M. Testa, Phys. Lett. B **245**, 585 (1990).
18. M. Luscher, Commun. Math. Phys. **105**, 153 (1986).
19. R. G. Edwards and B. Joo [SciDAC Collaboration], Nucl. Phys. Proc. Suppl. **140** (2005) 832 [arXiv:hep-lat/0409003].
20. C. McClendon, Jlab preprint, JLAB-THY-01-29 (2001).
21. S. R. Beane, P. F. Bedaque, K. Orginos and M. J. Savage [NPLQCD Collaboration], Phys. Rev. D **73**, 054503 (2006) [arXiv:hep-lat/0506013].
22. J. W. Chen, D. O'Connell, R. S. Van de Water and A. Walker-Loud, Phys. Rev. D **73**, 074510 (2006) [arXiv:hep-lat/0510024].
23. A. Walker-Loud, arXiv:hep-lat/0608010.
24. http://dirac.web.cern.ch/DIRAC/future.html
25. S. R. Beane, P. F. Bedaque, T. C. Luu, K. Orginos, E. Pallante, A. Parreno and M. J. Savage, arXiv:hep-lat/0607036.
26. S. R. Beane, P. F. Bedaque, A. Parreno and M. J. Savage, Nucl. Phys. A **747**, 55 (2005) [arXiv:nucl-th/0311027].
27. S. R. Beane, P. F. Bedaque, A. Parreno and M. J. Savage, Phys. Lett. B **585**, 106 (2004) [arXiv:hep-lat/0312004].
28. D. B. Kaplan, M. J. Savage and M. B. Wise, Nucl. Phys. B **534**, 329 (1998) [arXiv:nucl-th/9802075].
29. D. B. Kaplan, M. J. Savage and M. B. Wise, Phys. Lett. B **424**, 390 (1998) [arXiv:nucl-th/9801034].
30. S. R. Beane, P. F. Bedaque, M. J. Savage and U. van Kolck, Nucl. Phys. A **700**, 377 (2002) [arXiv:nucl-th/0104030].
31. S. Weinberg, Phys. Lett. B **251**, 288 (1990).
32. S. Weinberg, Nucl. Phys. B **363**, 3 (1991).
33. C. Ordoñez, L. Ray and U. van Kolck, Phys. Rev. C **53**, 2086 (1996) [arXiv:hep-ph/9511380].
34. P. Pennanen, C. Michael and A. M. Green [UKQCD Collaboration], Nucl. Phys. Proc. Suppl. **83**, 200 (2000) [arXiv:hep-lat/9908032].
35. M. S. Cook and H. R. Fiebig, arXiv:hep-lat/0210054.
36. D. Lee and T. Schafer, Phys. Rev. C **73**, 015202 (2006) [arXiv:nucl-th/0509018].

RESEARCH OPPORTUNITIES AT THE UPGRADED HIγS FACILITY

H. R. WELLER

Duke University & Triangle Universities Nuclear Laboratory (TUNL), Box 90308, Durham, North Carolina 27708-0308, USA.

The High Intensity γ-ray Source (HIγS) is a joint project between TUNL and the Duke Free Electron Laser Laboratory (DFELL). This facility utilizes intra-cavity back-scattering of the FEL light in order to produce intense γ-ray beams. An upgrade which will allow for the production of γ-rays up to energies of about 160 MeV having total intensities in excess of 10^8/sec is essentially completed. The primary component of the upgrade is a 1.2 GeV booster-injector which will make it possible to replace lost electrons at full energy. In addition, an upgrade of the present linear undulator to a helical system will provide nearly 100% linear and circularly polarized beams. The full system, including the booster injector, will be ready in 2007. The proposed experimental program includes low-energy studies of nuclear reactions of importance in nuclear astrophysics as well as studies of nuclear structure using the technique of nuclear resonance fluorescence (NRF). Future double-polarization experiments include a study of the Gerasimov-Drell-Hearn Sum Rule for the deuteron and ^{3}He, and an extensive Compton scattering program designed to probe the internal structure of the nucleon. A major focus of these studies will be the measurement of the spin-polarizabilities of the proton and the neutron. Studies at pion-threshold designed to observe Isospin-symmetry breaking effects are also being planned. A description of the anticipated facility following the present upgrades will be given in this talk, along with a description of some of the planned experiments.

1. Introduction

The major components of the HIγS facility are the OK-4 XUV FEL and the Duke 1.2 GeV electron storage ring with its 280 MeV linac injector. When operating in the two-bunch mode, the two electron bunches contained in the ring can be thought of as a *lasing* bunch and a *scattering* bunch. The lased photons from one electron bunch are reflected from the downstream mirror, then collide head on with the second electron bunch at the *collision point*, producing gamma rays.[1] The flux of γ rays can be increased by increasing the current in the target bunch and/or by operating with up to eight electron bunches. The present OK-4 FEL is being upgraded to a helical undulator system (OK-5[2]). Whereas OK-4 produces only linearly polarized beams, OK-5 will produce both linear and circularly polarized γ rays.[3]

The present system can support the HIγS operations in the *no loss* mode up to 20 MeV. Total beam intensities on the order of 10^6-to-10^8 γ/s are available between 2 and 20 MeV. For γ-ray energies above 20 MeV, the recoil electrons exceed the energy acceptance of the storage ring. This limitation was the main motivation for the HIγS upgrade project.[4] The primary component of the upgrade project is the addition of a 1.2 GeV Booster Injector, which will allow for *top-off mode* operation, ie. continuous replacement of lost electrons at full energy.

Figure 1 presents a schematic view of the upgraded HIγS facility. One of the first upgrades performed in this project was the design and installation

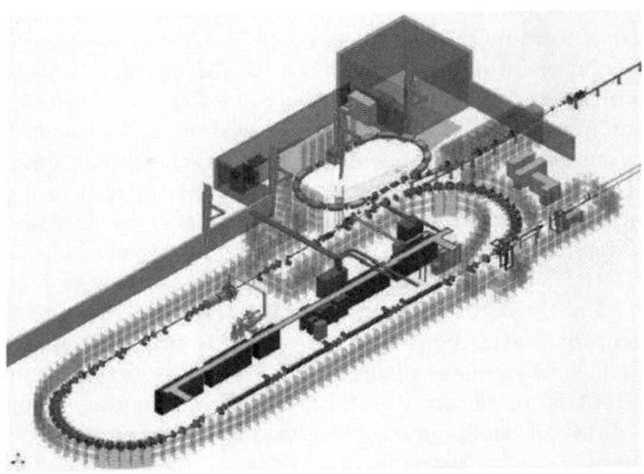

Fig. 1. The upgraded HIγS facility showing the Booster Injector, the Stroage Ring and the OK-5 system.

of a new RF System, especially designed to eliminate Higher-Order-Mode (HOM) damping. This damping had previously limited the amount of current we could store in the ring. The new RF system was designed and built in collaboration with scientists and engineers at the Budker Institute in Novosibirsk. It has been installed at HIγS, and is presently operational. The major aspect of the upgrade, however, is the 1.2 GeV Booster Injector, also shown in Fig. 1. A building extension was necessary to contain the Booster. Additional shielding has been installed to accommodate the radiation generated by the lost electrons. The upgrade is in its final stages, with research activities anticipated to resume in early 2007.

2. The Gerasimov-Drell-Hearn Sum Rule for the Deuteron

Measurements of the Gerasimov-Drell-Hearn integral on the proton and neutron to test the Gerasimov-Drell-Hearn (GDH) Sum Rule have been and are being performed at Mainz, LEGS, Bonn and TJNAF. The GDH sum rule for a nucleon is

$$\int_{\omega_\pi}^{\infty} (\sigma_P(\omega) - \sigma_A(\omega))\frac{d\omega}{\omega} = 2\pi^2 \left(\frac{\kappa_N\, e}{M_N}\right)^2, \tag{1}$$

where ω_π corresponds to the threshold energy for pion production from the nucleon, $\sigma_P(\sigma_A)$ is the total inelastic photon cross section when the nucleon and the circularly polarized photon spins are parallel (anti-parallel), and κ_N is the anomalous magnetic moment of the nucleon.[5,6] It was pointed out by Hosada and Yamamoto[7] and Gerasimov[6] that these arguments could be applied equally well to the deuteron. That is, the deuteron could be treated as the object of the sum rule rather than simply as a source of neutrons. The resulting *GDH* sum rule is given by

$$\int_{\omega_2}^{\infty} (\sigma_P(\omega) - \sigma_A(\omega))\frac{d\omega}{\omega} = 4\pi^2 \left(\frac{\kappa_d\, e}{M_d}\right)^2, \tag{2}$$

where ω_2 corresponds to the threshold not for pion production ($\approx$ 145 MeV) but for photodisintegration ($\approx$ 2.2 MeV), and M_d (κ_d) is the mass (anomalous magnetic moment) of the deuteron. The sum rule values for the proton, neutron, and deuteron are given in the table below:

Target	κ	$\int GDH$
p	1.79	204.0 μb
n	−1.91	232.0 μb
d	−0.14	0.6 μb

The GDH integral for the deuteron can be separated into two pieces

$$\int_{\omega_2}^{\infty} GDH_d = \int_{\omega_2}^{\omega_\pi} GDH_d + \int_{\omega_\pi}^{\infty} GDH_d = 4\pi^2 \left(\frac{\kappa_d e}{M_d}\right) = 0.6\mu\text{b}. \tag{3}$$

The second term on the right hand side will be measured at LEGS, Mainz and Bonn. Its value can be estimated from the GDH integrals for the nucleons under the assumption that the impulse approximation is valid (this assumption is useful to obtain this estimate but is not significant to the basic argument since what will actually be measured at Mainz, LEGS, Bonn, and TJNAF is the GDH integral for the deuteron). Thus,

$$\int_{k_\pi}^{\infty} GDH_d = (204\mu\text{b}) + (232\mu\text{b}) = 436\mu\text{b}. \tag{4}$$

Therefore, the previous equation predicts that

$$\int_{k_2}^{k_\pi} GDH_d \cong -436\mu\text{b}. \tag{5}$$

That is, if the assumptions underlying the GDH sum rules are generally valid, then asymmetries in the total cross sections dominated by high energy pion production and resonance excitation processes give a firm prediction for asymmetries in a very low-energy process, namely photodisintegration below pion threshold. This is a connection that merits testing.

The connection between the GDH sum rule for the nucleons and that for the deuteron is potentially more interesting. There are suggestions that the GDH sum rule may be violated for nucleons. If this turns out to be true, then the immediate question will be, *What is the origin of the violation?* Possible explanations include a failure of the GDH integral to converge, the need to use a subtracted dispersion relation in calculating the forward-angle Compton amplitudes, or a breakdown of the LETs. Measurements on the deuteron would help to differentiate among these possibilities. In summary, the measurement of the GDH integral for the deuteron will be a valuable complement to measurements on the nucleons which are currently underway or in preparation. No previous measurements of the asymmetries contained in the GDH integrals for the deuteron have ever been attempted at energies below pion threshold.

Figure 2 shows the results of a calculation[8] of the difference between the total cross section when the helicity of the incident photon and the spin of the target are parallel (σ_P) and that when they are antiparallel (σ_A).

The dominant contribution comes from the near threshold $^1S_0$ resonance. At higher energies this difference shows a surprisingly large sensitivity to relativistic effects, a result meriting serious investigation especially since the positive contribution it creates is important to cancel some of the negative contribution ($\sim$550 μb) which is present near threshold if the sum-rule integral is going to result in the theoretical value.

The HIGS facility has been used to perform measurements of the photodisintegration of the deuteron in the photodisintegration-threshold region using 100% linearly polarized γ rays on an unpolarized target. These results

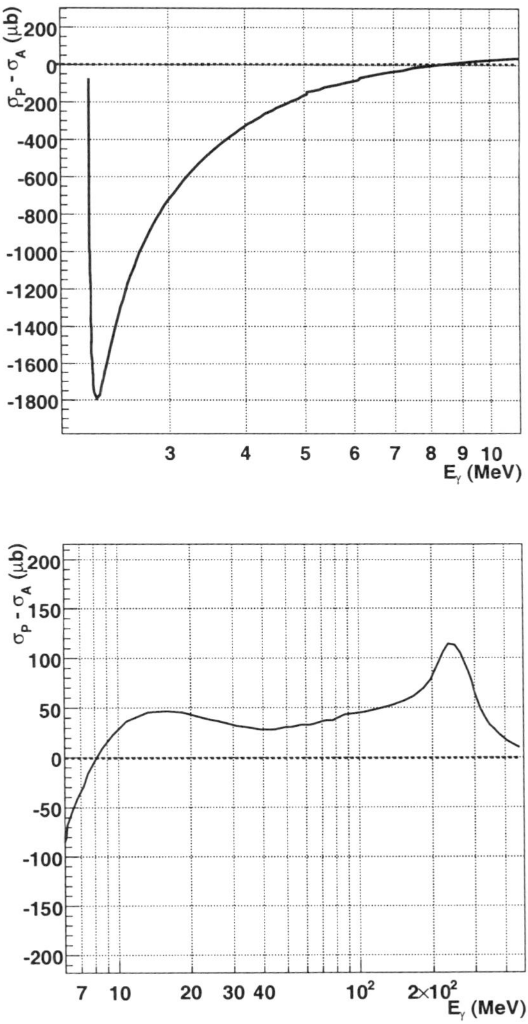

Fig. 2. The theoretical prediction of the GDH integrand for the deuteron (from Ref. [8]).

have provided an indirect determination of the GDH integrand below 10 MeV. This can be understood by noting that the term in the integrand of the GDH sum rule, $(\sigma_P - \sigma_A)$, can be written in terms of the contributing transition matrix elements. In the threshold region, there are only 5 matrix elements which are expected to contribute. Three of these are the E1 matrix elements. If we assume, as expected based on the small differences in the p-wave n-p scattering phase shifts at these energies, that the reduced E1 matrix elements are not j-dependent, then they do not contribute to the value of $(\sigma_P - \sigma_A)$. There are also two M1 matrix elements. One of these has the same quantum numbers as the ground state, and is therefore expected to not contribute to the M1 strength due to orthogonality. This result permits us to relate the M1 strength to the quantity $(\sigma_P - \sigma_A)$, the result being that $\sigma_P - \sigma_A$ = -3σ(M1). The linear analyzing powers from the HIγS measurements can be used to separate the M1 and E1 contributions to the cross section.[9,10] Another way of determining this is by means of polarized n-p capture (or, equivalently, by measuring the neutron's polarization in unpolarized photodisintegration of the deuteron). In this case the s-p phase (which can be obtained from n-p scattering data) is also required. All available data have been used to deduce the M1 cross section and hence the value of $(\sigma_P - \sigma_A)$, as shown in Fig 3 below. The integral of this strength amounts to about -620 μb.

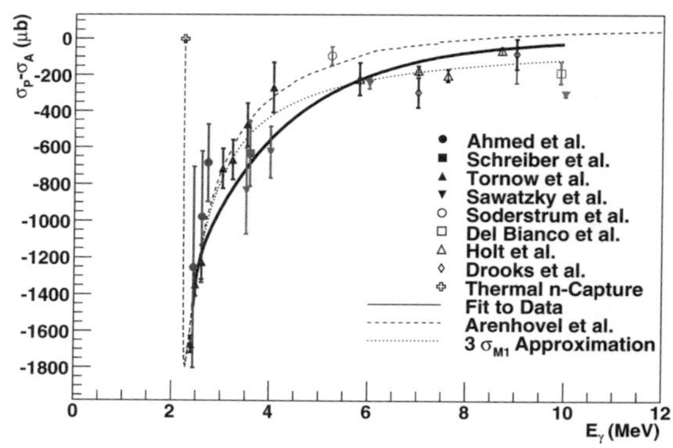

Fig. 3. Indirect measurement of the GDH integrand. The HIγS results are labeled: Ahmed et al., Schreiber et al., Tornow et al. and Sawatzky et al.

The assumptions which allowed us to extract the GDH integrand from these data break down at energies above 10 MeV where additional partial waves contribute. Theory predicts that the integrand becomes positive above 10 MeV as the result of a *relativistic* correction. The result is that the integral up to pion threshold is predicted to be -520 μb. A direct measurement of this integrand will begin at HIγS in late 2007. This experiment will utilize the circularly polarized beams from OK-5, the 88-neutron detector array *Blowfish*[10] (as shown in Fig. 4), and a frozen-spin polarized target (described below). It is expected that the integrand between 10 and 100 MeV will be measured to an accuracy of about 3% in about 400 hours.

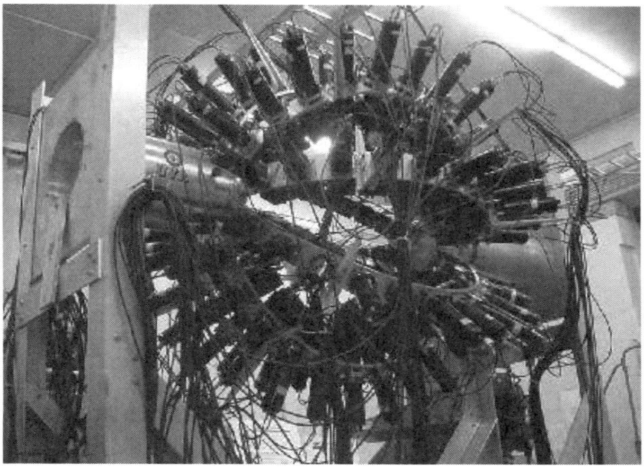

Fig. 4. The 88-neutron detector array: *Blowfish*.

3. The Compton @ HIGS Program

Nucleon polarizabilities are fundamental quantities related to the internal structure of the nucleon, describing the response of the nucleon to external electromagnetic fields. These quantities can be accessed through low-energy Compton scattering experiments. Compton scattering from nucleons at low energy is specified by low-energy theorems up to and including terms linear in photon energy. These terms are completely determined by the static properties of the nucleon, i.e., the nucleon charge, mass and magnetic moment. At next-to-leading order in photon energy there appear new structure constants which are related to the dynamical response of the nucleon's internal degrees of freedom. The electric (α) and magnetic (β) scalar polarizabilities, which describe the response of the nucleon to external electric and magnetic fields, respectively, enter in terms which are second order in photon energy. At the third order four new parameters, γ_1 to γ_4, the spin polarizabilities, appear. While spin polarizabilities do not have as intuitive a physical interpretation as the electric and magnetic polarizabilities, they can be thought of as describing the *stiffness* of the spin when it is subjected to external electric and magnetic fields of various multipolarities.

At present the magnetic polarizability of the proton is uncertain to more than +/-50%. As described in the original HIγS proposal,[4] a ~280 hour experiment using 100% linearly polarized γ-rays can reduce this uncertainty to ~5%. This is important not only in testing theories of this quantity, but also because it impacts our ability to extract accurate values of the spin-polarizabilities, which will be discussed below.

On the theoretical side, a new *industry* of effective field theory (EFT) has arisen in the past ten years (starting with the development of chiral perturbation theory) and great attention has been focused on Compton scattering calculations for the proton and the deuteron. Recently, there have been a series of EFT calculations[11–14] which suggest a new experimental approach to the problem of determining the isoscalar nucleon polarizabilities. These calculations have been performed using EFTs where the pions have been integrated out. Such pionless theories have been shown to be highly successful for systems whose characteristic momenta are below the pion mass.

We are proposing to perform a precision measurement of the differential cross section for elastic scattering of γ-rays from the deuteron at incident energies of 30-80 MeV. EFT calculations and a model independent sum rule will be used to determine the separate values of α_N and β_N from the differential cross section to an accuracy of $\sim$ 10%. These measurements are expected to provide neutron polarizability measurements for both α_n and β_n to an accuracy of better than ~20%. Presently, the statistical uncertainty alone is 65% in the case of β_n.[15]

Compared to the nucleon electric and magnetic polarizabilities, very little is known about nucleon spin polarizabilities ($\gamma_i, i = 1 - 4$). The only quantities which have been extracted so far are the forward and backward spin polarizabilities, γ_0 and γ_π: two independent linear combinations of γ_1, γ_2 and γ_4. In order to determine all four spin polarizabilities separately, two additional measurements which are sensitive to the nucleon's spin polarizabilities are needed. Spin polarizabilities of the nucleon can be probed directly via circularly polarized photons Compton scattered from a polarized nucleon target. The planned experimental setup at HIγS is as follows:

- Photon detection over a range in lab angles from 60° up to approximately 160°. Because the largest physics asymmetry is seen with transverse target polarization, the detector elements should be in the plane of the target polarization.

- Incident photon energies ranging from approximately 30 MeV (almost no spin polarizability effect) up to 140 MeV. The photon polarization should be circular.
- Data taking with longitudinal and transverse target polarizations.

The University of Virginia group is currently building a frozen-spin (butanol) polarized target for the GDH experiment at HIγS. This target will be similar to one under construction for JLab Hall B. The HIγS target will have a length of 10 cm, a diameter of 2.5 cm, a density of 1.5 g/cm^3, and a dilution factor of 7.4, which gives $\sim 4.3 \times 10^{23}$ polarized protons per cm^2. The polarization will be $\sim$90%. Although the GDH experiment doesn't require transverse polarization, this feature has been engineered into the GDH target design and will be used in the proton spin-polarizability measurements.

A proposal for the development of a large acceptance detection system (HINDA) was recently funded by the NSF through the MRI program.[16] This array will consist of eight large high-resolution NaI detectors surrounded by 3-inch thick segmented NaI shields (see Fig. 5).

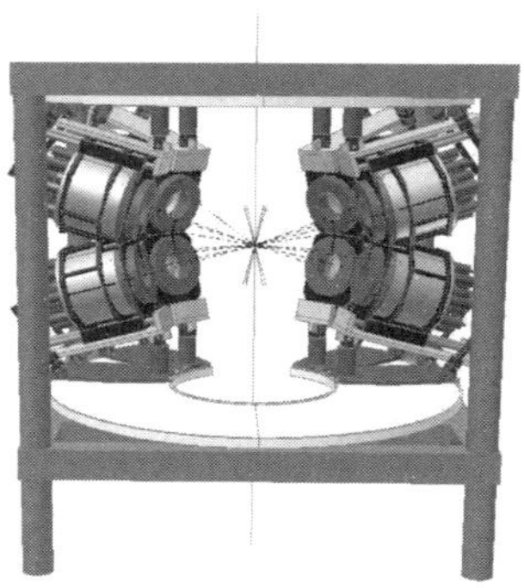

Fig. 5. The HINDA setup: Eight NaI detectors with segmented NaI shields.

Our studies indicate that the experimental sensitivity to the polarizabilities is maximized by measuring the helicity-dependent cross sections, not just the asymmetry. Also, the most effective test of the model-dependent calculations used to extract the polarizabilities is a measurement of the helicity-dependent cross sections. For these reasons we plan to measure cross sections in the experiment. It is expected that it will be possible to measure the photon flux to an accuracy of about 3%.

Figure 6 shows the helicity-dependent cross section for a longitudinally (upper) and transversely (lower) polarized proton target at a lab energy of 136 MeV. The projected error bars shown in the figure are also for a 100 hour measurement for each target spin orientation. Fig. 6 demonstrates that there is considerable sensitivity to the spin polarizabilities in the helicity-dependent cross sections below pion threshold, particularly at backward angles with the longitudinal polarization, and at mid to backward angles with transverse polarization.

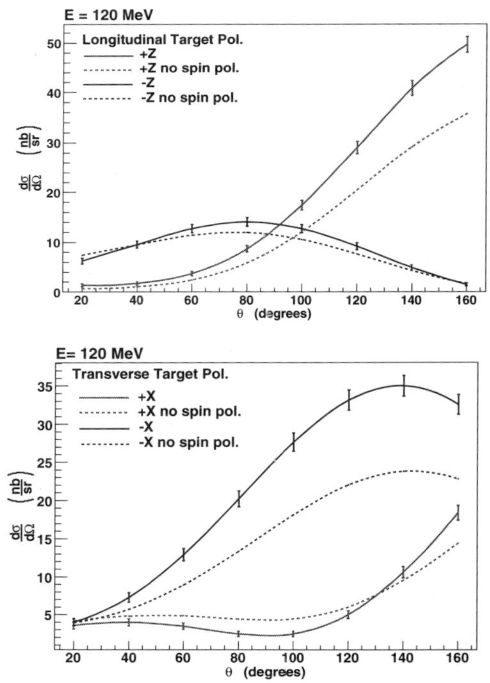

Fig. 6. Projected measurements as a function of photon angle at a center-of-mass energy of 120 MeV for a longitudinally and transversely polarized scintillation target, respectively. The solid and dotted curves indicate the results when the spin-polarizabilities are set to their nominal values or to zero, respectively.

We are also planning a Compton scattering experiment from a polarized ^{3}He target[17] at the elastic kinematics with circularly polarized photons. Sensitivities to the neutron spin polarizabilities from such measurements have been recently been evaluated.[18] Due to the rather small Compton cross-section, the proposed large acceptance HINDA system is essential for the success of this experiment. The polarized ^{3}He target is based on the principle of spin exchange between optically pumped alkali-metal vapor and noble-gas nuclei.[19] Such a target[17] has been built for the HIγS program at TUNL.

The static-spin polarizabilities of the nucleons have been calculated by several groups.[20–23] The earlier calculations were performed at order p^3 and ω^3, and were only sensitive to the isoscalar spin polarizabilities, so the calculated values were the same for the neutron and the proton. More recently,[20,21] the first nonvanishing contributions to the isovector part of the spin polarizabilities were calculated in a p^4 order calculation using heavy baryon chiral perturbation theory (HBCHPT). The results indicate that overall these *corrections* are almost an order of magnitude smaller than the isoscalar values with the exception of γ_1. The values of the neutron and proton spin polarizabilities obtained in Ref.[20] are presented in Table 1. Note that the values given here correspond to the structure part only: the non-structure pole term contributions have been removed.

Table 1. The calculated spin polarizabilities are given along with the anticipated statistical uncertanities expected from the HIγS measurements. The proton results assume 500 hours of beam time while the neutron results are based upon the assumption of a 1500 hour quassifree neutron Compton scattering experiment.

Proton	HIγS uncertainty	Neutron	HIγS uncertainty
$\gamma_1^P = 1.1$	± 0.25	$\gamma_1^n = 3.7$	± 0.4
$\gamma_2^P = -1.5$	± 0.36	$\gamma_2^n = -0.1$	± 0.5
$\gamma_3^P = 0.2$	± 0.24	$\gamma_3^n = 0.4$	± 0.5
$\gamma_4^P = 3.3$	± 0.11	$\gamma_4^n = 2.3$	± 0.35

4. Summary

The GDH and Compton research programs[19] described above represent just two examples of experiments being planned and developed for the upgraded HIγS facility. Another major program will focus on studies of photo-pion reactions on the proton at threshold energies. This program will attempt to measure Isospin and Chiral Symmetry breaking effects. It will also test and evolve the predictions of Chiral Perturbation Theory. A detailed discussion of the photo-pion threshold program can be found in the contribution to these Proceedings by Mohammad Ahmed.

Acknowledgements

The author is especially grateful to Drs. H. Gao and M. Ahmed for their help in preparing this manuscript. Many of the results presented here were derived from Ref.[19]

References

1. V. Litvinenko *et al.*, Phys. Rev. Lett., 78 (1997) 4569.
2. Y. Wu, private communication.
3. Henry R. Weller and Mohammad W. Ahmed, Modern Physics Letters. **A18**, No. 23, 1569-1590 (2003).
4. http://www.tunl.duke.edu/Local/proposal/higs/
5. S. D. Drell and A. C. Hearn, Phys. Rev. Lett. 16 (1966) 908.
6. S. B. Gerasimov, Sov. J. Nucl. Phys. 2 (1966) 430.
7. M. Hosada and K. Yamamoto, Prog. Theor. Phys. 36 (1966) 425.
8. H. Arenhövel *et al.*, Nucl. Phys. A631, 612c (1998)
9. E. Schreiber *et al.*, Phys. Rev. C61 (2000) 061604R.
10. B. Sawatsky *et al.*, private communication.
11. H.W. Griesshammer and G. Rupak, Phys. Lett. B529, 57 (2002).
12. G. Rupak, private communication (2003).
13. Jiunn-Wei Chen, Xiangdong Ji, and Yingchuan Li, Phys. Lett. **B620**, 33 (2005). nucl-th/0408003.
14. Jiunn-Wei Chen, Xiangdong Ji, and Yingchuan Li, Phys. Rev. C **71**, 044321 (2005); nucl-th/0408004.
15. K. Kossert *et al.*, Phys. Rev. Lett., 88, 162301 (2002) Phys. Rev. C **71**, 044321 (2005); nucl-th/0408004.
16. Steve Whisnant, James Madison University, private communication (2006).
17. K. Kramer, X. Zong, H. Gao *et al.*, to be submitted.
18. D. Phillips, D. Choudhury, A. Nogga, private communication.
19. "Compton scattering from nucleon and nuclear targets", H. Gao, H. Weller *et al.*, www.tunl.duke.edu/ mep/higs/compton.pdf
20. G.C. Gellas, T.R. Hemmert, and U.-G. Meissner, Phys. Rev. Lett. **85**, 14 (2000).
21. K.B. Vijay Kumar, J.A. McGovern, and M.C. Birse, Phys. Lett. B **479**, 167 (2000).
22. T.R. Hemmert, B.R. Holstein, J. Kambor, and G. Knöchlein, Phys. Rev. D **57**, 5746 (1998).
23. V. Bernard, N. Kaiser, and U.-G, Meissner, Int. J. Mod. Phys. E **4**, 193 (1995).

PART B

GOLDSTONE BOSON DYNAMICS

WORKING GROUP ON GOLDSTONE BOSON DYNAMICS: SUMMARY

G. COLANGELO* and S. GIOVANNELLA†

**Institute für Theoretische Physik der Universität Bern*
Sidlerstr. 5, 3012 Bern, Switzerland

† *Laboratori Nazionali di Frascati - INFN*
P.O. Box 13, 00044 Frascati (Rome) Italy

We summarize the activity in the Working Group dedicated to Goldstone Boson Dynamics

1. Introduction

The study of the chiral dynamics in the Goldstone Boson sector has a very long history and is by now a mature field. The activity in this working group witnesses this very clearly. In $\pi\pi$ scattering, e.g., the experiments measuring the scattering lengths are providing very stringent tests to theory predictions which are at the percent level. In kaon decays the precise understanding of the chiral dynamics allows the extraction of fundamental standard model (SM) parameters, like V_{us}, and sometimes provides hints of new physics effects. The decay $\eta \to 3\pi$, now measured with better accuracy, as reported in this working group, provides crucial information on a ratio of quark masses. In order to best exploit the experimental information chiral perturbation theory is often used together with other methods, like dispersion relations, large N_C expansion or lattice calculations – only in this manner one can reach the necessary level of precision. These few examples show that advances in this field, both in experiment as well as in theory, are important not only *per se*, for the challenge of understanding strong interactions, but also because they improve our knowledge of the standard model and even of what lies beyond it, at least in the form of constraints.

The progress of lattice calculations in the last few years – the most important being the removal of the quenching approximation – has made contact with phenomenology finally possible. This contact can be made best using chiral perturbation theory, because the extrapolation to the physical point in quark masses is better controlled if one uses analytic formulae. Chiral perturbation theory (ChPT) is extensively used also to remove other systematic effects of lattice calculations, like finite volume or finite lattice spacing effects. The latter are specific to the specific action one uses to formulate QCD on the lattice, and indeed chiral perturbation theory experts inside lattice collaborations have often developed effective theories adapted to their own way of simulating QCD on a computer. Some of these effective field theories have a remarkable level of sophistication.

The rest of this summary is organized as follows: we will first discuss separately the pions, the kaons and the etas and summarize the experimental results and the related theoretical presentations. We will then move up in mass and discuss the contributions related to chiral perturbation theory for heavy mesons. Section 6 is devoted to the lattice-related activity of the working group, and Sect. 7 to theoretical developments in effective field theories. In the last section we give our conclusions.

2. Pions

2.1. $\pi\pi$ *scattering*

Our knowledge of $\pi\pi$ scattering in the low energy region, based on chiral perturbation theory and dispersion relations is remarkably accurate.[1] In recent years a number of new experiments have measured the scattering lengths and are therefore providing sensitive tests of our knowledge of the interaction among pions.

A partial sample of 2003 data has been used by the NA48/2 collaboration to extract the $\pi\pi$ scattering length from two independent channels: $K^\pm \to \pi^+\pi^-e^\pm\nu$ ($K^\pm_{e4}$) and $K^\pm \to \pi^\pm\pi^0\pi^0$. A sample of 370,000 $K^\pm_{e4}$ decays is selected with a very small contamination (0.5%) from 3π background obtained with a π/e discriminant analysis. These events are used to perform a precise measurement of the decay form factors. Moreover, the $\delta = \delta_s - \delta_p$ variation along $M_{\pi\pi}$ is fitted using the numerical solution of the Roy equations.[7] The extracted value of a_0 is not in very good agreement with theory predictions – it will be interesting to see what the final value will be. The measurement of the scattering length parameters can also be obtained from the interpretation of the cusp observed in the $\pi^0\pi^0$ invariant mass of $K^\pm \to \pi^\pm\pi^0\pi^0$ events. A slope variation in the region $M^2_{00} = (2M_+)^2$, where M_+ is the mass of the charged pion, is clearly visible. Several checks on possible instrumental effects artificially creating the slope have been carried out. A fit to the M_{00} distribution allows one to extract $a_0 - a_2$ and the decay form factors. A non-zero quadratic term is found.[2]

The extraction of $a_0^0 - a_0^2$ from the cusp effect can be understood in very simple terms at the one-loop level on the basis of unitarity and analyticity, as proposed originally by Cabibbo.[3] The method of extraction has later been refined by Cabibbo and Isidori[4] by considering up to two loop diagrams, and again only unitarity and analyticity arguments. Within this framework it is unfortunately not clear how to include photons in the diagrams and to evaluate radiative corrections.

A step in this direction was described by Bastian Kubis, who presented a nonrelativistic field theory for the decay $K \to 3\pi$.[5] This is a remarkable example of how flexible and useful a tool the effective field theories are: the input on which the construction of this effective field theory is based is pure kinematics. In view of the masses of kaons and pions, the pions coming out of this decay move slowly – on the basis of this simple observation one can formulate a counting scheme (where the small quantities are the velocities of pions), write down a Lagrangian and check that the counting can be applied also to loops. Calculating the loop corrections automatically provides the effects due to unitarity which Cabibbo originally predicted. At one loop there is little one can gain with such a nonrelativistic field theory, but at two loops the advantages are enormous: instead of the complicated analysis of Cabibbo and Isidori, one can perform a straightforward calculation of two-loop diagrams. The comparison to the representation in[4] showed that some of the assumptions of Cabibbo and Isidori were unjustified. This gave rise to a difference in the representation of the amplitude at the two-loop level. Numerically, this difference appeared to be irrelevant. The second essential advantage of this effective field theory is that the inclusion of photons is straightforward.

The dispersive representation of the partial waves of $\pi\pi$ scattering, as given by Roy[6] – an essential ingredient of the calculation of the scattering lengths[1] – gives the real part of any partial waves in terms of integrals over the imaginary parts of all other partial waves. The integrals run only over the physical region ($s \geq 4M_\pi^2$) and extend up to infinity. To evaluate the dispersive integrals one therefore needs a representation of (the imaginary parts of) all the partial waves up to infinity. The latter is provided by Regge theory which, however, also needs some experimental input in order to pin down the parameters which appear in the representation. The determination of these parameters as used in[7] has been criticized in,[8] where an alternative representation has been proposed. In this Workshop we have heard two talks on these issues. Irinel Caprini has presented an overview of what is known about the Regge parameters for $\pi\pi$ scattering and discussed a general strategy for their determination. She has given preliminary numerical results and shown what is new with respect to previous determinations.[7,8] José Peláez has presented the results of two papers[9,12] where the $\pi\pi$ scattering amplitude has been analyzed in the region below 1.4 GeV. The spirit of these analyses is to use data and dispersion relations in order to determine the $\pi\pi$ scattering amplitude. The starting point is a parametrization which has the correct analyticity properties of the amplitude and uses a minimal number of parameters which are needed to fit the data. He discussed the differences they find in comparison to the representation in.[1]

Given the different approaches, the results do show some differences. Some of these can be clarified and work in this direction is in progress. One important point which we wish to emphasize is that in the analysis presented by Peláez chiral symmetry plays no role and so it is not surprising that, e.g., the values of the scattering lengths so obtained have larger uncertainties.

2.2. $\pi \longrightarrow \ell\nu\gamma$

The measurement of the radiative pion decay $\pi^+ \longrightarrow e^+\nu\gamma$ is one of the main physics goal of the PIBETA experiment at PSI. A high precision measurement of this rare decay (BR $\sim O(10^{-7})$) could give evidence of the deviation from the SM prediction, claimed to be observed by the ISTRA experiment,[10] although with low statistics. In this experiment a non zero value of the tensor current was detected, in contrast with SM expectations. This evidence was then confirmed by the first PIBETA data set (1999–2001),[11] with a sample of $\sim 40,000$ events of this kind coming from pions decaying at rest. In a specific kinematic region a 20% drop in the counting was found with respect to SM expectations. A further data taking (2004) was optimized to select events in the controversial region. The complete data set shows an overall agreement with the SM predictions, giving the most precise value of the vector and axial form factors to date together with the first measurement of their dependence on the momentum transfer. The upper limit on the tensor form factor ($O(10^{-4})$) rules out the previous measurements.

The hint of a discrepancy between SM and data in this channel stimulated a reevaluation of the SM prediction and in particular of the related estimate of the uncertainty. Vicent Mateu has presented work in this direction: they have evaluated the amplitude (in fact the underlying Green function) for the $\pi \to \ell\nu\gamma$ decay including contributions from vector resonances. When doing this it is important to check that the model has the correct behaviour at high energy, because this constrains the coupling of the resonances. Even if the discrepancy has now disappeared, this process constrains new physics contributions.

2.3. $\sigma(e^+e^- \to hadrons)$ *and* $(g-2)_\mu$

The measurement of e^+e^- cross sections in hadrons is an important ingredient for theory in the evaluation of $(g-2)_\mu$ and the running of $\alpha(s)$. Different from the traditional beam energy scan, a novel method has been introduced, first by KLOE and then by BABAR, to measure the energy dependence of the hadronic cross section by running at fixed beam energy. They take advantage of the "radiative return method" i.e. the beam energy is lowered by the emission of an initial state radiation (ISR) photon. In this way, all the mass range from the resonance (ϕ and Y respectively) down to the $\pi\pi$ threshold can be reached. At KLOE only the $\pi\pi$ channel can be studied which, however, corresponds to $\sim 70\%$ of the total hadronic contribution to a_μ. Two different samples are analized. The small angle one (SA) has the ISR photons produced below acceptance and reconstructed by missing energy and momentum. The large angle one (LA) instead requires ISR photon tagging in the detector. Using the first 150 pb^{-1}, the pion form factor dependence on $M^2_{\pi\pi}$ has been already been measured resulting in a determination of the hadronic contribution to a_μ with a 1.3% error. New measurements are in progress both improving the SA analysis through a normalization with the $e^+e^- \to \mu^+\mu^-$ channel and completing the study on the LA sample which allows one to cover the $M^2_{\pi\pi}$ region down to threshold. The BABAR experiment scans a wider energy region, from 4.5 GeV down to threshold, using events with tagged photons. Excellent momentum resolution, high fiducial efficiency and the hard momentum spectrum of the reconstructed particles allows one to reach systematic uncertainties from 5% up to 20%, depending on the channel. At the moment the 3 pions and 4/6 hadrons final states have been measured. For the last category, the accuracy is considerably improved for energies above 1.4 GeV and some channels have been measured for the first time. The $\pi^+\pi^-\pi^+\pi^-$ cross section has been also measured by the BES experiment in the 2.0-3.6 GeV range with the traditional energy scan method. The resulting shape agrees well with BABAR data.

Christopher Smith presented preliminary results of an evaluation of the hadronic vacuum polarization contribution below one GeV, based on analyticity, unitarity and chiral symmetry, and data of course. The idea is to fit the data on the electromagnetic pion form factor with a parametrization which respects analyticity and unitarity. The latter implies that, below the $\pi\omega$ threshold[a], the phase of the form factor coincides with the P wave $\pi\pi$ phase shift. All that is known about the latter plays then a crucial role also in the evaluation of the hadronic contributions to $(g-2)_\mu$. This is an excellent example of how important it is to know the $\pi\pi$ scattering amplitude as precisely as possible.

2.4. *Other experiments*

The PRIMEX experiment at JLAB has acquired a large sample of photoproduced π^0's with the aim of measuring the width $\Gamma(\pi^0 \to \gamma\gamma)$ with 1.5% accuracy, a factor of 10 improvement over current data. Such a precise measurement will allow a powerful test of the QCD axial anomaly prediction and of the NLO isospin symmetry breaking corrections. Theory predictions have a precision of 1% and are in agreement, although with a slightly higher central value (4%), with data. An improvement in the experimental precision will therefore tightly constrain the theory. PRIMEX has reconstructed the π^0 with high resolution ($\sigma_{\pi^0} = 1.3$ MeV) due to excellent calorimetry and tagged incident photons. The π^0 width is measured by extracting the Primakoff component from the measured cross section. At the moment, data has already been acquired and a preliminary analysis shows good understanding of the Compton component.

Another interesting test of a ChPT prediction is the study of the Q^2 dependence of π^0 electroproduction near threshold. Present data shows large disagreement versus theory[14] but with very large bins in Q^2. The proposed JLAB experiment E04-007 aims to determine this dependence with a thinner binning in Q^2. The construction of the detector is underway, $2^n d$ data taking is expected for fall 2008.

The pion polarizabilities describe how the pion reacts to slowly varying electromagnetic fields. In principle they should be reliably calculated in ChPT and recently a two-loop calculation has been performed.[15] Surprisingly, the chiral predictions are in disagreement with the experimental results, the latest of which come from Mainz and are very recent.[16] Kashe-

[a]This statement is a direct consequence of the bound obtained by Eidelman and Lukaszuk.[13]

varov has discussed the results of a dispersive evaluation of the pion polarizabilities which uses as input the $\gamma\gamma \to 2\pi$ cross section. These results agree with experiments. The authors claim that they have a large contribution from the σ meson (contributing to the $\gamma\gamma \to 2\pi$), and it is unclear whether the latter is properly accounted for in the chiral representation. This is a puzzle which certainly deserves further study.

3. Kaons

In the kaon sector, three different experiments reported interesting contributions to this conference: the fixed target NA48 and KTeV, with a large production rate able to pin down very rare decays, and KLOE at the DAΦNE ϕ-factory. Here, although with reduced statistics, pure beams of K_S or K_L are produced, well suited to absolute BR measurements.

The $K_L \to \pi e \nu \gamma$ decay gives rise to controversial results. To compare experiments and theory, the measurement requires identification of the radiative photon in a selected energy and angular range and a proper normalization with the $K_L \to \pi e \nu$ decay. KTeV has first precisely measured the branching fraction of this decay,[17] finding a marginal disagreement from the ChPT prediction. The new result from NA48/2[18] shows instead perfect agreement with theory. The preliminary number from KLOE agrees with the KTeV result.

The new measurement of the $K^+ \to \pi^+\pi^0\gamma$ decay from NA48/2 has shown for the first time evidence of the interference term, (INT) between the inner bremsstrahlung (IB) and the electric part of the direct emission (DE) amplitudes. Due to the very high statistics available and to a sophisticated analysis which greatly improves the background rejection, the cut on T^*, the kinetic energy of the π^+ on the kaon center of mass system, was lowered down to 0 to increase sensitivity to DE and INT terms. At the moment, with a sub-sample of 2003 data, the interference term is determined with a $\sim 30\%$ accuracy both on the statistical and systematic side.

The KTeV experiment presented several measurements on the radiative decays $K_L \to N\pi\gamma^{(*)}$, where different sources of photon emission are present: inner bremsstrahlung (charged modes only), direct emission - with multipole terms - and charged radius (virtual photons only). Since the inner bremsstrahlung is CP violating, the $\pi^+\pi^-\gamma$ final state is dominated by the M1 direct emission amplitude. The corresponding coupling and the overall DE fraction are measured with few % accuracy.[19] In the related mode $K_L \to \pi^+\pi^-e^+e^-$, with a virtual photon emission, the strength of the charged radius mode is also measured.[20] The neutral $\pi^0\pi^0\gamma$ final state probes higher order terms: the dominant term is the E2 direct emission process and the lowest ChPT contribution is $O(p^6)$.[21,22] Using the total KTeV/E799 data set one event is observed. The preliminary upper limit is getting close to the theoretical expectations. The decays $K_L \to \pi^+\pi^-\pi^0\gamma^{(*)}$ and $K_L \to \pi^0\pi^0\pi^0\gamma^{(*)}$ are also being studied. The BR measurements are in progress. There are no other experimental results on these modes, and only one theory prediction exists.[23]

A precise measurement of $K_S \to \gamma\gamma$ is an important test of ChPT since its decay amplitude can be evaluated at leading $O(p^4)$ order, providing an estimate of the BR with a few percent accuracy. The latest precise measurement of this decay rate (3% total error) from NA48 differs of about 30% from the ChPT prediction, suggesting a relevant contribution from higher order corrections. With its entire sample of 2.5 fb^{-1} KLOE aims to reach 5% accuracy on this measurement with a completely different systematics and, by using tagging, without any contamination from $K_L \to \gamma\gamma$ decay.

The decay K_{e3} allows the extraction of V_{us}, provided one knows the relevant form factor at $t = 0$. The decay $K_{\mu 3}$ allows a measurement also of the second form factor, the scalar $K\pi$ form factor. The latter satisfies a dispersion relation and can be represented *à la* Omnès: given its phase (which is related to the scalar $K\pi$ phase shift in the elastic region), one can obtain the form factor explicitly. Moreover, at the Callan-Treiman point $t = M_K^2 - M_\pi^2$ it satisfies a low-energy theorem: it can be expressed through the combination F_K/F_π modulo small corrections. If one implements this constraint in the Omnès representation, the integral over the phase shift becomes relatively insensitive to the (poorly known) high-energy part. Jan Stern and collaborators[24] proposed to use this representation when analyzing the data, i.e. to let the data determine the value at the Callan-Treiman point, treated as a free parameter. The recent NA48/2 data[25] give a value which disagrees with the low-energy theorem by much more than the expected size of the corrections. The interesting question is whether this might be due to new physics. A second experimental determination of the same parameter would be essential before taking this discrepancy more seriously. This could be provided by KTeV.

The presence of a low-lying resonance in the $K\pi$ scalar channel, the so-called κ is controversial as it is the σ in the $\pi\pi$ channel. The latter has now been established thanks to a new model-independent method based on the use of the Roy representation for complex values of s.[26] Bachir Moussallam has presented the results of an analogous analysis in the $K\pi$

channel[27] which also establishes the existence of the κ beyond any doubt and determines its parameters within small uncertainties.

4. Etas

Concerning the η and η' sector, new precise measurements are being produced by the KLOE and WASA experiments.

In KLOE, light mesons are generated via radiative decays of the ϕ. In the entire 2.5 fb^{-1} sample, 10^8 (10^6) η (η') mesons have been produced. These events are tagged by identifying the recoil monochromatic photon. The analysis of the first 450 pb^{-1} shows the possible reach of the ϕ factory. A Dalitz plot analysis of the $\eta \to \pi^+\pi^-\pi^0$ channel with unprecedented statistics of 1.4×10^6 events and negligible background allows one to extract the Q parameter,[28] related to the u-d quark mass difference. The η mass measurement is also in progress. In the η' sector an improved measurement of the ratio $R = BR(\phi \to \eta')/BR(\phi \to \eta)$ with 4% accuracy allows one to measure the pseudoscalar $\eta - \eta'$ mixing angle and to extract limits on the gluonium content of the η'.

WASA has run for a long period (1989-2005) at CELSIUS and is now being recommissioned at COSY in Jülich; the new experiment is expected to start in Feb 2007. Large samples of η in charged and neutral channels have been collected at CELSIUS. Analysis of the dinamics of $\eta \to 3\pi$ decay is in progress. The rare decay $\eta \to \pi^+\pi^-e^+e^-$ has been observed and the related BR measured with a 30% precision. At COSY a much larger sample of η and η' will be collected in the next years, aiming to achieve a comprehensive physics program in this sector.

5. Heavy Mesons

Any hadron whose mass does not vanish in the chiral limit represents a difficulty in ChPT because the heavy scale messes up the power counting. Hadrons containing heavy quarks, on the other hand, offer the advantage that in the infinite mass limit QCD acquires more symmetries. In transitions from a heavy quark to a different heavy quark, the light degrees of freedom do not perceive any change. Analogously, the light quark bound to a very heavy quark does not realize the difference to the case when there are two heavy quarks in the same hadron. The dynamics of the light degrees of freedom in these systems does give rise to visible effects and can be calculated with chiral techniques – technically the problem is similar to the treatment of the sector with baryon number one, where one has a stable heavy degree of freedom coupled to Goldstone Bosons. We had two talks about this subject: one by Roxanne Springer about $c\bar{s}$ parity partners. She showed mass formulae for these heavy hadrons which contain both one loop chiral corrections as well as $1/m_c$ counterterms and discussed the corresponding spectrum.[29] In the same framework she also discussed the electromagnetic decays of excited $c\bar{s}$ hadrons.[30]

Thomas Mehen discussed the doubly heavy hadrons: after introducing the formalism[31] he presented formulae for the mass splittings and for the electromagnetic decays.[32] He also discussed quenched and partially quenched versions of the formalism[33] and showed results again for the masses and electromagnetic decays.

6. Lattice

Among the lattice talks, three presented results of simulations with dynamical quarks for quantities like masses and decay constants. Claude Bernard presented the latest results from the MILC collaboration, the only one having three light quarks. As is well known, these simulations use staggered fermions and the fourth root trick to have the right number of quarks in the sea. The question of whether this lattice formulation corresponds to QCD in the continuum limit is still debated in the literature.[34] Independently of this, the formulation does imply an extra difficulty, because in order to extract the right continuum physics from the lattice data one has to necessarily rely on the appropriate effective theory at finite lattice spacing, as was discussed in great detail. The outcome of this analysis is remarkable: quite accurate numbers for the quark masses, the $N_f = 2$ and $N_f = 3$ condensate, decay constants and four low-energy constants.[35]

Leonardo Giusti gave a progress report on the simulations carried out in the group of Lüscher using Wilson fermions with two light flavours.[36] In this case the formulation of QCD on the lattice is straightforward and, provided one is close enough to the continuum limit, one can use the formulae of plain ChPT to make fits to the numerical simulations. The results for the pion mass and decay constant are very encouraging and will provide the best determination of $\bar{\ell}_3$ and a competitive one for $\bar{\ell}_4$. A complete study of all the systematic effects in these simulations is not yet available.

Gerrit Schierholz has shown the current status of the simulations of the QCDSF collaboration, which uses improved Wilson fermions, also with two dynamical fermions. There seems to be puzzling results about the pion decay constants, because the quark mass dependence cannot be fitted with ChPT formulae, and there seems to be a contradiction with the results of other groups. It is yet too soon to claim a clash, because the systematic effects have not yet been properly treated.

All these results show that the prospects for having soon a precision determination of fundamental QCD parameters and of the low energy constants controlling the quark mass dependence of masses and decay constants of the mesons are very good.

In fact Paulo Bedaque has shown that one can even determine a combination of low energy constants which enters in the scattering of pions. He presented the results of an evaluation of the $I = 2$ S wave pion scattering length based on the use of a mixed action: domain wall fermions are used for the valence quarks whereas the sea is made of staggered quarks as provided by the MILC configurations. First results have already been published[37] and he showed us preliminary numbers obtained on finer lattices.

The use of mixed actions is not quite without danger – like in many other cases, an estimate of the distortions introduced by a special action at finite lattice spacing can be obtained with the appropriate effective theory. Work in this direction has been presented by André Walker-Loud, who showed us that for the $I = 2$ scattering lengths with partial quenching there is nothing but a small, calculable effect.[38]

Hans Bijnens gave us an overview of the impressive work of his group in providing two-loop mass formulae in partially quenched ChPT.[39]

7. Theoretical developments

The construction and use of an effective field theory like ChPT is quite nontrivial. Indeed on the purely theoretical side there are also very interesting open questions, and some of these were also discussed in our Working Group. Maarten Golterman discussed the issue of resonance saturation for L_8 and showed that in the scalar channel there is no reason why one should be able to saturate this constant with only one (or a few) resonance.[40]

Martin Schmid gave a progress report of his project to match $SU(2)$ ChPT to $SU(3)$ at order p^6 and so determine the m_s-dependence of quantities like the quark condensate or the pion decay constant to or of the l_i's beyond leading order.

José Oller has presented a resummation of a certain class of loop diagrams in self energies. These diagrams are related to the rescattering of mesons in the S wave and are therefore potentially dangerous. This representation of the self-energies changes the values of the (ratios of) quark masses and low energy constants which are needed to describe the phenomenology.[41]

Andreas Fuhrer showed us how to obtain chiral logs to any arbitrary power (in the chiral limit) on the basis of analyticity and unitarity and that this is more efficient than the renormalization group equations.

Andrea Quadri discussed the problem of renormalization of the nonlinear sigma model applying techniques developed for gauge theories.[42]

8. Conclusions

The interplay among experiments, chiral perturbation theory, dispersion relations and other nonperturbative techniques, especially lattice calculations, make this a very lively field of research, with many interesting open problems. We look forward to the advances in this field until Chiral Dynamics 2009!

Acknowledgments

We wish to warmly thank the organizers of the Workshop for providing a perfect framework for the activities of the working group and for the excellent organization of the whole workshop. We also thank all the participants for their contributions and lively presence. This work has been supported by the Schweizerischer Nationalfonds.

References

1. G. Colangelo, J. Gasser and H. Leutwyler, Nucl. Phys. B **603** (2001) 125 [arXiv:hep-ph/0103088].
2. J. R. Batley *et al.*, Phys. Lett. B **633** (2006) 173.
3. N. Cabibbo, Phys. Rev. Lett. **93** (2004) 121801 [arXiv:hep-ph/0405001].
4. N. Cabibbo and G. Isidori, JHEP **0503** (2005) 021 [arXiv:hep-ph/0502130].
5. G. Colangelo, J. Gasser, B. Kubis and A. Rusetsky, Phys. Lett. B **638** (2006) 187 [arXiv:hep-ph/0604084].
6. S. M. Roy, Phys. Lett. B **36** (1971) 353.
7. B. Ananthanarayan, G. Colangelo, J. Gasser and H. Leutwyler, Phys. Rept. **353** (2001) 207 [arXiv:hep-ph/0005297].
8. J. R. Pelaez and F. J. Yndurain, Phys. Rev. D **68** (2003) 074005 [arXiv:hep-ph/0304067]; *ibid.* Phys. Rev. D **69** (2004) 114001 [arXiv:hep-ph/0312187].
9. R. Kaminski, J. R. Pelaez and F. J. Yndurain, Phys. Rev. D **74** (2006) 014001 [Erratum-ibid. D **74** (2006) 079903] [arXiv:hep-ph/0603170].
10. A. A. Poblaguev, Phys. Lett. B **238** (1990) 108.
11. D. Pocanic *et al.*, Phys. Rev. Lett. **93** (2004) 181804.
12. J. R. Pelaez and F. J. Yndurain, Phys. Rev. D **71** (2005) 074016 [arXiv:hep-ph/0411334].
13. S. Eidelman and L. Lukaszuk, Phys. Lett. B **582** (2004) 27 [arXiv:hep-ph/0311366].

14. T. Walcher, these proceedings.
15. J. Gasser, M. A. Ivanov and M. E. Sainio, Nucl. Phys. B **745** (2006) 84 [arXiv:hep-ph/0602234].
16. J. Ahrens *et al.*, Eur. Phys. J. A **23** (2005) 113 [arXiv:nucl-ex/0407011].
17. T. Alexopoulos *et al.*, Phys. Rev. D **71** (2005) 012001.
18. A. Lai *et al.*, Phys. Lett. B **605** (2005) 247.
19. E. Abouzaid *et al.*, Phys. Rev. D **74** (2006) 032004.
20. E. Abouzaid *et al.*, Phys. Rev. Lett. **96** (2006) 101801.
21. P. Heiliger and L.M. Sehgal, Phys. Lett. B **307** (1993) 182.
22. G. Ecker, H. Neufeld and A. Pich, Nucl. Phys. B **413** (1994) 321.
23. G. D'Ambrosio, G. Ecker, G. Isidori and H. Neufeld, Z. Phys. C **76** (xxx) 301.
24. V. Bernard, M. Oertel, E. Passemar and J. Stern, Phys. Lett. B **638** (2006) 480 [arXiv:hep-ph/0603202].
25. A. Lai *et al.* [NA48 Collaboration], CERN-PH-EP-2006-033
26. I. Caprini, G. Colangelo and H. Leutwyler, Phys. Rev. Lett. **96** (2006) 132001 [arXiv:hep-ph/0512364].
27. S. Descotes-Genon and B. Moussallam, arXiv:hep-ph/0607133.
28. B. V. Martemyanov and V. S. Sopov, Phys. Rev. D **71** (2005) 017501.
29. T. Mehen and R. P. Springer, Phys. Rev. D **72** (2005) 034006 [arXiv:hep-ph/0503134].
30. T. Mehen and R. P. Springer, Phys. Rev. D **70**, 074014 (2004) [arXiv:hep-ph/0407181].
31. S. Fleming and T. Mehen, Phys. Rev. D **73** (2006) 034502 [arXiv:hep-ph/0509313].
32. J. Hu and T. Mehen, Phys. Rev. D **73** (2006) 054003 [arXiv:hep-ph/0511321].
33. T. Mehen and B. C. Tiburzi, Phys. Rev. D **74** (2006) 054505 [arXiv:hep-lat/0607023].
34. S. R. Sharpe, PoS **LAT2006** (19??) [arXiv:hep-lat/0610094].
35. C. Bernard *et al.* [MILC Collaboration], arXiv:hep-lat/0609053.
36. L. Del Debbio, L. Giusti, M. Luscher, R. Petronzio and N. Tantalo, arXiv:hep-lat/0610059.
37. S. R. Beane, P. F. Bedaque, K. Orginos and M. J. Savage [NPLQCD Collaboration], Phys. Rev. D **73** (2006) 054503 [arXiv:hep-lat/0506013].
38. J. W. Chen, D. O'Connell, R. S. Van de Water and A. Walker-Loud, Phys. Rev. D **73** (2006) 074510 [arXiv:hep-lat/0510024].
39. J. Bijnens, N. Danielsson and T. A. Lahde, Phys. Rev. D **73** (2006) 074509 [arXiv:hep-lat/0602003].
40. M. Golterman and S. Peris, Phys. Rev. D **74** (2006) 096002 [arXiv:hep-ph/0607152].
41. J. A. Oller and L. Roca, arXiv:hep-ph/0608290.
42. R. Ferrari and A. Quadri, JHEP **0601** (2006) 003 [arXiv:hep-th/0511032].

RECENT RESULTS ON RADIATIVE KAON DECAYS FROM NA48 AND NA48/2

S. GOY LÓPEZ

Dipartimento di Fisica, Università di Torino,
Torino, I - 10125, Italy
E-mail: silvia.goy.lopez@cern.ch

The analysis of $K^0 \to \pi^\pm e^\pm \nu\gamma$ ($K^0e3\gamma$) and $K^\pm \to \pi^\pm\pi^0\gamma$ radiative decays will be discussed.

Keywords: Internal Bremsstrahlung, Direct Emission.

1. Ratio of $K^0e3\gamma$ with respect to K^0e3.

Using data from a dedicated run in 1999 at the CERN SPS, the NA48 collaboration has measured the branching ratio of $K^0e3\gamma$ with respect to inclusive K^0e3.[1]
To avoid collinear and infrared singularities cuts have been imposed on the photon energy (E^*_γ >30 MeV) and on the angle between the photon and the electron ($\theta^*_{e\gamma}$ >20^0) in the kaon center of mass system.
Monte Carlo simulation has been used to compute the detector acceptance and the selection efficiency for both channels. Radiative corrections were taken into account by modifying the PHOTOS[2] program package in order to reproduce the $\theta^*_{e\gamma}$ distribution obtained in data. The main source of systematic uncertainty is the knowledge of the kaon spectrum. The result is based on 19000 $K^0e3\gamma$ decays and 5.6×10^6 K^0e3 decays:

$$BR(K^0 \to \pi^\pm e^\pm \nu\gamma, E^*_\gamma > 30MeV, \theta^*_{e\gamma} > 20^0)/BR(K^0e3) = (0.964 \pm 0.008(stat)^{+0.011}_{-0.009}(syst))\%$$

This is in good agreement with theoretical predictions, but at variance with other experimental results.

2. The $K^\pm \to \pi^\pm\pi^0\gamma$ decay.

A subsample of 2003 NA48/2 data has been analyzed to measure the contribution of Direct photon Emission (DE) relative to Inner bremsstrahlung

(IB) and to detect the presence of interference (INT) between these two amplitudes in $K^{\pm} \to \pi^{\pm}\pi^{0}\gamma$ decays. About 124×10^{3} events have been reconstructed in the ranges $T^{*}_{\pi} < 80$ MeV and $0.2 < W < 0.9$, where T^{*}_{π} is the charged pion kinetic energy in the Kaon center of mass system and W^{2} is:

$$W^{2} = \frac{(p_{K}p_{\gamma})(p_{\pi}p_{\gamma})}{m_{\pi}^{2}m_{K}^{2}}$$

p_{K}, p_{π}, p_{γ} being the four-momenta of the kaon, charged pion and odd gamma, and m_{π}, m_{K} the charged pion and kaon masses.
An algorithm has been developed to reject $K^{\pm} \to \pi^{\pm}\pi^{0}\pi^{0}$ decays with two gammas overlapping in the detector. In addition, the reconstructed kaon mass is required to be within 10 MeV from its nominal value. These two conditions avoid the need of performing a lower cut on T^{*}_{π} , increasing the sensitivity to DE and INT terms with respect to previous measurements.[3] The upper cut $T^{*}_{\pi} < 80$ MeV rejects $K^{\pm} \to \pi^{\pm}\pi^{0}$ decays. After all cuts, the background contamination has been estimated to be less than 10^{-4}. Photon mistagging (i.e., choice of the wrong odd photon) has been kept smaller than the per mil level for all decay components.
The fractions of DE and INT with respect to IB present in data have been determined by fitting the proportions of simulated IB, DE and INT W distributions to reproduce the expetimental one, using a maximum likelihood method. The systematic error is dominated by the trigger efficiency. The results are:

$$\text{Frac(DE)} = (3.35 \pm 0.35_{stat} \pm 0.25_{syst})\%$$

$$\text{Frac(INT)} = (-2.67 \pm 0.81_{stat} \pm 0.73_{syst})\%$$

The correlation coefficient between these two parameters is -0.93.

References

1. A. Lai *et al.*, *Phys. Lett.* B **605** (2005) 247.
2. E. Barberio and Z. Was, *Comp. Phys. Comm.* B **79** (1994) 291.
3. R.J. Abrams *et al.*, *Phys. Rev. Lett.* **29** (1972) 1118.
   K.M. Smith *et al.*, *Nucl. Phys.* **B109** (1976) 173.
   V.N. Bolotov *et al.*, *Yad. Fiz.* **45** (1987) 1652
   S. Adler *et al.*, *Phys. Rev. Lett.* **85** (2000) 4856.
   M.A. Aliev *et al.*, *Phys. Lett. B* **554** (2003) 7.
   V.A. Uvarov *et al.*, *Phys. Atom. Nucl.* **69** (2006) 26.

CUSPS IN $K \to 3\pi$ DECAYS*†

BASTIAN KUBIS‡

HISKP, Universität Bonn, Nussallee 14–16, D-53115 Bonn, Germany
‡*E-mail: kubis@itkp.uni-bonn.de*

We present a non-relativistic effective field theory framework to analyze the cusp structure in $K \to 3\pi$ decays, which allows to construct a representation of the relevant amplitudes that contains the $\pi\pi$ scattering lengths in a straightforward manner.

It has been pointed out by Cabibbo and Isidori[1,2] that the pion mass difference generates a pronounced cusp in $K^+ \to \pi^+\pi^0\pi^0$ decays, an accurate measurement of which may allow one to determine the combination $a_0 - a_2$ of S-wave $\pi\pi$ scattering lengths to high precision. A first analysis of data based on this proposal has already appeared.[3] In order for this program to be carried through successfully, one needs to determine the structure of the cusp with a precision that matches the experimental accuracy.

Non-relativistic effective field theory is the appropriate systematic framework to analyze the structure of $K \to 3\pi$ amplitudes and their dependence on the $\pi\pi$ scattering lengths,[4] as the non-relativistic Lagrangian directly contains the parameters of the effective range expansion of the $\pi\pi$ scattering amplitude. This is in contrast to, e.g., chiral perturbation theory, where scattering lengths etc. are expanded in powers of the pion mass.[5] As the $\pi\pi$ scattering lengths are small, it is useful to perform a combined expansion in powers of the scattering lengths and a non-relativistic small parameter ϵ.

The power counting is set up such that pion three-momenta are counted as $\mathcal{O}(\epsilon)$; the kinetic energies are therefore of $\mathcal{O}(\epsilon^2)$, and consequently so is the mass difference $M_K - 3M_\pi$. In this way, the non-relativistic region covers the whole decay region. It is convenient to formulate the non-relativistic ap-

*Work done in collaboration with G. Colangelo, J. Gasser, and A. Rusetsky.

†Work supported in part by DFG (TR-16) and by EU I3HP (RII3-CT-2004-506078).

proach in a manifestly Lorentz-invariant/frame-independent manner, which can be achieved by employing a non-local kinetic-energy Lagrangian for the pion fields of the form

$$\mathcal{L}_{\rm kin} = \Phi^\dagger (2W)(i\partial_t - W)\Phi \ , \qquad W = \sqrt{M_\pi^2 - \triangle} \ . \tag{1}$$

Non-relativistic Lagrangians for $K \to 3\pi$ as well as for the $\pi\pi$ interaction are given by

$$\begin{aligned} \mathcal{L}_K &= \frac{G_0}{2}\big(K_+^\dagger \Phi_+ (\Phi_0)^2 + h.c.\big) + \frac{H_0}{2}\big(K_+^\dagger \Phi_- (\Phi_+)^2 + h.c.\big) + \ldots \ , \\ \mathcal{L}_{\pi\pi} &= C_x\big(\Phi_-^\dagger \Phi_+^\dagger (\Phi_0)^2 + h.c.\big) + \ldots \ , \end{aligned} \tag{2}$$

where the ellipses include higher-order derivative terms. The parameters G_0, H_0, C_x etc. have to be determined from a simultaneous fit of $K^+ \to \pi^+\pi^0\pi^0$ and $K^+ \to \pi^+\pi^+\pi^-$ amplitudes to experimental data. C_x in particular is proportional to $a_0 - a_2$ up to isospin breaking corrections.

The non-relativistic representation of the $K \to 3\pi$ amplitudes, in the combined expansion in $\pi\pi$ scattering lengths a and the non-relativistic parameter ϵ, has been given up to $\mathcal{O}(\epsilon^4, a\epsilon^5, a^2\epsilon^2)$.[6] This representation is valid to arbitrary orders in the quark masses. As the approach is based on a Lagrangian framework, constraints from analyticity and unitarity are automatically obeyed. Special care has to be taken for the representation of overlapping two-loop graphs, which have a particularly complicated analytical structure and which, for certain combinations of pion masses running in the loops, develop anomalous thresholds in the decay region.

An extension of the non-relativistic Lagrangian approach to include real and virtual photons is in principle straightforward. Photons modify the singularity structure at threshold at $\mathcal{O}(\alpha)$, a future analysis including such radiative correction effects is therefore mandatory.

References

1. N. Cabibbo, Phys. Rev. Lett. **93** (2004) 121801 [arXiv:hep-ph/0405001].
2. N. Cabibbo and G. Isidori, JHEP **0503** (2005) 021 [arXiv:hep-ph/0502130].
3. J. R. Batley *et al.* [NA48/2 Collaboration], Phys. Lett. B **633** (2006) 173 [arXiv:hep-ex/0511056].
4. G. Colangelo, J. Gasser, B. Kubis and A. Rusetsky, Phys. Lett. B **638** (2006) 187 [arXiv:hep-ph/0604084].
5. E. Gamiz, J. Prades and I. Scimemi, arXiv:hep-ph/0602023.
6. J. Gasser, B. Kubis and A. Rusetsky, in preparation.

RECENT kTeV RESULTS ON RADIATIVE KAON DECAYS

MICHAEL C. RONQUEST for the KTeV Collaboration

Department of Physics, University of Virginia, Charlottesville, Virginia, 22904
E-mail: mr7y@virginia.edu

This paper describes recent results from the KTeV project[1] on di-pion radiative kaon decays, which are interesting from a χPt standpoint, and tri-pion radiative kaon decays, which may reveal the existence of the CP violating charge radius process $K_L \to K_L\gamma^*$, similar to the CP conserving $K_L \to K_S\gamma^*$ transition seen in $K_L \to \pi^+\pi^-e^+e^-$.

Keywords: neutral kaon, radiative decay, CP violation

1. Photon Emission in $K \to N\pi\gamma$ and $K \to N\pi\gamma^*$

Both classes of decays receive contributions from three possible photon emission sources: Inner Bremsstrahlung where a photon is emitted from a charged particle, Direct Emission where the photon is emitted from the weak decay vertex, and finally Charge Radius Emission, where a virtual photon is emitted in the process $K \to K\gamma^*$.

Table 1. Photon emission sources in various decays

Decay	Inner Bremsstrahlung	Direct Emission	Charge Radius
$K_L \to \pi^+\pi^-\gamma$	CPV	✓	
$K_L \to \pi^0\pi^0\gamma$		✓	
$K_L \to \pi^+\pi^-\pi^0\gamma$	✓	✓	
$K_L \to \pi^+\pi^-\pi^0e^+e^-$	✓	✓	✓
$K_L \to \pi^0\pi^0\pi^0\gamma$		✓	
$K_L \to \pi^0\pi^0\pi^0e^+e^-$		✓	✓

2. $K_L \to \pi\pi\gamma$

KTeV's most recent radiative di-pion decay result[2] is from $K_L \to \pi^+\pi^-\gamma$. Using the models of Refs. 3 and 4 the KTeV experiment has determined:

$$|\widetilde{g_{M1}}| = 1.198 \pm 0.035(stat) \pm 0.086(syst) \quad (1)$$

$$a_1/a_2 = -0.738 \pm 0.007(stat) \pm 0.018(syst) GeV^2/c^2 \quad (2)$$

The decay $K_L \to \pi^0\pi^0\gamma$, an E2 multi-pole direct emission process[5] is a probe of chiral perturbation theory at $\mathcal{O}(p^6)$. KTeV has arrived at a *preliminary* limit of:

$$BR(K_L \to \pi^0\pi^0\gamma) < 2.32 \times 10^{-7} (90\% \; C.L.) \quad (3)$$

3. $K_L \to \pi\pi\pi\gamma$ and $K_L \to \pi\pi\pi\gamma^*$

In order to search for the CP-violating transition $K_L \to K_L\gamma^*$, KTeV has begun to search for tri-pion radiative decays, for which there have been no previous experimental and very few theoretical studies. Note that the contribution due to $K_L \to K_S\gamma^*$ is suppressed in these decays. KTeV has searched a portion of its data and has uncovered 2847 candidate $K_L \to \pi^+\pi^-\pi^0\gamma$ events. The expected branching ratio is[6] $(1.65 \pm 0.03) \times 10^{-4}$.

KTeV observes 132 candidate $K_L \to \pi^+\pi^-\pi^0 e^+e^-$ events in the same portion of its data, leading to a *preliminary* branching ratio of:

$$BR(K_L \to \pi^+\pi^-\pi^0 e^+e^-) = (1.60 \pm 0.18(stat) \pm ??) \times 10^{-7} \quad (4)$$

A search for $K_L \to \pi^0\pi^0\pi^0 e^+e^-$ using the entire rare decay KTeV dataset has not revealed any signal events.

A search for $K_L \to \pi^0\pi^0\pi^0\gamma$ is also in progress.

4. Conclusions

The KTeV collaboration has produced a new form factor measurement for the decay $K_L \to \pi^+\pi^-\gamma$ and will place new limits on the branching ratios of $K_L \to \pi^0\pi^0\gamma$ as well as $K_L \to \pi^0\pi^0\pi^0 e^+e^-$. In addition, first observations of the decays $K_L \to \pi^+\pi^-\pi^0\gamma$ and $K_L \to \pi^+\pi^-\pi^0 e^+e^-$ have been produced and form factor analyses of these modes are forthcoming.

References

1. A. Alavi-Harati *et al.*, *Phys. Rev.* **D67**, p. 012005 (2003).
2. E. Abouzaid *et al.*, *Phys. Rev.* **D74**, p. 032004 (2006).
3. L. M. Sehgal and J. van Leusen, *Phys. Rev. Lett.* **83**, 4933 (1999).
4. Y. C. R. Lin and G. Valencia, *Phys. Rev.* **D37**, p. 143 (1988).
5. P. Heiliger and L. M. Sehgal, *Phys. Lett.* **B307**, 182 (1993).
6. G. D'Ambrosio, G. Ecker, G. Isidori and H. Neufeld, *Z. Phys.* **C76**, 301 (1997).

THE $\pi\pi$ SCATTERING AMPLITUDE

J.R. PELÁEZ

Departamento de Física Teórica, II Facultad de Ciencias Físicas, Universidad Complutense de Madrid, E-28040, Madrid, Spain

We report on our precise determination of the $\pi\pi$ scattering amplitude obtained using Forward Dispersion Relations and fits to data. Very recently we have improved the fits above $K\bar{K}$ threshold and checked that it satisfies remarkably well the Roy equations.

Keywords: Dispersion relations, pion-pion scattering, Roy equations.

In a recent set of papers (see[1–3] for further details and references) we have obtained fits to data on $\pi\pi$ scattering for S0, P, S2, D0, D2 and F partial waves, checking first the Froissart-Gribov and the Olsson sum rules. These sum rules together with the model independent fits below $K\bar{K}$ threshold were also used to obtain precise values for threshold parameters.

In order to have a precise description of the high energy parts (1.1 GeV above thresholds) we performed a Regge analysis[4] of meson-meson, meson-nucleon and nucleon-nucleon data.

All these parametrizations together with their error analyses were then used[2] to check the often conflicting phase shift determinations of the S0 wave by means of a complete isospin set of Forward Dispersion Relations (FDR). Remarkably we found that some of the most popular phase shift analysis fail to satisfy FDR by many standard deviations. Just a few sets pass this stringent FDR test.

Next, and starting from a global fit[2,4] to different S0 analyses in the regions where they agree, we then constrained the different partial wave fits to satisfy FDR's up to 950 MeV, thus obtaining a precise description[2] of pion-pion scattering below $K\bar{K}$ threshold. Only the S0 and D2 waves changed sizably from their original unconstrained fits.

Very recently[3] we have updated our unconstrained fits above $K\bar{K}$ threshold and the Regge ρ exchange with more data and more flexible parametrizations, resulting in an overall improvement of all FDR's which are remarkably well satisfied up to 1.4 GeV, as can be seen in the Figure.

In this conference we have also reported on our check of the Roy equations, still in progress, that can be seen in the Figure, as well as on the ongoing data fits constrained to FDR's and the Roy equations, that will provide an extremely precise representation of $\pi\pi$ scattering up to 1.4 GeV.

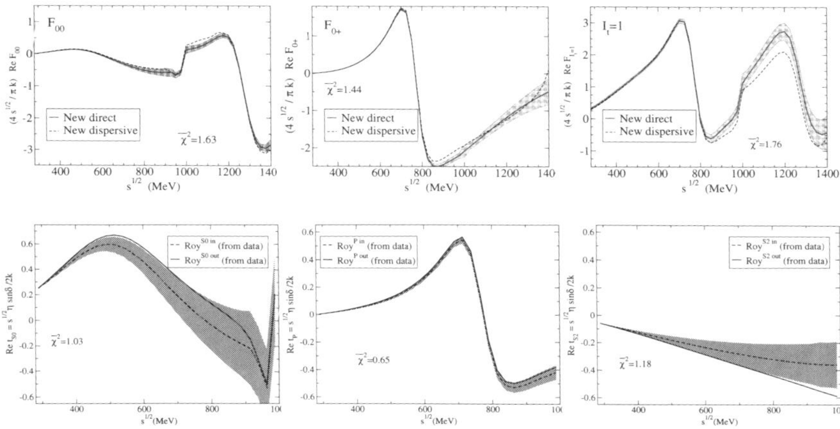

Fig. 1. Our data fits[2,3] satisfy very well the complete set of FDR and the Roy Equations below $K\bar{K}$ threshold and fairly well above. Here we compare the direct parametrization (direct or "in" curves) versus the integral representation (dispersive or "out"). Note that the curves come from data fits, without imposing FDR or the Roy equations, so that the agreement within uncertainties is even more remarkable.

References

1. J. R. Pelaez and F. J. Yndurain, Phys. Rev. D **68**, 074005 (2003)
2. J. R. Pelaez and F. J. Yndurain, Phys. Rev. D **71**, 074016 (2005)
3. R. Kaminski, J. R. Pelaez and F. J. Yndurain, Phys. Rev. D **74**, 014001 (2006) [Erratum-ibid. D **74**, 079903 (2006)]
4. J. R. Pelaez and F. J. Yndurain, Phys. Rev. D **69**, 114001 (2004)

DETERMINATION OF THE REGGE PARAMETERS IN THE $\pi\pi$ SCATTERING AMPLITUDE

I. CAPRINI

National Institute of Physics and Nuclear Engineering, Bucharest, R-077125 Romania
** E-mail: caprini@theory.nipne.ro*

An accurate Regge parametrization is required as input in solving the Roy equations for $\pi\pi$ scattering up to the maximum energy of 1.15 GeV. I report on recent work, done in collaboration with G. Colangelo and H. Leutwyler, on the determination of the Regge trajectories and residues entering the parametrization of the the $\pi\pi$ amplitudes for $\sqrt{s} > 1.7$ GeV and $-0.623\ \text{GeV}^2 \leq t \leq 0$.

Keywords: $\pi\pi$ scattering; Regge parametrization; Roy equations.

1. Introduction

Recent developments in Regge theory include the description of the Pomeron in QCD and updated global fits of πN, KN and NN high energy data.[1] For $\pi\pi$ scattering the recent interest in the Regge model is connected to the dispersive treatment of this process, based on the Roy equations, forward dispersion relations or sum rules.[2,3]

The standard contribution of a simple Regge pole is

$$T(s,t) \approx -\frac{\tau + e^{-i\pi\alpha(t)}}{\sin\pi\alpha(t)}\beta(t)\left(\frac{s}{s_1}\right)^{\alpha(t)}, \quad s \text{ large},\ t \leq 0, \tag{1}$$

where $\alpha(t)$ and $\beta(t)$ denote the trajectory and the residue, the signature $\tau = \pm 1$ for C-even (odd) trajectories and $s_1 = 1\,\text{GeV}^2$. Additional logarithmic dependence on s appears in the case of multiple poles (as appears to be the Pomeron), or for Regge cuts. In $\pi\pi$ scattering the dominant trajectories are the Pomeron P and the Reggeon f, which contribute to the amplitude $T^{(0)}(s,t)$ of isospin 0 in the t-channel, and ρ, which contributes to the amplitude $T^{(1)}(s,t)$. The exotic amplitude $T^{(2)}(s,t)$ receives contributions only from Regge cuts. Therefore, we have to determine three trajectories: $\alpha_P(t), \alpha_f(t)$ and $\alpha_\rho(t)$, and three residues: $\beta_P(t), \beta_f(t)$ and $\beta_\rho(t)$. A few results were presented already in Ref. 4. Here I shall describe the main ideas of our method.

2. Determination of trajectories and residues

For the trajectories and residues at $t = 0$ we use the global fits[1] of πN and NN total cross sections for $\sqrt{s} > 5$ GeV, combined with factorization of the residues. The Regge representation was then continued to lower energies, up to $\sqrt{s} = 1.7$ GeV, including preasymptotic corrections to match smoothly the partial wave expansions. The contributions of separate trajectories were extracted from proper combinations of $\pi^{\pm}p$, pp, $p\bar{p}$ and np amplitudes, using the fact that the C-odd terms (ρ, ω) change sign when a particle is replaced by its antiparticle, and the isospin-odd ones (a_2, ρ) change sign when p is replaced by n. The method allows an accurate determination of all trajectories and residues, except for $\beta_\rho(0)$: in this case the errors are large, since the contribution of ρ to nonflip NN amplitudes is small, and moreover one must rely on poor pn data. For a sharper determination of $\beta_\rho(0)$ we resort to the Olsson sum rule.[2] Unlike Refs. 3, we do not use input data on high energy $\pi\pi$ scattering, since their systematic errors are very large. We compare nevertheless our results with these data.

The slopes of the trajectories are rather well known from scattering data and Chew-Frautschi plots. The Pomeron residue $\beta_P(t)$ was determined using factorization and the shape of the difraction peaks in πN and NN data. The t-dependence of the f-residue was obtained from the requirement that $\beta_f(t)$ must vanish when $\alpha_f(t) = 0$, in order to avoid ghosts in Eq. (1). The residue $\beta_\rho(t)$ was calculated using a generalization of the Olsson sum rule.[2] Since the determinations from sum rules are based on a low energy input, the procedure was iterated until consistency of the framework was reached. A complete discussion of the method and the results will be presented in a forthcoming publication.

Acknowledgments: This work was supported in part by the Romanian Ministry of Education and Research, under Contract 2-CEx-11-92.

References

1. W.-M.Yao *et al*, *Journal of Physics* **G33**, 1 (2006).
2. B. Ananthanarayan, G. Colangelo, J.Gasser and H.Leutwyler, *Phys.Rept.* **353**, 207 (2001).
3. J. R. Peláez and F. J. Ynduráin, *Phys. Rev.* **D69** 114001 (2004);
   R. Kaminski, J. R. Peláez and F. J. Ynduráin, *Phys. Rev.* **D74**, 014001 (2006).
4. I. Caprini, G. Colangelo, H. Leutwyler, *Int.J.Mod.Phys.* **A21** 954, (2006).

e^+e^- HADRONIC CROSS SECTION MEASUREMENT AT DAΦNE WITH THE KLOE DETECTOR

The KLOE collaboration represented by PAOLO BELTRAME

Institut für Experimentelle Kernphysik, KIT - Universität Karlsruhe,
Postfach 3640, D-76021 Karlsruhe, Germany
E-mail: paolo.beltrame@iekp.fzk.de

The e^+e^- hadronic cross section is measured at the Frascati ϕ-factory DAΦNE with the KLOE detector using initial state radiation (*radiative return* method). Two different analyses have been developed for the channel $e^+e^- \to \pi^+\pi^-\gamma$, which differ by the Initial State Radiation (ISR) photon polar angle direction. The so-called small angle analysis, in which the photon is not tagged, has been published with a total error for the pion form factor of 1.3%. In a second approach the photon is tagged at large polar angles; this allows to cover the mass region below 0.35 GeV^2

Keywords: Radiative return; hadronic cross section, muon anomaly.

The precision measurement by the E821 collaboration[1] of the anomalous magnetic moment of the muon, $a_\mu = (11659208.0 \pm 6.3) \cdot 10^{-10}$, has led to renewed interest in accurate theoretical estimates of this quantity. The SM prediction is the sum of the QED, weak and hadronic contributions: (in 10^{-10} units) $a_\mu^{QED} = 11658470.4 \pm 0.3$, $a_\mu^{weak} = 15.4 \pm 0.2$ and $a_\mu^{had} \sim 700$.[4] The uncertainty on a_μ^{had} dominates the accuracy of the prediction. The hadronic correction to the photon propagator cannot be calculated by perturbative QCD for low energy, but the lowest-order hadronic contribution to a_μ can be expressed by means of a dispersion integral using the hadronic cross section, $\sigma(e^+e^- \to hadrons)$ as input.
The process $e^+e^- \to \pi^+\pi^-$ below 1 GeV accounts for $\sim 70\%$ of the dispersion integral for a_μ^{had}. Two recent measurements of $\sigma(e^+e^- \to \pi^+\pi^-)$, performed by CMD-2 and SND,[2,3] have been done with an energy-scan and claim accuracies of 0.9% and 1.3%, respectively. The final SM prediction for a_μ differs by 3.3σ with the measurement by E821.[4] Moreover, there is a rather strong disagreement between e^+e^--experiments and measurements using hadronic τ-decays, which can be used to extract the pion form factor

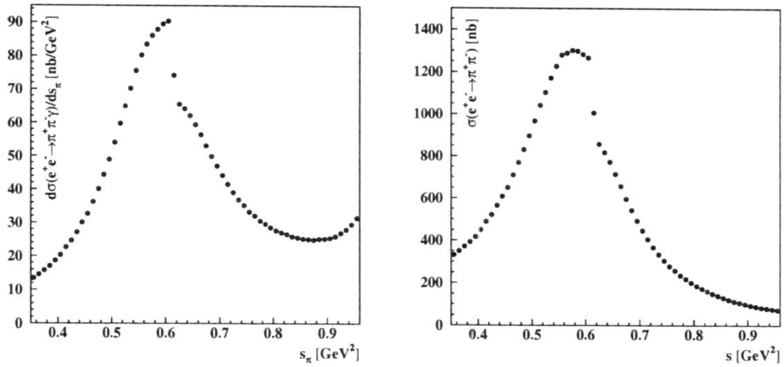

Fig. 1. *Left*: cross section for the process $e^+e^- \to \pi^+\pi^-\gamma$, with pions detected in apparatus and with $\theta_\gamma < 15°$. *Right*: bare cross section for $e^+e^- \to \pi^+\pi^-$.

via the conserved vector current theorem, taking into account appropriate isospin breaking corrections.

KLOE has developed a new and complementary way to determine the $e^+e^- \to \pi^+\pi^-$ cross section as a function of s_π, the squared center of mass energy of the $\pi\pi$ system. Although DAΦNE operates at fixed energy $W \sim m_\phi$, initial state radiation events (ISR) lower the available beam energy for the dipion system. Therefore, measuring the cross section for the process $e^+e^- \to \pi^+\pi^-\gamma$ and exploiting ISR, one can relate $d\sigma(\pi^+\pi^-\gamma, s_\pi)/ds_\pi$ to $\sigma(\pi^+\pi^-, s_\pi)$ with s_π from $2m_\pi$ to W. The two quantities are related by a radiator function obtained using the Monte Carlo generator PHOKHARA, which contains NLO QED corrections.[5]).
In the first measurement of KLOE the contribution from final state radiation (FSR) is largely suppressed by restricting the selection to events with the radiative photon at low polar angle ($\theta_\gamma < 15°$): in this region, ISR is the dominant contribution and FSR accounts for less than 1% over the entire s_π spectrum. The radiated photon is not detected, but s_π and θ_γ are reconstructed using the information from the charged particle tracks. The results of this *small angle* analysis have been published, with a total error of 1.3% (0.9%$_{exp}$ and 0.9%$_{theo}$), making KLOE the first experiment proving the feasibility and the reliability of the radiative return method.[6] The KLOE data are shown in Fig. 1. Calculating the dispersion integral, one obtains

$$a_\mu^{had-\pi\pi}(0.35 < s_\pi < 0.95 \text{ GeV}^2) = (388.7 \pm 0.8_{stat} \pm 3.5_{syst} \pm 3.5_{th}) \times 10^{-10}$$

An update on small angle analysis, using the data collected in 2002 (240 pb^{-1}), and normalizing with $\mu^+\mu^-\gamma$ events is on the way.
In order to access the threshold region, $s_\pi < 0.35$ GeV^2, which for kinematical reasons is not accessible with the previous selection criteria, KLOE is measuring the $e^+e^- \rightarrow \pi^+\pi^-\gamma$ cross section detecting the photon at *large* polar *angle* ($50° < \theta_\gamma < 130°$). In this condition there is a larger background due to the decay $\phi \rightarrow \pi^+\pi^-\pi^0$; moreover one has to correct for pure FSR events and the radiative decay $\phi \rightarrow f_0\gamma$ with $f_0 \rightarrow \pi^+\pi^-$. These latter processes must be evaluated by using proper phenomenological models.[7] The large angle analysis is close to being completed.
A further analysis, using a dedicated DAΦNE off-peak run, is on the way. The data sample at $\sqrt{s} = 1$ GeV^2, collected in 2006 with an integrated luminosity of 225 pb^-1, will give the possibility to reach the mass threshold region in background free conditions (namely, without ϕ decays into $\pi^+\pi^-\pi^0$ and $f_0\gamma$ channels), allowing ultimate precision for $\sigma^{\pi\pi}$ with the KLOE detector.

References

1. B. W. Bennet et al. [E821 Coll.], Phys. Rev. Lett. **92**, (2004) 161802
2. R. R. Akhmethsin et al. [CMD-2 Coll.], Phys. Lett. B **578**, (2004) 285
3. M. N. Achasov et al. [SND Coll.], hep-ex/0605013
4. S. Eidelman, Euridice Meeting, Kazimirez (August 27, 2006)
5. S. Binner, J. H. Kühn and K. Melnikov, Phys. Lett. B **459**, (1999) 279; G. Rodrigo, A. Gehrmann-De Ridder, M. Guilleaume and J. H. Kühn, Eur. Phys. J. **C22**, (2001) 81; G. Rodrigo, H. Czyż, J. H. Kühn and M. Szopa, Eur. Phys. J. **C24**, (2002) 71; H. Czyż, A. Grzelińska, J. H. Kühn and G. Rodrigo, Eur. Phys. J. **C27**, (2003) 563; H. Czyż, A. Grzelińska, J. H. Kühn and G. Rodrigo, Eur. Phys. J. **C33**, (2004) 333; H. Czyż, A. Grzelińska, J. H. Kühn and G. Rodrigo, Eur. Phys. J. **C39**, (2005) 411; H. Czyż and E. Nowak-Kubat, Phys. Lett. **B634**, (2006) 493
6. A. Aloisio et al. [KLOE Coll.], Phys. Rev. Lett. B **606**, (2005) 12
7. N.N. Achasov and V.V. Gubin, Phys. Rev. **D57**, (1998) 1987; G. Isidori, L. Maiani, M. Nicolaci and S. Pacetti, hep-ph/0603241; M. Boglione, M.R. Pennington, Eur. Phys. J. **C30**, (2003) 503

MEASUREMENT OF THE FORM FACTORS OF $e^+e^- \to 2(\pi^+\pi^-)$, $p\bar{p}$ AND THE RESONANT PARAMETERS OF THE HEAVY CHARMONIA AT BES.

HAIMING HU*

Institute of High Energy Physics, Beijing, 100049, P. R. China
**E-mail: huhm@ihep.ac.cn*

The preliminary results about the form factors for the $e^+e^- \to 2(\pi^+\pi^-)$, $p\bar{p}$ and the resonant parameters of the heavy charmonia $\psi(3070)$, $\psi(4040)$, $\psi(4160)$ and $\psi(4416)$ measured with BES are presented.

Keywords: form factor; resonant parameter.

1. Form factor for $e^+e^- \to 2(\pi^+\pi^-)$

The Born cross section of the $2(\pi^+\pi^-)$ final state is measured between $2.0-3.65$ GeV, the systematic error is about 10%, see Fig.1 (a), in which the result from *BABAR*[1] with the initial state radiation data are compared. The theoretical cross section and the form factor for $e^+e^- \to 2(\pi^+\pi^-)$ can be described in terms of the VMD model[2]. The form factor of $2(\pi^+\pi^-)$ is extracted by fitting the cross section, while the all parameters of the form factor are obtained. Fig.1 (b) and (c) show the form factor based on the results by CMD2, DM1, BES and *BABAR* respectively.

2. Form factor for $e^+e^- \to p\bar{p}$

The lowest order cross section of $e^+e^- \to p\bar{p}$ is measured at 10 energy points between 2.0 and 3.07 GeV [3]. Fig.2(a) plots the form factors in the processes $e^+e^- \to p\bar{p}$ and $p\bar{p} \to e^+e^-$ measured by some groups under the assumption of $|G_M| = |G_E| \equiv |G|$. The line shows the energy dependence of the form factor $|G(s)|$ by fitting all measurements, and the result is consistent with the pQCD prediction.

3. Resonant parameters of the heavy charmoia

Using the raw data for R value scan collected at BES in 1999[4], the resonant structure of the high excited ψ-family resonances, namely $\psi(3770)$, $\psi(4040)$,

$\psi(4160)$ and $\psi(4416)$, is fitted with the iterative method, and the values of the resonant parameters (mass M, total width Γ_{tot} and leptonic width Γ_{ee}) are obtained. In the fitting, the energy-dependence of total widthes, relative phase angles and the interference between the same decay final states are considered. In addition to the fitted errors, the uncertainties of the models are also included to estimate the systematic errors. Fig.2 (b) shows the new R values in heavy charmonia region, which are a little different from the old one reported in reference [4] due to the change of the resonant parameters and the correction factor for the initial state radiation.

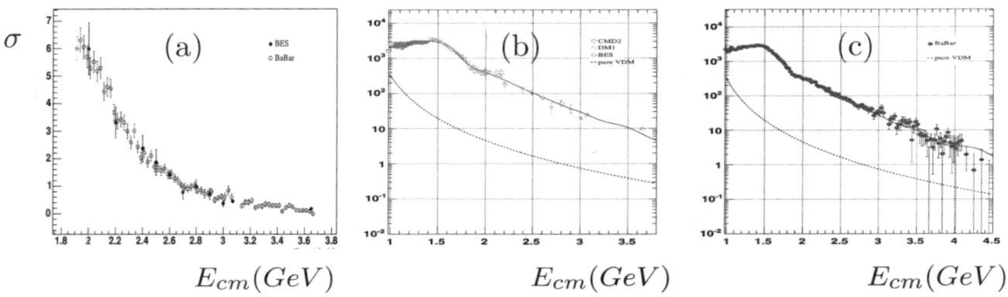

Fig. 1. (a) The Born cross section of $e^+e^- \to 2(\pi^+\pi^-)$ measured by BES (solid dots), and by *BABAR* (open dots) in unit of *nb*. (b) and (c) the form factor squared, the dots are the values extracted from the data by CMD2, DM1, BES and *BABAR*, the solid line is the fitted curve and the dash line represents the contribution of $\rho(770)$ only.

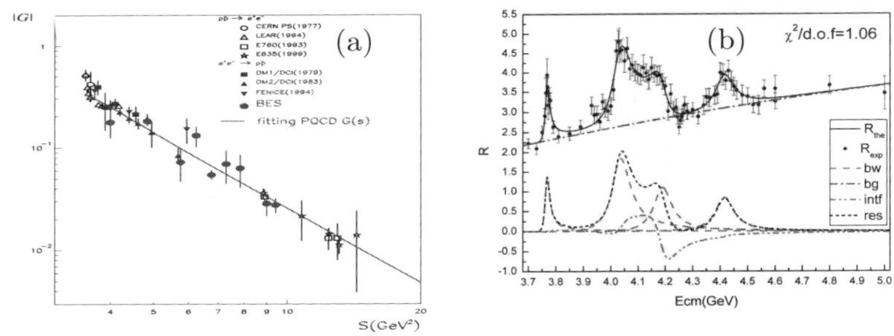

Fig. 2. (a) Form factors of $p\bar{p}$ measured by BES and other groups, the straight line illustrates the prediction by QCD. (b) The experimental and the theoretical predicted R value for the heavy charmonia structure; the curves for the continuum backgrounds, intrinsic Breit-Wigner cross section, and interferential contribution are shown respectively.

References

1. *BABAR* Collaboration, B.Aubert *et al.*, Phys. Rev. **D71** 052001 (2005).
2. N.N.Achasov and A.A.Kozhevnikov, Phys. Rev. **D55**·2663 (1997).
3. BES Collaboration, M.Ablikim *et al.* Phys. Lett. **B630** 14-20(2005)
4. BES Collaboration, J.Z.Bai *el al.*, Phys. Rev. Lett. **88**, 101802 (2002).

MEASUREMENT OF e^+e^- MULTIHADRONIC CROSS SECTIONS BELOW 4.5 GeV WITH *BABAR*

ACHIM DENIG Representing the *BABAR* collaboration

Institut für Experimentelle Kernphysik, Universität Karlsruhe (TH)
Postfach 3640, D-76021 Karlsruhe, Germany
E-mail: achim.denig@iekp.fzk.de

We present a summary of the hadronic cross section measurements performed with *BABAR* at the PEP-II collider via *radiative return*. *BABAR* has performed measurements of exclusive final states containing 3, 4 and 6 hadrons via this new method, as well as a measurement of the proton form factor.

1. Initial State Radiation Physics at *BABAR*

At the particle factories DAΦNE and PEP-II the hadronic cross section $\sigma(e^+e^- \rightarrow \text{hadrons})$ is measured over a wide energy range by *radiative return*[1,2] . In this new method only those events are considered, in which one of the beam electrons or positrons has emitted an initial state radiation (ISR) photon, lowering in such a way the effective invariant mass of the hadronic system. Precision measurements of the hadronic cross section are of utmost importance since they provide input to data-driven calculations of the hadronic contributions to the anomalous magnetic moment of the muon, a_μ, and of the running fine structure constant $\alpha(m_Z^2)$[3,4] . In this paper we present measurements of different exclusive final hadronic states in the mass range < 4.5 GeV, performed at the B-factory PEP-II ($\sqrt{s} = 10.6$ GeV) with the detector *BABAR*. At *BABAR* the ISR photon is required to be emitted at large polar angles with respect to the beam axis, allowing a kinematic closure of the event (*tagging*). Since the hadronic system is recoiling opposite to the ISR photon, a measurement of the cross sections with very high geometrical acceptance becomes possible. In order to extract the *non-radiative* cross section from the measured *radiative* cross section, one normalizes to a well-known radiator function[5] and to the PEP-II integrated luminosity, or - alternatively - to the yield of $e^+e^- \rightarrow \mu^+\mu^-\gamma$ events.

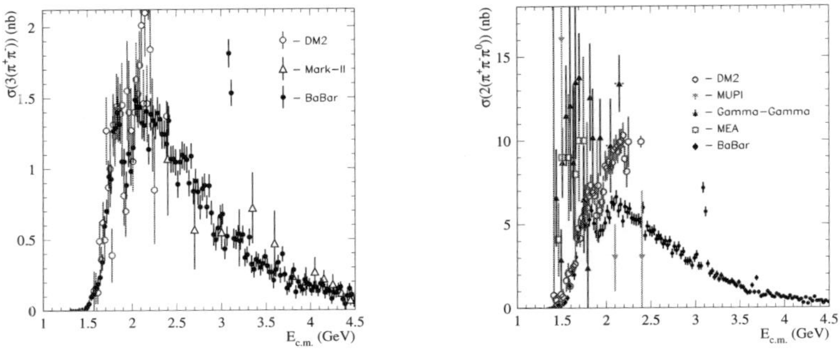

Fig. 1. The *BABAR* measurement of the cross section for $e^+e^- \to \pi^+\pi^-\pi^+\pi^-\pi^+\pi^-$ (left) and $e^+e^- \to \pi^+\pi^-\pi^+\pi^-\pi^0\pi^0$ (right) vs. $\sqrt{s}$ compared to the world data set.

2. Results

Three and four hadrons

BABAR has previously published measurements[6,7] of the $\pi^+\pi^-\pi^0$, $\pi^+\pi^-\pi^+\pi^-$, $K^+K^-\pi^+\pi^-$, $K^+K^-K^+K^-$ final states with better precision and coverage than all previous experiments, using 89fb^{-1} of data. The systematic accuracy of the 3π- and 4π-channels in the central mass region $1-2$ GeV is 5%. All states have been studied also in terms of their internal structures. In the $\pi^+\pi^-\pi^0$ analysis it was possible to improve significantly on the world's knowledge the excited ω states, while in the $\pi^+\pi^-\pi^+\pi^-$ channel a very strong contribution from the two-body mode $a_1(1260)\pi$ was identified. Preliminary results from a data sample of 232fb^{-1} are available for the $e^+e^- \to K^+K^-\pi^+\pi^-$ and $K^+K^-\pi^0\pi^0$ cross section[8] . In the $\phi(1020)f_0(980)$ intermediate two-body state a vector resonance-like structure is seen with a mass of $(2175 \pm 10_{\rm stat} \pm 15_{\rm syst})$ MeV and a width of $(58 \pm 16_{\rm stat} \pm 20_{\rm syst})$ MeV.

Six hadrons

The six-hadron process[9] has been studied in a data sample of 232fb^{-1} in the channels $e^+e^- \to 3(\pi^+\pi^-), 2(\pi^+\pi^-\pi^0)$ and $2(\pi^+\pi^-)K^+K^-$. The cross sections for the first two channels are shown in fig. 1; large improvements over existing measurements are seen, as well as a much wider coverage of the mass range. In the all-charged mode very little substructure has been found; a simulation containing one ρ^0 and four pions distributed according to phase space is adequate to describe the internal structure. On the contrary the partly neutral state shows a much more complex structure with signals for $\rho^0, \rho^\pm, \omega$ and η, and a substantial contribution from $\omega\eta$, which seems to be resonant. In both channels a structure at ca. 1900 MeV, which had previously been seen by DM2 and FOCUS[10] , is clearly visible. Fits to the $3(\pi^+\pi^-)$ and $2(\pi^+\pi^-\pi^0)$ spectra, assuming a resonant structure over a continuum shape, give consistent results for the mass M and width Γ of the structure. For the channel $3(\pi^+\pi^-)$ we find $M = (1880 \pm 30)$ MeV and $\Gamma = (130 \pm 30)$ MeV, for the channel $2(\pi^+\pi^-\pi^0)$ $M = (1860 \pm 20)$ MeV and $\Gamma = (160 \pm 20)$ MeV.

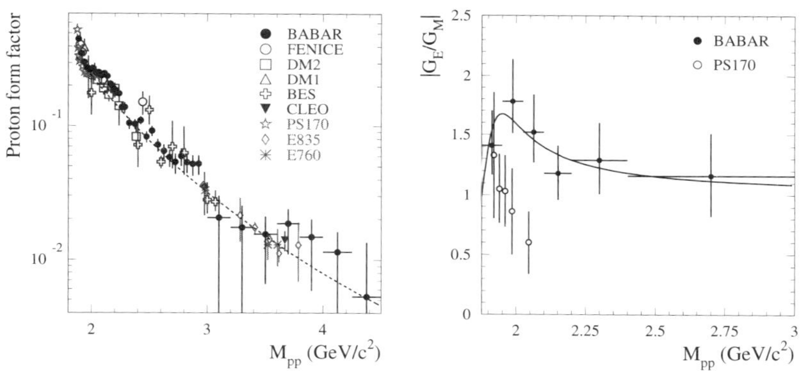

Fig. 2. Left: The *BABAR* measurement of the effective proton form factor and previous data obtained in e^+e^- and $p\bar{p}$ experiments. Right: The ratio of the electric and magnetic form factors $|G_E/G_M|$ as a function of $\sqrt{s}$.

Proton form factor

BABAR has also measured the cross section $e^+e^- \rightarrow p\bar{p}$ using 240fb^{-1} of data[11] ; the corresponding effective form factor is shown in fig. 2 (left), along with previous data from e^+e^- and $p\bar{p}$ experiments. We find an overall good consistency. The mass dependence shows a significant threshold enhancement, as well as two structures featuring sharp drops at 2.25 and 3.0 GeV, which illustrate the power of data from one single experiment over a wide range with no point-to-point uncertainties. Measuring the proton helicity angle θ_P in the $p\bar{p}$ rest frame, one can separate the ratio of the electric and magnetic form factor $|G_E/G_M|$, since both show a different functional behaviour in θ_P. The *BABAR* measurement of this ratio is shown in fig. 2 (right) for six different mass bins of $M_{p\bar{p}}$; a previous LEAR measurement[12] is in disagreement with *BABAR*. Our data shows a significant increase of the ratio $|G_E/G_M|$ towards threshold, while the two form factors are equal at larger masses.

3. Conclusions

Measurements of the hadronic cross section at PEP-II have improved the knowledge of the hadronic spectrum above 1 GeV. Thanks to the ISR-method, for the first time it becomes possible to cover the entire mass range of interest from threshold to 4.5 GeV in one single experiment. *BABAR* has not only performed precision measurements for exclusive hadronic states containing proton-antiproton, 3 pions and 4 and 6 hadrons, but has also measured 16 J/ψ and $\psi(2S)$ branching ratios, out of which 10 are better than the world average. Ongoing analyses are measuring the final states $\pi^+\pi^-$, K^+K^-, $\pi^+\pi^-\pi^0\pi^0$ and many more channels, which will further improve the standard model predictions for the muon anomaly a_μ and for the running fine structure constant $\alpha(m_Z^2)$.

References

1. A. Aloisio *et al.* [KLOE collaboration], Phys. Lett. **B606** (2005) 12
2. A. Denig, *The Radiative Return: a Review of Experiments*, Proceedings of the International Workshop on e^+e^- physics from ϕ to Ψ, Novosibirsk (Russia), Feb./March 2006; hep-ex/0611024
3. S. Eidelman and F. Jegerlehner, Z. Phys. **C67** (1995) 585
4. M. Davier, S. Eidelman, A. Höcker and Z. Zhang, Eur. Phys. J. **C31** (2003) 503
5. S. Binner, J.H. Kühn and K. Melnikov, Phys. Lett. B **459** (1999) 279; G. Rodrigo, A. Gehrmann-De Ridder, M. Guilleaume and J.H. Kühn, Eur. Phys. J.**C22**, (2001) 81; G. Rodrigo, H. Czyż, J.H. Kühn and M. Szopa, Eur. Phys. J. **C24** (2002) 71; H. Czyż, A. Grzelińska, J. H. Kühn and G. Rodrigo, Eur. Phys. J. **C27**, (2003) 563; H. Czyż, A. Grzelińska, J. H. Kühn and G. Rodrigo, Eur. Phys. J. **C33**, (2004) 333; H. Czyż, A. Grzelińska, J. H. Kühn and G. Rodrigo, Eur. Phys. J. **C39**, (2005) 411; H. Czyż and E. Nowak-Kubat, Phys. Lett. **B634** (2006) 493
6. B. Aubert *et al.* [*BABAR* collaboration], Phys. Rev. **D70** (2004) 072004
7. B. Aubert *et al.* [*BABAR* collaboration], Phys. Rev. **D71** (2004) 052001
8. E. Solodov, *ISR study at BABAR and the application to the R-measurement and hadron spectroscopy*, Proceedings of the International Conference ICHEP 2006, Moscow (Russia), July/Aug. 2006
9. B. Aubert *et al.* [*BABAR* collaboration], Phys. Rev. **D73** (2006) 052003
10. P.L. Frabetti *et al.* [E687 collaboration], Phys. Lett. **B514** (2001) 240
11. B. Aubert *et al.* [*BABAR* collaboration], Phys. Rev. **D73** (2006) 012005
12. G. Bardin *et al.* [PS170 collaboration], Nucl. Phys. **B411** (1994) 3

THE PION VECTOR FORM-FACTOR AND $(g-2)_\mu$

CHRISTOPHER SMITH

Institut für Theoretische Physik, Universität Bern, CH-3012 Bern, Switzerland.

The hadronic contribution to $(g-2)_\mu$ from states below 1 GeV is revisited. Imposing model-independent theoretical constraints while fitting the data is shown to decrease the theoretical error by a factor of two.

Keywords: Pion form-factor; $(g-2)_\mu$; $\alpha(M_Z)$.

1. Introduction

The hadronic vacuum polarization enters the theoretical predictions of the muon anomalous moment a_μ and the fine structure constant $\alpha\,(M_Z)$.

The prediction[1] $a_\mu^{th} = 11659180.4 \pm 5.1$ (units of 10^{-10}), including the LO hadronic piece $a_\mu^{hvp} = 689 \pm 5$, is about 3σ below experiment. New physics can easily account for this discrepancy. Hence it is important to decrease the error, especially for a_μ^{hvp}. Since low-energy dominates the a_μ^{hvp} dispersion integral, in terms of $\sigma\,(e^+e^- \to hadrons)$, our strategy towards improving the already very precise estimate is to supplement $e^+e^- \to \pi^+\pi^-$ data with model-independent constraints on the $\gamma^* \to \pi^+\pi^-$ form-factor.[2]

Knowledge of $\alpha\,(M_Z)$ is needed for precision study of the SM, as it enters the predictions for M_W, $\sin^2\theta_{eff}$,... The theoretical status is[1] $\alpha\,(M_Z)^{-1} = \alpha\,(0)^{-1}\,(1 - \Delta\alpha_{lep} - \Delta\alpha_{had} - \Delta\alpha_{top})$, $\Delta\alpha_{lep} = 3149.77$, $\Delta\alpha_{top} = -7.3 \pm 0.2$, $\Delta\alpha_{had} = 2768 \pm 22$ (units of 10^{-5}). Though $\Delta\alpha_{had}$ is again estimated from $\sigma\,(e^+e^- \to hadrons)$, only $\sim 10\%$ comes from low-energy in its dispersion integral. Still, as the precision improves for $\sigma\,(e^+e^- \to hadrons)$ in the $1 \to 5$ GeV range, reanalysis of the low-energy part is useful.

2. Theoretical constraints and fits to the data

The form factor $F_\pi\,(s)$ for $\gamma^*\,(s) \to \pi^+\pi^-$ is an analytic function of s, and can be expressed in terms of its phase $\delta_F\,(s)$ (Omnes representation).[2]

Below the opening of inelastic channels, Watson Theorem states that $\delta_F\,(s) = \delta_1^1\,(s)$, with $\delta_1^1\,(s)$ the $\pi\pi$ phase-shift in the $I = \ell = 1$ wave (including the ρ resonance). This phase-shift is obtained by solving the

Roy Equations for $\pi\pi$ scattering, which follow from unitarity and crossing-symmetry.[3]

Isospin breaking effects are important at the ρ peak ($\rho-\omega$ mixing), and are accounted for by explicitly giving to $F_\pi(s)$ the ω pole.[2] Finally, the various inelastic channels like $\gamma^* \to 4\pi, \pi^0\omega, ... \to \pi^+\pi^-$ are all encoded into a conformal polynomial.[2] A severe constraint on this piece is the Eidelman-Lukaszuk bound,[4] relating the inelasticity in $\pi\pi$ scattering, the inelastic phase $\delta_F - \delta_1^1$, and the cross-section for production of non-2π states. Experimentally, this cross-section is very small below the $\pi^0\omega$ threshold and the Watson Theorem is expected to hold up to $m_\pi + m_\omega$.

We can now fit the data using the constrained parametrization of $F_\pi(s)$. The CMD2[5] and SND[6] energy scans are found compatible among themselves and with the theoretical constraints. They lead to the precise values $a_\mu(2m_K) = 493.7\pm1.8$ and $\Delta\alpha_{had}(2m_K) = 327\pm1.2$. The radiative return data from KLOE[7] is not compatible with CMD2/SND, though it leads to similar values for $a_\mu(2m_K)$ and $\Delta\alpha_{had}(2m_K)$. From a theoretical perspective, we note that the behavior below the ρ peak as well as the ρ, ω masses are slightly different. Finally, $\tau^- \to \pi^-\pi^0\nu_\tau$ data[8] are also inconsistent with CMD2/SND, being significantly higher above the ρ peak.

3. Conclusion

Imposing model-independent theoretical constraints on the vector form-factor improves the precision of the evaluation of a_μ and $\alpha(M_Z)$. Further, this provides for meaningful tests of both experimental and theoretical techniques. Indeed, below 1 GeV, one has data from all three types of experiments: energy scans, radiative return and τ decays. Though for the moment only the former are compatible with our parametrization, the error in the determinations of $a_\mu(2m_K)$ and $\Delta\alpha_{had}(2m_K)$ is already reduced by a factor two.

Acknowledgements

Work supported by the Schweizerischer Nationalfonds.

References

1. K. Hagiwara, A. D. Martin, D. Nomura and T. Teubner, hep-ph/0611102.
2. G. Colangelo, Nucl. Phys. Proc. Suppl. **131** (2004) 185;
   G. Colangelo, H. Leutwyler, I. Caprini, C. Smith, in preparation.
3. B. Ananthanarayan *et al.*, Phys. Rept. **353** (2001) 207.

4. S. Eidelman and L. Lukaszuk, Phys. Lett. **B582** (2004) 27.
5. R. R. Akhmetshin *et al.* [CMD2], Phys. Lett. **B578** (2004) 285.
6. M. N. Achasov *et al.* [SND], J. Exp. Theor. Phys. **103** (2006) 380.
7. A. Aloisio *et al.* [KLOE], Phys. Lett. **B606** (2005) 12.
8. See e.g. S. Schael *et al.* [ALEPH], Phys. Rept. **421** (2005) 191.

PARTIALLY QUENCHED CHPT RESULTS TO TWO LOOPS

JOHAN BIJNENS[a]

[a] *Dept. of Theor. Physics, Lund University, Sölvegatan 14A, SE 22362 Lund, Sweden*

Keywords: Chiral Perturbation Theory; Lattice QCD; Meson Masses and Decay Constants

For lattice QCD calculations it is easy to distinguish between valence and sea quarks. This, together with the fact that valence quark masses can be changed with much less computational effort, makes it useful to perform so-called partially quenched studies where one has different properties for the valence and the sea quarks.

However, even with this, it still remains difficult to go to very low quark masses of the order of the physical up and down quark masses. The extrapolation to low quark masses is best performed within a theoretically sound framework. In the case of fully unquenched QCD, such a framework is Chiral Perturbation Theory (χPT). Here the present state of the art is two-loop calculations. An overview of these and a source of further references is Ref. 1. To do the same for the extrapolation from partially quenched lattice calculations, χPT must be extended to this case. A review of this and other uses of χPT for lattice QCD is Ref. 2.

One way to obtain the effect of treating valence and sea quarks differently in χPT is by adding ghost quarks, bosonic but with spin 1/2, with the same masses and couplings as the valence quarks. Due to their different statistics, closed ghost and valence quark loops cancel automatically as shown schematically in Fig. 1, achieving the desired effect of removing closed valence quark loops.

The work presented here involves extending the partially quenched (PQ) version of χPT also to two-loop order. One important fact that allows one to do this is that PQχPT can be formally obtained form n_F-flavor χPT by replacing traces by supertraces and the general number of flavors n_F by the number of sea quarks n_{sea}. This is true in the mesonic sector since all

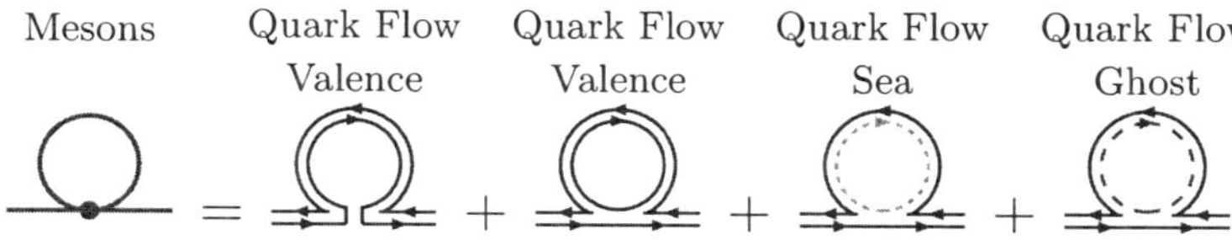

Fig. 1. A schematic view of a meson loop and the quark flows added to obtain the partial quenching effect.

operations used in constructing the Lagrangian and divergence structure go through.[3,4] This observation allows one to do a series of calculations for the masses and decay constants for two[5] and three[3,4,6] sea quark flavors at the two-loop level. Major technical problems to be dealt with were dealing with the extremely lengthy formulas due to the occurrence of many ratios of differences of quark masses and the presence of the double poles in the propagators allowed in partially quenched theories.

The behavior of the double poles to all orders in perturbation was studied via a direct resummation technique and then worked out explicitly to two-loop order for the simplest mass case.[7] This allows one to determine the eta mass at two-loop order from PQχPT. A more recent addition is the inclusion of photons.[8] For more details and references, please consult Refs. 3–8.

Acknowledgments

This work is supported by the EU contracts No. MRTN-CT-2006-035482 (FLAVIAnet) and No. RII3-CT-2004-506078 (HadronPhysics).

References

1. J. Bijnens, hep-ph/0604043, to be published in Prog. Part. and Nuc..
2. S. R. Sharpe, hep-lat/0607016.
3. J. Bijnens, N. Danielsson and T. A. Lähde, Phys. Rev. D**70**, 111503 (2004) [hep-lat/0406017].
4. J. Bijnens, N. Danielsson and T. A. Lähde, Phys. Rev. D**73**, 074509 (2006) [hep-lat/0602003].
5. J. Bijnens and T. A. Lähde, Phys. Rev. D**72**, 074502(2005)[hep-lat/0506004].
6. J. Bijnens and T. A. Lähde, Phys. Rev. D**71**, 094502(2005)[hep-lat/0501014].
7. J. Bijnens and N. Danielsson, Phys.Rev. D**74**,054503(2006)[hep-lat/0606017].
8. J. Bijnens and N. Danielsson, hep-lat/0610127.

PION-PION SCATTERING WITH MIXED ACTION LATTICE QCD

PAULO F. BEDAQUE

University of Maryland
College Park, MD 20742
E-mail:bedaque@umd.edu

We discuss the recent NPLQCD calculations of the $I = 2$ $\pi\pi$ scattering length using staggered sea and domain wall valence quarks. Particular enphasis is given to the chiral extrapolation down to realistic values of the quark mass.

Keywords: Pion-pion scattering; lattice QCD; Chiral perturbation theory

Lattice QCD calculations of hadron *interactions* are notoriously difficult. Among the difficulties is the fact that, contrary to most other observables, scattering amplitudes are not obtained by extrapolating the finite volume results to the infinite volume limit. In fact, it is the finite volume dependence of the energy levels of the two particle system that determines the scattering amplitude. The relation between scattering amplitudes and energy levels – known as the Luscher formula[1]– reduces, in the case where the scattering length a is much smaller than the size of the lattice L, to

$$\Delta E = -\frac{4\pi a}{mL^3}\left[1 + c_1\frac{a}{L} + c_2\left(\frac{a}{L}\right)^2 + \cdots\right] \tag{1}$$

where m is the mass of the particle, $c_1 = -2.8372, c_2 = 6.3752$ and $\Delta E = E - 2m$ is the difference between the lowest energy level of two particles and their rest mass. We see then that the signal sought after in the lattice calculation, namely ΔE, is a difference of two much larger quantities.

The Nuclear Physics with Lattice QCD collaborations (NPLQCD) was formed with the goal of computing hadron-hadron phase shifts using lattice QCD, specially the ones relevant for nuclear physics. As part of this program, the s-wave $\pi\pi$ scattering length in the isospin $I = 2$ channel was computed. A mixed action was used: an improved staggered quark action (asqtad) in the sea sector (with gauge configurations generated by MILC)

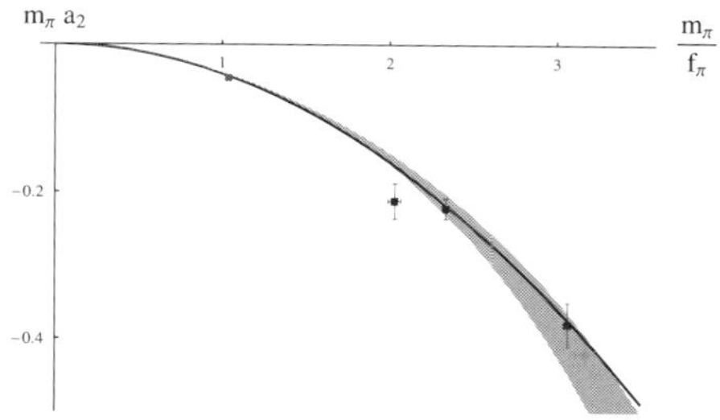

Fig. 1. $m_\pi a_2$ plotted as a function of f_π/m_π. The leftmost point (red) shows the experimental determination,[4] the rightmost one the calculation in 3 . The remaining (black) points correspond to the NPLQCD calculation. The (blue) band shows the range in the chiral extrapolations, as discussed in the text.

and domain wall quarks, with nearly perfect chiral symmetry, in the valence sector (with propagators generated by LHPC). The details are fully described in.[2] Three quark masses, corresponding to $m_\pi = 194, 348$ and 484 MeV at one lattice spacing, $b \approx 0.125$ fm, were used. The results are summarized in Fig. (1).

The data at the three values of the pion mass can then be used in conjunction with chiral perturbation theory (χPT) to extrapolate the results down to the realistic value of the pion mass. At next-to-leading order, the χPT prediction for the scattering length depends on one combination of Gasser-Leutwyler coefficients ($l_{\pi\pi}(\mu)$). Fitting the numerical data we determine $l_{\pi\pi}(m_\pi) = 3.3 \pm 0.3 \pm 0.6$, where the first error is statistical and the second systematic. From this value of $l_{\pi\pi}(m_\pi)$ we predict $m_\pi a_2$ at the realistic value of m_π to be $m_\pi a_2 = -0.0426 \pm 0.0006 \pm 0.0003 \pm 0.0018$, where the errors are, respectively, statistical, systematic and the systematic errors from the neglect of higher orders in χPT. This value is to be compared to the one extracted from the K(e4) decays $m_\pi a_2 = -0.0454 \pm 0.0031 \pm 0.0013 \pm 0.0008$.[4]

References

1. M. Luscher, Commun. Math. Phys. **105**, 153 (1986).
2. S. R. Beane et al, [NPLQCD Collaboration], Phys. Rev. D **73**, 054503 (2006).
3. T. Yamazaki *et al.* [CP-PACS Collaboration], Phys. Rev. D **70**, 074513 (2004).
4. S. Pislak *et al.*, Phys. Rev. D **67**, 072004 (2003).

MESON SYSTEMS WITH GINSPARG-WILSON VALENCE QUARKS

ANDRÉ WALKER-LOUD

Physics Department, University of Maryland,
College Park, MD 20742-4111, USA
E-mail: walkloud@umd.edu

Mixed action (MA) lattice simulations[1] are a new variant of lattice regularization which employ different discretization schemes for the *valence* and *sea* fermions. This allows one to simulate in the chiral regime with numerically expensive chiral fermions[2] in the valence sector, which satisfy the Ginsparg-Wilson (GW) relation[3] while using numerically cheaper fermions which violate chiral symmetry in the sea sector, for example Wilson or staggered fermions. The chiral extrapolation formulae determined from effective theories extended to include partial quenching (PQ)[4] and lattice spacing (a) artifacts[5] will generally involve not only the *physical* counterterms of interest, known as Gasser-Leutwyler coefficients in the case of mesons,[6] but will also involve *unphysical* counterterms corresponding to the finite lattice spacing and partial quenching effects. For MA theories with GW valence fermions, this is not the case; the extrapolation formulae for meson scattering processes through next-to-leading order (NLO) only involve the *physical* counterterms of interest provided one uses a lattice-physical (on-shell) renormalization scheme.[7–9]

It has recently been shown[7] that the MA $I = 2$ $\pi\pi$ scattering length, expressed in terms of leading order (LO) parameters has the following form at NLO,

$$m_\pi a_{\pi\pi}^{I=2} = -\frac{m_{uu}^2}{8\pi f^2}\left\{1+\frac{m_{uu}^2}{(4\pi f)^2}\left[4\ln\left(\frac{m_{uu}^2}{\mu^2}\right)+4\frac{\tilde{m}_{ju}^2}{m_{uu}^2}\ln\left(\frac{\tilde{m}_{ju}^2}{\mu^2}\right)-1+\ell'_{\pi\pi}(\mu)\right.\right.$$
$$\left.\left.-\frac{\tilde{\Delta}_{ju}^4}{6m_{uu}^4}-\frac{\tilde{\Delta}_{ju}^2}{m_{uu}^2}\ln\left(\frac{m_{uu}^2}{\mu^2}\right)\right]+\frac{\tilde{\Delta}_{ju}^2}{(4\pi f)^2}\,\ell'_{PQ}(\mu)+\frac{a^2}{(4\pi f)^2}\ell'_{a^2}(\mu)\right\}, \quad (1)$$

where the *valence-valence* meson mass is $m_{uu}^2 = 2B_0 m_u$, the mixed *valence-sea* meson mass is $\tilde{m}_{ju}^2 = B_0(m_u + m_j) + a^2\Delta_{\rm Mix}$ and the PQ parameter $\tilde{\Delta}_{ju}^2 = 2B_0(m_j - m_u) + a^2\Delta_{\rm sea}$ (which vanishes in the QCD limit). This expression depends on three unknown counterterms, only one of which is physical, $\ell'_{\pi\pi}(\mu)$, as well as the mixed meson mass shift, $a^2\Delta_{\rm Mix}$. If we express $m_\pi a_{\pi\pi}^{I=2}$ in terms of the lattice-physical parameters, the expression is identical in form to the physical expression up to a known additive shift,

$$m_\pi a_{\pi\pi}^{I=2} = -\frac{m_\pi^2}{8\pi f_\pi^2}\left\{1 + \frac{m_\pi^2}{(4\pi f_\pi)^2}\left[3\ln\left(\frac{m_\pi^2}{\mu^2}\right) - 1 - l_{\pi\pi}^{I=2}(\mu) - \frac{\tilde{\Delta}_{ju}^4}{6\,m_\pi^4}\right]\right\}, \quad (2)$$

and thus proportional to the *physical* counterterm, $l_{\pi\pi}^{I=2}(\mu)$. The *unphysical* contribution, $\tilde{\Delta}_{ju}^4/6m_\pi^4$, turns out to about 10% of the physical NLO contribution[7,9] in the recent lattice simulation of this quantity.[10]

This simplification is understood with the realization that every operator in a NLO MA Lagrangian relevant for external *valence-valence* mesons, incorporating either lattice spacing or PQ effects, amounts to a renormalization of one of the two operators in the LO Lagrangian.[9] The on-shell renormalization scheme absorbs all *unphysical* counterterm effects into the lattice-physical meson masses and decay constants, removing any explicit dependence upon these operators. These arguments hold for all meson scattering processes at NLO; non-zero momentum and $N > 2$ external mesons. Finally, the above scattering length, along with $m_K a_{KK}^{I=1}$, $\mu_{K\pi} a_{K\pi}^{I=3/2}$ and f_K/f_π, two of which have been computed with MA lattice QCD[11,12] share only two linearly independent counterterms, both of which are *physical*, allowing us to make a prediction[9] of $m_K a_{KK}^{I=1}$, and thereby test this MA formalism as well as the convergence of $SU(3)$ chiral perturbation theory.

References

1. O. Bar, G. Rupak and N. Shoresh, *Phys. Rev.* **D67**, p. 114505 (2003).
2. D. B. Kaplan, *Phys. Lett.* **B288**, 342 (1992).
3. P. H. Ginsparg and K. G. Wilson, *Phys. Rev.* **D25**, p. 2649 (1982).
4. C. W. Bernard and M. F. L. Golterman, *Phys. Rev.* **D49**, 486 (1994).
5. S. R. Sharpe and J. Singleton, Robert L., *Phys. Rev.* **D58**, p. 074501 (1998).
6. J. Gasser and H. Leutwyler, *Nucl. Phys.* **B250**, p. 465 (1985).
7. J.-W. Chen, D. O'Connell, R. S. Van de Water and A. Walker-Loud, *Phys. Rev.* **D73**, p. 074510 (2006).
8. D. O'Connell [hep-lat/0609046].
9. J.-W. Chen, D. O'Connell and A. Walker-Loud [hep-lat/0611003].
10. S. R. Beane, P. F. Bedaque, K. Orginos and M. J. Savage, *Phys. Rev.* **D73**, p. 054503 (2006).

11. S. R. Beane, P. F. Bedaque, K. Orginos and M. J. Savage [hep-lat/0606023].
12. S. R. Beane *et al.* [hep-lat/0607036].

LOW ENERGY CONSTANTS FROM THE MILC COLLABORATION

C. BERNARD

Department of Physics, Washington University,
Saint Louis, Missouri 63130, USA
E-mail: cb@wustl.edu

We present preliminary updates of results for QCD low energy constants.

Keywords: chiral dynamics, lattice QCD

The MILC Collaboration is in the process of updating its simulations[1] of QCD using three dynamical light flavors (u, d, s) of improved staggered quarks. New lattice data[2] includes a complete ensemble with average u, d mass $\hat{m} = 0.1m_s$ and lattice spacing $a \approx 0.09$ fm, and a partial ensemble at $\hat{m} = 0.4m_s$ and $a \approx 0.06$ fm, as well as several new coarser ensembles at $a \approx 0.15$ fm, with a range of sea quark masses.

Some of our preliminary results are:

$$\begin{aligned}
f_\pi &= 128.6 \pm 0.4 \pm 3.0 \text{ MeV}; &\quad f_K &= 155.3 \pm 0.4 \pm 3.1 \text{ MeV};\\
f_K/f_\pi &= 1.208(2)(^{+\ 7}_{-14}) \Rightarrow & |V_{us}| &= 0.2223(^{+26}_{-14});\\
f_\pi/f_2 &= 1.050(3)(10); & f_2/f_3 &= 1.09(4)(4);\\
2L_6 - L_4 &= 0.5(1)(2); & 2L_8 - L_5 &= -0.1(1)(1);\\
L_4 &= 0.1(2)(3); & L_5 &= 2.0(3)(2);\\
\langle \bar{u}u \rangle_2 &= -(\,276(2)(7)(5) \text{ MeV}\,)^3; & \langle \bar{u}u \rangle_3 &= -(\,247(10)(15)(4) \text{ MeV}\,)^3;\\
\langle \bar{u}u \rangle_2/\langle \bar{u}u \rangle_3 &= 1.38(15)(22); & &\\
m_s^{\overline{\rm MS}} &= 90(0)(5)(4)(0) \text{ MeV}; & \hat{m}^{\overline{\rm MS}} &= 3.3(0)(2)(2)(0) \text{ MeV};\\
m_u^{\overline{\rm MS}} &= 2.0(0)(1)(2)(1) \text{ MeV}; & m_d^{\overline{\rm MS}} &= 4.6(0)(2)(2)(1) \text{ MeV};\\
m_s/\hat{m} &= 27.2(0)(4)(0)(0); & m_u/m_d &= 0.42(0)(1)(0)(4)\,.
\end{aligned}$$

The errors are statistical, systematic (from the simulations), and, where relevant, perturbative (from two-loop perturbation theory[3]) and electromagnetic. The quantity f_2 (f_3) represents the three-flavor decay constant in the two (three) flavor chiral limit, and $\langle \bar{u}u \rangle_2$ ($\langle \bar{u}u \rangle_3$) is the corresponding condensate. The low energy constants L_i are evaluated at chiral scale m_η, and the condensates and masses are in the $\overline{\text{MS}}$ scheme at scale 2 GeV.

References

1. C. Aubin *et al.*, *Phys. Rev.* **D70**, p. 114501 (2004), hep-lat/0407028.
2. C. Bernard *et al.* (2006), hep-lat/0609053.
3. Q. Mason, H. D. Trottier, R. Horgan, C. T. H. Davies and G. P. Lepage, *Phys. Rev.* **D73**, p. 114501 (2006), hep-ph/0511160.

LATTICE QCD SIMULATIONS AT SMALL QUARK MASSES

GERRIT SCHIERHOLZ

Deutsches Elektronen-Synchrotron DESY
D-22603 Hamburg, Germany

and

John von Neumann-Institut für Computing NIC
Deutsches Elektronen-Synchrotron DESY
D-15738 Zeuthen, Germany

E-mail: Gerrit.Schierholz@desy.de

– For the QCDSF Collaboration –

Due to improvement in algorithms and computer performance unquenched simulations of Wilson-type fermions with lighter quark masses are now possible.[1,2] This enables us to make contact with chiral perturbation theory (ChPT) and the real world.

We report results for the pseudoscalar decay constants f_{PS}^{AB} at pion masses down to $O(300)$ MeV, using $N_f = 2$ nonperturbatively $O(a)$ improved Wilson fermions. For calculational details see.[3] The renormalzation constants and improvement coefficients of the axial vector current are computed nonperturbatively as well.

We first investigate to see if we are entering a regime, where chiral logarithms are becoming visible, and look at the ratio of nondegenerate, partially quenched decay constants[4]

$$R = \frac{f_{PS}^{VS}}{\sqrt{f_{PS}^{VV} f_{PS}^{SS}}}$$

(V: valence, S: sea quark). Our data displays chiral logarithms of about the expected size. Next we fit our data to the predictions of NLO ChPT. We obtain $\alpha_4 \approx -0.58$, $\alpha_5 \approx -0.45$. While α_4 is in reasonable agreement with other phenomenological estimates, α_5 is not. Using $r_0 = 0.5$ fm [$r_0 = 0.467$ fm] to set the scale, we find $f_{K^+}/f_{\pi^+} = 1.24$ [$f_{K^+}/f_{\pi^+} = 1.22$], to be compared with the experimental value 1.223, and $f_{\pi^+} = 76(4)(2)$ MeV [$f_{\pi^+} = 81(4)(2)$ MeV]. The first error is statistical, while the second error is due to the error in r_0/a.

References

1. M. Göckeler *et al.*, hep-lat/0610066.
2. M. Göckeler *et al.*, hep/lat/0610071.
3. M. Göckeler *et al.*, Phys. Rev. D57 (1998) 5562 [hep/lat/9707021].
4. S.R. Sharpe, Phys. Rev. D56 (1997) 7052, Erratum *ibid.* D62 (2000) 099901.

Lattice QCD simulations with two light dynamical (Wilson) quarks

LEONARDO GIUSTI*

CERN

Physics Department - TH Division

CH-1211 Geneva 23, Switzerland

**E-mail: Leonardo.Giusti@cern.ch*

Algorithmic and technical progress achieved over the last few years makes QCD simulations with light dynamical quarks much faster than before. As a result lattices with pions as light as 250–300 MeV can be simulated with the present generation of computers. In this talk I report on simulations of two-flavour QCD at sea-quark masses from slightly above to approximately 1/4 of the strange-quark mass, on lattices with up to 64×32^3 points and spacings from 0.05 to 0.08 fm. Physical sea-quark effects are clearly seen on these lattices, while the lattice effects appear to be quite small, even without O(a) improvement. A striking result is that the dependence of the pion mass on the sea-quark mass is accurately described by leading-order chiral perturbation theory up to meson masses of about 500 MeV. Details of the results can be found in:
"QCD with light Wilson quarks on fine lattices (I): First experiences and physics results", by L. Del Debbio et al. CERN-PH-TH-2006-198, hep-lat/0610059;
"QCD with light Wilson quarks on fine lattices (II): DD-HMC simulations and data analysis", by L. Del Debbio et al. CERN-PH-TH-2006-261.

DO WE UNDERSTAND the LOW-ENERGY CONSTANT L_8?

MAARTEN GOLTERMAN

Dept. of Physics, San Francisco State University, San Francisco, CA 94132, USA

In this talk, we raise some questions about how well we understand the relation between low-energy constants (LECs) and the hadronic spectrum, in particular for the case of L_8.[1]

We start with the dispersion relation for the difference of the scalar and pseudo-scalar two-point function (in the chiral limit; omit the pion pole)

$$\Delta\Pi(q^2 = -Q^2) = \frac{1}{2}\left(\Pi_S(q^2) - \Pi'_P(q^2)\right) = \frac{1}{2\pi}\int_0^\infty dt\, \frac{\mathrm{Im}\left(\Pi_S(t) - \Pi'_P(t)\right)}{t + Q^2} \,. \tag{1}$$

Saturating with an infinite number of zero-width resonances at large N_c, this leads to a sum rule

$$16B^2 L_8 = \Delta\Pi(0) = \sum_n^\infty \frac{F_S^2(n)}{M_S^2(n)} - \sum_n^\infty \frac{F_P^2(n)}{M_P^2(n)} \,, \tag{2}$$

which is then usually truncated to a finite number of resonances in practice. However, the sums over n in Eq. (2) are quadratically divergent, and this is a consequence of the fact that we interchanged the integration over t with the sum over n. It follows that if one uses this sum rule, for example, with one resonance in each channel, either the prediction of L_8 in terms of the resonance parameters will not be reliable, or, imposing the correct value of L_8, one expects to obtain poor estimates of the resonance parameters.

We studied this issue in more detail in a simple Regge-like model that satisfies the following properties:[1]

- Each resonance is given a finite width of order $1/N_c$ which grows proportionally with its mass, in a way consistent with the general analyticity properties of $\Delta\Pi(q^2)$;
- $\Delta\Pi(q^2)$ satisfies an unsubtracted dispersion relation (1);

- There is no multi-particle continuum, hence no scale dependence in L_8 (this simplification is not essential for our purposes).

This model predicts a spectral function which shows the "onset of the continuum," which is usually taken as justification for truncating the sum in Eq. (2) (see Fig. 1 in Ref. [1]). Nevertheless, making the same truncation in the model, one obtains incorrect results for L_8, which in the model can be exactly calculated. (Just as in QCD, one cannot interchange the integral over t with the sum over n.) Moreover, this exact value in the model is independent of N_c, suggesting that the problem has little to do with taking the large-N_c limit. The value of L_8 in the model comes in equal parts from the "continuum part" of the spectral function, suggesting that this contribution cannot be ignored. The only reason that we can think of that would make the situation different in QCD is the scenario that QCD (contrary to our simple model) would exhibit "parity doubling" in the higher-lying part of the scalar and pseudo-scalar meson spectrum. We note that, at this time, neither theoretical nor experimental evidence exists for this scenario.

We close with two observations. First, starting from a once-subtracted dispersion relation instead of Eq. (1), one can derive a well-defined sum rule at large N_c:

$$16B^2L_8 = \lim_{Q^2\to\infty} Q^2\left(\sum_n^{\infty} \frac{F_S^2(n)}{M_S^2(n)(M_S^2(n)+Q^2)} - (S\to P)\right) . \qquad (3)$$

Of course, the Q^2 limit may not be interchanged with the sum over n, lest we revert back to the ill-defined result (2). Second, in our model, we find that, if we fit the parameters of an approximation with only one resonance in each channel to L_8 and to the leading order OPE behavior, this approximation describes the function $\Delta\Pi(Q^2)$ very well for all (euclidean) values of Q^2. This is true even though the fitted resonance parameters turn out to be very different from the true values.

We thank Oscar Catà, Matthias Jamin, Toni Pich and Eduardo de Rafael for useful discussions. SP was supported in part by CICYT-FEDER-FPA2005-02211 and SGR2005-00916, and MG by the Generalitat de Catalunya under the program PIV1-2005 and by the US Dept. of Energy.

References

1. For more detail and references, see M. Golterman and S. Peris, Phys. Rev. D **74**, 096002 (2006) [arXiv:hep-ph/0607152].

QUARK MASS DEPENDENCE OF LECs IN THE TWO-FLAVOUR SECTOR

M. SCHMID*

Institute for Theoretical Physics, University of Bern
Sidlerstrasse 5, 3012 Bern, Switzerland
** email: schmidm@itp.unibe.ch*

For large values of m_s and suitably chosen external fields, $\mathbf{SU}(3)\times\mathbf{SU}(3)$ chiral perturbation theory reduces to $\mathbf{SU}(2)\times\mathbf{SU}(2)$. This relation can be used to determine the m_s-dependence of the low-energy constants in the two-flavour sector.

Keywords: Chiral symmetry; Chiral perturbation theory; Low-energy constants

Chiral perturbation theory[1–3] was successfully applied in the two-flavour sector [chiral $\mathbf{SU}(2)\times\mathbf{SU}(2)$] as well as in the three-flavour case [chiral $\mathbf{SU}(3)\times\mathbf{SU}(3)$]. In both frameworks, the low-energy constants (LECs) encode the heavy degrees of freedom not explicitly present as dynamical fields in the effective Lagrangian. As a result of this, the $\mathbf{SU}(2)\times\mathbf{SU}(2)$-LECs depend on the strange quark mass, while the ones in $\mathbf{SU}(3)\times\mathbf{SU}(3)$ are independent thereof.

If one limits the external momenta to values small compared to m_s and treats m_u and m_d as small in comparison to m_s, the degrees of freedom of the K- and η-mesons freeze out. Limiting the external fields to be the same as in the two-flavour theory, one can work out explicitly the strange quark mass dependence of the LECs in $\mathbf{SU}(2)\times\mathbf{SU}(2)$ from the effective action pertaining to $\mathbf{SU}(3)\times\mathbf{SU}(3)$. For example, for the pion decay constant F at $m_u = m_d = 0$, $m_s \neq 0$, the relation reads at one loop accuracy[3]

$$F = F_0\left\{1 - \frac{m_s B_0}{32\pi^2 F_0^2}\left[\ln\frac{m_s B_0}{\mu^2} - 256\pi^2 L_4^r\right] + O\big(m_s^2\big)\right\}.$$

Here, F_0, B_0, and L_4^r are LECs from $\mathbf{SU}(3)\times\mathbf{SU}(3)$, and the renormalization scale is denoted by μ. Relations of this type generate constraints on the possible values of the low-energy constants: information on LECs in $\mathbf{SU}(2)\times$

SU(2) may be translated into information on LECs in **SU**(3) × **SU**(3) and vice versa.[2,3]

In my talk, I discussed recent efforts[4] to establish the corresponding relations between the LECs at two-loop accuracy. Our method relies on calculating the local part of the effective action in **SU**(3) × **SU**(3) to two loops in the above mentioned limit. Loop functions containing K- or η-mesons can be expanded in the external momenta by use of heat-kernel techniques. Using the equation of motion finally allows one to identify the local part of the pertinent two-flavour effective action and to extract the relation between the LECs.

During our calculations we found a hitherto unknown linear relation between the basis elements of the $O(p^6)$ Lagrangian of **SU**(2) × **SU**(2), constructed in Ref.[5] This relation reduces the number of independent chiral polynomials at order p^6 by one.

References

1. S. Weinberg, *Physica* **A96**, 327 (1979).
2. J. Gasser and H. Leutwyler, *Ann. Phys.* **158**, 142 (1984).
3. J. Gasser and H. Leutwyler, *Nucl. Phys.* **B250**, 465 (1985).
4. J. Gasser, C. Haefeli, M.A. Ivanov, and M. Schmid, in preparation.
5. J. Bijnens, G. Colangelo and G. Ecker, *JHEP* **9902**, 020 (1999).

PROGRESS REPORT ON THE π^0 LIFETIME EXPERIMENT (PRIMEX) AT JLAB

D. E. McNULTY (for the PrimEx Collaboration*)

Laboratory for Nuclear Science,
Massachusetts Institute of Technology,
Cambridge, Mass 02139, USA
**E-mail: mcnulty@jlab.org*
www.jlab.org/primex

A precision measurement of the π^0 photo-production cross section from ^{12}C and ^{208}Pb nuclear targets has been made by the PrimEx Collaboration at Jefferson Lab using 4.9 to 5.5GeV photons tagged by the Hall B tagger facility. The experimental goal is to measure the $\pi^0 \to \gamma\gamma$ decay rate to an accuracy of 1.5% using the Primakoff component of the measured cross section. This represents an order of magnitude improvement over current world-data precision and will allow for powerful tests of the Axial Anamoly plus Chiral corrections—primarily from isospin-breaking and π^0, η, $\eta\prime$ mixing. In this presentation, the status of the π^0 lifetime analysis is given, including detector performance, γ-flux control, systematic checks, and experimental π^0 yields.

1. Motivation

The $\pi^0 \to \gamma\gamma$ decay rate is a fundamental prediction of QCD which gives insight into one of its most profound symmetry issues—namely, the Axial or Chiral Anomaly. It is this anomalous symmetry-breaking mechanism by which the $\pi^0 \to \gamma\gamma$ decay channel proceeds, and thus a measure of its rate or partial width, $\Gamma_{\gamma\gamma}$, represents a direct probe of the anomaly.

In the chiral limit, an *exact* expression for the π^0 decay amplitude can be formed resulting in the Leading Order (LO) prediction, $\Gamma_{\pi^0 \to \gamma\gamma} = 7.725 \pm 0.044$eV, using the current value for the pion decay constant. However, for non-zero quark masses, corrections to the decay amplitude give rise to the Next to Leading Order (NLO) $\Gamma_{\gamma\gamma}$ prediction. The current state of world-data on this subject is presented in Figure 1.

*Supported in part by NSF MRI grant PHY-0079840

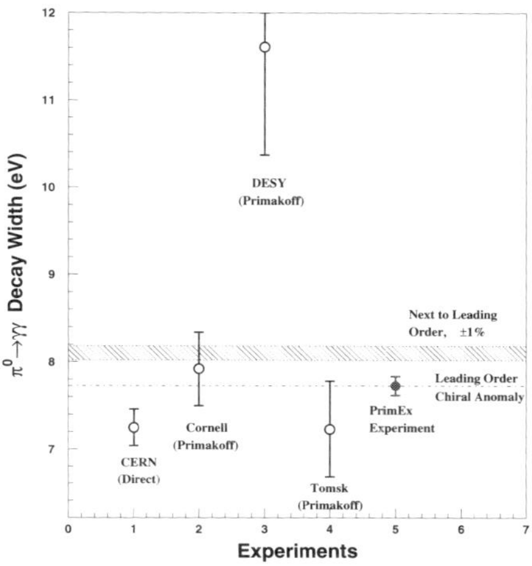

Fig. 1. Previous data on $\Gamma_{\pi^0 \to \gamma\gamma}$. PrimEx data-point is arbitrarily plotted at the LO value with the projected $\pm 1.5\%$ errorbar. The current PDG book value is $\Gamma_{\gamma\gamma} = 7.84 \pm 0.56$eV. The NLO prediction[1] is 8.1eV $\pm$ 1%. The precision of the PrimEx measurement will distinguish between LO and NLO predictions.

2. Experiment Design and Analysis Status

Use of the Hall B γ-tagging facility provides unprecedented γ-flux control $(1 - 2\%)$ which substantially reduces systematic errors associated with the flux normalization. This is combined with a new, high resolution calorimeter[2] designed to detect the two π^0 decay γ's. The precise energy and time resolution of the combination allow for a clean separation of signal and background events. Both e^+e^- pair-production and Compton scattering cross sections were measured for cross-checking the setup's ability to measure well known processes; both results are in excellent agreement with theory.

The Primakoff process is defined as π^0 photo-production from the Coulomb field of a nucleus. This implies an equivalent π^0 production and decay mechanism–which means the cross section $\propto$ lifetime. $\Gamma_{\gamma\gamma}$ is extracted from the data using a multi-parameter fit to the overall cross section measurement; preliminary π^0 cross section and lifetime results available soon.

References

1. J. L. Goity *et al.*, *Phys. Rev.* **D66**, p. 076014 (2002).
2. M. Kubantsev *et al.*, in *CALOR 2006 Proceedings*, (Chicago, June 2006).

DETERMINATION OF CHARGED PION POLARIZABILITIES

V. L. KASHEVAROV*

P.N. Lebedev Physical Institute, Leninsky Prospevct 53, Moscow, 119991, Russia

** E-mail: kashev@kph.uni-mainz.de*

The dipole and quadrupole polarizabilities of the charged pion have been found from the analyses of $\gamma\gamma \to \pi^+\pi^-$ reaction using dispersion relations with subtractions.

Keywords: pion polarizabilities, dispersion relations, sigma meson

The process $\gamma\gamma \to \pi^+\pi^-$ is described by the following invariant variables: $t = (k_1 + k_2)$, $s = (p_1 - k_1)^2$, $u = (p_1 - k_2)^2$, where $p_1(p_2)$ and $k_1(k_2)$ are the pion and photon four-momenta. The cross section of this process is defined through the helicity amplitudes M_{++} and M_{+-} as follows

$$\frac{d\sigma}{d\Omega} = \frac{1}{128\pi^2}\sqrt{\frac{(t-4\mu^2)}{t^3}}\left\{t^2|M_{++}|^2 + \frac{1}{16}t^2(t-4\mu^2)^2\sin^4\theta^*|M_{+-}|^2\right\},$$

where θ^* is the angle between the photon and the pion in the c.m.s. and μ is the $\pi^\pm$ meson mass. To determine the amplitudes M_{++} and M_{+-} we constructed dispersion relations (DRs) at fixed t with one subtraction at $s = \mu^2$, where the subtraction functions were determined with the help of the DRs with two subtractions. The subtraction constants are connected with the dipole and quadrupole polarizabilities.[1]

The cross section integrated over solid angle with $|\cos\theta^*| < 0.6$ was used to fit to all available experimental data in the energy region from threshold to 2.5 GeV (see references in work[1]). The DRs are saturated by the $\rho(770)$, $b_1(1235)$, $a_1(1260)$, and $a_2(1320)$ mesons in the s channel and σ, $f_0(980)$, $f_0(1370)$, $f_2(1270)$, and $f_2(1525)$ in the t channel. For σ meson parameters we used the values found by us in Ref.[2] The dipole and quadrupole polarizabilities were free parameters of this fit. The values of the polarizabilities found in our work[1] and the predictions of dispersion

	fit[1]	DSR[3]	ChPT[4]	
			one-loop	two-loop
$(\alpha_1 - \beta_1)$	$13.0^{+2.6}_{-1.9}$	13.60 ± 2.15	6.0	5.7 [5.5]
$(\alpha_1 + \beta_1)$	$0.18^{+0.11}_{-0.02}$	0.166 ± 0.024	0	0.16 [0.16]
$(\alpha_2 - \beta_2)$	$25.0^{+0.8}_{-0.3}$	25.75 ± 7.03	11.9	16.2 [21.6]
$(\alpha_2 + \beta_2)$	0.133 ± 0.015	0.121 ± 0.064	0	-0.001 [-0.001]

sum rules (DSR)[3] and ChPT[4] are listed below in the table. The numbers in brackets correspond to the order p^6 low energy constants taken from Ref.[5]

The most interesting result is the difference between the electric and magnetic dipole polarizabilities $(\alpha_1 - \beta_1)$. It agrees very well with results obtained in Serpukhov[6] and at MAMI[7] and with the DSR predictions. However, this value deviates substantially from the recent calculations in the framework of ChPT. One of the possible reasons for the small value of $(\alpha_1 - \beta_1)$ predicted by ChPT could be the omission of the contribution of the σ meson. As has been shown in Ref.,[3] this resonance gives the main contribution to DSRs for $(\alpha_1 - \beta_1)$.

The difference of the electric and magnetic quadrupole polarizabilities $(\alpha_2 - \beta_2)$ disagrees with the present two-loop ChPT calculations. One possible source of the disagreement is poor knowledge of low energy constants. Moreover, it should be noted that in this case the two-loop calculation generates nearly a 100% contribution as compared to one-loop result. Calculations of $(\alpha_{1,2}+\beta_{1,2})$ at order p^6 determine only the leading order term in ChPT. Therefore, contributions at p^8 could be essential, and considerably more work is required to put the ChPT prediction on a firm basis.

This research is part of the EU integrated initiative hadron physics project under contract number RII3-CT-2004-506078 and was supported in part by the Russian Foundation for Basic Research (Grant No. 05-02-04014).

References

1. L.V. Fil'kov and V.L. Kashevarov, Phys. Rev. C **73**, 035210 (2006).
2. L.V. Fil'kov and V.L. Kashevarov, Eur. Phys. J. A **5**, 285 (1999).
3. L.V. Fil'kov and V.L. Kashevarov, Phys. Rev. C **72**, 035211 (2005).
4. J. Gasser, M.A. Ivanov, and N.E. Sainio, Nucl.Phys., B **745**, 84 (2006).
5. J. Bijnens and J. Prades, Nucl. Phys. B **490**, 239 (1997).
6. Yu.M. Antipov *et al.*, Phys. Lett. **B121**, 445 (1983).
7. J. Ahrens *et al.*, Eur. Phys. J. A **23**, 113 (2005).

Proposed Measurements of Electroproduction of π^0 near Threshold using a Large Acceptance Spectrometer

R. A. LINDGREN for the Hall A Collaboration

Physics Department, University of Virginia, Charlottesville, VA 22904 USA

Approved experiment (E04-007) to conduct precision measurements of the $p(\mathbf{e}, e'p)\pi^0$ reaction to extract functions $\sigma_T + \epsilon_L\sigma_L$, σ_{TL} and σ_{TT} and asymmetry $A_{TL'}$ in a fine grid of ΔW near threshold and Q^2 in the range 0.04 to 0.14 $(GeV/c)^2$ at Jefferson Laboratory is discussed.

Repeated π^0 electroproduction measurements at $Q^2 > 0$ near threshold at Mainz[1] have shown strong disagreement with the predictions of Chiral Perturbation Theory. Since this theory is so firmly grounded in the symmetries of QCD, these possible violations are very fundamental and require substantiation by more than one experiment. A JLab approved experiment will be performed in Hall-A using the standard HRS spectrometer to detect the electron in coincidence with the proton detected in the BigBite spectrometer. The 90 msr acceptance of this spectrometer will enable us to make most measurements with the BigBite spectrometer in a single configuration thereby minimizing systematic uncertainties and will also provide substantial out of plane coverage for measurements of the fifth structure function. The structure functions $\sigma_T + \epsilon_L\sigma_L$, σ_{TL} and σ_{TT} will be extracted from the ϕ dependence of the differential cross sections and the fifth structure function asymmetry $A_{TL'}$ will be extracted from the beam polarization dependence. The results obtained in a fine grid of ΔW near threshold and Q^2 will provide a stringent test of chiral QCD and also provide valuable input to the MAID and SAID partial wave analysis codes.

Electron beams of 1.0 and 1.2 GeV will be used with scattered electrons detected at angles of 12.5, 14.7 and 16.5 degrees to obtain the necessary range of Q^2. The protons will be detected in the large acceptance BigBite spectrometer fixed at about 47 degrees. A liquid hydrogen target cell 15 cm in length and machined out of solid aluminum will be used with a 125 micron thick entrance and exit window. A helium bag will be used

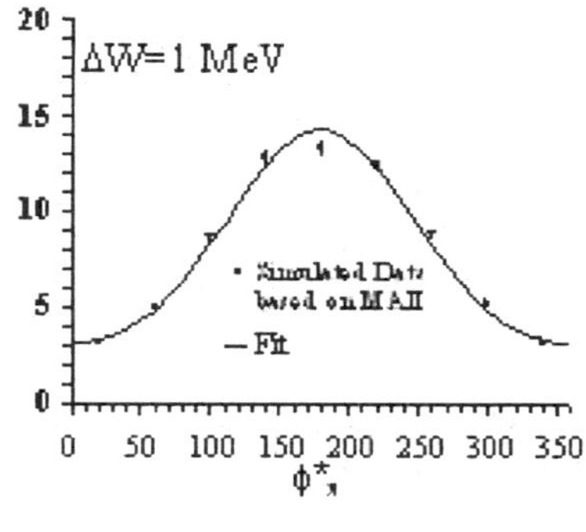

Fig. 1. A plot of the extracted five fold cross section from simulated data for a beam energy of 1.2 GeV at Q^2=0.04 - 0.05 $(GeV/c)^2$ and θ_π =90.

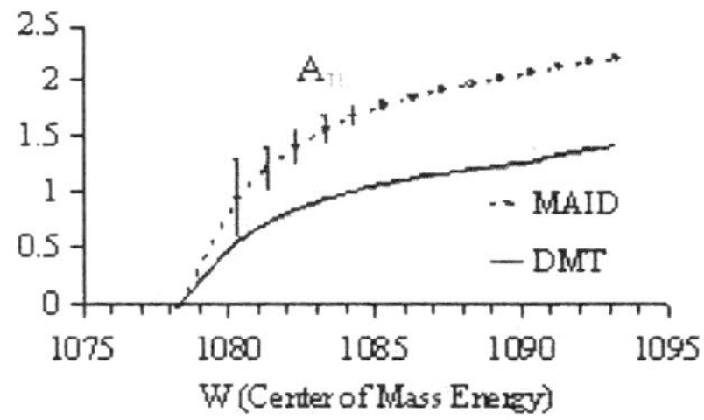

Fig. 2. A plot of the extracted asymmetry and error from simulation data based on MAID as a function of pion-nucleon center of mass energy for the same Q^2 as above. The DMT model is included for comparisons.

to minimize proton energy loss as the protons exit the target cell, pass through the spectrometer, and are finally detected in a set of multi-wire drift chambers optimized to detect low energy protons. Simulated data including the effects of the acceptance of BigBite are shown in Fig 1 for 1 MeV above threshold.

The fitted values of $\sigma_T + \epsilon_L \sigma_L$, σ_{TL} and σ_{TT} are 32.1 ± 0.46, -32.3 ± 0.73, 2.64 ± 0.48, respectively. Approximately 10,800 data points in twenty 1 MeV bins in ΔW, ten 0.01 $(GeV/c)^2$ bins in Q^2 and 180 bins in θ and ϕ center of mass will be obtained in the threshold region. Statistical errors on $\sigma_T + \epsilon_L \sigma_L$, σ_{TL} and σ_{TT} will be typically 2%, 3%, and 15% respectively. Measurements of the asymmetry of the fifth structure function, which is a longitudinal-imaginary / transverse-real interference, is shown in Fig 2.

Experimental results will be compared to calculations using chiral dynamics effective field theory and other models.

References

1. T. Walcher, Experiments at the MAINZ Microtron MAMI Confront Chiral Perturbation Theory, Chiral Dynamics 2006, Durham, NC

THE κ MESON IN πK SCATTERING

B. MOUSSALLAM

IPN, Université Paris-Sud, 91406 Orsay, France

A method is porposed for computing the S-wave πK scattering amplitude in the complex energy plane with controlled accuracy from the existing experimental data. The presence of a pole corresponding to a light, though highly unstable, scalar resonance is firmly established.

Low-energy $\pi\pi$ scattering provides crucial tests of the m_u, m_d chiral expansion (see Leutwyler's talk). πK scattering probes features of the expansion in m_s.[1] Properties of the chiral expansion are strongly influenced by the pattern of the light resonances.[2] Here, I discuss how a light (presumably) exotic resonance can be identified from the πK scattering amplitude. To date, very few exotic hadrons (i.e. multi-quarks, glueballs etc...) are known. Detection is difficult because they are usually expected to be very unstable. Recently, the existence of the $f_0(600)$ scalar meson (also called σ) was established from the $\pi\pi$ amplitude.[3]

Phase-shift analyses for πK scattering have been performed.[4] The $I = 1/2$ S-wave phase shift passes through $\pi/2$ at the energy $E \simeq 1.3$ GeV, which implies, according to standard scattering theory, that the lightest resonance in this channel should be the $K_0^*(1430)$. Yet, there has been suggestions of a lighter, but very wide, resonance.[5] A resonance may be defined to correspond to a pole of the S-matrix, for a complex value of the energy. This definition applies to resonances of any width. The problem then, is to devise a procedure for extrapolating the S-matrix to the complex energy plane with a controllable range of validity and precision. Rigorous results have been derived in the 50's from field theory concerning analyticity properties of scattering amplitudes (e.g.[6]) which apply to πK scattering. Based on these, using Cauchy representations, using also crossing-symmetry and the Froissart bound one can derive a so called Roy-Steiner type representation for the $l = 0$ $I = 1/2$ partial-wave. The integrals in this representation involve $\pi K \to \pi K$ $[f_l^I(s')]$ as well as $\pi\pi \to K\bar{K}$ ($g_l^I(t')$, via crossing) partial-waves, and kernels which are analytic except for cuts (and a simple pole in the case of $K_{00}^{1/2}$)

$$f_0^{\frac{1}{2}}(s) = \frac{1}{2}m_+ a_0^{\frac{1}{2}} + \frac{1}{12}m_+(a_0^{\frac{1}{2}} - a_0^{\frac{3}{2}})\frac{(s-m_+^2)(5s+3m_-^2)}{(m_+^2-m_-^2)\,s}$$
$$+\frac{1}{\pi}\int_{m_+^2}^{\infty} ds' \sum_{l=0}^{\infty}\left\{K_{0l}^{\frac{1}{2}}(s,s')\mathrm{Im} f_l^{\frac{1}{2}}(s') + K_{0l}^{\frac{3}{2}}(s,s')\mathrm{Im} f_l^{\frac{3}{2}}(s')\right\}$$
$$+\frac{1}{\pi}\int_{4m_\pi^2}^{\infty} dt' \sum_{l=0}^{\infty}\left\{K_{02l}^{0}(s,t')\mathrm{Im} g_{2l}^{0}(t') + K_{02l+1}^{1}(s,t')\mathrm{Im} g_{2l+1}^{1}(t')\right\}$$

This representation must be used in two steps. At first, for s real, it must be combined with elastic unitarity and analogous representations for $l = 1$ and $I = 3/2$ partial-waves. This constitutes a closed set of equations which allows one to construct the S and P waves in the low-energy region where experimental information is absent.[7] As a second step, one can let s become complex and compute the amplitude in the complex plane. The region of validity of this procedure can be rigorously derived from the singularity lines of Mandelstam's double spectral representation.[8] Finally, one can verify the presence of a complex pole by direct computation, at the following complex energy

$$\sqrt{s_0} = 658 \pm 13 + i(\,279 \pm 12)\ \mathrm{MeV} \tag{1}$$

This work and also ref.[3] show that it is possible to locate highly unstable resonances without relying on specific a priori assumptions or models. The σ and κ mesons, together with the $a_0(980)$ and the $f_0(980)$ are likely to form a flavour nonet. This nonet should be exotic since its mass pattern differs from the usual $Q\bar{Q}$ picture. This is confirmed by a study of how the masses behave upon varying N_c.[10]

References

1. B. Ananthanarayan and P. Büttiker, Eur. Phys. J. C **19** (2001) 517, B. Ananthanarayan, P. Büttiker and B. Moussallam, Eur. Phys. J. C **22** (2001) 133.
2. G. Ecker *et al.*, Nucl. Phys. B **321** (1989) 311.
3. I. Caprini, G. Colangelo and H. Leutwyler, Phys. Rev. Lett. **96** (2006) 132001.
4. D. Aston *et al.*, Nucl. Phys. B **296** (1988) 493.
5. E. Van Beveren*et al.*, Z. Phys. C **30** (1986) 615.
6. A. Martin, lecture notes in physics vol. 3, Springer-Verlag (1969).
7. P. Büttiker *et al* Eur. Phys. J. C **33** (2004) 409
8. S. Mandelstam, Nuovo Cim. **15** (1960) 658.
9. S. Descotes-Genon and B. Moussallam, `hep-ph/0607133`.
10. J. R. Pelaez, Mod. Phys. Lett. A **19** (2004) 2879.

STRANGENESS −1 MESON-BARYON SCATTERING in S-WAVE

JOSÉ ANTONIO OLLER*

Departamento de Física, Universidad de Murcia,
E-30071 Murcia, Spain
**E-mail: oller@um.es*
www.um.es/oller

We consider meson-baryon interactions in S-wave with strangeness −1. This is a sector populated by plenty of resonances interacting in several two-body coupled channels. We consider a large set of experimental data, where the recent experiments are remarkably accurate. This requires a sound theoretical description to account for all the data and we employ Unitary Chiral Perturbation Theory up to and including $\mathcal{O}(p^2)$. The spectroscopy of our solutions is studied within this approach, discussing the rise of the two $\Lambda(1405)$ resonances and of the $\Lambda(1670)$, $\Lambda(1800)$, $\Sigma(1480)$, $\Sigma(1620)$ and $\Sigma(1750)$. We finally argue about our preferred fit. We report about the references[3,4] and more details can be found there.

Keywords: Kaon-baryon interactions, Chiral Lagrangians, Multichannel scattering.

1. Introduction and Results

The study of strangeness −1 meson-baryon dynamics comprising the $\bar{K}N$ plus coupled channels, $\pi\Lambda$, $\pi\Sigma$, $\eta\Lambda$, $\eta\Sigma$ and $K\Xi$, has been renewed both from the theoretical and experimental sides. Experimentally, we have new exciting data like the increasing improvement in precision of the measurement of the α line of kaonic hydrogen accomplished recently by DEAR,[1] and its foreseen better determination, with an expected error of a few eV, by the DEAR/SIDDHARTA Collaboration.[2] On the theoretical side, there is also an on-going controversy whether the scattering data of $\bar{K}N$ and the DEAR measurement on kaonic hydrogen are compatible.[3–6] We show the reproduction of scattering data by the two kinds of fits A, fig.1, and fits B, fig.2, found in ref.[4] Both fits agree with the scattering data, but, while fits A reproduce the DEAR data on kaonic hydrogen as well, the fits B do not agree with the DEAR measurement.

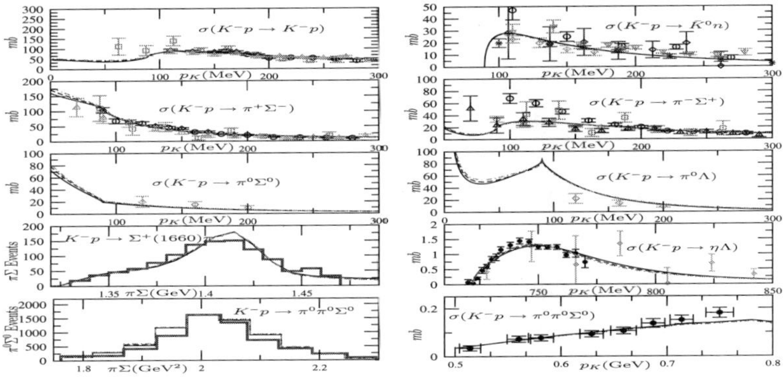

Fig. 1. Fits of type A.[4]

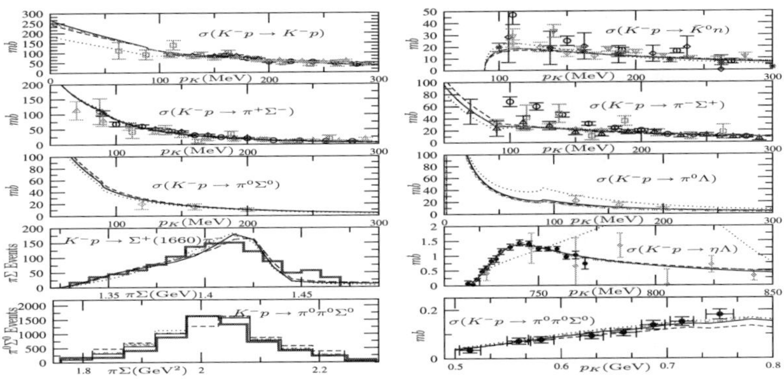

Fig. 2. Fits of type B.[4]

References

1. G. Beer *et al.* [DEAR Collaboration], Phys. Rev. Lett. **94**, (2005) 212302.
2. http://www.lnf.infn.it/esperimenti/dear/DEAR_RPR.pdf
3. J. A. Oller, J. Prades and M. Verbeni, Phys. Rev. Lett. **95**, 172502 (2005); Phys. Rev. Lett. **96** 199202 (2006); arXiv:hep-ph/0609065.
4. J. A. Oller, Eur. Phys. J. **A28**, 63 (2006).
5. U.-G. Meißner, U. Raha and A. Rusetsky, Eur. Phys. J. **C35**, 349 (2004).
6. B. Borasoy, R. Nissler and W. Weise, Phys. Rev. Lett. **94**, 213401 (2005); Ibid. **96**, 199201 (2006); Eur. Phys. J. **A25**; arXiv:hep-ph/0606108.

RESULTS on LIGHT MESONS DECAYS and DYNAMICS at KLOE

M. MARTINI On behalf of the KLOE collaboration

Laboratori Nazionali di Frascati,
Via E. Fermi 40, 00044, Frascati (Rm), Italy
E-mail: Matteo.Martini@lnf.infn.it

KLOE has finished its data taking collecting a whole sample of 2.5 fb^{-1} of e^+e^- collisions at $\sqrt{s} = M_\phi$. We present, for the η sector, the results for: $\eta \to \pi^+\pi^-\pi^0$ dynamics, $\eta - \eta'$ mixing, η mass measurement. For K physics we show the preliminary study of the $K_S \to \gamma\gamma$ decay and the preliminary result for $K_L \to \pi e \nu \gamma$.

Keywords: ϕ decays, η physics, η-η' mixing, K_S-K_L decays

1. The KLOE experiment

KLOE operates at DAΦNE, the Frascati ϕ factory. The detector consists essentially of a drift chamber,[1] DCH, surrounded by an electromagnetic calorimeter,[2] EMC, covering almost hermetically the solid angle.

2. Hadronic physics

The following items have been studied with the statistics of 450 pb^{-1} collected in 2001-2002.

- **Dynamics of $\eta \to \pi^+\pi^-\pi^0$.** This dynamics has been studied with a Dalitz plot analysis. The conventional variables X and Y are defined as: $X = \sqrt{3}(T_+ - T_-)/Q_\eta$, $Y = 3T_0/Q_\eta - 1$, where $Q_\eta = m_\eta - 2m_{\pi+} - m\pi^0$ and T_x are the kinetic energies of the particles. The measured distribution has been fitted by: $|A(X,Y)|^2 \simeq (1 + aY + bY^2 + cX + dX^2 + eXY + ...)$. C-parity conservation rules out odd terms in X; thus the c and e parameter should be zero (see Tab.1).
- **$\eta - \eta'$ mixing.** A clean sample of η' is selected by looking at the final state with two charged pions and seven photons. After background subtraction we observe $(3405 \pm 61 \pm 28)$ $\phi \to \eta'\gamma$ events. Normalizing to the

Table 1. Fitted parameter of $\eta \to \pi^+\pi^-\pi^0$ Dalitz plot

a	b	c	d	e	f
-1.072 ± 0.005	0.117 ± 0.006	0	0.047 ± 0.006	0	0.13±0.01

observed $\eta \to 3\pi^0$ decays in the same sample, we obtain a measurement of $R = \Gamma(\phi \to \eta'\gamma)/\Gamma(\phi \to \eta\gamma) = (4.79 \pm 0.09_{stat} \pm 0.20_{syst}) \times 10^{-3}$. This result well compares with our previous estimate[3] but is considerably improved in accuracy. Neglecting the possible gluonium admixture, we extract the pseudoscalar mixing angle in the flavour basis: $\phi_P = (41.4 \pm 0.3_{stat} \pm 0.7_{syst} \pm 0.6_{th})^\circ$.

- **η mass measurement.** The η mass is measured using events: $\phi \to \eta\gamma$ $(\eta \to \gamma\gamma)$. π and η are well separated by looking at different regions in the Dalitz plot of 3γ final state. The preliminary result obtained: $M(\eta) = (547822 \pm 5_{stat} \pm 69_{syst})KeV$. is compatible with the NA48 estimate.[4]

3. Kaon physics

- **Search for $K_S \to \gamma\gamma$ decays.** Precise measurement of this decay comes from NA48,[5] which shows a BR 30% larger than the $O(p^4)$ χ_{PT} expectation. In KLOE, we can perform the same measurement using a pure K_S beam. The background from $K_S \to 2\pi^0$ is separated using a kinematic fit and studying masses and angular distributions. Using a sample of 1.7 fb^{-1}, we obtain 607 ± 40 signal events. With the whole collected sample, KLOE well reach a statistical accuracy of 5% for this BR.
- **$K_L \to \pi e\nu\gamma$ decay.** The experimental measurements of this decay show a marginal disagreement with theoretical expectation. Using a sample of 2×10^6 K_{e3} decays, KLOE obtain a preliminary result: $R = BR(K_L \to \pi e\nu\gamma)/BR(K_L\pi e\nu) = 0.92 \pm 0.02_{stat} \pm 0.02_{syst}$, thus confirming the disagreement with theory.

References

1. M. Adinolfi *et al*, KLOE collaboration, Nucl. Inst. Meth. **488**, 51 (2002).
2. M. Adinolfi *et al*, KLOE collaboration, Nucl. Inst. Meth. **482**, 364 (2002).
3. A. Aloisio *et al*, KLOE collaboration, Phys. Lett. **B541**, 45 (2002).
4. A. Lai *et al*, Phys. Lett. **B533**, 196 (2002).
5. A. Lai *et al*, Phys. Lett. **B551**, 7 (2003).

STUDIES OF DECAYS OF η AND η' MESONS WITH WASA DETECTOR

A. KUPSC*
for CELSIUS/WASA and WASA-at-COSY Collaborations
Department of Particle and Nuclear Physics, Uppsala University, Uppsala, Sweden
**E-mail: kupsc@tsl.uu.se*

The status of η decay studies with the WASA detector at the CELSIUS storage ring (Uppsala, Sweden) and the continuation at the cooler synchrotron COSY storage ring (Jülich, Germany) is presented.

Keywords: Eta and Eta' meson decays.

1. Introduction

WASA is a universal, large acceptance detector system for photons and for charged particles built and operated at the CELSIUS storage ring in Uppsala (Sweden). The design was optimized for studies of η decays. After the shutdown of CELSIUS in 2005 the WASA detector has been moved and reassembled at COSY (Jülich, Germany) where the experimental studies of η decays will be continued and extended to η' decays.[1] The production experiments will start in the beginning of 2007. The programme of η decays at COSY will focus on rare decays with BR<10^{-5}. With the design luminosity of $10^{32}\text{cm}^{-2}\text{s}^{-1}$ a sensitivity up to 10^{-9} could be achieved for selected η decays involving photons and electrons. One of the aims is to observe the decay channel $\eta \to e^+e^-$ for the first time.

2. Results from CELSIUS/WASA

At CELSIUS two reactions were used to provide sources of η mesons for decay studies $pd \to {}^3\text{He}\eta$ at threshold and $pp \to pp\eta$ close to threshold.

In the $pd \to {}^3\text{He}\eta$ reaction the η meson is tagged in a very clean way and all decay channels are collected simultaneously.[2] In the CELSIUS/WASA experiment an internal deuterium pellet target and a proton beam was used. The ^{3}He nuclei recoiling at 0° were detected in a spectrometer. The accelerator quadrupole and dipole magnets were used to focus and deflect the ^{3}He onto a tagging telescope, which was placed inside the beam tube. With the luminosity 5×10^{30}cm^{-2}s^{-1} the production rate was about 1η event/s. The data sample used in the analysis corresponds to 5×10^5 η decay events. The decay products are detected in the central part of the WASA detector with an electromagnetic calorimeter, a plastic scintillator ΔE detector and a drift chamber made of mylar straw tubes placed in a 1 Tesla magnetic field produced by a thin walled (0.18 X_0) superconducting solenoid.[3] The data were used to extract the branching ratio of $\eta \to \pi^+\pi^-e^+e^-$ decay using $16.4\pm5.0_{stat}\pm2.0_{syst}$ identified events and the result is BR($\eta \to \pi^+\pi^-e^+e^-$)=$(4.6\pm1.4\pm0.5)\times10^{-4}$.[4] Samples of the $\eta \to e^+e^-\gamma$ and $\eta \to e^+e^-e^+e^-$ decay channels were extracted and the analysis is in progress. For the channel $\eta \to e^+e^-$ the hope is to obtain an upper limit for the branching ratio comparable with the present best value ($7.7\cdot10^{-5}$ 90% CL from CLEO II[5]).

The main drawback of this source of ηs is the rather low intensity and a way to study decay modes with lower branching ratios is to use the reaction $pp \to pp\eta$ at beam kinetic energies of 1350-1500 MeV. The η is tagged by measuring two outgoing protons in the forward part of WASA which consists of several layers of plastic scintillator hodoscopes and drift chambers covering scattering angles 3°-18°. For the trigger also information from the decay channels is also needed. During studies at CELSIUS a data sample of decays channels (mainly neutral), corresponding to 3×10^6 η events, was collected. One of the results is the determination of the quadratic slope parameter α for decay $\eta \to 3\pi^0$: $\alpha = -0.027 \pm 0.009(\text{stat})\pm0.01(\text{syst})$.[6] The uncertainty on the α value does not allow us to resolve the discrepancy between recent Crystal Ball[7] and KLOE[8] results. However already in the initial runs at COSY both the statistical and the systematical accuracy will be improved to match these results.

References

1. H. H. Adam *et al.*, *nucl-ex/0411038* (2004).
2. J. Berger *et al.*, *Phys. Rev. Lett.* **61**, 919 (1988).
3. J. Zabierowski *et al.*, *Phys. Scripta* **T99**, 159 (2002).
4. C. Bargholtz *et al.*, *submitted to Phys. Lett. B [hep-ex/0609007]* .
5. T. E. Browder *et al.*, *Phys. Rev.* **D56**, 5359 (1997).
6. C. Pauly, L. Demirors and W. Scobel, *Acta Phys. Slov.* **56**, 381 (2006).
7. W. B. Tippens *et al.*, *Phys. Rev. Lett.* **87**, p. 192001 (2001).
8. T. Capussela, *Acta Phys. Slov.* **56**, 341 (2005).

HEAVY QUARK-DIQUARK SYMMETRY AND χPT FOR DOUBLY HEAVY BARYONS

THOMAS MEHEN

Physics Department, Duke University,
Durham NC, 27707

Jefferson Laboratory
12000 Jefferson Ave.
Newport News VA, 23606
** E-mail: mehen@phy.duke.edu*

A chiral Lagrangian incorporating heavy quark-diquark symmetry is used to calculate doubly charm baryon electromagnetic decays and chiral corrections to doubly heavy baryon masses. Quenched and partially quenched versions of the theory allow one to derive chiral extrapolation formulae for lattice QCD simulations of doubly heavy baryons.

In a doubly heavy anti-baryon the heavy anti-quarks feel an attractive force when they are in the **3** of color $SU(3)$. If the heavy anti-quarks are the same flavor then Fermi statistics requires that they have spin one. They should be separated by a distance of $O(1/(m_Q v))$ which is small compared to $1/\Lambda_{\rm QCD}$. Thus, the light degrees of freedom in a doubly heavy antibaryon and a singly heavy meson orbit the same static source of color, with spin dependent interactions that are suppressed by $1/m_Q$, so properties of doubly heavy baryons and singly heavy mesons are related.

Savage and Wise[1] appealed to this physical picture to motivate a generalization of Heavy Quark Effective Theory[2] (HQET) for heavy diquarks. Combined with HQET, the theory enjoys a $U(5)$ symmetry which permutes the five spin states of the heavy quark and heavy diquark. Adding spin-symmetry breaking chromomagnetic couplings, one can relate the hyperfine splittings of the heavy mesons and doubly heavy baryons:

$$m_{\Xi^*} - m_{\Xi} = \frac{3}{4}(m_{P^*} - m_P)\,, \tag{1}$$

where $\Xi(\Xi^*)$ is the $J = \frac{1}{2}(\frac{3}{2})$ member of the ground state doubly heavy baryon doublet, and $P(P^*)$ is the pseudoscalar (vector) member of the ground state heavy meson doublet. Applied to the charm sector, this predicts a doubly charm baryon hyperfine splitting of 106 MeV, which is within $\approx 25\%$ of existing quenched lattice calculations.[3–5]

The recent observation by the SELEX collaboration of Ξ_{cc} candidates[6–8] has motivated recent theoretical work on doubly heavy baryons. Some aspects of the these states are difficult to interpret theoretically.[9,10] Also, these states were not observed in recent searches in e^+e^- colliders.[11,12] Both improved theoretical predictions and further experimental work are needed to clarify the situation.

An important theoretical issue is the use of the proper effective theory for the doubly heavy baryons. A hadron composed of two heavy quarks is characterized by several scales: m_Q, $m_Q v$, $m_Q v^2$ and $\Lambda_{\rm QCD}$, where v is the typical velocity of the heavy quarks within the bound state. The appropriate theory for these hadrons is not HQET but rather Non-Relativistic QCD[13–15] (NRQCD), which is formulated as an expansion in v. Our recent work[16] shows how the Savage-Wise Lagrangian emerges at the lowest order in NRQCD when a composite field for the diquark is introduced. Heavy quark-diquark symmetry predictions are also obtained in the pNRQCD approach to doubly heavy baryons.[17]

Recently we derived a generalization of Heavy Hadron Chiral Perturbation Theory[18–20] that combines heavy quark-diquark symmetry with chiral symmetry, allowing one to study the low energy interactions of doubly heavy baryons with Goldstone bosons and photons.[9] We obtain a new symmetry prediction for the electromagnetic decay of the Ξ^*. Including $1/m_Q$ corrections and fitting free parameters to the observed $D^* \to D\gamma$ decays, we predict[9]

$$\Gamma[\Xi_{cc}^{*++} \to \Xi_{cc}^{++} + \gamma] \approx \Gamma[\Xi_{cc}^{*++} \to \Xi_{cc}^{++} + \gamma] \approx 3\,\text{keV}\,. \tag{2}$$

This is the dominant decay mode of the Ξ_{cc}^*. The chiral Lagrangian can also be used to calculate chiral corrections to the masses of doubly heavy baryons.[9] Quenched and partially quenched versions of the theory can be used to develop extrapolation formulae for lattice simulations of doubly heavy baryons.[21] One interesting result of this analysis is that the quark-diquark symmetry relation receives small (< 10 MeV) nonanalytic chiral corrections in all of these theories.

References

1. M. J. Savage and M. B. Wise, Phys. Lett. B **248**, 177 (1990).
2. A. V. Manohar and M. B. Wise, Camb. Monogr. Part. Phys. Nucl. Phys. Cosmol. **10**, 1 (2000).

3. N. Mathur, R. Lewis and R. M. Woloshyn, Phys. Rev. D **66**, 014502 (2002).
4. R. Lewis, N. Mathur and R. M. Woloshyn, Phys. Rev. D **64**, 094509 (2001).
5. J. M. Flynn, F. Mescia and A. S. B. Tariq [UKQCD Collaboration], JHEP **0307**, 066 (2003).
6. M. Mattson *et al.* [SELEX Collaboration], Phys. Rev. Lett. **89**, 112001 (2002).
7. M. A. Moinester *et al.* [SELEX Collaboration], Czech. J. Phys. **53**, B201 (2003).
8. A. Ocherashvili *et al.* [SELEX Collaboration], Phys. Lett. B **628**, 18 (2005).
9. J. Hu and T. Mehen, Phys. Rev. D **73**, 054003 (2006).
10. T. D. Cohen and P. M. Hohler, Phys. Rev. D **74**, 094003 (2006).
11. R. Chistov *et al.* [BELLE Collaboration], arXiv:hep-ex/0606051.
12. B. Aubert *et al.* [BABAR Collaboration], Phys. Rev. D **74**, 011103 (2006).
13. G. T. Bodwin, E. Braaten and G. P. Lepage, Phys. Rev. D **51**, 1125 (1995) [Erratum-ibid. D **55**, 5853 (1997)].
14. N. Brambilla, A. Pineda, J. Soto and A. Vairo, Nucl. Phys. B **566**, 275 (2000).
15. M. E. Luke, A. V. Manohar and I. Z. Rothstein, Phys. Rev. D **61**, 074025 (2000).
16. S. Fleming and T. Mehen, Phys. Rev. D **73**, 034502 (2006).
17. N. Brambilla, A. Vairo and T. Rosch, Phys. Rev. D **72**, 034021 (2005).
18. M. B. Wise, Phys. Rev. D **45**, 2188 (1992).
19. G. Burdman and J. F. Donoghue, Phys. Lett. B **280**, 287 (1992).
20. T. M. Yan, H. Y. Cheng, C. Y. Cheung, G. L. Lin, Y. C. Lin and H. L. Yu, Phys. Rev. D **46**, 1148 (1992) [Erratum-ibid. D **55**, 5851 (1997)].
21. T. Mehen and B. C. Tiburzi, Phys. Rev. D **74**, 054505 (2006).

HHChPT APPLIED TO THE CHARMED-STRANGE PARITY PARTNERS

R. P. SPRINGER*

Department of Physics, Duke University,
Durham, NC 27708, USA
**E-mail: rps@phy.duke.edu*

We apply heavy hadron chiral perturbation theory (HHChPT) to study the properties of the even- and odd-parity charmed-strange mesons. We find that the experimental masses and electromagnetic decay rates are consistent with HHChPT. We consider the hypothesis that the even-parity states are molecules. An RG analysis shows that once the HHChPT parameter set overlaps with the parity doubling model, the parameters remain at those values. Finally, the implications for the bottom-strange system are described.

Keywords: heavy quarks, QCD, chiral perturbation theory, effective theories

The experimental observation of D_{s1} (2463 MeV, $J^P = 1^+$) and D_{s0} (2317 MeV, $J^P = 0^+$)[1,2] yielded the following interesting features: (a) The hyperfine splitting within a multiplet is the same for the even-parity states as for the odd-parity states. (b) The widths of the D_{s1} and D_{s0} are considerably narrower than predicted by most lattice calculations and potential models. This appears to be related to the fact that their masses are considerably lower than predicted; the only available channel is the isospin violating $D_{s1}(D_{s0}) \to D_s^*\pi(D_s\pi)$ rather than the expected $D_{s1}(D_{s0}) \to D^*K(DK)$ mode. (c) The SU(3) breaking between the even-parity charmed-strange mesons and their non-strange partners is anomalously small (10–50 MeV rather than the expected 100 MeV).

To investigate these issues we calculated electromagnetic (EM) decays and mass corrections to one loop order in HHChPT.[3] The Lagrange density and other details can be found in Ref. 4.

The EM calculation to $\mathcal{O}(Q^2)$, where $Q \sim m_s \sim m_K^2 \sim k_\gamma$, involves three unknown coefficients associated with EM operators, one of which turns out to be unimportant. The EM measurement is of branching fraction with respect to strong decays. Two additional coefficients are required to describe the strong decays to this order, but there is data from the non-strange system to fit them. We find that the measurement[2,5,6] of $1^+ \to 0^-\gamma$ and the limits[2,5] on $1^+ \to 1^-\gamma$ and $0^+ \to 1^-\gamma$ are consistent with HHChPT. If, on the other hand, we try to explain the EM measurements using a model where the D_{s1} and D_{s0} are bound states of D^*K and DK, respectively, we find that the relative strengths of the EM decays explicitly disagrees with data.

We calculated the masses of the multiplets to $\mathcal{O}(Q^3)$. We include $1/m_c$ corrections and SU(3) breaking effects to this order. Because there are 11 independent parameters and eight mass measurements we can only say that the data are consistent with HHChPT. More interestingly, we can investigate the stability of the parity-doubling model.[7] In this model, feature (a) is addressed explicitly; the axial coupling g' and mass splitting parameters of the even-parity states are set equal to those of the odd-parity states (axial coupling g) at tree level. Running these parameters in HHChPT shows that once the mass splitting parameters are equal and $|g| = |g'|$ they remain so. This shows that the parity doubling hypothesis is robust beyond tree level.

Under HHChPT, results from the charmed-strange mesons can be used to estimate the bottom-strange meson behavior by considering m_c/m_b corrections. We find that the analogue B_s states are below threshold to BK decay, and so should be light and narrow. We predict the hyperfine splitting in the strange sector to be on the order of 46 MeV. This is indeed observed for the odd-parity states. The even-parity B analogues have not yet been observed. Additional information in the b-meson spectrum, including decay observations, will be an important test of HHChPT predictions.

References

1. B. Aubert *et al.* [BABAR Collaboration], *Phys. Rev. Lett.* **90**, 242001 (2003).
2. D. Besson *et al.* [CLEO Collaboration], *Phys. Rev. D* **68**, 032002 (2003).
3. M. B. Wise, *Phys. Rev. D* **45**, 2188 (1992); G. Burdman and J. F. Donoghue, *Phys. Lett. B* **280**, 287 (1992); T. M. Yan, H. Y. Cheng, C. Y. Cheung, G. L. Lin, Y. C. Lin and H. L. Yu, *Phys. Rev. D* **46**, 1148 (1992) [Erratum-ibid. D **55**, 5851 (1997)].
4. T. Mehen and R. P. Springer, *Phys. Rev. D* **70**, 074014 (2004); *Phys. Rev. D* **72**, 034006 (2005).
5. K. Abe *et al.*, Phys. Rev. Lett. **92**, 012002 (2004);
6. P. Krokovny *et al.* [Belle Collaboration], Phys. Rev. Lett. **91**, 262002 (2003); B. Aubert *et al.* [BABAR Collaboration], Phys. Rev. Lett. **93**, 181801 (2004).
7. For a description in this context see, e.g., M.A. Nowak, M. Rho, and I. Zahed, *Phys. Rev. D* **48**, 4370 (1993); W.A. Bardeen and C.T. Hill, *Phys. Rev. D* **49**, 409 (1994).

STUDY OF PION STRUCTURE THROUGH PRECISE MEASUREMENTS OF THE $\pi^+ \to e^+\nu\gamma$ DECAY

D. POČANIĆ*, for the PIBETA Collaboration

Department of Physics, University of Virginia,
Charlottesville, VA 22904-4714, USA
** E-mail: pocanic@virginia.edu*
http://pibeta.phys.virginia.edu

We have measured radiative pion $\pi^+ \to e^+\nu\gamma$ (RPD), and muon $\mu^+ \to e^+\nu\bar{\nu}\gamma$ (RMD) decays, over broad phase space regions using the PIBETA detector. Preliminary results of fits to integral and differential RPD distributions yield the axial and vector pion form factors of $F_A = 0.0116(1)$, and $F_V = 0.0262(11)$, respectively, with a first ever evaluation of their q^2 dependence. We have evaluated $B(\mu^+ \to e^+\nu\bar{\nu}\gamma) = 4.40(9) \times 10^{-3}$ for $E_\gamma > 10\,\text{MeV}$ and $\theta_{e\gamma} > 30°$. We have extracted a new upper bound on the Michel decay parameter $\bar{\eta} \leq 0.033$ (at 68 % C.L.). Our results are in excellent agreement with the Standard Model.

Keywords: pion decays, meson properties, muon decays

1. The PIBETA Experiment and Its Motivation

Thanks to their theoretical simplicity, experimental study of rare decays of light mesons, the pion in particular, leads to improved precision of basic Standard Model (SM) parameters. Muon decays further benefit from the leading-order absence of hadrons. The PIBETA experiment[1,2] at the Paul Scherrer Institute (PSI), Switzerland, focused on the study of rare π and μ decays, primarily the pion beta decay,[3] $\pi^+ \to \pi^0 e^+\nu$, with data acquisition runs in 1999–2001 and 2004. Here we report preliminary results of the pion and muon radiative decay data analysis.

2. Radiative Pion Decay $\pi^+ \to e^+\nu\gamma$

The Standard Model with a pure V$-$A electroweak sector requires only two pion form factors, F_A and F_V, to describe the non-QED part of the RPD amplitude. The vector form factor is strongly constrained by the CVC hypothesis to $F_V = 0.0259\,(9)$. Analysis of our 1999-2001 data set comprising over 40 k RPD events yielded a fourfold improvement in precision of F_A, albeit with a large deficit of low-E_e events,[4] prompting a dedicated experiment in 2004. Detailed results of the 2004 run are available in Ref. 5. We are currently reanalyzing the combined 1999–2001 and 2004 data sets, comprising over 70 k RPD events, with these preliminary best-fit values:

$$F_A = 0.0116\,(1)\,, \quad F_V = 0.0262\,(11)\,, \quad \text{and} \quad a = 0.066\,(20)\,, \qquad (1)$$

where a accounts for the $q^2(\bar{e}\nu)$ dependence of $F_V(q^2) = F_V(0)(1 + a \cdot q^2)$, never before measured. Following the χPT ansatz of Ref. 6, we set $F_A(q^2) \equiv F_A(0)$. Our preliminary result provides the most accurate experimental determination to date of the pion polarizability:

$$\alpha_E = (6.24 \times 10^{-4}\,\text{fm}^3) \cdot \frac{F_A}{F_V} = (2.80 \pm 0.03) \times 10^{-4}\,\text{fm}^3\,. \qquad (2)$$

3. Radiative Muon Decay $\mu^+ \to e^+\nu\bar{\nu}\gamma$

The 2004 PIBETA data set accumulated more than 4×10^5 RMD events, which has enabled us to measure the branching ratio, with the restrictions $E_\gamma > 10\,\text{MeV}$ on the photon energy and $\theta_{e\gamma} > 30°$ on the positron–photon opening angle.[7] Our preliminary result is

$$B(\mu^+ \to e^+\nu\bar{\nu}\gamma) = [4.40 \pm 0.02\,(\text{stat.}) \pm 0.09\,(\text{syst.})] \times 10^{-3}\,. \qquad (3)$$

This represents a fourteenfold improvement in precision over the previous world average.[8] The best fit for the branching ratio is found for

$$\bar{\eta} = -0.084 \pm 0.050(\text{stat.}) \pm 0.034(\text{syst.})\,, \quad \text{or,} \quad \bar{\eta} < 0.033\,(68\%\,\text{C.L.})\,. \qquad (4)$$

All of the above results are in excellent agreement with the SM predictions. Both RPD and RMD will be studied further in the PEN experiment.[9]

References

1. D. Počanić *et al.*, *A Precise Measurement of the $\pi^+ \to \pi^0 e^+\nu$ Decay Rate*, PSI Experiment R-89-01, (Dec. 1991), `http://pibeta.phys.virginia.edu`.
2. E. Frlež *et al.*, *Nucl. Instrum. Meth. in Physics Research A* **526**, 300 (2004).
3. D. Počanić *et al.*, *Phys. Rev. Lett.* **93**, 181803 (2004).
4. E. Frlež *et al.*, *Phys. Rev. Lett.* **93**, 181804 (2004).
5. M. A. Bychkov, Ph.D. thesis, (University of Virginia, 2005).
6. C. Q. Geng, I-Lin Ho, and T. H. Wu, *Nucl. Phys. B* **684**, 281 (2004).
7. B. A. VanDevender, Ph.D. thesis (University of Virginia, 2005).
8. S. Eidelman *et al.*, *Phys. Lett. B* **592**, 1 (2004).
9. V. A. Baranov *et al*, *Precise Measurement of the $\pi^+ \to e^+\nu$ Branching Ratio*, PSI Expt. R-05-01, (Jan. 2006), `http://pen.phys.virginia.edu`.

EXCEPTIONAL AND NON-EXCEPTIONAL CONTRIBUTIONS TO THE RADIATIVE π DECAY

V. MATEU

Departament de Física Teòrica-IFC, Universitat de València-CSIC,
**E-mail: mateu@ific.uv.es*

We have studied the spin-one resonance dominated form factors governing the radiative decay of the π, within the framework of resonance chiral theory. We obtain predictions for their value at zero momentum transfer and also a description of their q^2 dependence. We also fit the possible new physics tensor coupling.

Keywords: Chiral lagrangians; $1/N_C$ expansions; Proceedings.

1. Main results and conclusions

For many years the radiative π decay (RPD) process ($\pi \to e\nu\gamma$) has been an open window for the speculation about new physics (NP).[1] The available experimental data[2] seemed to indicate a deficit of events in one region when compared to the Standard Model (SM) prediction.

The physical description of the process can be split into two pieces: the inner bremsstrahlung (IB) contribution contribution and the structure-dependent (SD) contribution, described by two form factors.

Still we lack a complete analysis of the process including q^2 dependence in the form factors, and this is the main purpose of this work. We will use resonance chiral theory (RχT) for the description of the form factors.[3]

Within the SM the SD contributions can be parametrized by a vector form factor and an axial-vector form factor. A tensor form factor would interfere destructively in the region in which the deficit of events occurs.

The axial-vector form factor has been already calculated.[4] The VVP Green Function so far has been calculated only with one multiplet of vector meson resonances and it is not able to fulfil at the same time all the short distance constraints.[5] Including a second multiplet, we can satisfy all the requirements and we obtain a reliable vector form factor :[3]

$$F_V(q^2) = -\frac{m_{\pi^+}}{3\sqrt{2}\,B_0\,F\,M_{V_1}^2\,M_{V_2}^2}\frac{c_{000} + c_{010}\,q^2}{(M_{V_1}^2 - q^2)\,(M_{V_2}^2 - q^2)} \tag{1}$$

where c_{000} is fixed by the anomaly and c_{010} is fitted from the $\pi^0 \to \gamma\gamma^*$ phenomenology.

Finally with this framework we calculate the tensor hadronic matrix element, usually expressed through the so called magnetic susceptibility χ and the vector meson couplings to vector and tensor currents currents.:[3]

$$\chi = -\frac{2}{M_V^2} \backsimeq -3.37\,\mathrm{GeV}^{-2}, \quad \frac{f_V^\perp}{f_V} = \frac{\langle\bar{q}q\rangle}{M_V F_V^2} \backsimeq 0.78(1). \tag{2}$$

With the new but preliminary data from the PIBETA collaboration[6] the missmatch now is gone, as can be seen in tab. 1. Nevertheless we fit the

$E_{e^+}^{\mathrm{min}}$(MeV)	E_γ^{min}(MeV)	$\theta_{e\gamma}^{\mathrm{min}}$	$R_{\mathrm{exp}}(\times 10^{-8})^6$	$R_{\mathrm{the}}{}^3$
50	50	–	2.655 ± 0.058	2.55(9)
10	50	40 °	14.59 ± 0.60	14.66(1)
50	10	40 °	37.95 ± 0.28	37.70(98)

value for the NP tensor coupling and get:[3]

$$f_T = (4 \pm 6) \times 10^{-4} \tag{3}$$

which is compatible with zero and with that dictated by SUSY.[1]

Acknowledgements

Work supported by EU FLAVIAnet (MRTN-CT-2006-035482) , the Spanish MEC (FPA2004-00996), Generalitat Valenciana (GRUPOS03/013 and GV05-164) and by ERDF funds from the EU Commission.

References

1. V. M. Belyaev and I. I. Kogan, Phys. Lett. B **280** (1992) 238.
2. A. A. Poblaguev, Phys. Lett. B **238** (1990) 108;
   E. Frlez *et al.*, Phys. Rev. Lett. **93**, 181804 (2004) [arXiv:hep-ex/0312029].
3. V. Mateu and J. Portolés, work in preparation
4. V. Cirigliano, G. Ecker, M. Eidemuller, A. Pich and J. Portoles, Phys. Lett. B **596** (2004) 96
5. M. Knecht and A. Nyffeler, Eur. Phys. J. C **21** (2001) 659;
   P. D. Ruiz-Femenia, A. Pich and J. Portolés, JHEP **0307** (2003) 003
6. See D. Pocanic proceedings of this workshop

LEADING CHIRAL LOGARITHMS FROM UNITARITY, ANALYTICITY AND THE ROY EQUATIONS

A. FUHRER*

Institute for Theoretical Physics, University of Bern,
3012 Bern, Switzerland
**E-mail: afuhrer@itp.unibe.ch*

The structure of leading logarithms in chiral perturbation theory was studied some time ago by Weinberg[1] and recently by Büchler and Colangelo.[2] Because the leading logarithms may generate sizeable numerical contributions to observables, it would be very interesting to know them to every order in the chiral expansion. We investigate this possibility for two specific Green functions in chiral perturbation with two flavours, in the chiral limit.

Keywords: Chiral symmetry; Chiral perturbation theory; Leading logarithms

At a given order in chiral perturbation theory, the term with the logarithm of the highest power is called *leading logarithm* (LL). In the following, we consider chiral perturbation theory for two flavours, in the chiral limit $m_u = m_d = 0$. Suppose that the LL of the partial wave t_0^0 of $\pi\pi$ scattering is known at chiral order $p^{2(N-2)}$. Unitarity of the S-matrix then determines the LL of the two point function of two scalar quark currents,

$$H(s) = i \int dx e^{ipx} \langle 0|TS^0(x)S^0(0)|0\rangle \; ; \; S^0 = \bar{u}u + \bar{d}d ; \; s = p^2 ,$$

and of the scalar form factor $F(s)$,

$$\langle 0|S^0(0)|\pi^i(p)\pi^k(p')\rangle = \delta^{ik} F(s) \; , \; s = (p+p')^2 ,$$

at the chiral order p^{2N} and $p^{2(N-1)}$, respectively.[3] For illustration, we note the pertinent relation between $H(s)$ and $F(s)$,

$$\text{disc}\, H(s) = \; i(2\pi)^4 \sum_n \delta^{(4)}(P_n - p)|\langle 0|S^0(0)|n\rangle|^2 = \; \frac{3i}{16\pi}|F(s)|^2 + \cdots .$$

Only two-pion intermediate states are relevant for the LL.[3] Furthermore, once the partial waves t_ℓ^I are known at tree level, invoking unitarity and the Roy equations[4] allows one to calculate the pertinent combination of LL of t_ℓ^I used in the unitarity relation for disc $F(s)$. The Roy equations are invoked, because unitarity does not provide sufficient information about the contribution generated by the left-hand cut in $t_\ell^I(s)$. We order the low energy expansion of $H(s)$ and $F(s)$ as

$$H(s) = \frac{B^2}{16\pi^2}\left\{P_0 + P_1 L + P_2 L^2 + \cdots\right\},$$
$$L \equiv \ln\left(-\frac{s}{\mu^2}\right),$$
$$F(s) = 2B\{T_0 + T_1 L + T_2 L^2 + \cdots\},$$

where the coefficients P_i and T_i are polynomials in $N = s/(16\pi^2 F^2)$. Up to five loops, the leading contributions to these polynomials are given by[3]

$$P_1 = -6, \qquad P_2 = 6N, \qquad P_3 = -\frac{61}{9}N^2, \qquad P_4 = \frac{68}{9}N^3,$$
$$P_5 = -\frac{140347}{16200}N^4, \quad T_1 = -N, \quad T_2 = \frac{43}{36}N^2, \quad T_3 = -\frac{143}{108}N^3, \quad T_4 = \frac{15283}{9720}N^4.$$

Higher order terms in T_i, P_i are omitted here, because they are suppressed in the chiral counting. Available one- and two-loop results[5–8] together with the renormalization group equation[2] provide a direct check on the polynomials P_i, T_i, with $i \leq 3$.

In the chiral limit, a straightforward use of the Roy equations produces infrared divergences, because $\mathrm{Im}t_\ell^I(s)$ behave like s^4 as $s \to 0^+$ for $\ell \geq 2$. In the calculation of P_5, the emerging divergence is of the form $s^4 \ln(M_\pi)$. However, this divergence does not affect the five loop LL, as we explicitly checked using the renormalization group equations for the partial wave t_0^0 at order p^8. We expect the method presented here to work to all orders.

References

1. S. Weinberg, PhysicaA **96** (1979) 327.
2. M. Buchler and G. Colangelo, Eur. Phys. J. C **32** (2003) 427.
3. M. Bissegger and A. Fuhrer, in preparation.
4. S. M. Roy, Phys. Lett. B **36** (1971) 353.
5. J. Gasser and H. Leutwyler, Annals Phys. **158** (1984) 142.
6. J. Gasser and U. G. Meissner, Nucl. Phys. B **357** (1991) 90.
7. J. Bijnens, G. Colangelo and P. Talavera, JHEP **9805** (1998) 014.
8. J. Bijnens, G. Colangelo, G. Ecker, J. Gasser and M. E. Sainio, Nucl. Phys. B **508** (1997) 263.

ALL ORDERS SYMMETRIC SUBTRACTION OF THE NONLINEAR SIGMA MODEL IN D=4

ANDREA QUADRI*

Physics Department, University of Milano,
via Celoria 16, I-20133 Milano, Italy
**E-mail: andrea.quadri@mi.infn.it*

The symmetric subtraction of the nonlinear sigma model in $D = 4$ is performed to all orders in the loop expansion by enforcing the functional equation associated with the invariance of the Haar measure under local left multiplication. This equation encodes a powerful hierarchy allowing one to fix all amplitudes involving at least one pion field in terms of those only containing insertions of the flat connection and the nonlinear σ-model constraint ϕ_0 (ancestor amplitudes).

The problem of the quantization of the nonlinear σ-model is strictly related to the program of giving a consistent structure to chiral perturbation theory.[1] Since a long time people realized that the nonlinear σ-model cannot be renormalized in a symmetric way (chiral symmetry) already at one loop. Some of the unwanted (chiral breaking) terms can be disposed of by redefinition of the fields (quartic divergences). However some divergent terms of the one-loop off-shell pion-pion scattering amplitude still violate chiral symmetry and can be reabsorbed by redefinition of the pion field only if derivatives are allowed. This strategy of removing the divergences never turned to a consistent program both for technical difficulty and for the impossibility of fixing the necessary finite subtractions.

From these previous experiences it is clear that the renormalization of the nonlinear σ-model cannot be achieved by using chiral-invariant counterterms only. It has been recently proposed[2,3] to quantize the nonlinear sigma model in $D = 4$ by embedding the pion fields in a flat connection in order to solve the long-standing problem of a symmetric removal of the divergences. The theory is defined by the functional equation expressing the invariance of the Haar measure in the path-integral under local left multiplication. This is a local functional equation which allows one to classify the whole structure of the divergences of the nonlinear σ-model in $D = 4$.

The recursive subtraction procedure of the poles in the D-dimensional amplitudes order by order in $\hbar$ is carried out in a way compatible with the functional equation. This means that only local functionals invariant under the linearized functional operator have to be subtracted. The complete classification of these invariants can be obtained by cohomological methods.[3] Their coefficients are uniquely fixed by the polar part of amplitudes only involving the external sources associated to the flat connection and the composite ϕ_0 field (ancestor amplitudes).

At any given order in the loop expansion only a finite number of divergent ancestor amplitudes exists. This establishes a weak power-counting theorem[3] for the nonlinear σ-model in $D = 4$. The polar part of the amplitudes involving at least one pion field is then completely determined by the invariants themselves upon projection on the relevant monomials. It depends on a finite number of parameters controlling the divergences of the ancestor amplitudes at any given order in the loop expansion, as a consequence of the weak power-counting theorem and of the symmetric subtraction.

Moreover it turns out that the functional equation encodes a hierarchy among the amplitudes: once the ancestor amplitudes are known, all the remaining amplitudes can be obtained by functional differentiation of the functional equation itself.

At higher orders in the loop expansion the bilinear term of the functional equation for the 1-PI generating functional yields non-trivial contributions depending on lower-order 1-PI Green functions. By using Quantum Action Principle arguments we have checked[4] up to two loop level that the subtraction procedure yields indeed finite amplitudes satisfying the functional equation (but the recursive method can be applied to all orders).

Applications of the proposed subtraction scheme range from the determination of the higher-order counterterms in chiral perturbation theory (including on-shell vanishing contributions) to the quantization of gauge theories based on nonlinearly realized gauge groups (in particular those implementing the Stückelberg mechanism for giving mass to non-abelian gauge bosons).

References

1. J. Gasser and H. Leutwyler, Annals Phys. **158**, 142 (1984).
2. R. Ferrari, JHEP **0508**, 048 (2005) [arXiv:hep-th/0504023].

3. R. Ferrari and A. Quadri, "A weak power-counting theorem for the renormalization of the nonlinear sigma model in four dimensions," arXiv:hep-th/0506220, to be published in Int. J. Theor. Phys.
4. R. Ferrari and A. Quadri, JHEP **0601**, 003 (2006) [arXiv:hep-th/0511032].

PART C

CHIRAL DYNAMICS IN FEW-NUCLEON SYSTEMS

WORKING GROUP SUMMARY: CHIRAL DYNAMICS IN FEW-NUCLEON SYSTEMS

H.-W. HAMMER

HISKP (Theorie), Universität Bonn, Nußallee 14-16, D-53115 Bonn, Germany
E-mail: hammer@itkp.uni-bonn.de

N. KALANTAR-NAYESTANAKI

Kernfysisch Versneller Instituut (KVI), University of Groningen, Groningen, The Netherlands
E-mail: nasser@kvi.nl

D. R. PHILLIPS

Department of Physics and Astronomy, Ohio University, Athens, OH 45701, USA
E-mail: phillips@phy.ohiou.edu

We summarize the findings of our Working Group, which discussed progress in the understanding of Chiral Dynamics in the $A = 2$, 3, and 4 systems over the last three years. We also identify key unresolved theoretical and experimental questions in this field.

Keywords: Chiral Dynamics; Few-Nucleon Systems

1. Introduction

This year is the tenth anniversary of the first quantitative treatment of the nuclear force based on chiral perturbation theory (χPT): the seminal paper of Ordóñez *et al.*[1] This paper employed Weinberg's proposal to expand the NN potential up to a given chiral order:

$$V = V^{(0)} + V^{(2)} + V^{(3)} + V^{(4)} + \ldots, \tag{1}$$

and then use the resulting potential in the Schrödinger equation and obtain the NN wave function.[2] However, in that same year Kaplan, Savage, and Wise raised serious questions about the consistency of such an approach.[3]

The paper of Ordóñez *et al.* analyzed NN scattering data using an NN potential which included all mechanisms up to next-to-next-to-leading order (N^{2}LO). This has now been extended to one higher order,[4,5] and the result is an energy-independent potential that describes NN data with an accuracy comparable to that of the "high-quality" NN potentials. However, questions remain about the consistency of the power counting used to derive this N^{3}LO potential.[6–8] These questions were discussed in a special "panel" format within our Working Group. The conclusions of this panel are summarized in Section 2.

At the same time lattice simulations have made significant strides in connecting these chiral potentials to QCD itself. The work of the NPLQCD collaboration is also discussed in Sec. 2, and it shows that NN scattering lengths are natural at quark masses corresponding to $m_\pi \geq 350$ MeV.[10] Chiral extrapolation from the physical point to these higher values of m_π shows that the numbers obtained in the simulations are consistent with the experimentally measured "unnaturally large" ($|a| \gg 1/m_\pi$) values:[9] $a^{{}^3S_1} = 5.4112(15)$ fm and $a^{{}^1S_0} = -23.7148(33)$ fm.

The existence of these large scattering lengths leads to "universal" features in the dynamics of few-nucleon systems. These are a consequence solely of the scale hierarchy $|a|m_\pi \gg 1$, and are independent of details of the nuclear force. There has been much recent progress on calculations in the universal effective theory without explicit pions that describes this physics. This provides the opportunity to compare precise, low-energy ($E < \frac{m_\pi^2}{M}$) scattering experiments in the three- and four-nucleon systems with a theory where calculations can be carried out with a comparable degree of control. Progress in this direction in both theory and experiment is discussed in Sec. 3.

At higher energies three-nucleon scattering experiments must be analyzed using theories in which the pion is an explicit degree of freedom. Recent experiments have produced a wealth of data, including single- and double-polarization observables, over a range of energies. Ideally one would analyze these data using nuclear forces derived from χPT. The great benefit of such an approach is the ability to derive three-nucleon (and, if desired, four- and five- and six-...) nucleon forces within the same framework, and to the same order in the chiral expansion as is used for the NN force. This is discussed in Sec. 4.

As with n-body forces, the operators that describe the interaction of external probes (electrons, photons, pions, ...) with the nucleus can also be expanded up to a chiral order that is consistent with the order used to analyze the NN potential. This "chiral effective theory" (χET) approach has now been applied to many reactions in the NN system, and calculations of reactions on the tri-nucleons are just beginning. These methods are particularly valuable in facilitating the extraction of neutron properties from experiments on multi-nucleon systems. Neutron polarizabilities are observables that have recently received particular attention in this regard. Progress in these areas is summarized in Sec. 5. In addition to the topics discussed explicitly there, related talks by F. Myhrer on $pp \to pp\pi^0$ and S. Nakamura on the renormalization-group behavior of weak current operators were also presented to the Working Group.

Finally, chiral perturbation theory is being used to analyze symmetry breaking in the NN system. New work on how parity-violating operators appear in the NN system provides a framework for the analysis of a new generation of experiments that probe hadronic parity violation. Isospin violation can also be included in the NN potential using χPT techniques, and this allows a careful examination of how isospin-violation effects appear in various reactions. These issues are discussed in Sec. 6.

2. The NN system: still fighting after all these years

2.1. *Connection to lattice QCD*

One of the major developments in the NN system since CD2003 is the first NN lattice calculation in full QCD.[10] The NPLQCD collaboration computed QCD correlation functions by using Monte Carlo techniques to evaluate the QCD path integral on a discrete Euclidean space-time lattice. This starts directly from QCD, thereby allowing truly "*ab initio*" calculations of nuclear physics observables. However, the computational effort required for these calculations increases sharply (i) with the number of valence quarks involved and (ii) as the light quark masses are lowered to approach physical values.

S. Beane reported on these results of the NPLQCD collaboration. Their hybrid calculations used configurations generated by the MILC collaboration including staggered sea quarks and they evaluated the NN correlator in this background using domain-wall valence quark propagators. Applying the Lüscher method, the spin-singlet and spin-triplet S-wave scattering lengths were extracted at three different pion masses between 350 and 590 MeV. At the lowest pion mass the values $a^{^3S_1} = (0.63 \pm 0.74 \pm 0.2)$ fm and $a^{^1S_0} = (0.63 \pm 0.50 \pm 0.2)$ fm were found. At this pion mass the errors from a chiral extrapolation to the physical m_π are still large, but chiral extrapolation from $m_\pi = 139$ MeV shows that the result is consistent with the experimental values, within (sizeable) error bars.

Thanks to continuing advances in computer power much progress in this direction can be expected over the next three years. Lattice computations of NNN correlators are a high priority since they will provide access to three-nucleon forces directly from QCD, but algorithmic and theoretical advances will be required in order for such calculations to become a reality.

2.2. *Panel on power counting for short-distance operators*

Fifteen years after Weinberg's seminal papers on a chiral effective theory description of nuclear physics, questions about this technique remain. The power counting (1) is quite successful phenomenologically, but various recent papers have questioned its validity.[6–8] As part of the working group, we had a panel discussion with U. van Kolck, U.-G. Meißner, and M. Pavon in order to try to shed some light on this important issue.

The key problem is how to renormalize the Schrödinger equation with (attractive) singular potentials. When the equation is solved there are two acceptable wave-function solutions at short distances. The "correct" linear combination must be fixed by physics input. This can be implemented by adding a contact potential in such channels. How many contact terms are required when the NN potential is computed to a given order in the chiral expansion? So far this question has only been intensively studied for the leading-order (LO) potential. That the inclusion of one such contact term renormalizes the $^3S_1$-$^3D_1$ channel at LO was established in Refs. 11–13. This vindicated Weinberg's counting in that channel. However, Ref. 12 confirmed that an additional m_π-dependent contact term must be considered in the $^1S_0$ channel for renormalization there.[3] The issue of how many contact terms to include in V in other channels, where Weinberg's counting does not indicate the presence of such a contribution at LO, is still a controversial one. Two main points of view on this issue were presented in the panel.

U. van Kolck argued that the necessity for the inclusion of contact terms beyond those present in Weinberg's counting is dictated by strict renormalization-group invariance of observables. Renormalization is achieved when the effects of the cutoff Λ are of the size of higher-order terms in the χET. But when Λ is varied over a wide range, unphysical bound states appear in the NN spectrum. The only way to then guarantee renormalization-group invariance of NN observables is to introduce additional contact terms which renormalize V. Typically the necessity for these extra operators to be included arises when Λ is larger than the breakdown scale of the theory, which is usually taken to be of order m_ρ. For example, in Ref. 7 renormalization of the $^3P_0$ phase shift for energies up to 100 MeV is achieved for $\Lambda \geq 8$ fm^{-1}, but only after a contact term that operates in this wave is included in the LO calculation. M. Pavon's results support this point of view, although the formulation employed in his work involved imposing boundary conditions on the radial wave function at a very short distance. He also used the requirement that wave functions corresponding to different NN energies should be orthogonal to derive restrictions on the maximum number of contact terms that can be included in a given channel.

The contrasting view was advocated by U.-G. Meißner: that calculations should only be carried out with a finite cutoff Λ which is of the order of the expected breakdown scale of the χET, m_ρ. Strict renormalization-group invariance is not required, but the cutoff should be varied around m_ρ in order to get a lower bound on the size of omitted short-distance physics. There is no point in considering $\Lambda \gg m_\rho$, since the error in the calculation is not expected to decrease in this regime.[14] In this case, the coefficients of short-distance operators scale as predicted by naive dimensional analysis with respect to Λ, and Weinberg's counting holds. In Ref. 15 Epelbaum and Meißner examined the behavior of NN phase shifts and compared to the results of Nogga *et al.*[7] They found that a result for the ^3P_J waves that deviates from the asymptotic result of Ref. 7 by less than the theoretical uncertainty of a LO calculation could be achieved for $\Lambda \sim 600$ MeV.

During the discussion no consensus was achieved between the panelists. Fundamental disagreement remained about whether it was necessary to consider cutoffs $\Lambda \gg m_\rho$. Meißner argued that Ref. 15 showed that for all practical purposes the range $\Lambda \lesssim m_\rho$ contained all useful information. A. Nogga pointed out that it would not have been possible to reach this conclusion—independent of data—in Ref. 15 if the cutoff-independent phase shifts of the analysis of Ref. 7 had not already existed. Following this argument, for each new order in V, an analysis which demands strict renormalization-group invariance may be necessary before we can identify regions in Λ-space where Weinberg's counting can be used in practice.

Meißner in particular, felt that no conclusions could be drawn about power counting for short-distance operators until their role in renormalization had been examined with V calculated to orders beyond leading. Some of the contact terms promoted to LO in the study of Ref. 7 appear in the NLO potential, and (almost) all of them are present by N^3LO. Therefore in N^3LO calculations employing Weinberg's counting may be consistent with the requirement of Λ-independence over a broad range. Such a resolution is possible because, for high enough J, phase shifts can be calculated in perturbation theory throughout the entire energy range for which χPT is valid. However, it should also be noted that as higher orders are considered the long-distance part of V becomes even more singularly attractive in some of the P- and D-waves discussed by the panel. Whether this leads to further difficulties for Weinberg's power counting remains to be seen. Indeed, issues associated with V at NLO and beyond can only be resolved by calculation, and the majority of Working Group participants felt that studies which examine such issues are very important.

2.3. *πNN coupling constant controversy resolved*

In the NN system, many measurements have been performed in the last 30 years. In 1993 the Nijmegen group performed a partial-wave analysis (PWA) of these data, and thereby obtained a statistically consistent database of over 4000 np and pp data.[16] In doing this they found a (charged) πNN pseudoscalar coupling of $\frac{g^2}{4\pi} = 13.54 \pm 0.05$. This disagreed with a dedicated measurement of backward-angle np scattering at Uppsala, which yielded a result consistent with the "old" higher value $\frac{g^2}{4\pi} \approx 14.4$.[17]

S. Vigdor gave an overview of a recent experiment aimed at resolving this discrepancy.[18] In this double-scattering experiment, the flux of the neutron beam was accurately determined by counting the outgoing protons in the reaction $^2\text{H}(p,n)2p$ that produced the beam. The results of this absolute np differential cross section measurement at backward angles and a lab energy of 194 MeV are consistent with the predictions of the Nijmegen PWA. The NN database therefore now provides an accurate and unambiguous determination of the $\mathcal{L}^{(3)}_{\pi N}$ LEC b_{19}, which is associated with the Goldberger-Treiman discrepancy.[19]

3. $p \ll m_\pi$: low-energy dynamics which isn't chiral

3.1. *N^2LO and beyond in the three-nucleon system*

One of the central goals of nuclear physics is to establish a connection to QCD and its spontaneously broken chiral symmetry. There are some properties of nuclear systems that are due to chiral dynamics. Other properties, however, are simply a consequence of the (apparently accidentally) large scattering lengths. Such properties are "universal": they occur in a wide variety of systems from atomic to nuclear to particle physics,[20] and can be studied in an effective field theory (EFT) with contact interactions only. In nuclear physics, this is a "pionless" theory, which can be applied to processes with typical momenta $p \ll m_\pi$ and no external pions. It allows for precise calculations of very low-energy processes and exotic systems with weak binding such as halo nuclei.

L. Platter gave an overview of the pionless theory and showed various applications from atomic and nuclear physics. He presented results from his recent calculations of nd scattering and the triton that included all effects up to second order in p/m_π (i.e. N^2LO in this EFT) using a subtractive renormalization scheme.[21,22] (For an overview of previous higher-order calculations in the pionless theory see Ref. 20.) Platter's results imply that only the two-body S-wave scattering lengths and effective ranges plus one datum from the three-body system (usually taken to be $a_{nd}^{(1/2)}$ or the triton binding energy) are needed to predict low-energy three-nucleon observables with an accuracy $\sim 3\%$. As a consequence, universal correlations such as the Phillips and Tjon lines, as well as an analogous correlation involving the triton binding energy and charge radius,[23] persist up to N^2LO.

For precise higher-order calculations, it is important to know at which order new three-body force terms appear.[24] A general classification of three-body forces in this theory was presented by H. Grießhammer. He also illustrated the practical usefulness of the pionless theory by showing precise predictions for a recent deuteron electrodisintegration experiment at S-DALINAC,[25] and threshold neutron-deuteron radiative capture.[26]

3.2. *Experimental input and output*

Low-energy neutron-deuteron (nd) scattering is characterized by two scattering lengths: one in the quartet ($J = 3/2$) and one in the doublet ($J = 1/2$) channel. A novel method for measuring the incoherent combination of these using polarized scattering was outlined by O. Zimmer. His experiment obtains the incoherent nd scattering length from the pseudomagnetic precession of neutrons in a mixture of polarized deuterons and protons. Combining with the well-known coherent scattering length yields an anticipated accuracy for $a_{nd}^{(1/2)}$ that is a factor of 10 better than the present value $a_{nd}^{(1/2)} = 0.65 \pm 0.04$ fm.[27] Since $a_{nd}^{(1/2)}$ is an input to pionless EFT calculations in the three- and four-nucleon sectors, this experiment should also lead to more accurate predictions from that theory.

With the ingredients of the calculations for the NNN system under control at low energies, theoretical attention is turning towards $4N$ systems. The four-body system provides a fertile laboratory for few-body physics since many features such as resonances and multiple thresholds are present there, but are absent for $A < 4$. $4N$ systems are also the simplest ones where amplitudes of isospin T=3/2 can be studied in the laboratory. T. Clegg presented a talk on recent measurements and plans for new measurements using the recently developed polarized ^{3}He target at TUNL in combination with polarized proton and neutron beams. In addition to analyzing powers, spin-correlation coefficients have been measured.

3.3. *The future*

The future of the pionless theory lies in the extension to larger systems and in more precise higher-order calculations. The new precise $3N$ and $4N$ data from TUNL provide a challenge for the pionless theory program. The best known example is the A_y puzzle which is a long-standing problem in few-nucleon physics. There are dramatic effects in the vector analyzing powers at low energies where the pionless theory is applicable. As emphasized by T. Clegg, these effects become more severe for higher target masses which underscores the necessity for a better understanding of the four-body system. Currently, only a leading-order calculation is available and the general power counting of four-body forces is not understood.[28] As a consequence, more theoretical work is required. Many more interesting calculations and experimental results pertinent to $4N$ systems can be expected before the next Chiral Dynamics workshop.

The extension to systems with $A > 4$ appears to be an opportunity for lattice implementations of the pionless theory. As discussed by D. Lee, the renormalization-group behavior of the three-body force was already verified and lattice techniques have successfully been applied to dilute neutron matter.[29,30] These techniques can also be extended to the theory with pions, and may provide novel ways to compute nuclear bound states from χPT.

4. Going higher in the NNN system

In the past fifteen years, experimental and theoretical developments in the NNN sector have taken place hand-in-hand. The numerical solution of Faddeev equations describing three-body systems has become routine, while high-precision experimental data have become available thanks to a new generation of experimental facilities. These include Bonn and Cologne in Germany, Kyushu in Japan, and TUNL in the US for low energies, and KVI in the Netherlands, RIKEN and RCNP in Japan, and IUCF in the US for intermediate energies. In this section, a summary of the measurements performed for the intermediate energies will be presented.

J. Messchendorp gave an overview of the experiments performed at KVI in the past 10 years on the elastic and break-up channels in proton-deuteron scattering. K. Sekiguchi presented data from Japan on the elastic and break-up channels in nucleon-deuteron scattering.[31] There is a persistent problem of disagreement between the elastic *pd* cross sections measured at KVI and RIKEN at 135 MeV, but the bulk of data (including analyzing powers) taken at both laboratories show unambiguously the need for an additional ingredient in the theory beyond standard NN potentials. In some cases the addition of phenomenological NNN forces brings the results into very good agreement with the data but elsewhere, particularly at higher energies, these three-nucleon forces do not resolve the discrepancies between the data and the calculations. Unfortunately, the χET calculations at N^{3}LO required to accurately predict NNN system observables for energies larger than 100 MeV/nucleon have not yet been carried out, and only calculations with phenomenological potentials are available for this energy range.

In order to understand the spin structure of three-body forces, double-scattering experiments with polarized deuteron beams have been performed at both KVI and RIKEN. Spin-transfer coefficients from the deuteron to the proton have been obtained with rather high precision for a large part of the phase space.[32] These measurements have been performed at lower energies, allowing a comparison with the N^{2}LO predictions from χET computed by Witala and collaborators.[33] Within the relatively large error bands of the calculations there is good agreement with the data, but some discrepancies remain. The need to go to higher orders in χET is obvious.

The progress being made in this direction was described in the talk of E. Epelbaum. As explained above, the N^{3}LO NN force already exists, and gives results in quite good agreement with NN data. Work on deriving the (parameter-free) N^{3}LO NNN force is ongoing. Results that have been obtained so far for certain classes of diagrams were presented. Completing the calculation of the full set of N^{3}LO NNN force diagrams and comparing N^{3}LO predictions with experimental data will be a key test of χET's usefulness in three-nucleon systems, and we anticipate results by CD2009. Epelbaum also outlined his derivation of a parameter-free $NNNN$ force, which enters the nuclear potential at N^{3}LO.[36] This is the first microscopic computation of a $NNNN$ force, and estimates of its impact on α-particle binding suggest a 200–400 keV effect.[37]

Returning to experimental results, both Sekiguchi and Messchendorp presented *pd* break-up data, which showed the rich phase space offered by this channel. Not surprisingly, the data supports the inclusion of NNN forces. For the kinematics where the relative energy of the two outgoing protons in the break-up reaction is small one expects sensitivity to Coulomb effects. These effects have now been explicitly included in the calculations of the Hanover-Lisbon group[34] and have been examined and shown to exist by the Cracow-KVI measurements.[35] Proton-deuteron break-up data can thus be analyzed in a theoretical framework which makes it clear which effects are due to the Coulomb interaction. Therefore, we are no longer restricted to examining only nd breakup data. This provides a much richer NNN database within which we can look for the effects of chiral dynamics.

Messchendorp noted that the coupled-channels calculations of the Hanover-Lisbon group, which treat the $\Delta(1232)$ as a dynamical degree of freedom, do reasonably well when compared to the measured data in all of the channels discussed above. This might point to the fact that calculations of these observables in χET should include an explicit $\Delta(1232)$—especially if they wish to address data at higher energies. This could improve the convergence of χET calculations, which at present show a sizeable shift between NLO and N^{2}LO (c.f. Ref. 1). This shift signals the impact of two-pion-exchange in both the NN and NNN forces. As such it is a place where χET gives concrete predictions for the impact of two-pion-exchange physics on nd and pd elastic and breakup data. More calculations that highlight the impact of chiral physics on these data are needed.

However, further opportunities for measurements of this physics appear dangerously limited. The database in the NNN system remains much poorer than that in the NN system, but many laboratories studying this physics have shut down in the past few years and more will cease operations in the near future.

5. Soft Photons and Light Nuclei

Electromagnetic reactions on light nuclei can also be calculated within χET. After using the chiral potential of Eq. (1) to generate a wave function $|\psi\rangle$ for the nucleus—and, in breakup reactions, a consistent wave function for the final scattering state $|\psi_f\rangle$—one derives the current operator, J_μ for that system, again up to a given chiral order. The matrix element $\mathcal{M} \equiv \langle\psi_f|J_\mu|\psi\rangle$ can then be computed.

5.1. *Electron-deuteron scattering*

Work on current operators for the NN system has been going on for more than 10 years now. Park, Min, Rho, and later collaborators have computed deuteron photo- and electrodisintegration as well as weak reactions using a hybrid approach where J_μ is sandwiched between wave functions obtained from the AV18 potential (see, e.g. Refs. 38,39). More recently, computations of elastic electron-deuteron scattering using both V and J_μ computed up to NLO[40] and N^2LO[41,42] in the chiral expansion have been carried out. At N^2LO these calculations seem to agree—within the combined theoretical and experimental error bars—with the preliminary BLAST data presented in the talk of R. Fatemi.

5.2. *Compton scattering in $A = 2$ and $A = 3$*

Compton scattering from nucleons probes chiral dynamics in novel ways that explore the interplay between the long-range pion cloud and short-range operators in chiral perturbation theory.[43] The spin-independent polarizabilities for the proton are now quite well established, but the equivalent quantities for the neutron are not as well constrained by existing experimental data. Better knowledge of α_n and β_n would provide more information on how chiral dynamics in Compton scattering expresses itself differently among the two members of the nucleon iso-doublet.

H. Grießhammer reported on recent advances in calculations of elastic γd scattering. He and his collaborators have developed a new power counting for γd, designed for photon energies $\omega \sim \frac{m_\pi^2}{M}$. In this regime, the usual χPT counting cannot be applied to the $\gamma NN \to \gamma NN$ operator, and resummation of certain classes of diagrams is mandatory. When this resummation is complete the calculation reproduces the Thomson limit for the zero-energy γd cross section.[44] This also reduces the theoretical uncertainty in the calculation due to different assumptions about short-distance physics. Thanks to these advances, as well as the inclusion of $\Delta(1232)$ effects that are important at $\omega \sim 100$ MeV and backward angles, α_n can now be extracted from the extant γd data with a precision of about 15%.[44]

This is particularly exciting in light of the new data anticipated from MAX-Lab. As described by K. Fissum in his talk, an experiment that is scheduled to begin there next summer will approximately quadruple the world data on γd scattering. Taken in concert with theoretical advances this should facilitate an extraction of spin-independent neutron polarizabilities at a precision comparable to that at which α_p and β_p are known.

D. Choudhury reported on the first calculation of Compton scattering from the Helium-3 nucleus. Her χET calculation shows that there is significant sensitivity to neutron spin polarizabilities in certain γ^3He double-polarization observables. There are plans to measure these at the HIγS facility. Within Choudhury's NLO calculation, a polarized ^{3}He nucleus behaves, to a very good approximation, as a polarized neutron. Neutron spin polarizabilities can also be probed in γd measurements at HIγS,[45] which provides an important cross-check on our ability to calculate "nuclear" effects in Compton scattering.

Further progress in χET calculations will reduce the theoretical uncertainties that arise when neutron polarizabilities are extracted from γd or γ^3He data. Meanwhile, accurate polarized and unpolarized data on these reactions are anticipated from MAX-Lab and HIγS. With lattice QCD beginning to make predictions[46,47] (albeit quenched ones) for baryon polarizabilities there is an exciting opportunity for χET to connect experimental data from few-nucleon systems to the results of lattice simulations.

5.3. *Photodisintegration*

A number of talks discussed photodisintegration of light nuclei as a probe of nuclear forces. W. Leidemann argued that this a particularly attractive possibility, because the Lorentz Integral Transform (LIT) is a technique by which bound-state methods can be used to take a given NN and NNN force and obtain predictions for the photodisintegration cross section.

The resultant predictions (with phenomenological force models)[48] for the photodisintegration of ^{4}He are in agreement with the data from Lund presented in the talk of K. Fissum.[49] They do not, however, agree with a recent experiment at RCNP in Japan which used a novel technique.[50] P. Debevec described an experiment that could be done at HIγS which, with careful control of systematics, could pin down this cross section with very small error bars. Fissum also anticipates a future experiment at MAX-Lab could significantly reduce the error bars obtained in Ref. 49.

All of these experiments bear on the height of the first peak in the γ^4He cross section. This quantity contains important information on nuclear dynamics, and as such can be used to constrain χET (and perhaps even pionless) descriptions of nuclear forces. A computation of ^{4}He photodisintegration within χET is therefore a high priority, as is a definitive experiment to resolve the discrepancies in the present ^{4}He photoabsorption data. Meanwhile, more detailed information on nuclear forces in general, and the χET NNN force in particular, is expected from measurements of three-body breakup (with neutron detection) of polarized ^{3}He by linearly-polarized photons. These experiments will take place at HIγS, and were described by X. Zong.

5.4. *The Future*

Few-body methods such as the LIT, combined with χET expansions for the nuclear potential and electromagnetic current operators, provide opportunities to confront χET predictions with existing data on electromagnetic reactions on light nuclei. But only a few groups world-wide are performing such calculations, so this opportunity may not be fully exploited. And as photon and electron machines shut down or increase their energy, the possibility to use real and virtual photons to probe chiral dynamics in light nuclei may disappear. Over the next few years the role of experimental facilities such as MAX-Lab and HIγS will be vital.

6. Frontiers in symmetry breaking

6.1. *Parity violation*

A discussion of hadronic parity violation experiments was presented by S. Page. Various probes in few-body systems were discussed. An experiment which is taking data presently is polarized neutron radiative capture on protons at LANSCE. By measuring the very small ($< 10^{-7}$) up-down γ-ray asymmetry, one can constrain the low-energy constants that represent parity-violation in the χET,[51] e.g. the parity-violating πNN coupling constant sometimes called $h_{\pi NN}$.

6.2. *Isospin violation*

The light-quark mass difference $m_u - m_d$ is only a small fraction of the total mass of the nucleon. But the large NN scattering lengths magnify this isospin-breaking to the point where it is a $\sim 10\%$ effect in $a_{nn} - a_{pp}$.

Theoretical extractions from pp data yield a strong proton-proton scattering length of $a_{pp} = -17.3 \pm 0.4$ fm.[52] However, up until now, the neutron-neutron scattering length, a_{nn}, has only been extracted from few-body data. Different experiments on nd breakup lead to numbers that disagree, as described in the talk of C. Howell. Over the next three years we can anticipate new data on a_{nn} from at least two sources: a (hopefully) definitive nd breakup experiment underway at TUNL, and a re-analysis of LAMPF $\pi^- d \rightarrow nn\gamma$ data, discussed in the talk of A. Gårdestig. In addition, the contribution of V. Lensky pointed out that $\gamma d \rightarrow nn\pi^+$ could provide a complementary a_{nn} extraction. Once HIγS attains energies above pion threshold such an experiment could perhaps be done there.

The NPLQCD collaboration has already examined the impact of $m_u - m_d$ on the neutron-proton mass difference.[53] If this calculation could be extended to NN correlators, and the impact of $m_u - m_d$ on $a_{nn} - a_{pp}$ predicted, it would allow us to confront lattice results with high-precision extractions of the scattering-length difference from few-nucleon systems.

7. Conclusion

All of this suggests an exciting future for Chiral Dynamics in these systems. An era of calculations of few-nucleon bound states and reactions that truly start from QCD is just beginning. The presentations at CD2006 allow us to foresee a future where lattice simulations provide constraints on the low-energy constants that appear in the χET, and few-body methods allow us to completely solve—even for systems with $A = 4$ and beyond—that effective theory up to a fixed order in the chiral expansion. The resultant computations, which involve a mix of lattice, effective theory, and traditional few-body techniques, can then be compared (including their theoretical uncertainties!) with the wealth of experimental data in the $A = 2, 3$ and 4 sectors. We look forward to significant progress in the formation of this linkage between the chiral dynamics of QCD and the behavior of few-nucleon systems by the time of CD2009.

Acknowledgments

We thank the conference organizers and group participants for their efforts over the three afternoons of the Working Group. We also acknowledge financial support from the US Department of Energy (DE-FG02-93ER40756, DP), Deutsche Forschungsgemeinschaft (SFB/TR-16, HWH), and Bundesministerium für Bildung und Forschung (Grant no. 06BN411, HWH).

References

1. C. Ordóñez, L. Ray and U. van Kolck *Phys. Rev. C* **53**, 2086 (1996).
2. S. Weinberg, *Nucl. Phys. B* **363**, 3 (1991).
3. D. B. Kaplan, M. J. Savage and M. B. Wise, *Nucl. Phys. B* **478**, 629 (1996).
4. D. R. Entem and R. Machleidt, *Phys. Rev. C* **68**, 041001 (2003).
5. E. Epelbaum, W. Glöckle and U.-G. Meißner, *Nucl. Phys. A* **747**, 362 (2005).
6. D. Eiras and J. Soto, *Eur. Phys. J. A* **17**, 89 (2003).
7. A. Nogga, R.G.E. Timmermans and U. van Kolck, *Phys. Rev. C* **72**, 054006 (2005).
8. M.C. Birse, *Phys. Rev. C* **74**, 014003 (2006).
9. R. W. Hackenburg *Phys. Rev. C* **73**, 044002 (2006).
10. S.R. Beane, P.F. Bedaque, K. Orginos and M.J. Savage, *Phys. Rev. Lett.* **97**, 012001 (2006).
11. T. Frederico, V. S. Timoteo and L. Tomio, *Nucl. Phys. A* **653**, 209 (1999).
12. S. R. Beane, P. F. Bedaque, M. J. Savage and U. van Kolck, *Nucl. Phys. A* **700**, 377 (2002).
13. M. Pavon Valderrama and E. Ruiz Arriola, *Phys. Rev. C* **72**, 054002 (2005).
14. G. P. Lepage, arXiv:nucl-th/9706029.

15. E. Epelbaum and U.-G. Meißner, arXiv:nucl-th/0609037.
16. V. G. J. Stoks, R. A. M. Klomp, M. C. M. Rentmeester, and J. J. de Swart, *Phys. Rev. C* **48**, 792 (1993).
17. J. Rahm *et al.*, *Phys. Rev. C* **57**, 1077 (1998).
18. M. Sarsour *et al. Phys. Rev. C* **74**, 044003 (2006).
19. J. Gasser, M. E. Sainio and A. Svarc, *Nucl. Phys. B* **307**, 779 (1988).
20. E. Braaten and H.-W. Hammer, *Phys. Rep.* **428**, 259 (2006).
21. L. Platter and D.R. Phillips, arXiv:cond-mat/0604255.
22. L. Platter, *Phys. Rev. C* **74**, 037001 (2006).
23. L. Platter and H.-W. Hammer, *Nucl. Phys. A* **766**, 132 (2006).
24. H. W. Grießhammer, *Nucl. Phys.* **A760**, 110 (2005).
25. P. von Neumann-Cosel *et al.*, *Phys. Rev. Lett.* **88**, 202304 (2002).
26. H. Sadeghi, S. Bayegan, and H. W. Grießhammer, arXiv:nucl-th/0610029.
27. B. van den Brandt *et al.*, *Nucl. Instrum. Meth. A* **526**, 91 (2004).
28. L. Platter, H.-W. Hammer and U.-G. Meißner, *Phys. Lett. B* **607**, 254 (2005).
29. B. Borasoy, H. Krebs, D. Lee and U.-G. Meißner, *Nucl. Phys. A* **768**, 179 (2006).
30. D. Lee, arXiv:cond-mat/0606706.
31. K. Ermisch *et al.*, *Phys. Rev. C*, **71**, 064004 (2003).
32. K. Sekiguchi *et al.*, *Phys. Rev. C*, **69**, 054609 (2004).
33. H. Witala *et al.*, *Phys. Rev. C* **73**, 044004 (2006).
34. A. Deltuva, A. C. Fonseca and P. U. Sauer, *Phys. Rev. C* **73**, 057001 (2006).
35. S. Kistryn *et al.*, *Phys. Lett. B* **641**, 23 (2006).
36. E. Epelbaum, *Phys. Lett. B* **639**, 456 (2006).
37. D. Rozpedzik *et al.*, arXiv:nucl-th/0606017.
38. T. S. Park, D. P. Min and M. Rho, *Phys. Rev. Lett.* **74**, 4153 (1995).
39. T. S. Park *et al.*, *Phys. Rev. C* **67**, 055206 (2003).
40. M. Walzl and U. G. Meißner, *Phys. Lett. B* **513**, 37 (2001).
41. D. R. Phillips, *Phys. Lett. B* **567**, 12 (2003).
42. D. R. Phillips, arXiv:nucl-th/0608036.
43. J. A. McGovern, these proceedings, and references therein.
44. R. P. Hildebrandt, H. W. Griesshammer and T. R. Hemmert, arXiv:nucl-th/0512063.
45. D. Choudhury and D. R. Phillips, *Phys. Rev. C* **71**, 044002 (2005).
46. J. Christensen, W. Wilcox, F. X. Lee and L. M. Zhou, *Phys. Rev. D* **72**, 034503 (2005).
47. F. X. Lee, L. Zhou, W. Wilcox and J. Christensen, *Phys. Rev. D* **73**, 034503 (2006).
48. D. Gazit *et al.*, *Phys. Rev. Lett.* **96**, 112301 (2006).
49. B. Nilsson *et al.*, Phys. Lett. B **626**, 65 (2005), and arXiv:nucl-ex/0603030.
50. Y. Nagai, these proceedings.
51. S. L. Zhu, C. M. Maekawa, B. R. Holstein, M. J. Ramsey-Musolf and U. van Kolck, *Nucl. Phys. A* **748**, 435 (2005).
52. R. B. Wiringa, V. G. J. Stoks and R. Schiavilla, *Phys. Rev. C* **51**, 38 (1995).
53. S. R. Beane, K. Orginos and M. J. Savage, arXiv:hep-lat/0605014.

POWER COUNTING IN NUCLEAR CHIRAL EFFECTIVE FIELD THEORY

U. VAN KOLCK
Department of Physics, University of Arizona,
Tucson, AZ 85745, USA
E-mail: vankolck@physics.arizona.edu

The renormalization of singular potentials leads to a new power counting in nuclear chiral effective field theory.

Keywords: renormalization; effective field theory.

Scattering amplitudes involving several nucleons with momenta $Q \sim m_\pi \ll M_{QCD}$ (where m_π is the pion mass and M_{QCD} the characteristic QCD scale) can be derived[1,2] from an effective field theory (EFT) with pions, nucleons and delta isobars, which is a generalization of chiral perturbation theory (ChPT). Short-range physics is arbitrarily separated between pion loops and contact interactions through a cutoff, on which amplitudes do not depend. The issue is power counting: which contact terms appear at each order to ensure cutoff independence, or equivalently, how the contact interactions scale.

Weinberg[1] proposed a power counting based on naive dimensional analysis (NDA), where derivative and quark-mass interactions scale with Q/M_{QCD}. The leading long-range interaction is one-pion exchange (OPE), $\mathcal{O}(1/f_\pi^2)$ (where f_π is the pion decay constant). Purely nucleonic intermediate states contribute to nuclear amplitudes factors of $\mathcal{O}(m_N Q/4\pi)$ (where m_N is the nucleon mass), an enhancement of $\mathcal{O}(4\pi m_N/Q)$ over intermediate states in standard ChPT. For momenta $Q \sim 4\pi f_\pi^2/m_N$ OPE has to be treated non-perturbatively. Because loops include factors of the large m_N in the numerator, contact interactions do not necessarily obey NDA.

Pion exchange produces singular potentials between two nucleons. In spin-singlet channels OPE consists of a (S-wave) delta function and an m_π^2/r Yukawa term. Renormalization of the delta function requires a chiral-symmetric contact interaction with a cutoff-dependent coefficient. The interference of the delta and Yukawa functions generates further cutoff dependence that can only be removed by a chiral-breaking contact interaction,[3] which in Weinberg's power counting does not appear at this order.

In spin-triplet channels OPE has a tensor component with a $1/r^3$ singularity. When it is repulsive, the wavefunction goes fast to zero at the origin, and the problem is well defined. However, in channels where it is attractive, the two solutions of the Schrödinger equation oscillate wildly at small distances, and the singular potential is not sufficient to fix the phase of the wavefunction, which determines low-energy observables. One needs one piece of short-range physics in order to make the problem cutoff independent,[4] that is, one contact interaction in every wave where tensor OPE is attractive,[5] while in Weinberg's power counting it only occurs in the $^3S_1$-$^3D_1$ channel. Fortunately OPE does not need to be iterated in all waves, because of the centrifugal barrier. A finite set of short-range parameters are driven by pion parameters, effects of derivatives scaling[5] as $Q/l f_\pi$ (where l is the angular momentum), not Q/M_{QCD}.

Because they are suppressed by powers of Q/M_{QCD}, subleading interactions can be treated in (distorted-wave) perturbation theory. The most effective organizational scheme employs[6] an explicit delta-isobar field.[2]

We have thus shown[3,5] that Weinberg's power counting is inconsistent, and conjectured[5] that the new power counting sketched above is the correct foundation of chiral EFT in nuclear systems.

Acknowledgments

This research was supported in part by the U.S. Department of Energy and by the Alfred P. Sloan Foundation.

References

1. S. Weinberg, *Phys. Lett. B* **251**, 288 (1990); **295**, 114 (1992).
2. P.F. Bedaque and U. van Kolck, *Ann. Rev. Nucl. Part. Sci.* **52**, 339 (2002).
3. S.R. Beane *et al.*, *Nucl. Phys. A* **700**, 377 (2002).
4. S.R. Beane *et al.*, *Phys. Rev. A* **64**, 042103 (2001).
5. A. Nogga, R.G.E. Timmermans, and U. van Kolck, *Phys. Rev. C* **72**, 054006 (2005); A. Nogga, plenary talk at this conference.
6. V.R. Pandharipande, D.R. Phillips, and U. van Kolck, *Phys. Rev. C* **71**, 064002 (2005).

ON THE CONSISTENCY OF WEINBERG'S POWER COUNTING*

ULF-G. MEIßNER‡
HISKP, Universität Bonn, D-53115 Bonn, Germany
and
IKP, Forschungszentrum Jülich, D-52425 Jülich, Germany
‡*E-mail: meissner@itkp.uni-bonn.de*

I review the arguments in favor of Weinberg's power counting for chiral effective nuclear field theory.

Keywords: Chiral Lagrangians, nuclear forces, power counting

In Ref.[1] the consistency of Weinberg's power counting[2] for chiral nuclear effective field theory was challenged. At leading order, the effective potential consists of the static one-pion exchange and two S-wave contact interactions with two corresponding LECs. This potential is then used in a regularized LS-equation, with a typical regulator like $f_R(p) = \exp(-p^4/\Lambda^4)$ and Λ the cut-off. It was argued that to achieve cut-off independence at this order, one should promote some selected contact interactions that in W-counting appear in higher orders to leading order (LO) (called N-counting). In Ref.[3] we have critically examined these statements and arrived at the following conclusions (keeping always in mind that such studies based on the LO potential are of limited significance):

[1] The LS-equation with the point-like static OPEP is UV divergent, requiring an infinite number of counter terms for $\Lambda \to \infty$. This also means that treating the high partial waves in perturbation theory implicitly introduces an UV cutoff, so that this formal limit appears useless for explicit calculations.

[2] Many forms of the one-pion exchange potential exist that differ in their short-range part (e.g. choosing covariant or non-relativistic normalization).

*Work supported in part by DFG (TR-16), by BMBF (06BN411) and by EU I3HP (RII3-CT-2004-506078).

All these are equivalent from the EFT point of view since the short-distance physics can not be resolved.
[3] It is natural to choose a *finite* cutoff – any value is acceptable *if the error introduced is within the theoretical uncertainty at this order* (as stressed long ago by Lepage[4]). We have indeed shown that this is the case for most partial waves if the cutoff is $\Lambda \sim 2 \ldots 3\,\mathrm{fm}^{-1}$.
[4] A small value of the cutoff offers computational advantages, the corresponding effective potential is easy to use in few- and many-body calculations.
[5] Most importantly, the Weinberg scheme when applied to *observables* seems not to be in conflict with the data (except for the analyzing power A_y at very low energy - but this is improved at the next order).
[6] Other advantages of the Weinberg scheme are: (i) the extension to systems with 3-, 4-, ... nucleons is straightforward and extremely successful, see e.g. the talk by Epelbaum; (ii) at a given order, one has less parameters (LECs), that means higher predictive power as in for N-counting; (iii) physical scales are naturally included, especially when using spectral function regularization; (iv) the quark mass expansion of the nuclear forces can be done straightforwardly; and (v) these forces are amenable to a lattice formulation, see e.g. Dean Lee's talk.
[7] We have argued that the extension of the N-counting scheme to higher orders is plagued with conceptual problems, as stressed also by Pavon Valderrama in his talk in this working group. Clearly, the authors of Ref.[1] are challenged to provide a viable and accurate alternative to the Weinberg approach.

For a more detailed discussion of these and further arguments, please consult Ref.[3] and the literature cited therein.

Acknowledgements

I thank the organizers for a superb job and Evgeny Epelbaum for sharing his insights into the topics discussed here.

References

1. A. Nogga, R. G. E. Timmermans and U. van Kolck, Phys. Rev. C **72**, 054006 (2005) [arXiv:nucl-th/0506005].
2. S. Weinberg, Nucl. Phys. B **363**, 3 (1991).
3. E. Epelbaum and U.-G. Meißner, arXiv:nucl-th/0609037.
4. G. P. Lepage, arXiv:nucl-th/9706029.

RENORMALIZATION OF SINGULAR POTENTIALS AND POWER COUNTING

M. PAVÓN VALDERRAMA*

Departmento de Física Atómica, Molecular y Nuclear, Universidad de Granada, Granada, E-18071, Spain

**E-mail: mpavon@ugr.es*

We analyze the renormalization of the nucleon–nucleon interaction at low energies in coordinate space for both one and two pion exchange chiral potentials. The singularity structure of the long range potential and the requirement of orthogonality respectively determines, once renormalizability is imposed, the minimum and maximum number of counterterms allowed in the effective description of the nucleon–nucleon interaction in a non-perturbative context.

Keywords: Renormalization; NN interaction; Two Pion Exchange.

The nucleon–nucleon (NN) interaction can be better understood if we take into account that there is a separation of scales between long and short range physics, which is the basis of the Effective Field Theory (EFT) formulation of nuclear forces.[1] The long range piece of the interaction, V_L, is given by pion exchanges, and its form is unambiguously determined by the imposition of chiral symmetry, while the short range piece V_S is a zero-range potential (i.e. $V_S(r) = 0$ for $r > 0$) which represents the NN contact terms. This last piece is regularization-dependent and determines the number of free parameters, or counterterms, of the EFT description.

In Weinberg's power counting these potentials can be written as an expansion in terms of the dimensionality of the contributions, $V_{NN} = V_{LO} + V_{NLO} + \dots$, where V_{LO}, V_{NLO}, and so on, represent increasing order contributions to the NN potential. Standard EFT wisdom states that the power countings of the long and short distance potentials are independent, regardless of the renormalization procedure, although there is a lively discussion on the field about this issue,[2–6] and, as it is shown below, if one takes the long distance potential seriously from $r > 0$ to infinity, this cannot be the case. One easy example is provided by the $^1S_0$ neutron–proton scattering state at NNLO, which can be described by the s-wave

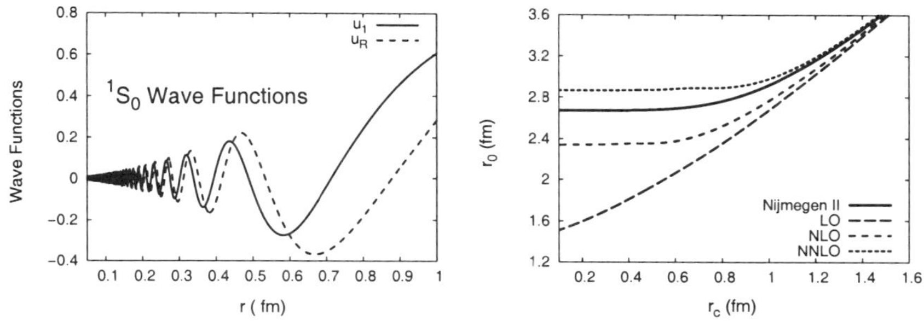

Fig. 1. (Left panel) Two zero-energy, linearly independent wave functions for the $^1S_0$ singlet channel at NNLO; u_1 and u_r respectively behave as 1 and r for $r \to \infty$. (Right Panel) Value of the effective range r_0 as a function to the cut-off for the same channel and different orders; r_0 is computed by $r_0(r_c) = 2\,[\int_0^\infty (1 - r/\alpha_0)^2\, dr - \int_{r_c}^\infty u_0^2\, dr]$, with $\alpha_0 = -23.74$fm the scattering length.

reduced Schrödinger equation, i.e. $-u'' + M_N V_L(r)\, u(r) = k^2\, u(r)$, with V_L the long range potential at NNLO, which is attractive, and displays a $-1/r^6$ singularity near the origin. As a second order differential equation, the Schrödinger equation has two linearly independent solutions, so how does one choose the physical solution? The regularity condition $u(0) = 0$, equivalent to the assumption that there is no short distance physics, cannot be applied for this case, since any solution $u(r)$ is regular at the origin,[2,5] as shown in Fig. (1). Then we are forced to use a boundary condition near the origin, equivalent to adding a counterterm to the theory.

And how many more counterterms can be added? According to Weinberg's power counting, there are two counterterms for the $^1S_0$ wave at NNLO. But if we take into account orthogonality between different energy solutions, we find out that the boundary condition must be energy independent, meaning that one can only have one counterterm[2,5] (any extra counterterm breaking orthogonality). So one is led to this alternative when removing the cut-off. The prediction of the effective range for the singlet, $r_0 = 2.86$ fm (the experimental value is 2.77 ± 0.05 fm), can serve as an orthogonality test.

References

1. See e.g. P. F. Bedaque and U. van Kolck, Ann. Rev. Nucl. Part. Sci. **52**, 339 (2002) for a review and references therein.
2. M. Pavon Valderrama and E. Ruiz Arriola, Phys. Rev. C **72**, 054002 (2005)
3. A. Nogga, R. G. E. Timmermans and U. van Kolck, Phys. Rev. C **72**, 054006 (2005)

4. M. C. Birse, Phys. Rev. C **74**, 014003 (2006)
5. M. Pavon Valderrama and E. Ruiz Arriola, Phys. Rev. C **74**, 054001 (2006)
6. E. Epelbaum and U. G. Meissner, arXiv:nucl-th/0609037.

THE CHALLENGE OF CALCULATING BARYON-BARYON SCATTERING FROM LATTICE QCD

SILAS R. BEANE

Department of Physics, University of New Hampshire, Durham, NH 03824-3568.

We discuss recent progress in determining baryon-baryon scattering parameters using fully-dynamical lattice QCD calculations.

Lattice QCD predictions of scattering parameters in the two-nucleon system would be a significant milestone toward rigorous calculations of nuclear properties and decays. One quenched lattice QCD calculation of the nucleon-nucleon (NN) scattering lengths[1] has been attempted. The S-wave NN scattering lengths are extremely large as compared to the sizes of state-of-the-art lattices: $a^{({}^1S_0)} \sim -23.714$ fm and $a^{({}^3S_1)} \sim +5.425$ fm. While naively this is a concern, it was recently shown that exact solutions to Lüscher's formula[2] for the energy-levels of the NN system on a lattice will allow for the extraction of scattering parameters from lattice calculations with volumes that are much smaller than the anomalously-large scattering lengths.[3] Moreover, while these lattice volumes are still unrealistically large at the physical quark masses, the size of the lattices required decreases with increasing quark mass, as the cutoff of the EFT is set by the pion mass. Therefore, for this problem, it is a definite advantage to work with unphysical lattice quark masses.

In a recent study,[4] the NPLQCD collaboration has performed the first full-QCD calculation of the S-wave NN scattering lengths. At the pion masses used in these calculations, the NN scattering lengths were found to be of natural size in both channels, and much smaller than the $L \sim 2.5$ fm lattice spatial extent. The lowest pion mass calculated ($\sim$ 350 MeV) is at the upper limit of where we expect the EFT describing NN interactions to be valid. With only one lattice point within the EFT, a prediction for the scattering lengths at the physical pion mass is not possible: the experimental values of the scattering lengths are still required for an extrapolation to the chiral limit and naive dimensional analysis (NDA) is still required to select only those operator coefficients that are consistent with perturbation theory. As the NN signals improve with increased statistics, a lattice QCD prediction of the scattering parameters should become possible. As a consequence of studying the NN systems on the lattice, we will be able, for the first time, to look for a bound-state of two nucleons at the available lattice quark masses. This will require looking at several distinct lattice volumes in order to distinguish bound states from scattering states and may require a program of lattice generation specifically suited to this problem. The difficulties in separating bound states from scattering states in the NN system is discussed in some detail in Ref.[3] Naively it may seem that the deuteron is too diffuse an object to be studied using lattice QCD methods as enormous lattices would be required to contain the deuteron. That this is actually not the case was pointed out originally by Lüscher[2] and demonstrated in detail for the NN system in Ref.[3] The basic point is that the range of the interaction, which is roughly given by the pion Compton wavelength, and not the scattering length, sets the relevant length scale which should be compared to the size of the lattice.

Study of the interactions of hyperons with nucleons and nuclei is an exciting area of nuclear physics. Hyperon-nucleon interactions influence the structure and energy-levels of hypernuclei. As the nucleon chemical potential increases with density, hyperons are also expected in the core of neutron stars. Therefore, a basic input in studies of the Equation of State of dense stellar matter is the hyperon-nucleon (and hyperon-hyperon) interaction. However, these interactions are quite poorly known in comparison with the interactions between nucleons. An investigation of these interactions using lattice QCD is currently underway. Scattering lengths of natural size are expected as there are probably no ΛN bound states near threshold. In Ref.,[5] an EFT was used to compute the scattering length and effective range for elastic ΛN.

References

1. M. Fukugita, Y. Kuramashi, M. Okawa, H. Mino and A. Ukawa, *Phys. Rev.* **D52**, 3003 (1995).
2. M. Luscher, *Commun. Math. Phys.* **105**, 153 (1986).
3. S. R. Beane, P. F. Bedaque, A. Parreno and M. J. Savage, *Phys. Lett.* **B585**, 106 (2004).
4. S. R. Beane, P. F. Bedaque, K. Orginos and M. J. Savage, *Phys. Rev. Lett.* **97**, p. 012001 (2006).
5. S. R. Beane, P. F. Bedaque, A. Parreno and M. J. Savage, *Nucl. Phys.* **A747**, 55 (2005).

PRECISE ABSOLUTE NP SCATTERING CROSS SECTION AND THE CHARGED π NN COUPLING CONSTANT

S. E. VIGDOR

Dept. of Physics, Indiana University, Bloomington, IN 47405, U.S.A.

A tagged medium-energy neutron beam has been used in a precise measurement of the absolute *np* back-scattering differential cross section. The results resolve significant dis-crepancies within the *np* database, bearing on the charged π *NN* coupling constant value.

Discrepancies among different experiments have led to a drastic pruning of cross section measurements for medium-energy *np* scattering from the database used in partial wave analyses (PWA)[1] and to debates concerning the value and extraction methods for the charged π *NN* coupling constant.[2] The rejected data include nearly all of the most recent results,[3,4] which differ systematically in angular dependence from earlier Los Alamos measurements[5] that dominate the data retained.

We have carried out a new experiment,[6] utilizing a tagged neutron beam to maintain tight control of systematic errors, to resolve the experimental discrepancies. The beam was produced in the IUCF Cooler ring by the 2H(p,n)2p reaction induced by a cooled, stored 202.5 MeV proton beam on a deuterium gas jet target. Detection of the two low-energy recoil protons in an array of double-sided silicon strip detectors determined the energy and angle of each produced neutron with resolutions σ $E \approx 60$ keV and σ angle ≈ 2 mrad. The tagged neutrons impinged on carefully matched secondary targets of CH2 and C, permitting accurate subtraction of backgrounds from quasifree scattering and non-target sources, and minimizing reliance on kinematic cuts to identify free scattering. Forward protons from the secondary targets were detected in a large-acceptance array of scintillators and wire chambers. Tagging was essential to the experimental approach, providing many internal crosschecks and calibrations;[6] no *extra* work was required to provide an *absolute* cross section standard for medium-energy neutron-induced reactions.

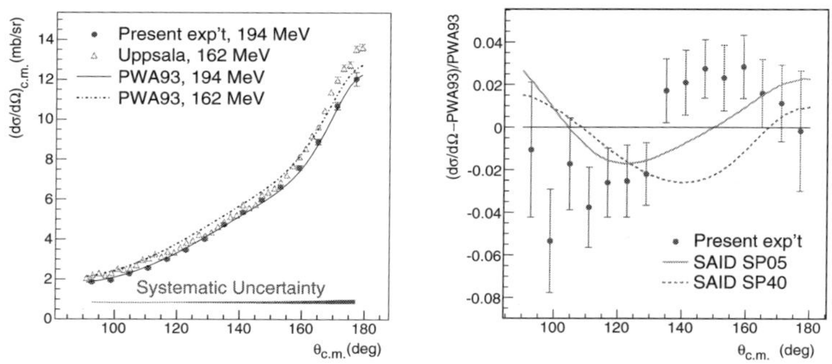

Fig. 1. Left: absolute differential cross section from the present experiment, compared with data from[2] and PWA calculations at two relevant energies. The plotted error bars are statistical (including from background subtraction). The net systematic uncertainty (shaded band, see[6] for details) is dominated by an absolute scale uncertainty of $\pm$ 1.5%. Right: relative differences of the present results and two recent PWA solutions from Nijmegen PWA93, all at En = 194 MeV.

Results of the experiment are compared to previous data and to several PWA's in Fig. 1 The results deviate systematically from the earlier Uppsala [2] and PSI[3] data, especially at large c.m. angles. They agree relatively well with the Nijmegen PWA93 calculations,[1] and hence support the corresponding π NN coupling constant value $g_\pi^2/4\pi \cong 13.5$. 2-3% deviations in angular distribution shape (Fig. 1(b)) from PWA93 are comparable to deviations among different PWA's, and may reflect the need for phase shift refinement. Our careful study of systematics suggests that quoted PWA uncertainties should be treated cautiously, as most systematic errors for previous np experiments are under-reported in the literature and generally ignored in the PWA fits.

Acknowledgments

This work was carried out with support from the U.S. National Science Foundation under grant numbers NSF-PHY-9602872, 0100348 and 0457219.

References

1. V.G.J. Stoks *et al.*, *Phys. Rev.* **C48**, 792 (1993); M.C.M. Rentmeester, R.G.E. Timmermans and J.J. deSwart, *Phys. Rev.* **C64**, 034004 (2001).
2. Proc. Workshop on *Critical Issues in the Determination of the Pion-Nucleon Coupling Constant,* ed. J. Blomgren, *Phys. Scripta* **T87** (2000).
3. J. Rahm *et al., Phys. Rev.* **C57**, 1077 (1998).

4. W. Hrster *et al.*, *Phys. Lett.* **B90**, 367 (1980); J. Franz *et al.*, *Phys. Scripta* **T87**, 14 (2000).
5. B.E. Bonner *et al.*, *Phys. Rev. Lett.* **41**, 1200 (1978).
6. M. Sarsour *et al.*, *Phys. Rev. Lett.* **94**, 082303 (2005) and *Phys. Rev.* **C74**, 044003 (2006).

PROBING HADRONIC PARITY VIOLATION USING FEW NUCLEON SYSTEMS

S. A. PAGE*
Department of Physics & Astronomy, University of Manitoba
Winnipeg, Manitoba, Canada R3T 2N2
**Email: spage@cc.umanitoba.ca*

Fifty years since the discovery of parity violation in weak interactions, the field of hadronic parity violation still harbors a number of mysteries that remain to be resolved. Their origin likely lies in the complex interplay between strong and weak interactions. Current experimental work is focussed on a set of extremely challenging measurements in two and few-nucleon systems, while recent theoretical developments involving effective field theory ideas has led to the formulation of a new framework for interpreting experimental data.

Keywords: Hadronic Weak Interaction; Parity Violation

1. Experimental Situation and Future Prospects

Experimentally, the strangeness-conserving hadronic weak interaction can be isolated via nuclear processes that violate parity, thereby eliminating the much larger effects of the parity-conserving strong and electromagnetic interactions. Since 1957, a large body of such experimental data has been acquired, with definitive measurements of parity-violating (PV) observables ranging from 10^{-7} level helicity asymmetries measured in proton-proton scattering, to greatly enhanced 10^{-1} level asymmetries in low energy polarized neutron transmission experiments through heavy nuclei, as reviewed e.g. in references[1,2] .

Until recently, the conventional theoretical framework for interpreting experimental data has been a meson exchange model, detailed by the seminal work of Desplanques, Donoghue, and Holstein (DDH).[3] In this model, the PV nucleon-nucleon (NN) interaction is dominated by the exchange of the pion and two lightest vector mesons (ρ and ω), and its strength is characterized by seven PV meson-nucleon couplings: h_π^1, $h_\rho^{0,1,2}$, $h_\rho^{1\prime}$ and $h_\omega^{0,1}$, where the superscript indicates the isospin. DDH provided theoretical "reasonable ranges" ($\pm \simeq 100\%$) and "best values" for the h^i_M using SU(6) symmetry, constraints from non-leptonic hyperon decay data, and the quark model to estimate the experimentally unconstrained terms. Experimental results from nuclear and hadronic PV measurements have been analyzed using the DDH framework, yielding inconsistent results for the π and heavier weak meson-nucleon couplings.

Recently, a new framework for interpreting hadronic parity violation has been formulated in Reference[4] using effective field theory ideas. Two versions of the EFT are useful: for energies well below the pion mass, the EFT contains only four-nucleon operators and five "low-energy constants": $\lambda_s^{0,1,2}$, λ_t, and ρ_t; while at higher energies, the the pion becomes dynamical and three additional constants associated with π-exchange effects appear at lowest order: h^1_π, along with a second parameter in the EFT potential, k^{1a}_π, and a new meson-exchange current operator characterized by $\bar{C}_\pi$.

On the experimental side, attention is focusing on measurements of PV in few nucleon systems, which offer the cleanest theoretical interpretation. The few-body PV observables of interest include: (longitudinally) polarized $\mathbf{p}p$ and $\mathbf{p}\alpha$ scattering below 50 MeV, yielding asymmetries A_z^{pp} and $A_z^{p\alpha}$; radiative np capture at low energy, with both polarized and unpolarized neutrons, yielding the photon asymmetry A^d_γ and circular polarization P^d_γ respectively (or alternatively, the asymmetry A^γ_L in $\gamma d \to np$); PV neutron spin rotation about the momentum direction upon transmission through ^{4}He, yielding $d\phi^{n\alpha}/dz$; and finally, radiative capture of polarized neutrons on deuterium at threshold, $\mathbf{n}d \to t\gamma$, yielding the photon asymmetry A^t_γ. Of these few body observables, the longitudinal asymmetries with polarized protons have been measured to high precision; the NPDGamma experiment[5] is currently underway at LANSCE, and the ^{4}He spin rotation experiment is underway at NIST. Immediate prospects for improvements in neutron beam measurements are excellent at the new fundamental physics beamline FNPB at the SNS, due to reach high intensity production by 2008. For details of the experimental situation and further discussion, see the recent review[1] and references therein.

References

1. Ramsey-Musolf, M.J. and Page, S.A., Ann. Rev. Nucl. Part. Sci. **56** (2006) 1
2. Adelberger, E.G. and Haxton, W.C., Ann. Rev. Nucl. Part. Sci. **35** (1985) 501
3. Desplanques, B., Donoghue, J.F. & Holstein, B.R., Ann. Phys. **124** (1980) 449
4. Zhu, S.L. *et al.*, Nucl. Phys. A **748** (2005) 435
5. Gericke, M.T. *et al.*, Nucl. Instrum. Meth. A **540** (2005) 428

EXTRACTING THE NEUTRON-NEUTRON SCATTERING LENGTH FROM NEUTRON-DEUTERON BREAKUP

C. R. Howell*

Physics Department, Duke University,
and Triangle Universities Nuclear Laboratory,
Durham, NC 27708, USA

This article provides a brief summary of the values for the $^1S_0$ neutron-neutron scattering length obtained from neutron-deuteron breakup cross-section measurements. The results of recent kinematically incomplete and complete experiments are reported, and a description of a new measurement at the Triangle Universities Nuclear Laboratory (TUNL) is given.

Keywords: neutron-neutron scattering; neutron-deuteron breakup; nn; nd.

1. Summary of a_{nn} Measurements using nd breakup

There are two primary reactions used to determne a_{nn}, pion-deuteron (π^--d) capture and neutron-deuteron (nd) breakup. The results obtained from these two reactions differ substantially, and there are significant discrepancies between the nd breakup results.[1] The focus of this paper is on measurements using nd breakup. The value of a_{nn} is determined by fitting the measured cross section for the neutron-neutron (nn) final-state interaction (FSI) in nd breakup with a theoretical model of the three-nucleon reaction.[2] The cross section for the nn FSI can be measured with sufficient accuracy to determine a_{nn} to better than $\pm$ 0.5 fm. However, the values for a_{nn} obtained in different nd breakup experiments do not have this level of consistency. Two types of nd breakup measurements have been used to determine a_{nn}, kinematically incomplete (KI) and kinematically complete (KC).

In KI experiments the energy spectrum of the recoil proton emitted near zero degree is measured. The average value of a_{nn} obtained from KI

---

*This work was supported in part by the U.S. Department of Energy, Office of High Energy and Nuclear Physics, under Grant No. DE-FG02-97-ER41033.

measurements from 1964 through 1986 is -18.7 $\pm$ 02 fm.[1] This average is mostly determined by the measurements reported in Refs. 3 and 4 and is consistent with the average value of -18.6 $\pm$ 0.4 fm obtained from π^--d experiments.[5] However, a recent KI measurement[6] is about 2 fm (2.4 sd) lower than the average value from previous KI nd breakup measurements.

In KC experiments the detectors are set up in either coincidence geometry (CG) or recoil geometry (RG). In the CG the momenta of the two final-state neutrons are measured along with the energy of the recoil proton, while in the RG the momenta of the proton and one of the final-state neutrons are measured. The average value of a_{nn} obtained in KC experiments from 1965 through 1980 is -16.9 $\pm$ 0.3 fm,[1] which is dominated by the CG and RG measurements reported in Refs. 7 and 8, respectively. The most recent measurement made in RG gives a value of a_{nn} = -16.2 $\pm$ 0.3 fm[9] in contrast to the value of a_{nn} = -18.7 $\pm$ 0.7 fm[10] obtained in the most recent CG measurement.

2. New Measurements at TUNL

To help resolve the discrepancy between the most recent determinations of a_{nn} using the nd breakup reaction, new cross-section measurements of the nn FSI in nd breakup have been made at TUNL at an incident neutron energy of 19.0 MeV. The measurements are made simultaneously in both the CG and RG using the same neutron beam, and a value for a_{nn} will be determined to an accuracy better than $\pm$ 0.7 fm from each measurement independently. Data analysis is expected to be completed by the end of 2007. Preliminary results from both measurements are consistent with a_{nn} = -17.5 $\pm$ 1 fm.

References

1. C.R. Howell, Proceedings of the X International Seminar on Interactions of Neutrons with Nuclei, Dubna, Russia, ISINN-10 (2003), p. 71.
2. W. Glöckle *et al.*, Phys. Reports **274** (1996) 107.
3. S. Shirato *et al.*, Nucl. Phys. **A215** (1973) 277.
4. N. Koori *et al.*, Conf. on Few Body Systems in Part. and Nucl. Physics, p.406, T. Sasakaw *et al.* (eds.) Tohoku Univ. 1986; S. Shirato *et al.*, *ibid*, p. 412.
5. C.R. Howell *et al.*, Phys. Lett. **B444** (1998) 252.
6. W. von Witsch *et al.*, Phys. Rev. **C74** (2006) 014001.
7. B. Zeitnitz *et al.*, Nucl. Phys. **A149** (1970) 449.
8. W. von Witsch *et al.*, Phys. Lett. **B80** (1979) 187.
9. V. Huhn *et al.*, Phys. Rev. Lett. **85** (2000) 1190.
10. D.E. González Trotter *et al.*, Phys. Rev. Lett. **83** (1999) 3788.

EXTRACTION OF a_{nn} FROM $\pi^- d \to nn\gamma$

A. GåRDESTIG

Department of Physics and Astronomy, University of South Carolina, Columbia, SC 29210, U.S.A.
E-mail: anders@physics.sc.edu

I present a calculation of the $\pi^- d \to nn\gamma$ reaction to third order in chiral perturbation theory. The short-distance physics of this reaction can be constrained by relating it to several important low-energy weak reactions. The theoretical error in a_{nn} extracted from this reaction can thus be reduced by a factor larger than three to ± 0.05 fm.

Keywords: neutron-neutron scattering length, radiative pion capture

Since there are no neutron targets, the neutron-neutron scattering length (a_{nn}) can only be accessed using indirect methods, with neutrons detected in phase space regions sensitive to a_{nn}. Measurements of a_{nn} using $nd \to nnp$ suffer from unresolved discrepancies between experiments.[1] Experiments using the final state of $\pi^- d \to nn\gamma$ are more consistent and dominate the presently accepted value of $a_{nn} = -18.59 \pm 0.4$ fm.[2,3] The latter extractions used theory models by Gibbs, Gibson, and Stephenson and de Téramond et al.,[4] both with a theoretical error of ± 0.3 fm. I will show how chiral perturbation theory can reduce this error considerably.[5–7]

The $\pi^- d \to nn\gamma$ reaction is dominated by well-known single-nucleon photon-pion amplitudes. Two-nucleon diagrams occur first at $\mathcal{O}(Q^3)$ (where $Q \sim m_\pi$ is a small energy/momentum) but nevertheless cause a small, but significant, change in the shape of the $\pi^- d \to nn\gamma$ spectrum. The wave functions for the bound and scattering states are calculated from the known asymptotic behavior, integrated in from $r = \infty$ using the one-pion-exchange potential. At short distances, the unknown short-distance physics is parametrized and regularized by matching the wave functions at some radius $r = R$ (1.4 fm $< R <$ 3 fm) to a spherical well solution for $r < R$. Details about the calculation and the wave functions can be found in Ref. 5.

The result at $\mathcal{O}(Q^3)$ is sensitive to the value of R, indicating that some unknown short-distance physics is at play (left panel of Fig. 1). As shown

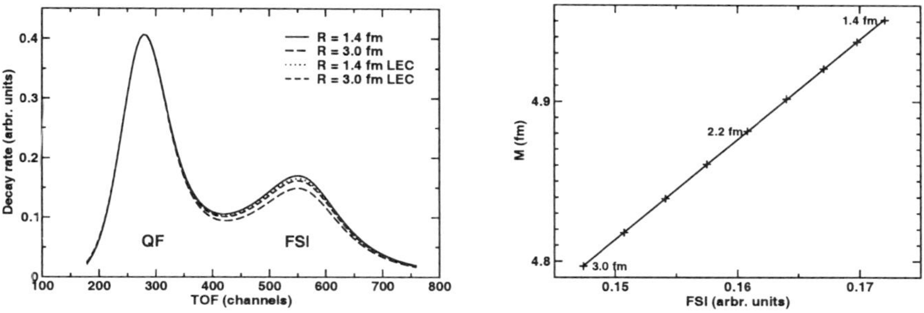

Fig. 1. (Left panel) Neutron time-of-flight spectrum for $\pi^- d \to nn\gamma$, without and with the LEC as indicated. The labels indicate where the reaction is dominated by quasi-free (QF) and final-state-interaction (FSI) kinematics. (Right panel) The relation between the GT matrix element and the FSI peak height, for varying R as indicated.

in Refs. 6,7, the $\pi^- d \to nn\gamma$ reaction is sensitive to the same short-distance physics as several important weak reactions, e.g., solar *pp* fusion and tritium beta decay. Calculations confirm this. Figure 1 (right panel) shows a linear relation between the $\pi^- d \to nn\gamma$ FSI peak height and the Gamow-Teller (GT) matrix element of *pp* fusion. This indicates that both reactions can be simultaneously renormalized by *one* low-energy constant (LEC). The same LEC appears in the chiral three-nucleon force (3NF) with similar kinematics, enabling extracting (part of) the 3NF from two-nucleon systems.

Once the LEC is included the theoretical error due to short-distance physics can be reduced significantly, as is obvious from Fig. 1. Including other sources of errors,[5] the theoretical error in a_{nn} from fitting the entire spectrum is ± 0.3 fm, while fitting only the FSI peak gives ± 0.05 fm.[6] For further details and discussions about future work, see Refs. 5–7.

This work was supported in part by DOE grant No. DE-FG02-93ER40756 and NSF grant No. PHY-0457014.

References

1. See, e.g., C. R. Howell, these proceedings.
2. C. R. Howell et al., Phys. Lett. B **444**, 252 (1998).
3. R. Machleidt and I. Slaus, J. Phys. G **27**, R69 (2001).
4. W. R. Gibbs, B. F. Gibson, and G. J. Stephenson, Jr., Phys Rev. C **11**, 90 (1975); **16**, 327 (1977); G. F. de Téramond, Phys. Rev. C **16**, 1976 (1977); G. F. de Téramond and B. Gabioud, *ibid.* **36**, 691 (1987).
5. A. Gårdestig and D. R. Phillips, Phys. Rev. C **73**, 014002 (2006).
6. A. Gårdestig and D. R. Phillips, Phys. Rev. Lett. **96**, 232301 (2006).
7. A. Gårdestig, Phys. Rev. C **74**, 017001 (2006).

THE THREE- AND FOUR-BODY SYSTEM WITH LARGE SCATTERING LENGTH

L. PLATTER

Department of Physics & Astronomy, Ohio University, Athens, OH 45701, USA

We dicuss recent results for nuclear three- and four-body systems obtained within the effective theory field theory with contact interactions alone.

The seminal paper by Bedaque, Hammer and van Kolck[1] applied the effective field theory (EFT) with contact interactions alone to the three-body system. In this work a three-body force was introduced at leading order to renormalize observables to deal with the strong cutoff-dependence of observables which arises due to the use of contact interactions in the two-body sector. Since then, this EFT has been used to calculate various observables in atomic and nuclear three-body systems and was able to explain well-known correlations between different observables as a result of a large scattering length in the two-body system. Different conclusions have been achieved regarding the order at which an energy-dependent three-body force appears. Bedaque *et al.*[2] introduced a three-body force at next-to-next-to-leading order (NNLO). It was found recently[3] that within a subtractive scheme[4,5] no energy dependent three-body force has to be included at this order. A particularly interesting implication of this result is that the triton binding energy can be calculated to one order higher than previously thought. In[6] we calculated the triton binding energy to be 8.1 Mev, 8.2 MeV and 8.54 MeV at LO, NLO and NNLO respectively which should be compared 8.48 MeV which is the experimentally observed value of the triton binding energy. As an additional step towards an understanding of many-body forces we considered the four-body system with large scattering length.[7] We computed the binding energies of the four-boson and four-nucleon system and analyzed their cutoff dependence. It was found that observables are cutoff-independent which implies that no four-body force is needed at

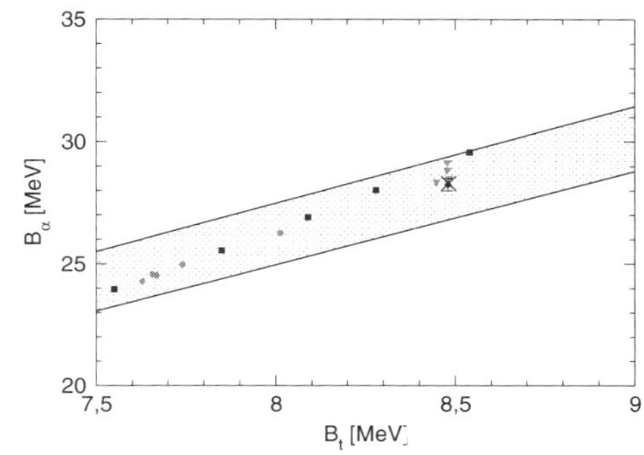

Fig. 1. The correlation between the triton and the α-particle binding energies. The lower (upper) line shows our leading order result using a_s and B_d (a_s and a_t) as two-body input. The grey dots and triangles show calculations using phenomenological potentials without or including three-nucleon forces, respectively[8]. The squares show the results of chiral EFT at NLO for different cutoffs while the diamond shows the N^2LO result[9,10]. The cross shows the experimental point.

leading order for the renormalization of observables. Therefore, the familiar linear correlation between the triton and α-particle binding energy can be understood as a result of the large scattering length in the two-body sector which requires exactly one three-body input parameter at LO in the R/a expansion. In Fig.1 we show the Tjon band generated by using the triton binding energy as a three-body input and using the singlet scattering length with either the deuteron binding energy or triplet scattering length as two-body input parameters.

References

1. P. F. Bedaque, H.-W. Hammer, and U. van Kolck, Phys. Rev. Lett. **82**, 463 (1999).
2. P. F. Bedaque, G. Rupak, H. W. Griesshammer and H. W. Hammer, Nucl. Phys. A **714**, 589 (2003).
3. L. Platter and D. R. Phillips, Few-Body Systems *in press.*
4. I.R. Afnan and D.R. Phillips, Phys. Rev. C **69**, 034010 (2004).
5. H.-W. Hammer and T. Mehen, Nucl. Phys. A **690**, 535 (2001).
6. L. Platter, Phys. Rev. C **74**, 037001 (2006).
7. L. Platter, H.-W. Hammer and U.-G. Meißner, Phys. Lett. B **607**, 254 (2005).
8. A. Nogga, H. Kamada and W. Glöckle, Phys. Rev. Lett. **85**, 944 (2000).
9. E. Epelbaum, A. Nogga, W. Glöckle, H. Kamada, U.-G. Meißner and H. Witala, Phys. Rev. C **66**, 064001 (2002).
10. E. Epelbaum, H. Kamada, A. Nogga, H. Witala, W. Glöckle and U.-G. Meißner, Phys. Rev. Lett. **86**, 4787 (2001).

3N AND 4N SYSTEMS AND THE A_y PUZZLE*

T. B. CLEGG

Department of Physics & Astronomy, University of North Carolina, Chapel Hill, NC 27599-3255 , USA, and Triangle Universities Nuclear Laboratory, Durham, NC 27708-0308, USA

In recent years, a major theoretical effort has sought, starting with modern NN potentials (e.g Nijmegen, CD Bonn, AV-18) and including a 3N interaction (e.g. Urbana IX), to calculate scattering observables in 3N and 4N systems. Recently it has been shown that calculations in coordinate space and. momentum space agree to within 1% for observables in p+d scattering.[1] Theoretical attention is turning now toward similar calculations in 4N systems. Four-N nuclei are a fertile 'theoretical laboratory' because these systems are the lightest with resonant states and thresholds, and the simplest where amplitudes of isospin T=3/2 can be studied.

In both 3N and 4N systems, excellent agreement is seen between theoretical calculations and some experimental observables, most notably the cross section and tensor analyzing powers in elastic scattering.[2,3] However, significant differences exist in this same system at low bombarding energies for the vector analyzing power, A_y. These differences are not removed, and can even be enhanced, by the addition of a 3N interaction in the theoretical calculations. At the scattering angle where the analyzing power peaks, the fractional differences between experiment and theory $\frac{(expt-theory)}{theory}$ are $\sim$ 20% for p+d and n+d, and $\sim$ 40% for p+^{3}He. New, highly accurate data from TUNL between 1.6 and 5.5 MeV for both[3] p+^{3}He and n+^{3}He have only confirmed differences first seen nearly 20 years ago. The origin of the differences has been traced[4] to incorrect splitting in the theoretical l=1 phase shifts. These differences disappear for energies above 50 MeV.

**Work supported in part by the US Department of Energy under Grant # DE-FG02-97ER41041

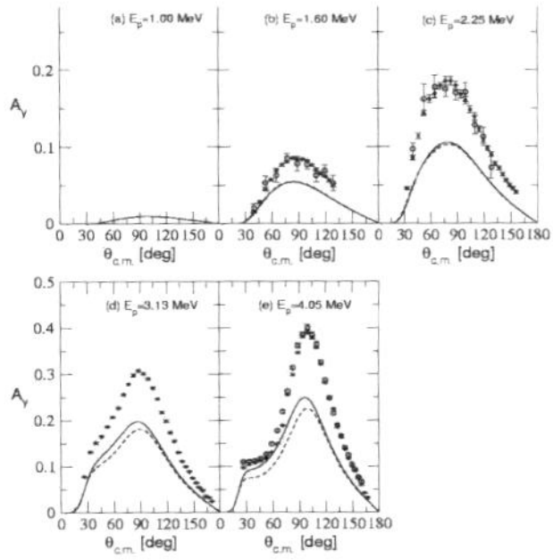

Fig. 1. Analyzing power data for p+^{3}He scattering from Ref[1] compared with calculations made with (dotted line) and without (solid line) a 3N interaction

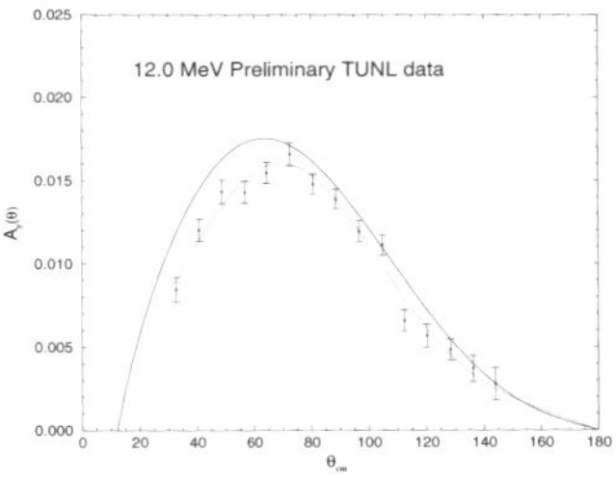

Fig. 2. Preliminary TUNL measurements[5] of A_y for n+p scattering at 12 MeV. The dotted curve is a polynomial fit to data, while the solid curve shows the prediction using the Nijmegen potential.

One possible explanation for this 'A_y puzzle' is that the input NN potentials used in all the theoretical calculations may be poorly defined at these low scattering energies due to lack of A_y data. Another possibility is that the correct 3N interaction may not yet have been found. Or is something being left out in the calculations? No agreement exists about the origin of the problem.

1. New Experimental Measurements Underway at TUNL

n+p Scattering–Recent very careful A_y measurements at 12 MeV by G. Weisel *et al.* show significant differences from values predicted by the Nijmegen potential, causing some concern about the validity of that accepted potential model standard at these low energies.

n+d Scattering–Other recent measurements by G. Weisel *et al.* of A_y at E_n = 19 and 22.5 MeV to map out the energy dependence of the disappearance of the A_y puzzle.

n+^{3}He Scattering– New A_y measurements underway by J. Esterline *et al.* between E_n = 1.6 and 5.5 MeV will be compared with theoretical calculations mentioned above and with similar data3 recently measured for p+^{3}He scattering.

p+^{3}He Scattering– Daniels *et al.* are utilizing a polarized ^{3}He target for measuring the target analyzing power A_{0y} and the spin correlation coefficients A_{xx} and A_{yy} in p+^{3}He scattering at energies between 2 and 5.5 MeV. These data will be used in a new energy-dependent phase shift analysis similar to that of George *et al*[6] to try to determine p+^{3}He phase shifts uniquely in this low energy range.

References

1. A. Deltuva, *et al.,* Phys. Rev. **C71**, 064003 (2005).
2. C. R. Brune *et al.*, Phys Rev **C63,** 44013 (2001).
3. B. M. Fisher *et al.*, Phys. Rev. **C74,** 034001 (2006).
4. M. H. Wood, *et al.*, Phys. Rev. C65, 034002 (2002).
5. M. D. Barker *et al.*, Phys. Rev. Lett. **48,** 918 (1982).
6. E. A. George and L. D. Knutson, Phys Rev. **C67**, 027001 (2003).

RECENT PROGRESS IN NUCLEAR LATTICE SIMULATIONS WITH EFFECTIVE FIELD THEORY

D. LEE

Department of Physics, North Carolina State University,
Raleigh, NC 27695, USA
E-mail: dean_lee@ncsu.edu

This proceedings article summarizes recent work presented at Chiral Dynamics 2006 on nuclear lattice simulations with chiral effective field theory for light nuclei. This work has been done in collaboration with Buḡra Borasoy, Evgeny Epelbaum, Hermann Krebs, and Ulf-G. Meißner.

Keywords: nuclear lattice simulations, effective field theory, light nuclei

1. Introduction

One unique feature of the lattice effective field theory approach is the ability to study in one formalism both few- and many-body systems as well as zero- and nonzero-temperature phenomena. A large portion of the nuclear phase diagram can be studied using exactly the same lattice action with exactly the same operator coefficients. Another feature is the computational advantage of many efficient Euclidean lattice methods developed for lattice QCD and condensed matter applications. This includes the use of Markov Chain Monte Carlo techniques, Hubbard-Stratonovich transformations,[1,2] and non-local updating schemes such as a hybrid Monte Carlo.[3] A third feature is the close theoretical link between nuclear lattice simulations and chiral effective field theory. One can write down the lattice Feynman rules and calculate lattice Feynman diagrams using precisely the same action used in the non-perturbative simulation. Since the lattice formalism is based on chiral effective field theory, we have a systematic power-counting expansion, an a priori estimate of errors for low-energy scattering, and a clear theoretical connection to the underlying symmetries of QCD.

The first study combining lattice methods with effective field theory for low-energy nuclear physics looked at infinite nuclear and neutron matter at nonzero density and temperature.[4] Recently nuclear lattice simulations were used to study the triton at leading order in pionless effective field theory with three-nucleon interactions.[5] The approach we summarize here is based on chiral effective field theory at leading order. This lattice formalism was also used to study neutron matter at nonzero temperature.[6]

In the work reported at Chiral Dynamics 2006 we considered the physics of instantaneous one-pion exchange and the leading-order S-wave contact interactions for light nuclei. This was implemented on the lattice using pions and auxiliary fields. At a lattice spacing of $(100\ \text{MeV})^{-1}$ we determined the two contact interaction coefficients using the deuteron binding energy and the $^1S_0$ scattering length using Lüscher's finite volume formula.[7–9]

In our first attempt we found large cutoff errors associated with the leading-order contact interactions. We therefore considered an improved action which included higher-derivative interactions to reduce cutoff effects. This was implemented as a one-parameter Gaussian broadening of the contact interactions. The extra parameter was determined by the average of the effective ranges for the $^1S_0$ and $^3S_1$ phase shifts.

We calculated binding energies, radii, and density correlations for the deuteron, triton, and helium-4. We also probed the computational scaling in systems with up to eight nucleons. We found that for light nuclei simulations, complex phase oscillations were suppressed by approximate $SU(4)$ Wigner symmetry. Also we found that for fixed volume and fixed number of Monte Carlo configurations the scaling in CPU time was almost linear in the number of nucleons.

References

1. J. Hubbard, *Phys. Rev. Lett.* **3**, 77 (1959).
2. R. L. Stratonovich, *Soviet Phys. Doklady* **2**, 416 (1958).
3. S. Duane, A. D. Kennedy, B. J. Pendleton and D. Roweth, *Phys. Lett.* **B195**, 216 (1987).
4. H. M. Müller, S. E. Koonin, R. Seki and U. van Kolck, *Phys. Rev.* **C61**, p. 044320 (2000).
5. B. Borasoy, H. Krebs, D. Lee and U.-G. Meißner, *Nucl. Phys.* **A768**, 179 (2006).
6. D. Lee, B. Borasoy and T. Schaefer, *Phys. Rev.* **C70**, p. 014007 (2004).
7. M. Lüscher, *Commun. Math. Phys.* **105**, 153 (1986).
8. S. R. Beane, P. F. Bedaque, A. Parreno and M. J. Savage, *Phys. Lett.* **B585**, 106 (2004).
9. R. Seki and U. van Kolck, *Phys. Rev.* **C73**, p. 044006 (2006).

FEW-BODY STUDIES AT KVI

J. G. MESSCHENDORP

KVI, University of Groningen,
Zernikelaan 25, 9747 AA, Groningen,
The Netherlands

The KVI facility allows a detailed study of few-nucleon interactions below the pion-production threshold exploiting polarized proton and deuteron beams up to energies of 190 MeV. In this contribution, some recent results are discussed and interpreted exploiting rigorous Faddeev calculations and predictions derived from chiral-perturbation theory.

Keywords: Three-Nucleon Forces; Proton-Deuteron Scattering

One of the experimental programs at KVI focuses on obtaining high-precision data in the few-nucleon scattering processes below the pion-production threshold. The goal is to study the details of the nucleon-nucleon and three-nucleon interactions through a comparison with predictions from state-of-the-art effective nucleon-nucleon potentials and models based on a chiral-symmetry expansion. For this purpose, cross sections and analyzing powers are measured in few-nucleon scattering processes. The focus of the few-body program at KVI is mainly oriented towards understanding the three-nucleon system by exploiting $p + d$ and $d + p$ reactions with polarized proton and deuteron beams. Different final states can be observed which include the elastic, break-up, and radiative capture reactions.

The simplest channel is the elastic $p + d$ scattering. A systematic study of this reaction has been carried out in which high-precision cross sections and analyzing powers have been determined for several bombarding energies up to 190 MeV using the Big-Bite Spectrometer (BBS). The data show that effective models based on a well-understood two-nucleon interaction are not sufficient to describe the three-nucleon pd system.[1,2] Including the most modern type of three-nucleon potentials in the calculations resolves only a part of the discrepancies. Recently, an experiment has been carried out with the BBS to determine spin-transfer coefficients in the elastic $\mathbf{d} + p \rightarrow \mathbf{p} + d$ reaction.[3] Strikingly, the predictions based on the most precise nucleon-nucleon and three-nucleon potentials do not give a consistent picture with respect to the measurements. Similar deficiencies have been observed in preliminary results from a high precision measurement of vector and tensor analyzing powers in deuteron-proton elastic scattering.[4] These data were obtained at RIKEN, where the $^{12}C(\mathbf{d}, \alpha)^{10}B(2^+)$ reaction at $0°$ was exploited for an unambiguous determination of the incident beam polarization.

Pioneering measurements of the $d+p$ break-up reaction at E_d=130 MeV using the SALAD detector are being analyzed by collaborators from Jagellonian University in Cracow and the University of Silesia in Katowice. These studies give access to differential cross sections and tensor-analyzing powers for protons scattered towards forward angles. Various kinematical regions have meanwhile been published[5] and compared to rigorous Faddeev calculations and prediction from chiral-perturbation theory. In general, three-nucleon effects have been observed in these observables as well. Recently, a new 4π hadron detector, carrying the name "Big Instrument for Nuclear-polarization Analysis" (BINA), has been commissioned at KVI. This apparatus allows us to study the break-up reaction for an almost 4π coverage in phase space. Preliminary results have meanwhile been obtained in proton-deuteron scattering using a polarized proton beam of 190 MeV. Data at various other energies have been collected as well. In the near future, BINA will also be employed to study the four-nucleon system via deuteron-deuteron elastic and inelastic scattering processes.

The experimental studies of scattering observables in the three-nucleon system have gained a significant amount of popularity in the last decade. For a large part, this is due to enormous theoretical developments in this field of physics, which is now close to providing an unambiguous data analysis. In addition, there is still a large need for more experimental data to refine our present knowledge of nuclear forces. Several laboratories, including KVI, are presently providing precision data in the three-nucleon scattering processes. Unfortunately, the number of contributing institutes is decreasing in time, which could endanger the progress in this exciting field of physics for the near future.

References

1. K. Ermisch et al., Phys. Rev. Lett. **86**, 5862 (2001).
2. K. Ermisch et al., Phys. Rev. C **68**, 051001 (2003).
3. H. Amir-Ahmadi et al., submitted to Phys. Rev. Lett. (2006).
4. H. Mardanpour et al., submitted to Eur. Phys. J. A (2006).
5. St. Kistryn et al., Phys. Rev. C **68**, 054004 (2003).

Results of Three Nucleon Experiments from RIKEN

KIMIKO SEKIGUCHI*
RIKEN
Wako, Saitama 351-0198, JAPAN
**E-mail: kimiko@ribf.riken.jp*

The measurements of deuteron–proton scattering with polarized deuteron beams at RIKEN are presented. The data at 70 MeV/A are compared with the Faddeev calculations based on the chiral effective field theory potential.

Keywords: ^{1}H($\mathbf{d}$, d)^{1}H at 70–135 MeV/A, ^{1}H($\mathbf{d}$, $\mathbf{p}$)^{2}H at 135 MeV/A, ^{1}H($\mathbf{d}$, $\mathbf{p}p$)n at 135 MeV/A, iT_{11}, T_{20}, T_{21}, T_{22}, $K_{ij}^{y'}$

For the last decade we have intensively performed precise measurements of deuteron(d)–proton(p) scattering at intermediate energies ($E \sim$ 100 MeV/nucleon) with polarized deuteron beams at RIKEN to study the three nucleon force (3NF) properties.[1] Precise cross section data for elastic dp scattering at 135 MeV/A have shown large disagreement between data and rigorous Faddeev calculations with the meson theoretical approach to nucleon–nucleon (NN) forces.[2] Combination of the NN forces and the 2π-exchange 3NF removes this discrepancy, showing the necessity of the 3NF. However insufficient descriptions have been found in the spin observables. The chiral effective field approach[3] is providing new 3NF contributions that will help to explain those spin observables.

All experiments were performed using the magnetic spectrograph SMART system. The polarized deuteron beams accelerated by the AVF and Ring cyclotrons bombarded a liquid hydrogen or polyethylene target. Either the scattered deuteron or the recoil proton was momentum analyzed by the SMART depending on the scattering angle and detected at the focal plane. For the polarization transfer measurement, the polarizations of the scattered protons were measured with the focal-plane polarimeter. The observables we measured are (1) the cross sections and all deuteron analyzing powers (iT_{11},T_{20},T_{21},T_{22}) for dp elastic scattering at 70, 100, 135 MeV/A,

(2) d to p polarization transfer coefficients for dp elastic scattering at 135 MeV/A, and (3) all deuteron analyzing powers and polarization transfer coefficients for coplanar configuration of $^1\mathrm{H}(\mathbf{d},\mathbf{p}p)n$ at 135 MeV/A. Thus we have performed the experiments mainly at 135 MeV/A. Note the experiments on dp elastic and breakup channels have also been performed at 135 MeV/A at IUCF and KVI. A lot of data are now available at 135 MeV/A, providing testing grounds for the theory.

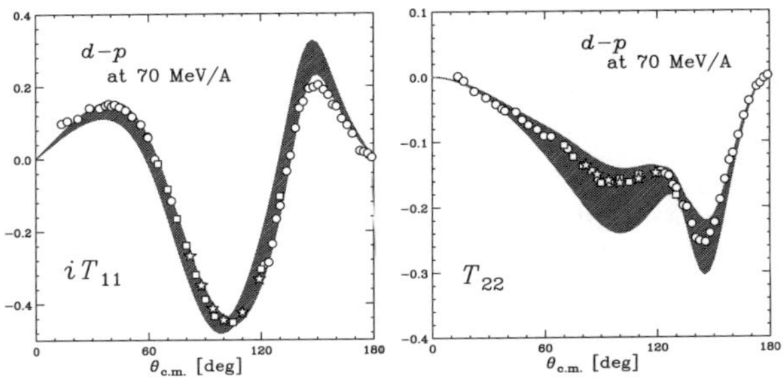

Fig. 1. Analyzing powers for elastic dp scattering at 70 MeV/A.

At present the predictions of the chiral effective approach are only available below 100 MeV/A. Parts of the data for dp scattering at 70 MeV/A are compared with the Faddeev calculations based on the $\mathrm{N^2LO}$ chiral potential[4] in Fig. 1. The data are well described by the calculations. Three nucleon force effects are considered to be enhanced relative to the NN force contributions at higher energies, and thus it would be interesting to see how this theoretical framework, which shows the hierarchy of many nucleon forces in a consistent way, works at 135 MeV/A.

The experiments were performed under the collaboration with RIKEN, University of Tokyo, RCNP, CYRIC, Tokyo Institute of Technology, Kyushu University, Saitama University, and KVI.

References

1. N. Sakamoto *et al.*, Phys. Lett. B **367**, 60 (1996), H. Sakai *et al.*, Phys. Rev. Lett. **84**, 5288 (2000), K. Sekiguchi *et al.*, Phys. Rev. C **65**, 034003 (2002), *ibid* **70**, 014001 (2004), Phys. Rev. Lett **95**, 162301 (2005).
2. R. Machleidt, Phys. Rev. C **63**, 024001 (2001), R. B. Wiringa *et al.*, *ibid* **51**, 38 (1995), V. G. J. Stoks *et al.*, *ibid* **49**, 2950 (1994).
3. For example, E. Epelbaum, Prog. Part. Nucl. Phys. **57**, 654 (2006).
4. E. Epelbaum, private communications.

A New Opportunity to Measure the Total Photoabsorption Cross Section of Helium**

P. T. DEBEVEC

Department of Physics, University of Illinois at Urbana-Champaign, 1110 W. Green St., Urbana, IL 61801

The total photoabsorption cross section for photon energies below the pion threshold gives a global view of the response of a nucleus to a certain aspect of the electromagnetic probe, since the interaction is dominated by the electric dipole operator. Thus at low energies the characteristic feature of the nuclear response is the strongly energy dependent response of the giant dipole resonance, and at higher energies the characteristic feature is a largely energy independent response of the quasi-deuteron process. These features are well-established for nuclei from lithium to the actinides. Missing from this list are the He isotopes for which the experimental situation is highly uncertain.

1. The Mainz experiments

The total photoabsorption cross sections on light nuclei from lithium to calcium were measured at Mainz in the 70's by the absorption methodc.[1] The Mainz accelerator, a low-duty factor linac, was used to produce a bremsstrahlung beam. The relative flux of the collimated bremsstrahlung beam as a function of energy was determined by counting the forward angle Compton-scattered electrons produced by the bremsstrahlung radiator in a magnetic spectrometer with a scintillator hodoscope in its focal plane. The bremsstrahlung beam then passed through a long, cylindrical sample in which atomic and nuclear interactions removed photons from the beam. After the sample the relative flux of the beam, now depleted by the absorption in the sample, was again measured by an identical forward-angle Compton electron spectrometer. The ratio of the transmitted flux to the incident flux uniquely determines the product of the total photoabsorption cross section and the areal density of the sample. The strength of this method is that only the absolute efficiency of the two flux measurements

** This work supported in part by the U. S. National Science Foundation.

need not be determined. Only the stability of the two relative flux measurements is required. The weakness of this method is that the nuclear absorption is only a small component of the total photon absorption cross section, the atomic photon absorption processes dominating. The ratio of the two relative cross sections then must be determined with high statistical precision so that the subtraction of the calculated atomic cross section from the measured total absorption cross section yields a nuclear absorption cross section with the desired precision. Taking into account both statistical and systematic errors, the precision of the Mainz measurements varied from approximately 2% for the giant dipole region of the lightest nuclei in their survey to approximately 10% near pion production threshold for the heaviest nuclei in their survey. Mainz did not extend their survey beyond Ca, due to the increase in the atomic absorption cross section with increasing atomic number. Mainz also did not measure the cross section on He, due to the smallness of the absorption cross. The Mainz setup could only accommodate a maximum target length of two meters, which, for liquid He, was very far from the optimal target length.[2]

2. Status of ^{4}He photodisintegration data

It should be practical to determine the total nuclear photoabsorption cross section of ^{4}He by summing the five channels. The Q-values of three two-body channels, ^{3}H + p, ^{3}He + n, d + d, are -19.9 MeV, -20.6 MeV and -23.8 MeV, respectively. The contribution from the d + d channel is negligible, since the electric dipole transition is highly forbidden. (The cross sections for the two-body channels can be measured in both photodisintegration and capture reactions.) The Q-values of the three-body channel, d + p + n, and the four-body channel, p + n + p + n, are -26.1 MeV and -28.3 MeV, respectively. These channels contribute significantly only above approximately 60 MeV. The inconsistent state of the experimental data, especially for photon energies between 20 and 30 MeV, has been commented on in both recent and not so recent theoretical and experimental works. It seems hardly useful to give yet another rendition of this history.[3]

At a photon energy at which the real part of the photon scattering amplitude vanishes, the nuclear photoabsorption cross section can be obtained model independently from the photon scattering section. This energy is approximately 25 MeV, and from their photon scattering data, Wells et al. obtain a 4% determination of the photoabsorption cross sectionc.[4]

3. Theoretical interest

Calculational techniques for nuclear few-body systems are highly developed. With these techniques exact wave functions for bound and continuum states can be obtained for Hamiltonians that have either phenomenological or theoretical two-body and three-body forces. A wide range of observables have been calculated with these techniques, and with one particular technique, the Lorentz integral transform, the total photoabsorption cross section can be calculated. Calculations with this technique have been done for both the helium and lithium isotopes. For ^{4}He these calculations[5] show a peak in the total cross section at approximately 26 MeV, a giant dipole resonance. Interesting in these calculations is the effect of the three-body force, which above 50 MeV has effects as large as 35%, and at the peak of the cross section effects of the order of 5%. Experiment is unable to confront these calculations due to the inconsistent data set.

4. A new opportunity

The HIGS facility at the Duke FEL storage ring offers a new opportunity to measure the ^{4}He photoabsorption cross section by the attenuation method. The HIGS facility can produce a photon intensity of more than 10^5 Hz with an energy definition of the order of 1% with energies of up to approximately 100 MeV.[6] Depending on the particular optical klystron, the photons are either 100% linear or circularly polarized. (The photon polarization is unimportant in the attenuation method. Much of the proposed program of the HIGS facility, however, exploits the photon polarization.) In its initial low-energy operation this facility has made highly precise photon asymmetry measurements for deuteron photodisintegration,[7] photon scattering measurements on ^{16}O, and photodisintegration measurements on ^{3}He. HIGS produces its photons in a free electron laser. The photon beam is intrinsically highly collimated due to the properties of the free electron laser. The photon beam is transported in air to a shield area in which nuclear physics experiments are done. There is ample space for a long target in the region between the free electron laser and the target room. The target room itself can accommodate a target of up to five meters in length. Thus the HIGS facility then represents a unique opportunity to measure the ^{4}He photoabsorption cross section by the attenuation method.

References

1. J. Ahrens, *et al.*, Nucl. Phys. **A251,** 479 (1975).

2. J. Ahrens, private communication.
3. See, for example, T. Shima, *et al.*, Phys. Rev. **C72**, 044004 (2005), for recent results and a comparison to previous results.
4. D. P. Wells, *et al.*, Phys. Rev. **C46**, 449 (1992).
5. Doron Gazit, *et al.*, Phys. Rev. Lett. **96**, 112301 (2006).
6. Henry R. Weller and Mohammad W. Ahmed, Mod. Phys. Lett. **A18**, 1569 (2003).
7. E. C. Schreiber, *et al.*, Phys. Rev. **C61**, 061604(R) (2000).

THREE-BODY PHOTODISINTEGRATION OF ^{3}He WITH DOUBLE POLARIZATIONS

X. ZONG

Triangle University Nuclear Laboratory and
Physics Department, Duke University,
Durham, NC 27708, U.S.A
** E-mail: xz12@phy.duke.edu*

The paper describes a planned first measurement of a spin-dependent asymmetry from three-body photodisintegration of ^{3}He. We propose to use a linearly polarized photon beam at an incident photon energy of 12.8 MeV on a high-pressure ^{3}He gas target which is transversely polarized to the reaction plane. The 3-body photodisintegration process will be studied by detecting the neutron from the 3-body breakup channel. Such a double-polarizations asymmetry has been predicted to be sensitive to 3-body force.

1. Introduction

The study of the three-nucleon system is of fundamental importance to nuclear physics[1] since knowledge of cross section can shed light upon the nuclear ground-state structure and thus allows a deeper understanding of the underlying three-body froce. With the availability of polarized beams and polarized targets,[2] it becomes possible to investigate the additional spin-dependent quantities.[3] This has spurred our experiment which will measure the spin-dependent asymmetry from 3-body photodisintegration of ^{3}He.

2. Polarized ^{3}He target

A new polarized ^{3}He target is being developed for the High Intensity Gamma Source (HIγS) at Duke FEL. The cell usually contains 7-9 amagats of ^{3}He gas, and is separated into three parts, pumping chamber where the gas is polarized, target chamber where the gas interacts with the photon beam and transfer tube which connects these two. The technique we use to polarize ^{3}He is the spin exchange optical pumping (SEOP), i.e. the gas is polarized by using spin-exchange with polarized rubidium vapor. Rubidium atoms can be polarized by optical pumping using circularly polarized photons. The ^{3}He nuclei is polarized through a hyperfine interaction with the rubidium electrons.

3. The Experiment

We plan to carry out measurement of spin-dependent asymmetry from three-body photodisintegration of ^{3}He employing the above target and a linearly polarized photon beam with energy of 12.8 MeV. The ^{3}He nuclear spin will be aligned transverse to the incident photon momentum direction, perpendicular to the reaction plane. The three-body photodisintegration process will be studied by detecting the neutron from the three-body breakup channel and right-left asymmetry will be formed between detectors placed symmetrically with respect to the incident photon beam.

4. Experimental Projection

In projecting the asymmetry measurement, the assumptions we have made are as follows: a ^{3}He target thickness of 8×10^{21} cm^{-2}(40 cm long) with a target polarization of 40%, a 100% linearly polarized photon beam at an intensity of 5×10^7/second. We plan to use the existing neutron counters in TUNL. Each neutron counter is 5.3 inches in diameter and the distance between the neutron counter and the center of the target is 15.75 inches. The average neutron counter efficiency is 35%. The total beam time assumed in the projection is 100 hours. The statistical error is around 0.04[3].[4] The experiment is scheduled to be running in Duke FEL in spring, 2007.

Acknowledgement

This work is supported by the U.S. Department of Energy under contract number DE-FG02-03ER41231 and the School of Arts and Science at Duke University.

References

1. W. Glöckle *et al.*, Phys. Rep. **274**, 107 (1996).
2. B. Blankleider *et al.*, Phys. Rev. C **29**, 538 (1984).
3. H.Witala, private communication.
4. A.Deltuva *et al.*, Phys. Rev. C **71**, 054005, and private communication.

LARGE TWO-PION-EXCHANGE CONTRIBUTIONS TO THE $pp \to pp\pi^0$ REACTION

F. MYHRER

Department of Physics and Astronomy, University of South Carolina, Columbia, SC 29208, U.S.A. E-mail: myhrer@physics.sc.edu

The large two-pion exchange amplitudes are calculated in HBχPT and their net contribution to the reaction cross section is large.

Refs.[1,2] evaluated this reaction at threshold in HBχPT. They found that the impulse approximation (I.A.) and one-pion-exchange (Resc) diagrams interfere destructively resulting in a very small reaction cross section. According to Lee and Riska[3] a model explanation of the measured reaction cross section near threshold requires the contributions from heavy meson (σ and ω) exchange in addition to the one-pion-exchange. The σ-meson-exchange is more properly described by correlated two-pion-exchange. This knowledge prompted a HBχPT study of two-pion-exchange (TPE) contributions to the $pp \to pp\pi^0$ reaction amplitude.[4]

In Weinberg chiral counting the TPE contributions are of higher chiral order in HBχPT. However, it was shown by Ref.[4] that some TPE amplitudes are as large or larger than the lower chiral order Resc contribution. At threshold the typical momentum is $p \sim \sqrt{m_\pi m_N}$. This large momentum prompted Cohen *et al.*[2] to propose a momentum counting rule, reviewed in Ref.[5] According to this counting, one finds that the Resc diagram is higher order in p/Λ compared to some "dominant" TPE diagrams, and this counting agrees with the numerical evaluations of the TPE diagrams of Ref.[4] One drawback with the momentum counting is that the sum of the diagrams in each "momentum order" no longer is independent of the definition of the pion field. Hanhart and Kaiser (HK)[6] used momentum counting to evaluate the "leading" momentum behavior of the dominant TPE diagrams. HK also found that diagram II in Ref.[4] should have opposite sign (which we confirmed), and they found that the sum of the leading momentum behavior of the three dominant TPE diagrams cancel. We will present results from Ref.[4] and a recent calculations[7] which show that the "sub-leading" parts of a dominant TPE diagram gives a contribution comparable to the Resc amplitude.

The TPE transition operators (TO) were evaluated analytically by Ref.[4] in HBχPT. When these operators are sandwiched between phenomenological determined distorted NN wave functions, the momentum integrals convergence slowly.[7] This slow convergence can be understood when we adopt the threshold fixed kinematics approximation (FKA). Imposing FKA on the analytic expressions for the TO given in Ref.,[4] we make an asymptotic expansion in the two-nucleon momentum transfer ($|\mathbf{k}| = |\mathbf{p} - \mathbf{p}\,'| \to \infty$). The TO matrix T of the TPE diagrams is of the form

$$T = \left(\frac{g_A}{f_\pi}\right) (\mathbf{\Sigma} \cdot \mathbf{k})\, t(p, p', x)$$

where $x = \hat{p} \cdot \hat{p}'$. The *asymptotic* momentum behavior for $t(p,p',x)$ is $t(p,p',x) \sim t_1\ |\mathbf{k}| + t_2\ \ln[\Lambda^2/|\mathbf{k}|^2] + t_3 + \delta t(p,p',x)$, where $\delta t(p,p',x)$ is $\mathcal{O}(|\mathbf{k}|^{-1})$, and the amplitudes t_i, $i = 1,2,3$ are known analytic expressions for each diagram. The t_1 amplitude is the dominant TPE amplitude of HK.[6]

Amplitude K =	I	II	III	IV	V	VI	VII
R_K	$-.70$	-6.70	-6.70	9.50	0.18	0.14	2.65
$t_1 \quad \propto$	$-$	-2	-1	$+3$	$-$	$-$	$-$

In the table the row marked R_K gives the values of the ratio of TO to the Resc amplitude in the plane wave approximation for the seven amplitudes of Ref.[4] As indicated in the last row of the table, marked t_1, the leading momentum terms of the TO from diagrams II, III and IV sum to zero, confirming HK's result.[6] The non-cancellation of the dominant amplitudes can however be inferred from the R_K row since the ratios II:III:IV are not -2:-1:3 but roughly -2:-2:3. When we remove t_1, we find that the sum of the two-pion-exchange amplitudes is larger than the Resc amplitude.[7]

References

1. B.-Y. Park *et al.*, Phys. Rev. **C 53**, 1519 (1996)
2. T.D. Cohen *et al.*, Phys. Rev. **C 53**, 2661 (1996)
3. T.-S.H. Lee and D.O. Riska, Phys. Rev. Lett. **70**, 2237 (1993)
4. V. Dmitrašinović *et al.*, Phys. Lett. **B 465**, 43 (1999).
5. C. Hanhart, Phys. Rep. **397**, 155 (2004)
6. C. Hanhart and N. Kaiser, Phys. Rev. **C 66**, 054005 (2002).
7. Y. Kim, K. Kubodera, F. Myhrer and T. Sato, in preparation.

TOWARDS A SYSTEMATIC THEORY OF NUCLEAR FORCES

E. EPELBAUM*

IKP (Theorie), Forschungszentrum Jülich, D-52425 Jülich, Germany
and
HISKP (Theorie), Universität Bonn, D-53115 Bonn, Germany

*E-mail: e.epelbaum@fz-juelich.de

I review recent progress in understanding the structure of the nuclear forces based on chiral effective field theory.

Keywords: Nuclear forces, effective field theory, chiral perturbation theory

Understanding nuclear reactions and various properties of nuclei requires a detailed knowledge of the interactions between the nucleons. This is a very old but still actual problem in nuclear physics. An important breakthrough occurred in the early 1990s with Weinberg's proposal to derive nuclear forces using chiral perturbation theory.[1] This field-theory-based approach is systematic, offers a well-defined expansion parameter corresponding to the low external momenta of the nucleons and is linked to QCD via chiral symmetry. The major difficulty that needs to be overcome in the few-nucleon sector and is absent in the Goldstone boson and single-nucleon sectors is the inapplicability of perturbation theory at the S-matrix level even in the low-energies region. This is obvious from the existence of shallow bound states. The breakdown of perturbation theory can be traced back to the enhancement of the iterative contributions to the amplitude arising from energy denominators corresponding to purely nucleonic intermediate states.[1] A natural way out of this difficulty proposed by Weinberg is to apply chiral effective field theory (EFT) to derive the kernel of the corresponding dynamical equation, which can be viewed as an effective nuclear potential. The scattering amplitude is then obtained as solution of the dynamical equation.

A lot of progress has been made towards the practical realization of Weinberg's program in the past one-and-a-half decades. Nuclear forces have been worked out at lowest orders in the chiral expansion using old-fashioned time-ordered perturbation theory,[2,3] a Feynman-diagram-based approach[4] and the method of unitary transformation;[5] See[6] for a recent review. The most advanced studies[7,8] of the two-nucleon (2N) system at next-to-next-to-next-to-leading order (N^3LO) in the chiral expansion demonstrate the ability of the novel chiral forces to provide an accurate description of the low-energy data, comparable to the one of the modern semi-phenomenological potentials. At this order in the chiral expansion, the 2N force receives contributions from one- two- and three-pion exchange and from the corresponding contact interactions. While the two-pion exchange contributions turn out to be very important,[4] the leading three-pion exchange potential is negligible.[9] Isospin-breaking and relativistic corrections have also been taken into account; see[6] and references therein for more details.

Promising results have also been obtained for systems with three and more nucleons which have so far been studied up to next-to-next-to-leading order (N^2LO), see e.g.[10,11] At this order, one needs to account for the chiral three-nucleon (3N) force which depends on two unknown low-energy constants. These two parameters can be fixed from few-nucleon observables like e.g. the ^{3}H, ^{4}He binding energies[11] or the doublet neutron–deuteron scattering length.[10] While most of the N^2LO results for 3N scattering observables and the spectra of light nuclei are in a reasonable agreement with the data, certain problems such as e.g. the A_y-puzzle remain. It is, therefore, important to extend these studies to N^3LO. A first step in this direction has recently been done in,[12] where the corresponding four-nucleon force has been worked out. The derivation of the 3N force at N^3LO is in progress.

References

1. S. Weinberg, Phys. Lett. **B251**, 288 (1990); Nucl. Phys. **B363**, 3 (1991).
2. C. Ordóñez, L. Ray, and U. van Kolck, Phys. Rev. **C53**, 2086 (1996).
3. U. van Kolck, Phys. Rev. **C49**, 2932 (1994).
4. N. Kaiser, R. Brockmann, and W. Weise, Nucl. Phys. **A625**, 758 (1997).
5. E. Epelbaum, W. Glöckle, and U.-G. Meißner, Nucl. Phys. **A637**, 107 (1998).
6. E. Epelbaum, Prog. Part. Nucl. Phys. **57**, 654 (2006).
7. D. R. Entem and R. Machleidt, Phys. Rev. **C68**, 041001 (2003).
8. E. Epelbaum, W. Glöckle, and U.-G. Meißner, Nucl. Phys. **A747**, 362 (2005).
9. N. Kaiser, Phys. Rev. **C61**, 014003 (2000); Phys. Rev. **C62**, 024001 (2000).
10. E. Epelbaum *et al.*, Phys. Rev. **C66**, 064001 (2002).
11. A. Nogga *et al.*, Phys. Rev. **C73**, 064002 (2006).
12. E. Epelbaum *et al.*, Phys. Lett. **B639**, 456 (2006).

AB INITIO CALCULATIONS OF ELECTROMAGNETIC REACTIONS IN LIGHT NUCLEI

W. LEIDEMANN*

Department of Physics, The George Washington University, Washington DC 20052, USA and INFN (GC di Trento), Italy
E-mail: leideman@science.unitn.it

The electromagnetic break-up of nuclei is calculated *ab initio in* the mass range from A=2–7. The nuclear interaction is described by conventional NN and 3N potential models. Full final state interaction is taken into account; for A>2 nuclei this is achieved via the Lorentz integral transform (LIT) method. It is pointed out that the wealth of polarization observables in A=2 electro-disintegration is an ideal testing ground for comparing conventional and chiral calculations. For A> 2 nuclei it is shown that the LIT method allows a proper treatment of the complicated many-body continuum state interaction.

Keywords: polarization observables; electromagnetic break-up; LIT.

Discussion

Deuteron electrodisintegration offers a rich spectrum of 324 linearly independent polarization observables. In a recent review[1] many different theoretical aspects are discussed and theoretical results of a non-relativistic potential theory calculation with meson exchange and isobar currents as well as relativistic contibutions are shown for some selected observables. A comparison of this theory with experiment is published elsewhere (for a list of references see Ref. 1). Generally quite good agreement between theory and experiment is found, but there is also an interesting exception with a rather large discrepancy.[2] The theoretical result was confirmed within a pionless chiral calculation (see contribution of H. Griesshammer), thus further experimental investigations will be necessary. Since an experimental determination of a large number of $d(e, e')$ polarization observables at various kinematical settings is out of question, it would be of great impor-

*on leave of absence from Dipartimento di Fisica, Università di Trento, I-38050 Povo (Trento), Italy tance to have more systematic comparisons between conventional and chiral calculations. Such a study could possibly point out crucial observables.

The LIT method[3] reduces a continuum state problem to a much easier to solve bound-state-like problem. Nonetheless the full continuum state interaction is taken into account rigorously. It has been applied to the following electromagnetic break-up calculations: (1) inclusive (e,e') and photoabsorption of ^{3}H/^{3}He with realistic nuclear forces;[4–6] (2) inclusive[7,8] and exclusive[9] photoabsorption and (e, e') of ^{4}He with semirealistic NN potentials; (3) ^{4}He total photoabsorption and photonuclear sum rules with realistic nuclear forces,[10] (4) total photoabsorption of ^{6}He, ^{6}Li and ^{7}Li with semirealistic NN potentials.[11,12] Here we mention particularly the following results: (i) inelastic longitudinal (e, e') form factor of ^{3}He at higher momentum transfer leading to a recommendation of a specific reference frame in order to minimize relativistic effects;[6] (ii) ^{6}He total photoabsorption,[11] the cross section exhibits two separate peaks (interpretation in a many-body picture: oscillations of surface neutrons relative to the α-core (soft mode) and of neutrons relative to protons (giant dipole resonance)); (iii) ^{4}He photoabsorption with a realistic nuclear force (AV18+UIX),[10] the 3N-force reduces the peak of the giant dipole resonance quite considerably, nonetheless the full result shows a pronounced peak confirming a calculation with a semirealistic NN interaction.[8]

References

1. H. Arenhövel, W. Leidemann and E. L. Tomusiak, *Eur. Phys. J. A* **23**, 147 (2005).
2. P. von Neumann-Cosel *et al.*, *Phys. Rev. Lett.* **88**, 202304 (2002).
3. V. D. Efros, W. Leidemann and G. Orlandini, *Phys. Lett. B* **338**, 130 (1994).
4. V. D. Efros, W. Leidemann, G. Orlandini and E. L. Tomusiak, *Phys. Lett. B* **484**, 223 (2000); *Phys. Rev. C* **69**, 044001 (2004).
5. J. Golak *et al.*, *Nucl. Phys. A* **707**, 365 (2002).
6. V. D. Efros, W. Leidemann, G. Orlandini and E. L. Tomusiak, *Phys. Rev. C* **72**, 011002(R) (2005).
7. V. D. Efros, W. Leidemann and G. Orlandini, *Phys. Rev. Lett.* **78**, 432 (1997).
8. V. D. Efros, W. Leidemann and G. Orlandini, *Phys. Rev. Lett.* **78**, 4015 (1997); N. Barnea *et al. Phys. Rev. C* **63**, 057002 (2001).
9. S. Quaglioni *et al.*, *Phys. Rev. C* **69**, 044002 (2004); *Phys. Rev. C* **72**, 064002 (2005); D. Andreasi *et al.*, *Eur. Phys. J. A* **27**, 47 (2006).
10. D. Gazit *et al.*, *Phys. Rev. Lett.* **96**, 112301 (2006); nucl-th/0610025; N. Barnea, W. Leidemann and G. Orlandini, *Phys. Rev. C* **74**, 034003 (2006).
11. S. Bacca *et al.*, *Phys. Rev. Lett.* **89**, 052502 (2002); *Phys. Rev. C* **69**, 057001 (2004).
12. S. Bacca *et al.*, *Phys. Lett. B* **603**, 159 (2004).

ELECTRON SCATTERING FROM A POLARIZED DEUTERIUM TARGET AT BLAST

R. FATEMI for the BLAST Collaboration
Massachusetts Institute of Technology
Cambridge, MA 02139, USA

The deuteron, the simplest nucleus, is an ideal arena for testing models of the nucleon-nucleon (NN) interaction and exploring the low-energy behavior of Quantum Chromodynamics (QCD). The Bates Large Acceptance Spectrometer (BLAST) Experiment[1] was designed to use both a tensor and vector polarized deuterium target[2] in combination with a longitudinally polarized electron beam for the simultaneous measurement of the vector (A_{ed}^V) and tensor (A_d^T) asymmetry. These asymmetries allow for the determination of the analyzing powers, T_{20} and T_{21}, and spin correlation coefficients, T_{10} and T_{11}, from which the elastic form factors can then be extracted.[3]

The average spin angle of the BLAST target for the 2004(2005) run was oriented at a $32^o(47^o)$ with respect to the beamline. This unique configuration allows for the spin asymmetries:

$$A_{ed}^V = \sqrt{\frac{3}{2}}(\cos\theta_d T_{10} - \sqrt{2}\sin\theta_d \cos\phi_d T_{11}) \tag{1}$$

$$A_d^T = \frac{\sqrt{3}(\cos^2\theta_d - 1)}{\sqrt{8}} T_{20} - \frac{\sqrt{3}\sin 2\theta_d \cos\phi_d}{2} T_{21} + \frac{\sqrt{3}\sin^2\theta_d \cos 2\phi_d}{2} T_{22} \tag{2}$$

to be measured simultaneously for the scenario where the Q^2 vector is parallel and perpendicular to the spin angle, therefore permitting the independent determination of T_{20}, T_{21}, T_{10} and T_{11} from the same dataset.

Parameterization III by Abbott *et al.*[4] of the world form factor data is used to subtract the small T_{22} contribution from the measured A_d^T and to determine the tensor polarization at the two lowest Q^2 bins. The BLAST analyzing powers T_{20} and T_{21}, plotted in Figs. 1 and 2,[5] are in good agreement with the existing world data while providing comparable or improved statistical precision between 2 and 4 fm^{-1}.

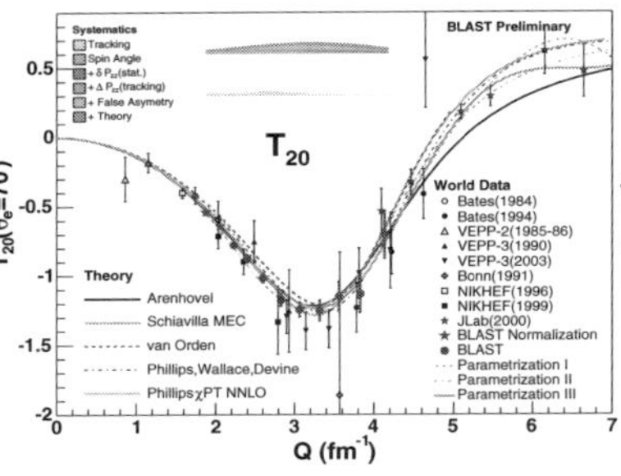

Fig. 1. Measured tensor analyzing power T_{20} compared to existing data and theoretical calculations

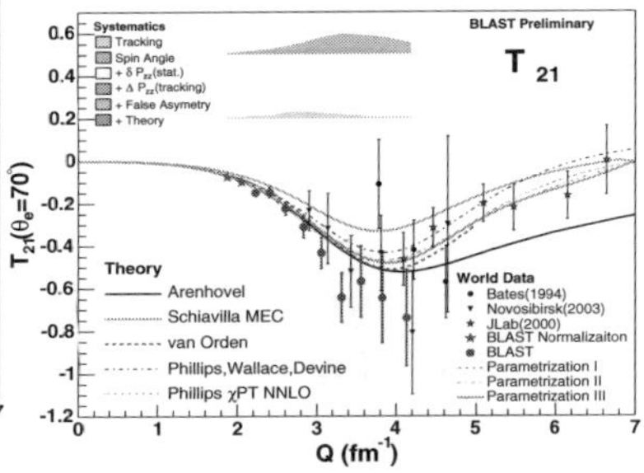

Fig. 2. Measured tensor analyzing power T_{21}.

Analogously, the measured vector asymmetries allow for the first time the extraction of the spin coefficients T_{11} and T_{10}, shown in Fig. 3.[6] While the statistical precision for the vector observables is lower than for T_{20}, the BLAST results are again in close agreement with calculations from the parameterizations of world data by Abbott[4] and predictions from theory.

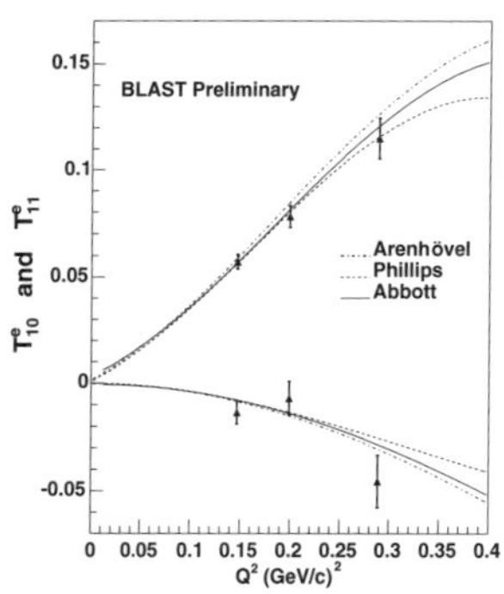

Fig. 3. Measured spin correlation coefficients T_{10} (neg) and T_{11} (pos)

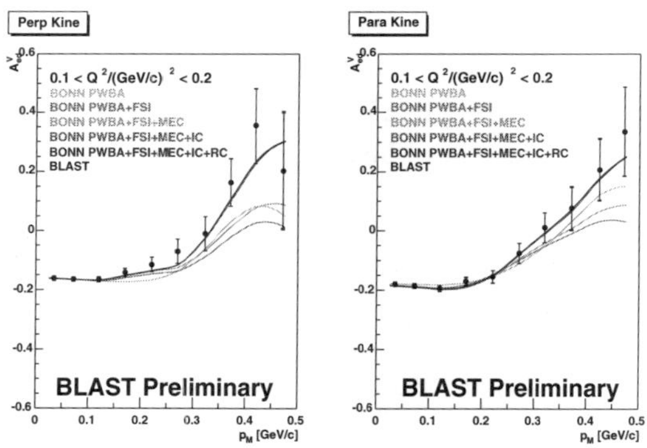

Fig. 4. Quasi-elastic proton scattering from a vector polarized deuterium target

Quasi-elastic proton scattering from vector polarized deuterium provides additional, but complementary, information about NN interactions, particularly at higher values of missing momentum (p_M). The vector asymmetry, $A^V_{ed}(p_M)$, is negative and flat if the deuteron orbital angular momentum is set to zero in the Plane Wave Impulse Approximation (PWIA). However, several effects, such as the D-state contribution, final state interactions (FSI), meson exchange currents (MEC), isobar contributions (IC), and relativistic effects (RC) can cause the asymmetry to change sign. Figure 4 shows the behavior of the BLAST A_{ed}^{V}[7] as a function of p_M for $Q^2 = 0.1 - 0.2\ (GeV/c)^2$ for both the perpendicular and parallel orientations. The agreement between data and predictions from the Plane Wave Born Approximation[8] (PWBA), a model which incorporates the NN exchange effects with the PWIA framework, is excellent and clearly shows the strong role these effects play for $p_M > 0.2\ GeV/c$. The dependence of the PWBA curves on various potentials, such as Born, Paris and V18, was tested and shown to be negligible within the statistical error of the data.

References

1. D. Hasell *et al.*, The BLAST Experiment, to be published.
2. D. Cheever et al., Nucl. Instr. Meth. **A556**, 410 (2006).
3. R. Redwine, see plenary proceedings from this workshop.
4. D. Abbott, Eur. Phys. J., **421-427** (2000)
5. C. Zhang, Ph.D. thesis, Massachusetts Institute of Technology (2006).
6. P. Karpius, Ph. D. thesis, University of New Hampshire (2005).
7. A. Maschinot, Ph.D. thesis, Massachusetts Institute of Technology (2005).
8. Arenhövel *et al.*, Phys.Rev. **C46** (1992).

NEUTRON–NEUTRON SCATTERING LENGTH FROM THE REACTION $\gamma d \to \pi^+ nn$

V. LENSKY*

Institut für Kernphysik (Theorie), Forschungszentrum Jülich,
Jülich, D-52425 Germany
**E-mail: v.lensky@fz-juelich.de*

We argue on how to extract the neutron–neutron scattering length from the reaction $\gamma d \to \pi^+ nn$. We provide a calculation of the differential cross-sections for this reaction within ChPT up to the order 5/2. We show that the error in the extraction of the scattering length induced by uncertainties of the NN wave functions is of the order of 0.1 fm.

Keywords: Pion photoproduction near threshold, neutron–neutron scattering length, chiral lagrangians

A precise determination of the neutron-neutron scattering length a_{nn} is important for an understanding of the effects of charge symmetry breaking in nucleon–nucleon forces. However a direct measurement of a_{nn} in a scattering experiment is practically impossible at the moment due to the absence of a free neutron target. For this reason, the commonly used value for a_{nn} is obtained as a result of analysis of reactions where there are three particles in the final state, *e.g.* $\pi^- d \to \gamma nn$, $nd \to pnn$. There is a large spread in the results for a_{nn} obtained by various groups. For instance, analyses of the reaction $nd \to pnn$ give significantly different values for a_{nn}, specifically $a_{nn} = -16.1 \pm 0.4$ fm[1] and $a_{nn} = -18.7 \pm 0.6$ fm,[2] whereas the value obtained from the reaction $\pi^- d \to \gamma nn$ is $a_{nn} = -18.5 \pm 0.3$ fm.[3] At the same time, for the proton-proton scattering length, which is directly measured, analysis gives $a_{pp} = -17.3 \pm 0.4$ fm[4] after correction for electromagnetic effects. This means that even the sign of $\Delta_a = a_{pp} - a_{nn}$ is not fixed.
Here we discuss a possibility to determine a_{nn} from differential cross-sections in the reaction $\gamma d \to \pi^+ nn$. Our calculation is based on the recent work of Ref.[5] in which the transition operator for the reaction $\gamma d \to \pi^+ nn$ was calculated up to order $\chi^{5/2}$ with $\chi = m_\pi / M_N$, where $m_\pi (M_N)$ is the pion (nucleon) mass. Half-integer powers of χ in the expansion arise from

three–nucleon cuts. Results of Ref.[5] showed a very good agreement with the experimental data for the total cross-section (the main theoretical error came from the uncertainty in the leading photoproduction multipole E_0^+). In order to estimate the theoretical uncertainty which arises from the NN wave functions, one has to use NN wave functions constructed within the same ChPT framework — this is what a consistent field theory calls for. In our present analysis we used NNLO NN wave functions constructed in Ref.[6]

We show that one can extract the value of a_{nn} fitting the shape of the FSI peak in the differential cross-section. In this case the (large) uncertainty due to E_0^+ is largely suppressed. At the same time there is a suppression of the quasi-free pion production at certain angles. We show that these angular configurations are preferred for extraction of a_{nn} since the suppression of quasi-free production leads to a smaller sensitivity to the NN wave function. Special care is taken to estimate the uncertainty that comes from the latter. We find as an estimate for this uncertainty $\delta a_{nn}^{\text{wave function}} \approx 0.1$ fm. Taking a conservative estimate for the uncertainty of the transition operator, one ends up with an estimate for the total uncertainty of order of $\delta a_{nn} \approx 0.15$ fm. At the same time, using different angular configurations for the analysis could potentially decrease the systematic uncertainties. We argue that the reaction $\gamma d \to \pi^+ nn$ (along with an alternative way, the reaction $\pi^- d \to \gamma nn$[7]) appears to be a good tool for the extraction of a_{nn}.

Acknowledgments

I thank V. Baru, E. Epelbaum, J. Haidenbauer, C. Hanhart, A. Kudryavtsev, and U.-G. Meißner for a very fruitful collaboration that lead to the results presented.

References

1. V. Huhn *et al.*, Phys. Rev. Lett. **85**, 1190 (2000); Phys. Rev. C **63**, 014003 (2000)
2. D. E. Gonzales Trotter *et al.*, Phys. Rev. C **73**, 034001 (2006)
3. C. R. Howell *et al.*, Phys. Lett. B **444**, 252 (1998)
4. R. B. Wiringa, V. G. J. Stoks, and R. Schiavilla, Phys. Rev. C **51**, 38 (1995)
5. V. Lensky, V. Baru, J. Haidenbauer, C. Hanhart, A. Kudryavtsev, and U.-G. Meißner, Eur. Phys. J. A **26**, 107 (2005)
6. E. Epelbaum, Prog. Part. Nucl. Phys. **57**, 654 (2006)
7. A. Gårdestig, D. R. Phillips, Phys. Rev. C **73**, 014002 (2006)

RENORMALIZATION GROUP ANALYSIS OF NUCLEAR CURRENT OPERATORS

SATOSHI X. NAKAMURA

Theory Group, TRIUMF, 4004 Wesbrook Mall, Vancouver, BC V6T 2A3, Canada
E-mail: snakamura@triumf.ca

We review our study of the Wilsonian renormalization group (WRG) analysis for nuclear current operators. We apply the WRG method to axial-current operators derived from various approaches and obtain the unique effective low-energy operator.

Keywords: renormalization group; effective field theory; few-nucleon system

In Wilsonian renormalization group (WRG) analysis, one integrates out high-energy modes and examines the evolution of interactions. We apply the WRG analysis to nuclear operators such as nuclear potentials and current operators. It is known that various nuclear potentials equally well reproduce all of the data below the pion production threshold, while they appear quite different in describing the short range part. As a result of the model-space reduction using the WRG equation, all the potentials converge to a single effective low-momentum potential.[1] Moreover, a parameterization of the single potential becomes the NEFT-based operator which, by construction, does not depend on modeling the small scale physics.

In evaluating an amplitude of an electroweak process in few-nucleon systems, nuclear current operators are necessary ingredients as well as the nuclear force. The current operators based on different approaches have quite different behaviors in the short-range part, however, all of them give essentially the same reaction rates for low-energy reactions, *e.g.*, solar-neutrino reactions on the deuteron.[2] This implies that we can obtain a single effective current operator through the WRG analysis.

We derive a WRG equation for the current operator and use it to reduce the model space of currents from the models or NEFT with the pion (EFT(π)).[3] We are specifically concerned with the exchange axial-current operators relevant to the solar neutrino-induced breakup of the deuteron.

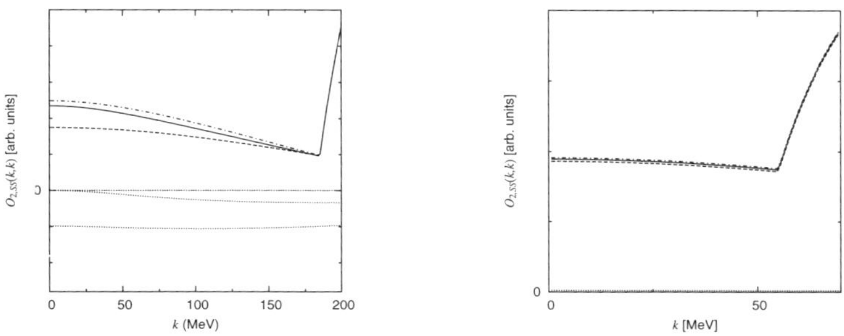

Fig. 1. Effective two-body current operators for initial S- to final S-wave state evaluated with the cutoff $\Lambda = 200$ MeV (left panel) and 70 MeV (right panel).

In Fig. 1, by reducing the model space, we find the evolution of the bare two-body operators (lower three curves) to the effective ones (upper three ones). However, a model dependence still remains among the effective operators at $\Lambda = 200$ MeV (left panel). This is because even the one-pion range mechanism is model dependent. These effective currents are further evolved up to $\Lambda = 70$ MeV (right panel). With this resolution of the system, the model dependence among the currents is not seen any more, and thus we obtain the unique effective operator.

Furthermore, we simulate the effective two-body current with $\Lambda = 70$ MeV using the EFT($\not\pi$)-based parameterization. Except for "jump-up" structures in Fig. 1, due to the bare one-body current contribution, the two-parameter fit yields an almost perfect simulation. Therefore, one can obtain the EFT($\not\pi$)-based operator from the models or EFT(π) in this way.

SA is supported by Korean Research Foundation and The Korean Federation of Science and Technology Societies Grant funded by Korean Government (MOEHRD, Basic Research Promotion Fund).

References

1. S.X. Nakamura, Prog. Theor. Phys. **114**, 77 (2005); M.C. Birse *et al.*, Phys. Lett. B **464**, 169 (1999); S.K. Bogner *et al.*, Phys. Lett. B **576**, 265 (2003).
2. M. Butler *et al.*, Phys. Rev. C 63, 035501 (2001); S. Nakamura *et al.*, Nucl.Phys. A 707, 561 (2002); S. Ando *et al.*, Phys. Lett. B 555, 49 (2003).
3. S.X. Nakamura and S. Ando, Phys. Rev. C **74**, 034004 (2006).

RECENT RESULTS AND FUTURE PLANS AT MAX-LAB

K. G. FISSUM*

Department of Physics, Lund University, Lund, SE-221 00, Sweden
** E-mail: kevin.fissum@nuclear.lu.se*
www.maxlab.lu.se/kfoto/index.html

Recent results obtained for the ^{4}He(γ, n) reaction prior to the facility upgrade are presented. Future plans for measuring Compton scattering from the deuteron at the upgraded facility are discussed.

Keywords: tagged photons, ^{4}He(γ, n), Compton scattering, deuteron.

1. Recent results for the ^{4}He(γ, n) reaction

During the past decade, 10 – 80 MeV tagged photons[1] have been available at MAX-lab. In this experiment,[2,3] tagged photons from $23 < E_\gamma < 70$ MeV were directed toward a liquid ^{4}He target, and neutrons were identified using pulse-shape discrimination and the time-of-flight technique in liquid-scintillator arrays. Seven-point angular distributions were measured for fourteen photon energies. For $E_\gamma < 29$ MeV, the angle-integrated data (see Fig. 1) are significantly larger than the values recommended[4] in 1983.

2. Future plans for Compton scattering from the deuteron

With the recent successful commissioning of the new energy-upgraded facility at MAX-lab, 10 – 225 MeV tagged photons are available. The COMPTON@MAX-lab Collaboration[a] is working to dramatically increase the available data for elastic Compton scattering from the deuteron, with the goal being to provide significant constraints upon the various models used to extract the electric and magnetic polarizabilities of neutron. A 40

[a]Lund University, Duke University, University of Edinburgh, The George Washington University, University of Glasgow, University of Göttingen, University of Illinois, University of Kentucky, University of Kharkov, Mount Allison University, and the University of Saskatchewan.

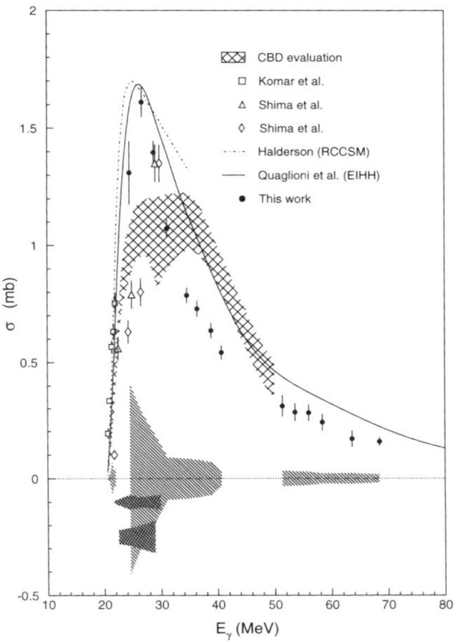

Fig. 1. The ^{4}He(γ, n) cross section: filled circles – present data; open squares[5] – Komar *et al*; open triangles[6] and open diamonds[7] – Shima *et al*; CBD evaluation[4] – hatched band; dashed-dotted line – RCCSM calculation;[8] and solid line – EIHH calculation.[9]

– 110 MeV tagged-photon beam will be used, and scattered photons will be detected simultaneously in three large NaI detectors (BUNI, CATS, and DIANA; each with ~2% energy resolution at 100 MeV) placed at angles from 30° to 150°. The experimental effort has commenced, with first measurements upon the deuteron anticipated to occur in late 2006.

References

1. J. O. Adler *et al.*, *Nucl. Instrum. Methods Phys. Res. A* **388**, p. 17 (1997).
2. B. Nilsson *et al.*, *Phys. Lett.* **B626**, p. 65 (2005).
3. B. Nilsson *et al.*, accepted for publication in Phys. Rev. C. in 2006.
4. J. R. Calarco *et al.*, *Phys. Rev. C.* **27**, p. 1866 (1983).
5. R. J. Komar *et al.*, *Phys. Rev. C.* **48**, p. 2375 (1993).
6. T. Shima *et al.*, *Nucl. Phys.* **A687**, p. 127c (2001).
7. T. Shima *et al.*, *Phys. Rev. C.* **72**, p. 044004 (2005).
8. D. Halderson, *Phys. Rev. C.* **70**, p. 034607 (2004).
9. S. Quaglioni *et al.*, *Phys. Rev. C.* **69**, p. 044002 (2004).

Nucleon Polarizabilities from Deuteron Compton Scattering, and Its Lessons for Chiral Power Counting

HARALD W. GRIEßHAMMER*

Center for Nuclear Studies, Department of Physics,
*The George Washington University, Washington DC, USA. *E-mail: hgrie@gwu.edu*

Polarizabilities measure the global stiffness of the nucleon's constituents against displacement in an external electro-magnetic field. We examined them in elastic deuteron Compton scattering $\gamma d \to \gamma d$ for photon energies between zero and 130 MeV in Chiral Effective Field Theory χEFT with explicit $\Delta(1232)$ degrees of freedom, see Refs.[1,2] for details and better references. An excellent tool to identify the active low-energy degrees of freedom are the *dynamical polarizabilities*, defined by a multipole decomposition of the structure part of the Compton amplitude at fixed energy. Unique signals allow one to study the temporal response of each constituent.

For example, the strong, para-magnetic N-to-$\Delta(1232)$ transition induces a strong energy-dependence which is pivotal to resolve the "SAL-puzzle", see Fig. 1: While all previous analyses of the SAL-data at 95 MeV extracted vastly varying nucleon polarizabilities, χEFT with an explicit $\Delta(1232)$ captures correctly both normalisation and angular dependence of the data without altering the static (namely zero-energy) polarizabilities.

A consistent description must also give the correct Thomson limit, i.e. the exact low-energy theorem which is a consequence of gauge invariance. Its verification is straight-forward in the 1-nucleon sector, where the amplitude is perturbative. In contradistinction, all terms of the leading-order Lippmann-Schwinger equation of NN-scattering, including the potential (and hence one-pion exchange), must be of order Q^{-1} when all nucleons are close to their non-relativistic mass-shell, to accomodate the shallow bound-state.[3] In a consistent power-counting, all NN-rescattering processes between photon-absorption and emission must thus be included. Our Green's function approach embeds these diagrams to guarantee the Thomson limit, which is however statistically significant only below 70 MeV, see Fig. 1. Up to next-to-leading order, the only unknowns are contributions to the polarizabilities from short-distance Physics, leading to two energy-independent parameters. The iso-scalar Baldin sum rule is in excellent agreement with our 2-parameter fit, serving as input to model-independently determine the iso-scalar, spin-independent dipole polarizabilities of the nucleon at zero energy from all Compton data below 100 MeV:

$$\alpha^s_{E1} = 11.3 \pm 0.7_{\text{stat}} \pm 0.6_{\text{Baldin}} \pm 1_{\text{th}} \; , \; \beta^s_{M1} = 3.2 \mp 0.7_{\text{stat}} \pm 0.6_{\text{Baldin}} \pm 1_{\text{th}}$$

(in 10^{-4} fm^3). We estimate the theoretical uncertainty to be ± 1 from typical higher-order contributions in the 1- and 2-nucleon sector. Dependence on the NN-potential or deuteron wave-function used is virtually eliminated with the correct Thomson limit. Comparing this with our analysis of all proton Compton data below 170 MeV by the same method,

$$\alpha^p_{E1} = 11.0 \pm 1.4_{\text{stat}} \pm 0.4_{\text{Baldin}} \pm 1_{\text{th}} \; , \; \beta^p_{M1} = 2.8 \mp 1.4_{\text{stat}} \pm 0.4_{\text{Baldin}} \pm 1_{\text{th}} \; ,$$

we conclude that the proton and neutron polarizabilities are to this leading order identical within (predominantly statistical) errors, as predicted by χEFT. More and better data from MAXlab (Lund) will lead to a more precise extraction, allowing one to zoom in on the proton-neutron differences.

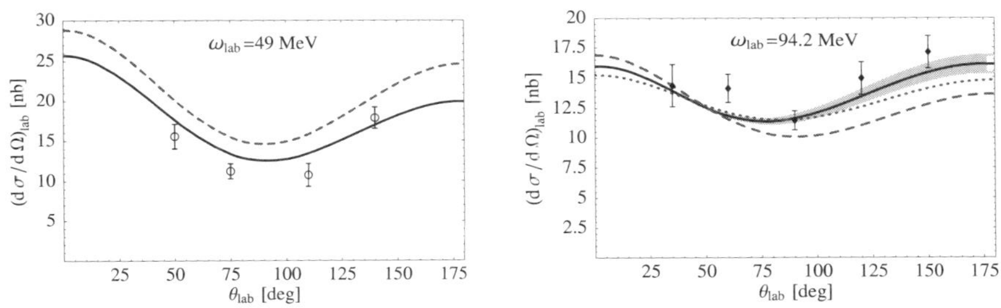

Fig. 1. Left: Example of prediction using proton polarizabilities with (solid) and without (dashed) NN-rescattering in intermediate states. Right: Example of 1-parameter fit result using the Baldin sum rule for the deuteron (solid, with stat. uncertainty), compared to χEFT without explicit $\Delta(1232)$ ($\mathcal{O}(p^3)$, dashed) and to a fit[4] at $\mathcal{O}(p^4)$ ($\alpha^s_{E1} = 11.5$, $\beta^s_{M1} = 0.3$, dotted). From Ref.[1]

References

1. R.P. Hildebrandt, H.W. Grießhammer, T.R. Hemmert: [nucl-th/0512063], subm. Phys. Rev. C .
2. R.P. Hildebrandt, PhD thesis TU München Dec. 2005 [nucl-th/0512064].
3. H.W. Grießhammer, forthcoming.
4. S.R. Beane, M. Malheiro, J.A. McGovern, D.R. Phillips, U. van Kolck: Phys. Lett. B567 (2003), 200; err. B607 (2005), 320; Nucl. Phys. A747 (2005), 311.

COMPTON SCATTERING ON HE-3

D. CHOUDHURY*

Department of Physics and Astronomy, Ohio University, Athens, Ohio 45701, USA
*E-mail: * choudhur@phy.ohiou.edu; † phillips@phy.ohiou.edu*

We present first calculations for Compton scattering on ^{3}He. The objective of the calculation is an extraction of the neutron polarizabilities.

1. Introduction

Our goal is to devise ways to extract the neutron polarizabilities. Direct experiments on the neutron are not possible due to the lack of free neutron targets and this encouraged physicists to look at other avenues to extract information about the neutron polarizabilities. Elastic Compton scattering on ^{3}He is one such avenue and here we report on the first calculations of this process. Our results indicate that $\gamma\,^3$He scattering is indeed a promising way to extract the neutron polarizabilities.

2. The Calculation and Results

The irreducible amplitudes for the elastic scattering of real photons from the NNN system are first ordered and calculated in Heavy Baryon Chiral Perturbation Theory (HBχPT) to $\mathcal{O}(Q^3)$– they are the same as those computed in Ref. 1. These amplitudes are then sandwiched between the nuclear wavefunctions to finally obtain the scattering amplitude–

$$\mathcal{M} = \langle \Psi_f | \hat{O} | \Psi_i \rangle. \tag{1}$$

Using these amplitudes we calculate the differential cross-section (dcs) or the double polarization observables, Δ_z or Δ_x.[2]

Fig. 1 shows the differential cross-section in the center of mass (c.m.) frame at 80 MeV. The left panel corresponds to calculations at different orders– $\mathcal{O}(Q^3)$, impulse approximation (IA) which is actually $\mathcal{O}(Q^3)$ but does not include two-body currents, and $\mathcal{O}(Q^3)$; and the right panel to varying the value of β_n around its $\mathcal{O}(Q^3)$ predicted value.

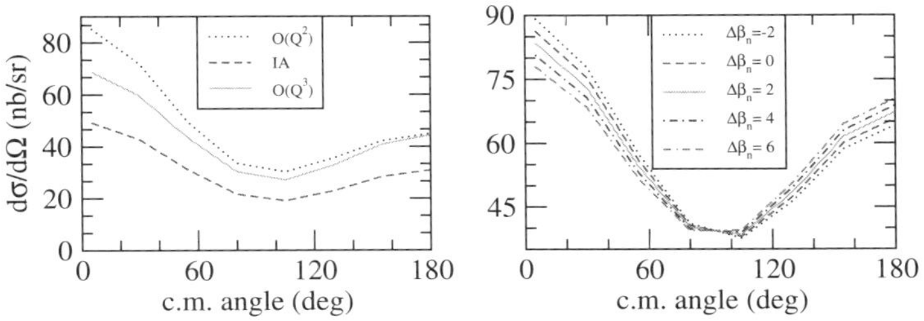

Fig. 1. The differential cross-section vs. c.m. angle at 80 MeV. Left panel shows $\frac{d\sigma}{d\Omega}$ calculated to different orders. Right panel shows the sensitivity of $\frac{d\sigma}{d\Omega}$ to $\Delta\beta_n(\times 10^{-4}\text{fm}^3)$.

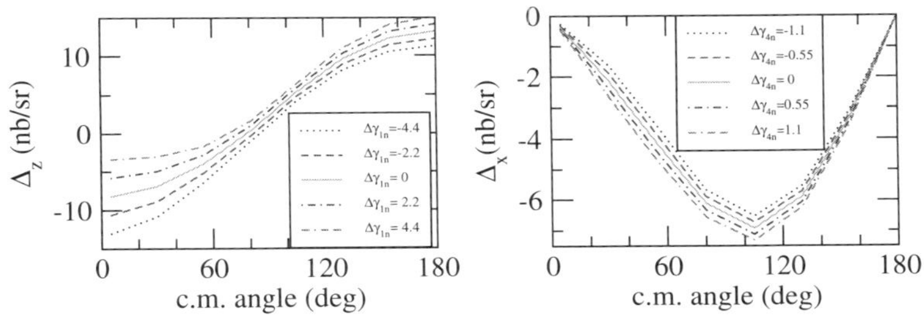

Fig. 2. Figure showing the sensitivity of the double-polarization observable, Δ_z to $\Delta\gamma_{1n}(\times 10^{-4}\text{fm}^4)$ and of Δ_x to $\Delta\gamma_{4n}(\times 10^{-4}\text{fm}^4)$ at 120 MeV in the c.m. frame.

Fig. 2 shows the variation in Δ_z (at 120 MeV in the c.m. frame) when $\Delta\gamma_{1n}(\times 10^{-4}\text{fm}^4)$ is varied and that in Δ_x when $\Delta\gamma_{4n}(\times 10^{-4}\text{fm}^4)$ is varied.

Both Figs. 1 and 2 suggest sizeable sensitivity to the neutron polarizabilities in the $\gamma\,^3$He dcs and the double polarization observables. For a more detailed description of the calculation and results please refer to Ref. 2.

References

1. S. R. Beane, M. Malheiro, D. R. Phillips, and U van Kolck, Nucl. Phys. **A656**, 367 (1999).
2. D. Choudhury, D. R. Phillips, and A. Nogga, (in preparation).

PART D

HADRON STRUCTURE AND MESON-BARYON INTERACTIONS

SUMMARY OF THE WORKING GROUP ON HADRON STRUCTURE AND MESON-BARYON INTERACTIONS

G. FELDMAN

Department of Physics, George Washington University, Washington, DC 20052

T.R. HEMMERT

Physik Department T39, TU Munchen, D-85747 Garching, Germany

In this Working Group (WG) on "Hadron Structure and Meson-Baryon Interactions" of *Chiral Dynamics 2006* there were 24 oral presentations in 6 different topical sections:

- **ChEFT and Lattice QCD**
- **ChEFT and Baryon Form Factors**
- **Nucleon Compton Scattering and Polarizabilities**
- **Theoretical Developments**
- **Chiral Meson-Baryon Interactions at Finite Density**
- **Information on Baryon Structure from Inelastic Reactions**

The talks had been selected based on the philosophy that new experimental and theoretical developments which could provide new insights into our understanding of baryon structure since the *Chiral Dynamics 2003* workshop should be represented in this WG. Given the vast variety of possible experimental probes, theoretical machinery and the richness of possible observables in low-energy baryon structure physics, the organizers chose to identify one talk per subject. All the speakers in this WG therefore had been asked not to cover only their own research work, but also to provide the audience with an overview of recent work from other research groups active in the respective sub-field. In this way, the speakers collectively helped to present the state-of-the-art development in this field to the community.

ChEFT and Lattice QCD (4 talks): The interface between Chiral Effective Field Theory (ChEFT) and Lattice QCD has produced a large amount of new research work since *Chiral Dynamics 2003*. Lattice data for a large variety of baryon structure properties are now available from simulations with *dynamical fermions* for quark masses corresponding to "effective" pion masses $m_\pi < 600$ MeV, with many groups working towards the region of $m_\pi \approx 350$ MeV.

For such "light" pion masses, quantitative control over the influence of the size of the simulation volume L becomes more and more important. In the baryon sector, examples were shown (*G. Schierholz*), where the L-dependence observed in the simulations can be quantitatively understood with the help of ChEFT. For mesonic observables, the agreement between ChEFT predictions and Lattice QCD was reported to be less precise. More work is needed to establish the L-dependence for $m_\pi \approx m_K$ near $L = 2$ fm in this sector.

While it is desirable to use large lattices in the new simulations with "small" quark masses, possible effects from discretization artifacts of the lattice spacing a also need to be kept under control. In principle, ChEFT can provide the theoretical machinery needed to study such a-dependent effects (*B. Tiburzi*). However, for each action used in the lattice simulations, a specific low-energy theory needs to be developed to study the restoration of continuum physics.

At this point in time, there exist relatively few studies on the properties of strange baryons within lattice QCD. The necessary chiral extrapolation formulae for 3-flavor QCD are more involved than those in the more popular 2-flavor case (*S. Simula*). Nevertheless, such studies carry the promise of access to interesting physics outside the usual 2-flavor simulation regime.

Chiral extrapolation studies for baryon properties are typically performed to finite order in perturbation theory, fitting some parameters to lattice data and constraining other parameters from analyses of scattering experiments or phenomenology. Each chiral extrapolation "curve" should therefore be accompanied by a set of "uncertainty bands" to facilitate a proper judgment of its validity (*B. Musch*). Many "uncertainty bands" shown in the literature are not compliant with mathematical error analysis, but rather seem to reflect the prejudices of their authors.

ChEFT and Baryon Form Factors (4 talks): The role of Goldstone Boson dynamics in baryon form factors is one of the "classic" topics in this field. Perturbative ChEFT calculations exist for a momentum transfer $Q^2 < 0.4\ (\mathrm{GeV/c})^2$ and new high precision data in this low-momentum region have also been reported in recent years. The interpretation of these results is still ongoing.

When dividing the experimental data on nucleon form factors by the dipole parameterization, recent high precision measurements (*C. Crawford*) from Bates (using the BLAST detector) suggest deviations/structures around the dipole expectation at low values of Q^2. While it is appealing to attribute the origin of these structures to the pion cloud of the nucleon, one would like to see more theoretical calculations in ChEFT that address the mechanisms by which such signatures are dynamically generated.

New experimental results in the Δ resonance region for pion electroproduction at low Q^2 have been reported from MAMI and JLab, which complement earlier data from Bates using the OOPS spectrometer system (*S. Stave*). The extracted Q^2-dependences of the E2/M1 and C2/M1 ratios are in good agreement with the predictions from ChEFT and can be interpreted in terms of N-Δ transition form factors. The existing leading-one-loop ChEFT calculations for this process should be extended to the next order to judge the stability of the predicted non-linear Q^2-dependences in this transition. It is hoped that future experiments at MAMI and/or JLab will exploit target and recoil polarization data to provide more stringent constraints for testing ChEFT and the various phenomenological and pion-cloud models.

Lattice QCD has now produced a wealth of results for nucleon form factors as well as for N-Δ transition form factors related to the E2/M1 and C2/M1 ratios (*C. Alexandrou*). In addition, first results have recently become available for N-Δ axial-vector form factors which can be related to experimental parity-violating asymmetries. For future progress, it will be essential to push these simulations to the regime of low $Q^2 \ll 0.5$ (GeV/c)^2 in order to directly utilize ChEFT for the extrapolation to the physical pion mass. Dipole extrapolation methods in Q^2 that have been used previously can only be viewed as an intermediate step, as they tend to wash out the interesting structures at low Q^2 that are possibly connected to pion dynamics (*C. Crawford*).

Parity-violating electron scattering experiments at MAMI and JLab (G0 and HAPPEX) are now in a position to provide results on the strange form factors of the nucleon (*F. Benmohktar*) at several values of Q^2, both for the electric and the magnetic form factors. Preliminary data suggest the possibility of a non-null strange quark contribution to the form factors. While ChEFT cannot predict the absolute strength of the strange form factors, several calculations regarding the Q^2-dependence exist in the literature. In light of the recent experimental progress, these calculations should be updated and a global fit to all existing strangeness data should be attempted, in order to judge the role of kaon dynamics at low Q^2 in these form factors.

Nucleon Compton Scattering and Polarizabilities (4 talks): Compton scattering on the nucleon provides an experimentally difficult but theoretically "clean" window on the structure of the nucleon at femtometer scales. With the upgrade of HIGS (at Duke University) and the continued effort at MAX-Lab (in Lund, Sweden), this field is expected to see plenty of new challenges in the next few years, both for experimentalists and for theorists.

The measured experimental differential cross sections for Compton scattering on a nucleon contain much more information than simply the static values for the electric and magnetic polarizabilities of the nucleon. Dynamical polarizabilities can be extracted via a newly developed projection operator formalism (*B. Pasquini*) which enables one to study the energy dependence (*i.e.* dispersive effects) of the polarizabilities of the nucleon, including spin-dependent terms as well. This machinery is able to distinguish between different theoretical calculations that might agree on the magnitude of the static polarizabilities, but differ in their predictions for the energy dependence, providing a direct signal of the active degrees of freedom in this energy range.

Recent data for Virtual Compton Scattering (VCS) on the proton both at very low $Q^2 = 0.05$ (GeV/c)^2 (Bates) and at high $Q^2 = 1.0$-1.9 (GeV/c)^2 (JLab) have provided new information on the VCS response functions of the nucleon (*C. Hyde-Wright*). The new Bates data are consistent with the pioneering experiment from MAMI (~ 0.4 (GeV/c)^2), providing a coherent picture in good agreement with the ChEFT predictions for $Q^2 < 0.4$ (GeV/c)^2. The extraction of generalized polarizabilities from these response functions leads to the conclusion that the observed Q^2-dependence behaves significantly different from a dipole-like form factor! However, it must be noted that the discrepancy between ChEFT and Dispersion Theory calculations for the generalized spin polarizabilities of the nucleon still remains. A proposed polarized VCS experiment on the proton at MAMI is expected to shed new light on this issue in the next few years.

Lattice QCD has recently started to deliver results on the static values of a large number of hadron polarizabilities (*F. Lee*). These quantities are obtained on the lattice by calculating the mass/energy shift of the hadrons due to their interaction with an external electric or magnetic field. Thus far, the results are preliminary, but they show a great deal of promise, and spin polarizabilities are also considered to be within reach. However, at the present time, the quark mass dependences observed in the lattice simulations for many polarizabilities have not yet been understood from the point of view of ChEFT. More theoretical work is needed here to clear up this important issue!

Polarized Compton scattering on the proton and on light nuclei (deuterium and 3He) is the new frontier in low-energy Compton scattering (*H. Gao*). These processes can provide access to the elusive spin polarizabilities of the nucleon, which should display, according to the predictions of ChEFT, an interference between long- and short-distance physics different from the one observed in spin-independent polarizabilities. It is hoped that the high intensity polarized photon facility at HIγS will soon become fully operational in order to mount the first polarized Compton experiment on a polarized proton, which has been discussed and anticipated in this WG since *Chiral Dynamics 1997*! Neutron spin polarizabilities will also be studied using a polarized 3He target that has been developed at TUNL for use at HIγS.

Theoretical Developments (4 talks): The continued development of theoretical machinery has been part of the *Chiral Dynamics* workshops since the first meeting in 1994. The activity in this direction since *Chiral Dynamics 2003* in the field of single-baryon physics has focused on the development of covariant schemes for nucleons and Δ's and of non-perturbative methods for the Goldstone Boson dynamics around a baryon.

Covariant BChPT has seen a revival in the past few years, mainly among European groups. It carries the hope of providing results in a perturbation series which, under some circumstances, is better behaved than the one from a non-relativistic scheme like HBChPT. It is expected that the previously used IR-scheme will be superseded (*T. Gail*) by EOMS or Modified-IR due to some recently discussed inconsistencies.

Calculations for baryon scattering observables with 3 active flavors are known to be difficult in ChEFT. The naive perturbative treatment of the kaon-cloud dynamics around a baryon with a method like HBChPT is now known to be inadequate for many observables. The development of non-perturbative methods for SU(3) ChEFT in single-baryon physics (*J. Oller*) is very promising in this sector. Such calculations are also needed for the study of the possibility of kaon condensation (*P. Camerini*).

Goldstone Boson dynamics dominates the structure of baryons only at low energies and for small momentum transfers. On the other hand, state-of-the-art quark model approaches like the "Schwinger-Dyson-Equations Ansatz" are expected to contain the correct physics at intermediate energies and perform well in the cross-checks with Lattice QCD. It is therefore one of the challenges for theoretical hadron physics in the next few years to systematically incorporate the known chiral dynamics at low energies into such models (*C. Roberts*), in order to bridge the gap in many observables (*e.g.* the nucleon form factors) between the chiral dynamics at low Q^2 and the dynamical constituent quark (correlation) physics at intermediate Q^2.

One of the motivations in the early days of ChEFT was the hope of gaining theoretical access to rare physical processes which allow for precision tests of the Standard Model. When the program of this WG was compiled, the calculation of the branching ratio for a radiative decay of a neutron was only a theoretical prediction. However, by the time of the *Chiral Dynamics 2006* workshop, a first comparison with data from NIST has been reported, which agrees reasonably well with the prediction from ChEFT (*S. Gardner*). In order to make this process competitive with other determinations of CKM elements and the V-A structure of the weak interactions, the experiment should be upgraded to determine the polarization of the final-state photon.

Chiral Meson-Baryon Interactions at Finite Density (3 talks): The mass of a particle that lives at the femtometer scale is strongly influenced by the quantum mechanical fluctuations allowed by the strong interaction in its environment. Besides their masses, hadrons are in fact expected to change all of their basic properties when embedded in a nuclear medium of finite density, due to the additional fluctuation channels not present in "free space". Given that the spatial extent of such nuclear matter is constrained by the size of a nucleus in present-day earth-bound experiments, the main difficulty in this field lies in the proper identification of such medium-induced effects from experimental signals.

The properties of vector mesons (*i.e.* the change of the observed mass and width compared to the known values from "free space") produced during a nuclear reaction have played a large role in this field. It is hoped that an upcoming experiment (*M. Kotulla*) on recoil-free vector meson production in carbon at ELSA will provide a very clean experimental scenario, where the majority of the vector mesons decay *inside* the nucleus.

The $\bar{K}$-N interaction at low energies is very interesting due to the interference between Goldstone Boson dynamics and nearby resonances. It has been speculated that this system, when put into the nuclear medium, could lead to the existence of deeply bound K-nucleus states. Mainly utilizing K- beams, several experimental groups throughout the world have been searching for such states (*P. Camerini*). At the moment, the reported evidence is not yet conclusive.

A systematic approach within ChEFT for calculations at finite nuclear matter density is still being developed (*D. Furnstahl*). At the moment, the relevant nucleon potentials are calculated within a variety of approaches ranging from purely perturbative calculations to renormalization group analyses. The emerging organizational principle ("power counting") for ChEFT in nuclear matter is likely to be different from the one used in "free space" calculations due to a different hierarchy of scales at finite density. Plenty of new insights can be expected in this area for the next *Chiral Dynamics 2009* workshop!

Information on Baryon Structure from Inelastic Reactions (5 talks): In the past, single-pion and double-pion photo/electroproduction near threshold has been the main focus in this subfield. More recently, the analysis of total inclusive electroproduction experiments and their relationship to chiral dynamics has also started to play a prominent role.

Impressive experimental progress on the extraction of the moments of spin structure functions of the nucleon at low momentum transfer has been reported from JLab (*A. Deur*). At this point in time, the theoretical understanding of the observed Q^2-dependences within the framework of ChEFT only works qualitatively (*e.g.* location of the zero-crossing in the first moment of g_1), with lots of room for improvement (*e.g.* depth of the observed minimum in the first moment of g_1). This is one of the areas in which the convergence pattern of HBChPT results appears questionable, triggering investigations in covariant BChPT. Explicit inclusion of $\pi\Delta$ intermediate states at next-to-leading one-loop order appears mandatory in future calculations, as most existing ChEFT calculations so far only account for πN fluctuations.

Lattice QCD has reported a wealth of results for the first 3 moments of the Generalized Parton Distribution functions (GPDs) of the nucleon (*D. Richards*). The main surprise is still the sizable discrepancy between the "large" lattice QCD value for the isovector moment x compared to the "small" value of x extracted from experiments at the physical point. Interestingly, the lattice "data" from different groups which utilize *different* lattice actions are basically in agreement on the "large" value, showing only a very weak quark-mass dependence down to the lightest quark masses available at present. On the other hand, the observed "plateau" of the lattice data for isovector x cannot extend to significantly smaller quark masses because the expected chiral curvature – which hopefully can connect the lattice results with the known value at the physical point – cannot be arbitrarily steep (with "natural size" couplings). It is therefore hoped that a "bending down" of the lattice results will actually be observed in the near future!

Many groups have worked on the chiral extrapolation functions for the moments of nucleon GPDs (*J-W. Chen*), both at finite and at infinite volume. However, most calculations so far have only been performed to leading-one-loop order, making it difficult to judge the theoretical stability of the results obtained. The experience from the "zeroth-order" moments like g_A or the magnetic moment of the nucleon suggests that the calculated leading-order quark-mass dependence for the higher moments so far is only under control for quark masses near the physical point. More calculations beyond the leading non-analytic quark-mass dependence are urgently needed, before the wealth of information on nucleon GPDs provided by lattice QCD (D. Richards) can be analyzed on a quantitative level.

The discrepancy between the Q^2-dependence of single-pion electroproduction multipoles predicted by ChEFT and the value reported from a new experiment at MAMI still remains (*M. W. Ahmed*). Given the low value of Q^2, ChEFT should be capable of correctly describing the multipoles involved, so this continued discrepancy (*Chiral Dynamics 2003-2006*) is quite surprising and definitely needs to be resolved. Ideally one would like to see an extraction of all 6 s-wave and p-wave multipoles as a function of energy and of momentum transfer. It is hoped that the BIGBITE experiment at JLab can contribute to an extended experimental data base in this most elementary inelastic reaction! The upgrade of the HIγS facility in the next few years to photon energies above pion threshold also carries the hope to finally measure the imaginary part of the s-wave multipole and to study the associated cusp structure predicted by ChEFT, an idea that has been around since the first *Chiral Dynamics* workshop in 1994!

In contrast to single-pion electroproduction at finite Q^2, the theoretical situation at the real photon point $Q^2 = 0$ seems to be well under control. A consistent picture between phenomenological models, Dispersion Theory and perturbative ChEFT calculations in the s-wave and p-wave multipoles is emerging (*L. Tiator*). Covariant BChPT allows for an easier mapping to Dispersion Theory than the HBChPT framework. An extension of such a combined analysis to finite Q^2 is challenging but necessary in light of the reported discrepancies (M. W. Ahmed).

The organizers would like to thank all the speakers and the participants in this WG for their professional contributions and the focused discussions. The authors of this Summary Report would also like to thank Alan Nathan for his valuable assistance in selecting speakers and topics during the early planning stages of this workshop.

Regarding the overall organization of this working group, it is our opinion that the new strategy of including lattice QCD talks in this WG focused on baryon structure has worked well and should be continued at *Chiral Dynamics 2009*. Looking back at the proceedings of the previous *Chiral Dynamics* workshops, we note that next to "classic" topics of this WG like "Baryon Form Factors", "Nucleon Polarizabilities" or "Single-Pion Electroproduction", new aspects such as the "Quark-Mass Dependence of Baryon Observables", "Density Dependence of Hadronic Properties" or "Nucleon Structure from the Viewpoint of GPDs" have developed into active research areas involving a large number of groups across the globe. We are therefore confident that the "Hadron Structure and Meson-Baryon Interactions" working group will continue to be an important meeting point for the community at the next *Chiral Dynamics 2009* workshop in Bern.

FINITE VOLUME EFFECTS: LATTICE MEETS CHPT

GERRI SCHIERHOLZ

Deutsches Elektronen-Synchrotron DESY
D-22603 Hamburg, Germany

and

John von Neumann-Institut für Computing NIC
Deutsches Elektronen-Synchrotron DESY
D-15738 Zeuthen, Germany

E-mail: Gerrit.Schierholz@desy.de

– For the QCDSF Collaboration –

In lattice calculations of hadronic observables one faces the problem that the results have to be extrapolated in a controlled way to the infinite volume and the chiral limit. A suitable tool for this task is chiral perturbation theory (ChPT).

In Fig. 1 I show a fit of ChPT to the nucleon mass in finite volume[1] and its extrapolation to the infinite volume and chiral limit. We find a remark-

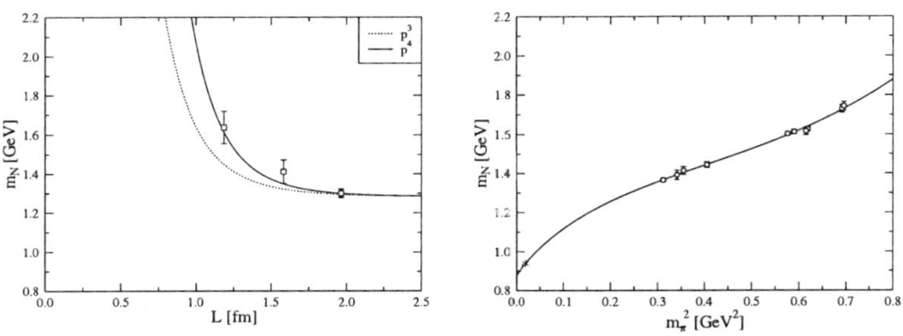

Fig. 1. Left figure: Volume dependence of the nucleon mass for $m_\pi \approx 550$ Mev together with a fit of ChPT. Right figure: The nucleon mass extrapolated to the infinite volume and the resulting chiral extrapolation.

ably consistent picture of the finite volume and pion mass dependence of the nucleon mass, in spite of the fact that our pion masses are rather large and the applicability of ChPT at these mass values is far from obvious.

A similarly consistent result was found for the axial vector charge of the nucleon.[2] In Fig. 2 I show the same sort of fit and extrapolation for g_A.

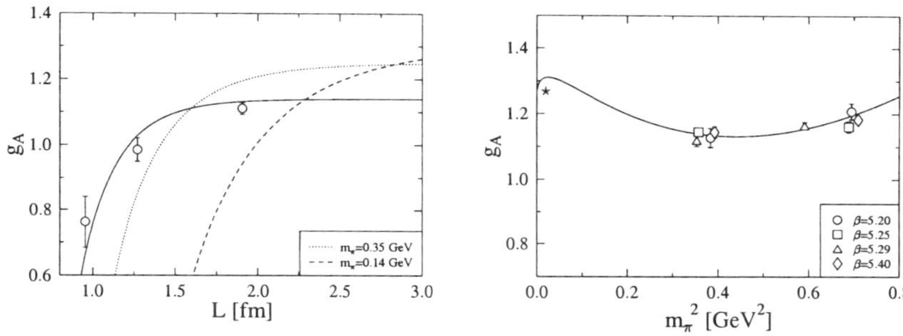

Fig. 2. Left figure: Volume dependence of g_A for $m_\pi \approx 550$ Mev together with a fit of ChPT. Right figure: g_A extrapolated to the infinite volume and the resulting chiral extrapolation.

To corroborate these results, lattice data at smaller pion masses will be needed. Simulations at pion masses of $O(300)$ MeV are in progress.[3]

References

1. A. Ali Khan *et al.*, Nucl. Phys. B689 (2004) 175 [arXiv:hep-lat/0312030].
2. A. A. Khan *et al.*, arXiv:hep-lat/0603028, to be published in Phys. Rev. D.
3. M. Göckeler *et al.*, arXiv:hep-lat/0610066; arXiv:hep-lat/0610071; arXiv:hep-lat/0610118.

LATTICE DISCRETIZATION ERRORS IN CHIRAL EFFECTIVE FIELD THEORIES

B. C. TIBURZI*

Department of Physics, Duke University, Box 90305, Durham, NC 27708
**E-mail: bctiburz@phy.duke.edu*

We give a brief review of how to include lattice discretization effects in chiral perturbation theory. Focusing on the hybrid actions used by LHP[1] and NPLQCD[2] collaborations, continuum extrapolations are discussed.

Keywords: Lattice QCD, Chiral Perturbation Theory, Discretization

Simulating QCD on a spacetime lattice provides a first principles method for strong interaction physics. As a numerical technique, we must address how to connect lattice data to nature. Spontaneous symmetry breaking complicates extraction of infinite volume physics as propagation of Goldstone modes is highly volume dependent. Low-energy effective theories (EFTs), such as chiral perturbation theory (χPT), can address these long-range effects. The lattice discretization leads to short-range effects, and somewhat surprisingly χPT can be extended to address discretization errors too.

Near the continuum, $a^{-1} \ll \Lambda_{\text{QCD}}$, the lattice action can be described by a continuum effective field theory known as the Symanzik action

$$S_{\text{Symanzik}} = S_0 + aS_1 + a^2 S_2 + \ldots .$$

Operators in this EFT respect the symmetries of the lattice action, and higher dimensional terms are suppressed by powers of the ultraviolet cutoff a^{-1}. The crucial observation is that solutions to the fermion doubling problem generally lead to some form of chiral symmetry breaking at finite lattice spacing. Thus in the chiral limit, the masses of Goldstone modes will no longer vanish and short-range discretization effects will then enter long-range physics. One then builds χPT for the Symanzik action to incorporate explicit symmetry breaking from discretization effects.[3,4] The result is an EFT with a dual expansion in quark mass and lattice spacing.

Each lattice fermion has a differing χPT. For example, unimproved Wilson quarks break chiral symmetry at $\mathcal{O}(a)$ via the Pauli term $\overline{\psi}\sigma_{\mu\nu}G_{\mu\nu}\psi$. Wilson χPT has been developed to $\mathcal{O}(a^2)$[3,5–7] and nucleons have been included.[8–10] Staggered quarks retain a remnant of chiral symmetry at finite lattice spacing, but discretization leads to the breaking of taste symmetry for which staggered χPT has been developed for the continuum extrapolation.[4,11,12] Hybrid lattice actions have gathered increased recently attention and their chiral theories too have been constructed.[13–15]

Current hybrid actions use valence quarks with good chiral symmetry (such as domain wall) on a sea of staggered quarks. At finite a no symmetry connects the valence and sea Symanzik operators. Mixed bilinears of the form $(\overline{\psi}_v\gamma_\mu\psi_v)(\overline{\psi}_s\gamma_\mu\psi_s)$ lead to an $\mathcal{O}(a^2)$ additive mass renormalization of pions formed from one valence and one sea quark. The pion charge radius depends only on valence-sea meson masses at one-loop. If quark masses continue to be lowered nearing the physical value, the chiral logarithm will not be discernible at finite a.[16] Similar behavior appears in other observables, such as the nucleon mass,[17] the pion decay constant,[14] and the $I = 2$ $\pi\pi$ scattering length.[18] As the chiral limit is approached, discretization can spoil the chiral curvature or enhance partial quenching effects.

Chiral EFTs formulated for the Symanzik action provide a systematic approach to determine lattice spacing artifacts. For lighter quarks, one may see acute effects from discretization rather than chiral curvature.

References

1. J. W. Negele, *Lattice QCD and nucleon spin structure, these proceedings* .
2. M. J. Savage, *Few-body lattice calculations, these proceedings* .
3. S. R. Sharpe and R. L. Singleton, *Phys. Rev.* **D58**, p. 074501 (1998).
4. W.-J. Lee and S. R. Sharpe, *Phys. Rev.* **D60**, p. 114503 (1999).
5. G. Rupak and N. Shoresh, *Phys. Rev.* **D66**, p. 054503 (2002).
6. O. Bär, G. Rupak and N. Shoresh, *Phys. Rev.* **D70**, p. 034508 (2004).
7. S. Aoki, *Phys. Rev.* **D68**, p. 054508 (2003).
8. S. R. Beane and M. J. Savage, *Phys. Rev.* **D68**, p. 114502 (2003).
9. D. Arndt and B. C. Tiburzi, *Phys. Rev.* **D69**, p. 114503 (2004).
10. B. C. Tiburzi, *Nucl. Phys.* **A761**, 232 (2005).
11. C. Aubin and C. Bernard, *Phys. Rev.* **D68**, p. 034014 (2003).
12. C. Aubin and C. Bernard, *Phys. Rev.* **D68**, p. 074011 (2003).
13. O. Bär, G. Rupak and N. Shoresh, *Phys. Rev.* **D67**, p. 114505 (2003).
14. O. Bär *et al.*, *Phys. Rev.* **D72**, p. 054502 (2005).
15. M. Golterman, T. Izubuchi and Y. Shamir, *Phys. Rev.* **D71**, p. 114508 (2005).
16. T. B. Bunton *et al.*, *Phys. Rev.* **D74**, p. 034514 (2006).
17. B. C. Tiburzi, *Phys. Rev.* **D72**, p. 094501 (2005).
18. J.-W. Chen *et al.*, *Phys. Rev.* **D73**, p. 074510 (2006).

SU(3)-BREAKING EFFECTS IN HYPERON SEMILEPTONIC DECAYS FROM LATTICE QCD

SILVANO SIMULA

INFN, Sezione Roma 3, Via della Vasca Navale 84, I-00146 Roma, Italy
E-mail: simula@roma3.infn.it

We present the first quenched lattice QCD study of all the vector and axial form factors relevant for the hyperon semileptonic decay $\Sigma^- \to n\ \ell\ \nu$.

Recently it has been shown[1] that SU(3)-breaking corrections to the $K \to \pi$ vector form factor can be determined from lattice simulations with great precision, allowing to reach the percent level of accuracy in the extraction of V_{us} from $K_{\ell 3}$ decays. An independent way to extract V_{us} is provided by hyperon semileptonic decays, and Ref.[2] has shown that it is possible to extract the product $|V_{us} \cdot f_1(0)|$ at the percent level from experiments, where $f_1(0)$ is the vector form factors at zero-momentum transfer.

In Ref.[3] we have performed a lattice QCD study of SU(3)-breaking corrections in all the vector and axial form factors relevant for the $\Sigma^- \to n\ \ell\ \nu$ decay. Though the simulation has been carried out in the quenched approximation, our results represent the first attempt to evaluate hyperon f.f.'s using a non-perturbative method based only on QCD.

For each f.f. we have studied its momentum and mass dependencies, obtaining its value extrapolated to $q^2 = 0$ and at the physical point. Our results are collected in Table 1.

It can be seen that the SU(3)-breaking corrections to $f_1(0)$ have been determined with great statistical accuracy in the regime of the simulated quark masses, which correspond to pion masses above 0.7 GeV. The magnitude of the errors reported in Table 1 is mainly due to the chiral extrapolation and to the poor convergence of the Heavy Baryon Chiral Perturbation Theory.[4] Though within large errors the central value of $f_1(0)$ arises from a partial cancellation between the contributions of local terms, evaluated on the lattice, and chiral loops. This may indicate that SU(3)-breaking corrections on $f_1(0)$ are moderate, giving support to the analysis of Ref.[2]

Table 1. Lattice results[3] for the vector and axial form factors at $q^2 = 0$ for the $\Sigma \to n$ transition. The errors do not include the quenching effect.

$f_1(0)$	$-0.988 \pm 0.029_{\text{lattice}} \pm 0.040_{\text{HBChPT}}$
$g_1(0)/f_1(0)$	-0.287 ± 0.052
$f_2(0)/f_1(0)$	-1.52 ± 0.81
$f_3(0)/f_1(0)$	-0.42 ± 0.22
$g_2(0)/f_1(0)$	$+0.63 \pm 0.26$
$g_3(0)/f_1(0)$	$+6.1 \pm 3.3$

The ratio $g_1(0)/f_1(0)$ is found to be negative and consistent with the value adopted in Ref.[2] In the limit of exact SU(3) symmetry we obtain $[g_1(0)/f_1(0)]_{SU(3)} = -0.269 \pm 0.047$. This means that SU(3)-breaking corrections are moderate also on this ratio, though it is not protected by the Ademollo-Gatto theorem against fist-order corrections.

The weak electric form factor g_2 is found to be non-vanishing because of SU(3)-breaking corrections. Our result for $g_1(0)/f_1(0)$ combined with that of $g_2(0)/f_1(0)$ are nicely consistent with the experimental result from Ref.[5] Our findings favor the scenario in which $g_2(0)/f_1(0)$ is large and positive with a corresponding reduced value for $|g_1(0)/f_1(0)|$ with respect to the conventional assumption $g_2(q^2) = 0$ based on exact SU(3) symmetry.

Future lattice QCD studies of the hyperon semileptonic transitions should remove the quenched approximation and lower the quark masses as much as possible in order to reduce the impact of the chiral extrapolation. The accuracy of the ratios $f_2(0)/f_1(0)$, $f_3(0)/f_1(0)$, $g_2(0)/f_1(0)$ and $g_3(0)/f_1(0)$ may be improved by implementing twisted boundary conditions for the quark fields (see Ref.[6]), which allows to reach values of the momentum transfer closer to $q^2 = 0$. Finally, the use of smeared source and sink for the interpolating fields as well as the use of several, independent interpolating fields may help in increasing the overlap with the ground-state signal, particularly at low values of the quark masses.

References

1. D. Becirevic *et al.*, *Nucl. Phys. B* **705**, 339 (2005); *Eur. Phys. J. A* **24S1**, 69 (2005).
2. N. Cabibbo *et al.*, *Ann. Rev. Nucl. Part. Sci.* **53**, 39 (2003).
3. D. Guadagnoli *et al.*, hep-ph/0606181; *Nucl. Phys. Proc. Suppl.* **140**, 390 (2005); *PoS LAT2005*, 358 (2005).
4. G. Villadoro, *Phys. Rev. D* **74**, 014018 (2006).
5. S. Y. Hsueh *et al.*, *Phys. Rev. D* **38**, 2056 (1988).
6. D. Guadagnoli *et al.*, *Phys. Rev. D* **73**, 114504 (2006).

UNCERTAINTY BANDS FOR CHIRAL EXTRAPOLATIONS

B. U. MUSCH*

Physics Department T39, Technische Universität München,
85747 Garching, Germany
** speaker; E-mail: bmusch@ph.tum.de*

Lattice QCD results can be extrapolated to physical quark masses using Chiral Perturbation Theory. Here, treatment of statistical and systematic errors is discussed. One way of estimating the theoretical uncertainty is to perform a matching of the theory to a framework with an additional degree of freedom. This is illustrated for the nucleon axial vector coupling constant g_A by effectively integrating out the leading $\Delta(1232)$ contribution.

Keywords: Chiral Perturbation Theory; Lattice QCD; extrapolation; errors; uncertainties; nucleon axial vector coupling constant; delta resonance;

1. Introduction

Due to limited computer power, present lattice QCD calculations still make some unrealistic assumptions. Most prominently they exhibit an unphysically large value of the pion mass m_π. Extrapolation to the physical regime can be achieved using Chiral Perturbation Theory (χPT), which parametrizes the quark mass dependence in terms of universal low energy constants (LECs). As an example, we analyze state of the art lattice data for the nucleon axial vector coupling constant using a fit function $g_A^{\mathrm{SSE}}(m_\pi, f_\pi^0, g_A^0, \Delta, c_A, g_1, C^{\mathrm{SSE}})$. The latter has been calculated to leading one loop order in the Small Scale Expansion (SSE), which treats nucleons, pions, and $\Delta(1232)$ as explicit degrees of freedom.[1] Numerical estimates of f_π^0, c_A, and $\Delta \equiv m_\Delta^0 - m_N^0$ have been deduced from phenomenology. We can extract the other LECs from a fit to lattice data points $(m_{\pi,i}\,,\, y_i \pm \Delta y_i)$, where the statistical errors Δy_i come from the lattice Monte Carlo method. Here, we also constrain g_A^{SSE} to pass through the physical point. This talk presents techniques to perform the error analysis associated with the extrapolation.

2. Statistic and Systematic Errors[2,3]

g_A^0 and g_1 are determined from the minimum $\chi^2_{\rm opt}$ of $\chi^2 \equiv \sum_i (g_A(m_{\pi_i}, g_A^0, g_1) - y_i)^2/\Delta y_i^2$. In Fig. 1 (left) the statistical error band at 68% confidence level emerges from drawing all functions $g_A^{\rm SSE}(m_\pi)$ for LECs within the confidence region, obtained from the condition $\chi^2 \leq \chi^2_{\rm opt} + 2.30$. Varying the fixed LECs within phenomenologically acceptable bounds leads to the systematic envelope in Fig. 1 (left). In general, correlations among parameters are important and should be taken into account.

3. Theoretical Uncertainty from Resonances[4]

An alternative theory framework, Heavy Baryon χPT, does not include $\Delta(1232)$ degrees of freedom explicitly. We get an estimate of the size of expected theoretical errors for this theory by integrating out the $\Delta(1232)$ in the SSE framework. In practice we expand $g_A^{\rm SSE}(m_\pi)$ in powers of m_π/Δ and plot the result at different truncation orders as in Fig. 1 (right). Our analysis suggests that for g_A theoretical errors in the HBχPT framework get out of control for pion masses larger than about 300 MeV.

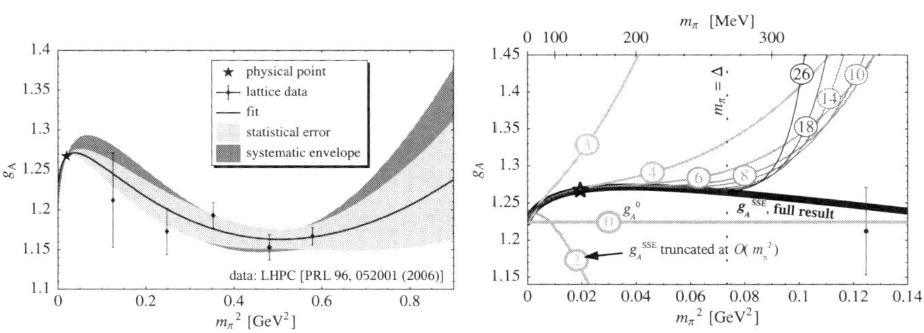

Fig. 1. g_A in SSE at leading one-loop order. Left: Statistic and systematic error band. Right: Integrating out the $\Delta(1232)$ resonance. Labels n of the curves give the order m_π^n at which the series is truncated.

B. M. acknowledges support by BMBF and DFG Emmy Noether-program.

References

1. T. R. Hemmert, M. Procura and W. Weise, *Phys. Rev.* **D68**, 075009 (2003).
2. B. Musch, diploma thesis, hep-lat/0602029, TU München (2005).
3. M. Procura *et al.*, *Phys. Rev.* **D73**, 114510 (2006).
4. M. Procura, B. Musch, T. R. Hemmert and W. Weise, *hep-lat/0610105* (2006).

UPDATE OF THE NUCLEON ELECTROMAGNETIC FORM FACTORS

C. B. CRAWFORD*

Department of Physics, University of Tennessee,
Knoxville, TN 37996, USA
**E-mail: chris2@lns.mit.edu*

Research on the nucleon electromagnetic form factors has been revitalized by new class of experiments exploiting spin degrees of freedom. A summary of recent progress at both high and low Q^2 is presented.

Keywords: Electromagnetic form factors; nucleon structure; electron scattering.

Recent activity at high Q^2 has been prompted by the observation of a decrease in $\mu G_E^p/G_M^p$ at high Q^2 up to $Q^2 = 5.6$ (GeV/c)^2.[1] This is at odds with the unpolarized result $\mu G_E^p/G_M^p \approx 1$, verified by an L-T separation with improved detector acceptance systematics.[2] An independent polarization experiment using the reaction ^{1}H(e, e′)p was unable to confirm either the FPP or unpolarized result.[3] However, the discrepancy has been nearly resolved by calculations of the two-photon exchange (TPE) contribution, with the intermediate nucleon in the ground state[4] or highly excited states.[5] Experiments are approved to directly measure the TPE amplitude.[6] For the neutron, G_E^n has been measured to $Q^2 = 1.45$ (GeV/c)^2 from the reactions ^{2}H(e, e′n)[7] and ^{2}H(e, e′n).[8] G_E^n was measured at lower Q^2 from ^{3}He(e, e′n),[9] and data up to $Q^2 = 3.4$ (GeV/c)^2 from a separate experiment are currently being analyzed.[10] Results of G_M^n at $Q^2 = 0.5$–4.5 (GeV/c)^2 from the ratio of ^{2}H(e, e′n) to ^{2}H(e, e′p) unpolarized cross sections with simultaneous in-situ detector efficiency calibration[11] are forthcoming. Recent theoretical activities include, among others, updates of dispersion-theoretical[12] and extended VMD[13] models, a Poincare covariant quark core model augmented by a pseudoscalar meson cloud,[14] and a covariant spectator quark model.[15]

There is renewed interest in the region Q^2=0.1–0.6 GeV^2 due to phenomenological observations of structure due to effects of the pion cloud.[16] A comprehensive mapping of the form factors in this region has been carried out by the BLAST experiment, with longitudinally polarized electrons scattered from an internal polarized ABS hydrogen target. In addition to the usual advantages of double-polarization measurements, systematic errors were further reduced by having an isotopically pure, windowless, highly polarized target. The large acceptance spectrometer was capable of measuring multiple reactions at all Q^2 values simultaneously. By orienting the target spin angle at 45° with respect to the beam, the elastic ^{1}H(e, e′p) asymmetry was measured at different kinematics in the left and right sectors to extract $\mu G_E^p/G_M^p$ independent of beam and target polarization, significantly improving the world precision of G_E^p and G_M^p in this region.[17] Using vector polarized deuterium, G_E^n was measured from the quasielastic ^{2}H(e, e′n) channel, and G_M^n from the inclusive ^{2}H(e, e′) channel. In both cases the product of beam and target polarizations was obtained from ^{2}H(e, e′p). The BLAST results were consistent with the aforementioned findings.[16]

Complementary to electron scattering, at low Q^2, there is an ongoing experiment at PSI[18] to measure $\langle r_p \rangle$ to 0.1% accuracy from the Lamb shift of muonic hydrogen, which is 200 times more sensitive to $\langle r_p \rangle$ than that of ordinary hydrogen. There is also a neutron interferometry experiment at NIST[19] to measure $\langle r_n^2 \rangle$ to 0.3% accuracy from the e-n scattering length, extracted from the dynamical phase shift through the Bragg condition.

The author wishes to thank Haiyan Gao for helpful discussions. This work is supported in part by the U.S. Department of Energy.

References

1. V. Punjabi *et al.*, *Phys. Rev. C* **71**, 055202 (2005).
2. I.A. Qattan *et al.*, *Phys. Rev. Lett.* **94**, 142301 (2005).
3. M.K. Jones, *et al.*, *Phys. Rev. C* **74**, 035201 (2006).
4. P.G. Blunden, W. Melnitchouk, J.A. Tjon, *Phys. Rev. C* **72**, 034612 (2005).
5. Y.C. Chen, *et al.*, *Phys. Rev. D* **72**, 013008 (2005).
6. JLab E04-116, E04-019, E05-017; Arrington *et al.*, nucl-ex/0408020 (2004).
7. B. Plaster *et al.*, *Phys. Rev. C* **73**, 025205 (2006).
8. G. Warren *et al.*, *Phys. Rev. Lett.* **92**, 042301 (2004).
9. D.I. Glazier *et al.*, *Eur. Phys. J.* **A24**, 101 (2005).
10. JLab E02-013.
11. W.K. Brooks, J.D. Lachniet, *Nucl. Phys.* **A755**, 261 (2005).
12. H.-W. Hammer and Ulf-G. Meissner, *Eur. Phys. J.* **A20**, 469 (2004).
13. R. Bijker and F. Iachello, *Phys. Rev. C* **69**, 068201 (2004).
14. A. Hoell *et al.*, *Nucl. Phys.* **A755**, 298 (2005).
15. F. Gross, P. Agbakpe, *Phys. Rev. C* **73**, 015203 (2006).
16. A. Faessler, *et al.*, *Phys. Rev. D* **73**, 114021 (2006).
17. C.B. Crawford *et al.*, nucl-ex/0609007 (2006).
18. A. Antognini *et al.*, *A.I.P. Proc.* **796**, 253 (2005).
19. F.E. Wietfeldt *et al.*, nucl-ex/0509018 (2005).

N and N to Δ TRANSITION FORM FACTORS FROM Lattice QCD

C. ALEXANDROU

Department of Physics, University of Cyprus,
P.O. Box 20537, CY-1678 Nicosia, Cyprus
E-mail: alexand@ucy.ac.cy

We present recent lattice QCD results on nucleon form factors and N to Δ transition form factors. We predict the parity violating asymmetry in N to Δ and check the off-diagonal Goldberger-Treiman relation.

We present the evaluation, within lattice QCD, of fundamental physical quantities of the nucleon-Δ system. On the theoretical side, providing a complete set of form factors and coupling constants constitutes an important input for model builders and for fixing the parameters of chiral effective theories. On the experimental side, there is ongoing effort to measure accurately these quantities. Examples are the recent polarization experiments, of the electric, G_E, and magnetic, G_M, nucleon form factors, and the accurate measurements on the electric and scalar quadrupole multipoles and the magnetic dipole in N to Δ transition as well as the ongoing experiment to measure the parity violating asymmetry in N to Δ. Using state-of-the-art-lattice techniques we obtain results with small statistical errors for pion masses, m_π, in the range 600-360 MeV. We use two flavors of dynamical Wilson fermions and domain wall valence quarks (DWF) on MILC configurations to study the role of pion cloud contributions. Results obtained with dynamical Wilson fermions and DWF are in agreement showing that lattice artifacts are under control.

In this work only the isovector nucleon form factors are evaluated since isoscalar contributions involved quark loops that are technically difficult to calculate. In Fig. 1 we display the momentum dependence of the ratio of the isovector electric to magnetic form factors for the lightest pion mass namely 410 MeV in the quenched theory and 380 MeV for dynamical Wilson fermions.[1] The lattice results are in agreement but higher than experiment.

Given the weak quark mass dependence of this ratio for the quark masses used in this work a linear extrapolation in m_π^2 to the physical limit fails to reproduce experiment. On the other hand, the isovector magnetic moments extracted from lattice results using a dipole Ansatz are well described by chiral effective theory[2] that includes explicitly the Δ. As can be seen in Fig. 1 the extrapolated value of the magnetic moment is in agreement with experiment. This is consistent with the fact that lattice results are closer to experiment for the magnetic than for the electric form factor.

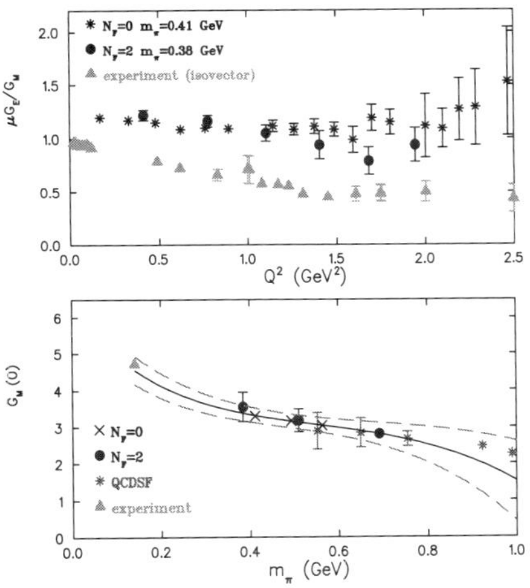

Fig. 1. Top: $\mu G_E/G_M$ with $\mu = 4.71$ versus $Q^2 = -q^2$ for the lightest pion for quenched and dynamical Wilson fermions. Bottom: The magnetic moment versus m_π.

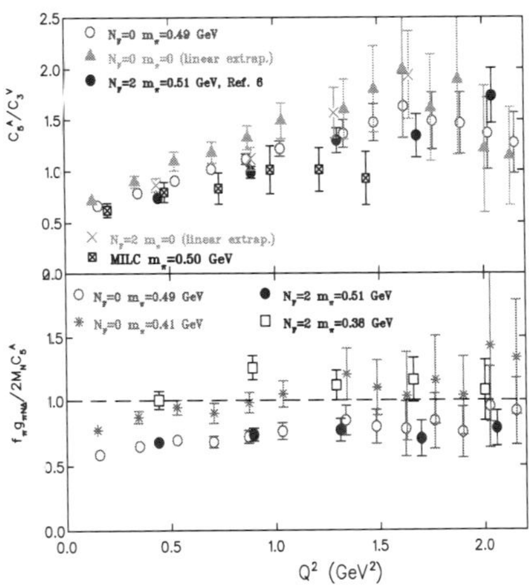

Fig. 2. Top: C_5^A/C_3^V versus Q^2 for Wilson and DWF fermions at $m_\pi \sim 500$ MeV and, at the physical limit for Wilson fermions. Bottom: The ratio R_{GT} for Wilson fermions versus Q^2 for the two smallest pion masses.

We evaluate the three electromagnetic Sachs and four axial Adler form factors for the N to Δ transition. We show in Fig. 2 the ratio C_5^A/C_3^V, which is the analogue of g_A/g_V and determines, to a first approximation, the parity violating asymmetry[3] . The off-diagonal Goldberger-Treiman relation implies that the ratio $R_{GT} = \frac{f_\pi g_{\pi N\Delta}(Q^2)}{2M_N C_5^A(Q^2}$, is one, where $g_{\pi N\Delta}(Q^2)$ is determined from the matrix element of the pseudoscalar density $< \Delta^+|\bar{\psi}(x)\gamma_5 \frac{\tau^3}{2}\psi(x)|p >$ and f_π is the pion decay constant. As can be seen in Fig. 2 this ratio approaches unity as m_π decreases from ~ 500 MeV to $\sim 410(380)$ MeV for quenched (dynamical) Wilson fermions.

References

1. C. Alexandrou *et al.*, Phys. Rev. D **74** 034508 (2006).
2. T. R. Hemmert and W. Weise, Eur. Phys. J. A **15**,487 (2002).
3. C. Alexandrou, hep-lat/0608025, C. Alexandrou *et al.*, hep-lat/0607030.

THE $\gamma^* N \to \Delta$ TRANSITION AT LOW Q^2 AND THE PIONIC CONTRIBUTION

S. STAVE* †

Department of Physics, LNS, MIT,
Cambridge, MA 02139, USA

and an

MIT/Mainz A1/Athens Collaboration

The first excited state of the proton, the Δ, can be reached through a magnetic spin flip of one of the quarks or through quadrupole amplitudes which indicate a deviation from spherical symmetry. The quark models using the color hyperfine interaction underestimate the size of the quadrupole amplitudes by an order of magnitude. Chiral symmetry breaking leads to a pion cloud around the proton and Δ. Calculations which include the effects of the pion cloud do a much better job of describing the data. New data taken at the Mainz Microtron fill in gaps in the low Q^2, long distance region where the pion cloud is expected to dominate. The latest multipole ratios will be shown and compared with models and recent lattice and chiral perturbation theory results. The new data continue to confirm the presence of non-spherical components in the N/Δ system and test the various calculations and models.

Since Chiral Dynamics 2003, several items have been accomplished: New chiral effective field theory calculations with errors by Gail and Hemmert and Pascalutsa and Vanderhaeghen,[1,2] new lattice QCD calculations,[3] studies of the effect of background amplitudes by the authors and, the subject of this article, new data taken at Mainz $Q^2 = 0.06$ to 0.20 (GeV/c)^2. New data are also available from Jefferson Lab at $Q^2 = 0.16$ to 0.36 (GeV/c)^2 but are not public yet.

See Stave *et al.* and Sparveris *et al.*[4] for a more complete discus-

*Current address: TUNL/Dept. of Physics, Duke University, Durham, NC 27708, USA
†E-mail: stave@tunl.duke.edu

sion of the experimental and theoretical details. Figure 1 shows our new CMR $= \mathrm{Re}(S_{1+}^{3/2}/M_{1+}^{3/2})$ data points compared with existing data, models, and calculations. The new data differ greatly from the quark model calculations and show the importance of including the effect of the pion cloud. The lattice QCD calculation deviates from the data at the lower Q^2 values. Calculations indicate that this discrepancy can be fixed using a chiral extrapolation.[1] Finally, while no other model is able to describe the data exactly, the models which dynamically calculate the effect of the pion cloud follow the same trend as the data further indicating the presence and importance of the pion cloud.

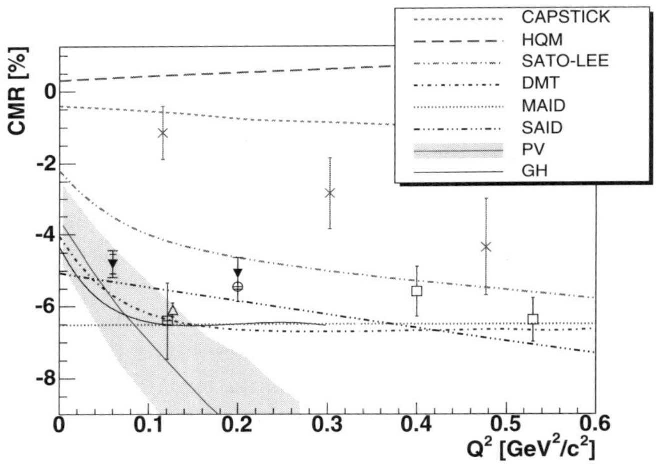

Fig. 1. The low Q^2 dependence of the CMR at $W = 1232$ MeV for the $\gamma^* p \rightarrow \Delta$ reaction. The ▼ symbols are our data points and include the experimental and model errors added in quadrature. Other data: Mainz photon point (Beck) ○, CLAS (Joo) □, Bates (Sparveris) △, Mainz (Elsner) ⊕, Mainz (Pospischil) ⊞. Calculations: lattice QCD with linear pion mass extrapolations of Alexandrou *et al.* ×, chiral perturbation calculations of Pascalutsa and Vanderhaeghen (PV) and Gail and Hemmert (GH). See Stave *et al.* and Sparveris *et al.*[4] for full references.

References

1. V. Pascalutsa and M. Vanderhaeghen, *Phys. Rev.* **D73**, p. 034003 (2006).
2. T. A. Gail and T. R. Hemmert (2005).
3. C. Alexandrou *et al.*, *Phys. Rev. Lett.* **94**, p. 021601 (2005).
4. S. Stave *et al.*, (2006), nucl-ex/0604013; Sparveris *et al.*, to be published.

STRANGE QUARK CONTRIBUTIONS TO THE FORM FACTORS OF THE NUCLEON

FATIHA BENMOKHTAR (for the G^0 collaboration)

Department of Physics, University of Maryland,
College park, MD 20742, USA
** E-mail: fatiha@jlab.org*

A brief summary of the experimental program to extract strange quark contributions to the nucleon's electromagnetic properties is reported.

Keywords: Strange quark; Form factors.

1. Introduction

The contributions of strange quarks to nucleon properties have been studied in several observables. The are contributions to the momentum from deep inelastic neutrino scattering, to the spin with polarized deep inelastic electron scattering and to the mass with pion-nucleon scattering. In order to extract the contribution of strange quarks to the ground state charge and magnetization distributions of the nucleon, several Parity Violating (PV) electron scattering experiments have been carried out during the last decade. These experiments involve measurement of the helicity dependent cross section of elastically scattered polarized electrons from an unpolarized target. Combining several experiments at different kinematics along with existing data on the nucleon's electromagnetic form factors will allow the separation of the contributions of up, down and strange quarks to these distributions.

2. Experimental status

Four major experimental programs have been underway to access the Q^2 range of 0.1-1$(\text{GeV/c})^2$. The SAMPLE experiment at Bates[1] determined the asymmetry from H and D targets for backward angles at Q^2=0.04 and 0.1 $(\text{GeV/c})^2$, allowing the separation of the magnetic (G^s_M) and axial (G^e_A) form factors. In the PVA4 experiment at MAMI,[2] the asymmetry on the proton at forward angles for $Q^2 = 0.1$ and 0.23 $(\text{GeV/c})^2$ was determined. Measurements at backward angles with H and D targets are presently underway at Q^2=0.23 $(\text{GeV/c})^2$, with expected completion in the next few years. The HAPPEX collaboration at Jefferson Laboratory, determined the PV asymmetry from the proton at forward angles for $Q^2 = 0.1$ and 0.48 $(\text{GeV/c})^2$, and from ^{4}He at $Q^2 = 0.1$, which when combined allow separate extratcions of G_E^s and G_M^s. New results have recently been reported,[5] which have significantly tightened the constraints on G_M^s and G_E^s at low momentum transfer. A third HAPPEX program at Q^2=0.6 $(\text{GeV/c})^2$ is scheduled to run in 2008.

The G0 experiment at Jefferson Laboratory, is a program to perform the complete separation of the electric, magnetic and axial form factors for two different momentum transfers, 0.23 and 0.6 $(\text{GeV/c})^2$. The forward angle asymmetry results, for electron scattering angles between 7 and 15° corresponding to a Q^2 range of 0.1-1 $(\text{GeV/c})^2$, were published in 2005.[4] For these, a toroidal spectrometer was used to detect the recoiling protons from a liquid hydrogen target. The results show a non trivial Q^2 dependence which may indicate two contributions of opposite sign. The G0 data are in agreement with other experiments for the common Q^2 values (0.1, 0.23 and 0.48 $(\text{GeV/c})^2$).

The backward angle phase of the G0 experiment is presently underway, corresponding to an electron scattering angle of about 110°. Data are being collected with hydrogen and deuterium targets for beam energies of 362 and 687 MeV. The toriodal spectrometer was turned around, and scattered electrons are detected using the original detector system augmented by an array of cryostat exit detectors and aerogel Čerenkov counters that allow separation of elastic and inelastic scattering as well as rejection of pions. This phase of the experiment also allows the first measurement in the neutral-weak process of the N-Δ axial transition form factor. Data collection will continue through early 2007.

References

1. E.J. Beise, M.L. Pitt, & D.T. Spayde, Prog. Part. Nucl. Phys., **54** (2005)289.
2. F. Maas *et al.*, Phys. Rev. Lett. **93** (2004) 022002.
3. K.A. Aniol, Phys. Lett. **B 635** (2006) 275.
4. D. Armstrong *et al.*, Phys. Rev. Lett. **95**, 092001 (2005).
5. B. Michaels, these proceedings, A. Acha *et al.*, nucl-ex/0609002.

DYNAMICAL POLARIZABILITIES OF THE NUCLEON

B. PASQUINI

Dipartimento di Fisica Nucleare e Teorica, Università degli Studi di Pavia, and INFN, Sezione di Pavia, Pavia, Italy
E-mail: pasquini@pv.infn.it

We introduce the concept of dynamical polarizabilities in real Compton scattering off the nucleon as a new tool to test and interpret predictions about the low energy degrees of freedom inside the nucleon. We show results for the energy dependence of these dynamical polarizabilities both from Dispersion Theory and from leading one-loop Chiral Effective Theory.

1. Introduction

Static nucleon polarisabilities gauge the stiffness of the nucleon against an external electromagnetic field, parametrizing the part of the real Compton scattering (RCS) amplitude at zero energy which is beyond the contribution from the pole terms, i.e. the interaction of the photon with a pointlike nucleon with anomalous magnetic moment κ. Dynamical nucleon polarizabilities are the energy dependent generalization to RCS, and thus provide more information about the low-energy effective degrees of freedom inside the nucleon. Here, I sketch their definition and interpretation, referring to Refs.[1–3] for more details.

2. Definition of dynamical polarizabilities and results

By performing a multipole expansion of the non-pole part of the Compton amplitude, we can parametrize the response of the nucleon at leading order in terms of six polarizabilities: two spin-independent ones ($\alpha_{E1}(\omega)$ and $\beta_{M1}(\omega)$) for electric and magnetic transitions which do not couple to the nucleon spin; and, in the spin sector, two diagonal spin-polarizabilities ($\gamma_{E1E1}(\omega)$, $\gamma_{M1M1}(\omega)$), and two off-diagonal spin polarizabilties ($\gamma_{E1M2}(\omega)$ and $\gamma_{M1E2}(\omega)$). These dynamical polarizabilities are functions of the photon energy ω, and in the limit $\omega \to 0$ reproduce the static polarizabilities. Their energy dependence contains detailed information about dispersive effects caused by internal relaxation, baryonic resonances and mesonic production thresholds. This is clearly seen in Fig. 1, where we show a comparison between a Dispersion Relation (DR) analysis,[4] leading-one loop heavy baryon Chiral Perturbation Theory (ChPT) as well leading-one loop Small Scale Expansion (SSE) (with an error band associated with the uncertainty in three energy independent parameters of the theory fitted to unpolarized Compton scattering data). In α_{E1} and γ_{E1E1} all three approaches agree remarkably well, displaying a pronounced cusp at the pion production threshold. In β_{M1} the SSE approach and the dispersion analysis both predict a rapidly increasing paramagnetic response of the nucleon as function of the photon energy. This feature is due to the strong $\gamma N\Delta$ $M1$ excitation which in heavy baryon ChPT only gradually gets built up via higher order terms. The Δ-pole contribution is visible also in γ_{M1M1}, but it does not rise as dramatically as in the case of β_{M1}. The mixed spin polarizabilities are quite small, with an energy dependence similar in different approaches, but rather different values at zero energy. Concerning the possibility to extract these six dipole polarizabilities from unpolarized and polarized RCS experiments, we refer to the analysis of Refs.[3,5]

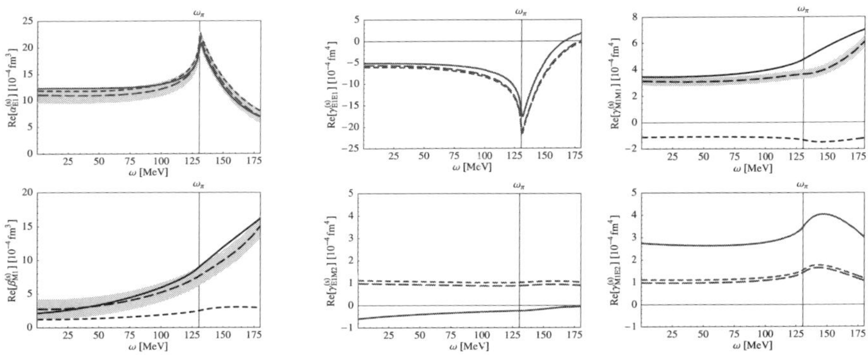

Fig. 1: Results for the dipole polarisabilities within DR analysis (solid) and χEFT with (long dashed + band from fit errors) and without (short dashed) explicit Δ. ω_π denotes the one-pion production threshold.

References

1. H. W. Grießhammer and T. R. Hemmert, *Phys. Rev. C* **65**, 045207 (2002).
2. R. P. Hildebrandt, et al., *Eur. Phys. J. A* **20**, 329 (2004).
3. R. P. Hildebrandt, et al., *Eur. Phys. J. A* **20**, 293 (2004).
4. D. Drechsel, B. Pasquini, M. Vanderhaeghen, *Phys. Rep.* **378** (2003), 99.
5. D. Drechsel, B. Pasquini, M. Vanderhaeghen, in preparation.

HADRON MAGNETIC MOMENTS AND POLARIZABILITIES IN LATTICE QCD

FRANK X. LEE*

Center for Nuclear Studies, Physics Department, George Washington University, Washington, DC 20052, USA

**E-Mail: fxlee@gwu.edu*

Results are presented for the magnetic moments and polarizabilities for a variety of hadrons in the background field method.

In the background field method, the mass shift in the presence of small fields can be expanded as a polynomial in the case of magnetic field B, $\delta m(B) = m(B) - m(0) = c_1 B + c_2 B^2 + c_3 B^3 + \cdots$ The linear coefficient is related to magnetic moments and the quadratic coefficient is related to the magnetic dipole polarizability (β). In practice, we separate the odd- and even-powered terms by computing the mass shifts both in the field B and its reverse $-B$ for each value of B. In order to place a magnetic field on the lattice, we construct an analogy to the continuum case. The fermion action is modified by the minimal coupling prescription $D_\mu = \partial_\mu + gG_\mu + qA_\mu$ where q is the charge of the fermion field and A_μ is the vector potential describing the background field. On the lattice, the prescription amounts to a modified link variable $U'_\mu = U_\mu U_\mu^{(B)}$. Choosing $A_y = Bx$, a constant magnetic field B can be introduced in the z-direction. Then the phase factor is in the y-links $U_y^{(B)} = \exp(iqa^2 Bx)$.

Using standard lattice technology (Wilson action, 24^4 quenched lattice at $\beta = 6.0$, 150 configurations), we have computed the magnetic moments of the baryon octet and decuplet,[2] as a function pion mass down to about 500 MeV. The results show that the method is robust, and provide an alternative to the form factor method.[4,5] Only mass shifts are required. We also computed electric polarizabilities for a number of neutral particles,[1] most of which are new. For magnetic polarizability,[3] we did a sweeping calculation for about 30 particles covering the entire baryon octet and decuplet plus selected mesons.[3] Aside from the proton and neutron, all other results for magnetic polarizability are predictions. The results are compared with available values from experiments and other theoretical calculations. Overall, our results are consistent with experimental observations, but systematic errors due to the lattice must be assessed before more definite comparisons can be made. As far as the cost is concerned, it is equivalent to 11 standard mass-spectrum calculations using the same action (5 values of the parameter η to provide 4 non-zero magnetic fields, 5 to reverse the field, plus the zero-field to set the baseline). This factor can be reduced to 7 if only two non-zero values of magnetic field are desired.

The interaction of a hadron with external electric and magnetic fields can be characterized by

$$H = -\frac{1}{2}\alpha E^2 - \frac{1}{2}\beta B^2 - \frac{1}{2}\left(\alpha_{E\nu}\dot{\mathbf{E}}^2 + \beta_{M\nu}\dot{\mathbf{B}}^2\right) - \frac{1}{12}\left(\alpha_{E2}E_{ij}^2 + \beta_{M2}B_{ij}^2\right)$$
$$-\frac{1}{2}\left(\gamma_{E1}\sigma\cdot\mathbf{E}\times\dot{\mathbf{E}} + \gamma_{M1}\sigma\cdot\mathbf{B}\times\dot{\mathbf{B}} - 2\gamma_{M1E2}E_{ij}\sigma_i B_j + 2\gamma_{E1M2}B_{ij}\sigma_i E_j\right), \quad (1)$$

where the dots mean a time derivative and E_{ij} and B_{ij} are spatial derivatives. In addition to α and β which enter at order $\mathcal{O}(\omega^2)$ in low-energy Compton scattering amplitude, there are eight more polarizabilities: four spin polarizabilities (γ_{E1}, γ_{M1}, γ_{M1E2}, γ_{E1M2}) at order $\mathcal{O}(\omega^3)$; two dispersion corrections to the dipole polarizabilities ($\alpha_{E\nu}$, $\beta_{M\nu}$); and two quadrupole polarizabilities (α_{E2}, β_{M2}) at order $\mathcal{O}(\omega^4)$. They describe hadron internal structure in increasingly finer detail, but our knowledge on them is fairly limited. Lattice QCD offers an unique opportunity to make predictions and to guide experiments. On the lattice, each one can be isolated by carefully measuring mass shifts using specifically-designed time and space-varying electric and magnetic fields. Computations are under way to extract these polarizabilities on the lattice. There also exists work to account for finite-volume effects in such lattice calculations.[6]

This work is supported in part by U.S. Department of Energy under grant DE-FG02-95ER40907. The computing resources at NERSC and JLab have been used.

References

1. J. Christensen, W. Wilcox, F.X. Lee, L. Zhou, Phys. Rev. D 72, 034503 (2005).
2. F.X. Lee, R. Kelly, L. Zhou, W. Wilcox, Phys. Lett. B 627, 71 (2005).
3. F.X. Lee, L. Zhou, W. Wilcox, J. Christensen, Phys. Rev. D 73, 034503 (2006).
4. S. Boinepalli, D. B. Leinweber, A. G. Williams, J. M. Zanotti, J. B. Zhang, hep-lat/0604022
5. V. Pascalutsa, M. Vanderhaeghen, and S.N. Yang, hep-ph/0609004.
6. W. Detmold, B.C. Tiburzi, A. Walker-Loud, Phys. Rev. D 73, 114505 (2006).

SPIN-DEPENDENT COMPTON SCATTERING FROM ^{3}He AND THE NEUTRON SPIN POLARIZABILITIES

HAIYAN GAO

Triangle Universities Laboratory and Department of Physics
Duke University, Durham, NC 27708, U.S.A.
**E-mail: gao@phy.duke.edu*

A new experiment is being planned at the High Intensity Gamma Source (HIγS) at Duke Free Electron Laboratory for first measurements of the spin-dependent asymmetries from elastic Compton scattering of circularly polarized photons from a high-pressure polarized ^{3}He gas target. The new experiment will allow for an extraction of neutron spin polarizabilities.

1. Introduction

Nucleon polarizabilities can be accessed through low energy Compton scattering experiments. Compton scattering from nucleon at low energy is specified by low-energy theorems up to and including terms linear in photon energy. These terms are completely determined by the nucleon charge, mass and its anomalous magnetic moment. At next to leading order in photon energy, new structure constants: the electric (α) and the magnetic (β) scalar polarizabilities appear. At the third order, four new parameters, γ_1 to γ_4, the spin polarizabilities appear. Very little is known about nucleon spin polarizabilities. The only quantities that have been extracted from experiments so far are the forward and backward spin polarizabilities, γ_0 and γ_π, which are two independent linear combinations of γ_1, γ_2 and γ_4.

Spin polarizabilities of the nucleon can be probed directly via circularly polarized photons Compton scattered from a polarized nucleon target.[2,3] We plan to carry out a double polarization experiment with the use of circularly polarized photons incident on a polarized target, where the target spin direction is either aligned along the beam direction, or in the scattering plane transverse to the beam direction.

2. The experiment

A polarized ^{3}He nucleus is very useful in probing the neutron electromagnetic and spin structure because of the unique spin structure of the ^{3}He ground state. We are planning a new experiment on Compton scattering experiment from a polarized ^{3}He target at the elastic kinematics. The polarized elastic Compton scattering from a polarized ^{3}He target allows for an extraction of the neutron spin polarizabilities as shown in a recent calculation.[4] The advantage is a larger Compton scattering cross-section and a relatively straightforward theoretical calculation of the Compton scattering process compared with quasifree Compton scattering process from ^{3}He. The complication is that the extraction of the neutron spin polarizabilities will be sensitive to our knowledge of the proton spin polarizabilities. Another experiment[5] on polarized Compton scattering with circularly polarized photons from a polarized proton target is also being planned at the HIγS facility which will determine the proton spin polarizabilities. This experiment is expected to run before the polarized ^{3}He Compton experiment.

We plan to carry out the aforementioned double polarization Compton scattering experiment at the HIγS facility at Duke Free-electron Laser Laboratory with an incident photon energy around 100 MeV. The Compton scattered photons will be detected using a large acceptance, to-be-upgraded NaI detector array.[1] The polarized ^{3}He target is based on the principle of spin-exchange optically pumping.[6] The design is similar to that used in electron scattering experiments in Hall A at Jefferson Lab (JLab),[7] but with a much larger target cell diameter due to the larger photon beam size. With a density of 2.75×10^{20} atoms/cm^3, and a target cell length of 40 cm, the target thickness will be 1.0×10^{22} atoms/cm^2. A target polarization of $\sim 40\%$ has been achieved for the HIγS polarized ^{3}He target.[8]

The neutron spin polarizabilities extracted from the planned experiment together with the proton spin polarizabilities extracted from the proton experiment[5] will allow for tests of theoretical predictions of the nucleon spin polarizabilities.

Acknowledgment

We thank D. Choudhury, M. Fujiwara, H.W. Griesshammer, T.R. Hemmert, R. Hildebrandt, J.A. McGovern, A. Nathan, D.R. Phillips, and H. Weller for stimulating discussions. This work is supported by the U.S. Department of Energy under contract number DE-FG02-03ER41231 and the School of Arts and Science at Duke University.

References

1. H. Gao and H. Weller *et al.* on "Compton scattering from nucleon and nuclear targets", http://www.tunl.duke.edu/~mep/higs
2. J. McGovern, private communication.
3. R.P. Hildebrandt *et al.*, Eur. Phys. J. **A20**, 329 (2004).
4. D. Choudhury *et al.* in this proceedings; nucl-th/0611032.
5. R. Miskimen, private communication.
6. M. A. Bouchiat *et al.*, Phys. Rev. Lett. **5**,373 (1960); N. D. Bhaskar *et al.*, Phys. Rev. Lett. **49**, 25 (1982); W. Happer *et al.*, Phys. Rev. **A29**, 3092(1984).
7. J. Alcorn, *et al.*, Nucl. Instru. Method A522, 294-346 (2004).
8. X. Zong *et al.*, to be submitted.

CHIRAL DYNAMICS FROM DYSON-SCHWINGER EQUATIONS

C. D. ROBERTS

Physics Division, Argonne National Laboratory
Argonne, IL, 60439, USA

A strongly momentum-dependent dressed-quark mass function is basic to QCD. It is central to the appearance of a constituent-quark mass-scale and an existential prerequisite for Goldstone modes. Dyson-Schwinger equation (DSE) studies have long emphasised these facts and are a natural way to exploit them.

Contemporary results from the world's experimental hadron facilities impact dramatically on our understanding of the strong interaction.[1] Theory requires flexible tools, which can rapidly provide an intuitive understanding of information in hand and simultaneously anticipate its likely consequences. Models, parametrisations and truncations of QCD play this role. Prominent amongst these are the DSEs and truncations thereof, and herein we will intimate recent progress.

The gap equation is primary. Its dynamical chiral symmetry breaking solution (DCSB) only has a chiral expansion on a measurable domain of current-quark mass.[2] Within this domain a perturbative expansion in the current-quark mass around the chiral limit ($\hat{m} = 0$) is uniformly valid for physical quantities. That fails at $\hat{m}_{\rm cr} \sim 60$–$70\,$MeV; i.e., a mass $m_{0^-} \sim 0.45\,$GeV for a flavour-nonsinglet 0^- meson constituted of equal mass quarks.

For chiral dynamics it is singularly important that a nonperturbative, systematic and symmetry preserving DSE truncation exists.[3] That enables proof of Goldstone's theorem in QCD[4] and yields

$$f_{\pi_n}\, m_{\pi_n}^2 = 2m(\zeta^2)\, \rho_{\pi_n}(\zeta^2)\,, \tag{1}$$

wherein f_{π_n} is the leptonic (pseudovector) decay constant for a pseudoscalar meson and ρ_{π_n} is the pseudoscalar analogue. ($n = 0$ denotes the lowest-mass pseudoscalar and increasing n labels bound-states of increasing mass.) The

Gell-Mann–Oakes-Renner relation is a corollary of (1). Importantly, (1) is equally valid for a meson containing one heavy-quark; i.e., a heavy-light system,[5] and also heavy-heavy mesons.[6] Equation (1) entails[7] that for $\hat{m} = 0$, $f^0_{\pi_n} \equiv 0\,, \forall n \geq 1$; viz., Goldstone modes are the only mesons to possess a nonzero leptonic decay constant in the chiral limit when chiral symmetry is dynamically broken. The decay constants vanish for all other pseudoscalar mesons on this trajectory, e.g., radial excitations. NB. In the absence of DCSB, the leptonic decay constant of *all* such pseudoscalar meson vanishes when $\hat{m} = 0$; namely, $f^0_{\pi_n} \equiv 0\,, \forall n \geq 0$. The consequences of the axial-vector Ward-Takahashi identity for two-photon decays of pseudoscalar mesons have been explored.[8] Amongst other things it has been proven that when $\hat{m} = 0$ the leading-order ultraviolet power-law behaviour of the transition form factor for excited state pseudoscalar mesons is $O(1/Q^4)$.

The study of scalar mesons is naturally important. Modern DSE analyses of the u, d-quark scalar meson provide results that are compatible with a picture of the lightest 0^{++} as a bound state of a dressed-quark and -antiquark supplemented by a material pion cloud.[9] Meson, baryon and dressed-quark σ-terms have also been studied.[9,10] Consequences of symmetries are manifest and the behaviour of the dressed-quark σ-term shows that the essentially dynamical component of chiral symmetry breaking decreases with increasing current-quark mass.

Acknowledgments

This work was supported by US Department of Energy, Office of Nuclear Physics, contract no. DE-AC02-06CH11357.

References

1. J. Arrington, C.D. Roberts and J.M. Zanotti, "Nucleon electromagnetic form factors," nucl-th/0611050.
2. L. Chang, Y.-X. Liu, M. S. Bhagwat, C. D. Roberts and S. V. Wright, "Dynamical chiral symmetry breaking and a critical mass," nucl-th/0605058.
3. M. S. Bhagwat, A. Höll, A. Krassnigg, C. D. Roberts and P. C. Tandy, *Phys. Rev.* **C 70**, 035205 (2004); and references therein.
4. P. Maris, C. D. Roberts and P. C. Tandy, *Phys. Lett.* **B 420**, 267 (1998).
5. M. A. Ivanov, Yu. L. Kalinovsky and C. D. Roberts, *Phys. Rev.* **D 60**, 034018 (1999).
6. M. S. Bhagwat, A. Krassnigg, P. Maris and C. D. Roberts, "Mind the gap," nucl-th/0612027.
7. A. Höll, A. Krassnigg and C. D. Roberts, *Phys. Rev.* **C 70** (2004) 042203(R).
8. A. Höll, A. Krassnigg, P. Maris, C. D. Roberts and S. V. Wright, *Phys. Rev.* **C 71** (2005) 065204.

9. A. Höll, P. Maris, C. D. Roberts and S. V. Wright, "Schwinger functions and light-quark bound states, and sigma terms," nucl-th/0512048.
10. V. V. Flambaum, A. Höll, P. Jaikumar, C. D. Roberts and S. V. Wright, *Few Body Syst.* **38**, 31 (2006).

RADIATIVE NEUTRON β–DECAY IN EFFECTIVE FIELD THEORY

S. GARDNER

Department of Physics and Astronomy, University of Kentucky,
Lexington, KY 40506-0055, USA
E-mail: gardner@pa.uky.edu

I consider radiative β–decay of the neutron in heavy baryon chiral perturbation theory. Nucleon-structure effects not encoded in the weak coupling constants g_A and g_V enter at the $\mathcal{O}(0.5\%)$-level, making a sensitive test of the Dirac structure of the weak currents possible.

Radiative neutron beta-decay enters the radiative corrections to neutron beta-decay and is necessary to realizing precision tests of the Standard Model. Here we study this process in its own right,[1] (i) to realize an alternative determination of g_V and g_A, (ii) to study the hadron matrix elements in $\mathcal{O}(1/M)$, as the same enter in muon radiative capture,[2] and (iii) to test the Dirac structure of the weak current, through the circular polarization of the associated photon.[3,4] I report on a systematic analysis of neutron radiative β–decay in the framework of heavy baryon chiral perturbation theory and in the small scale expansion (SSE), including all terms in $O(1/M)$, i.e., at next–to–leading order (NLO) in the small parameter ϵ,[1] where ϵ contains all the small external momenta and quark (pion) masses, relative to the heavy baryon mass M; $\Delta(1232)$–nucleon mass splitting enters as well. In this way the recoil-order corrections can be calculated in a controlled way[1,2] — only the leading bremsstrahlung terms have been computed previously.[3]

The resulting photon energy spectrum $d\Gamma/d\omega$ is shown in Fig. 1,[1] as is the spectrum in leading-order in $\mathcal{O}(1/M)$.[3] The recoil-order corrections are no larger than $\mathcal{O}(0.5\%)$; the SSE contribution is itself of $\mathcal{O}(0.1\%)$. Neutron radiative β-decay has recently been observed;[5] our results are consistent with experiment. We find a branching ratio of $2.85 \cdot 10^{-3}$ for $\omega \in [0.015\,\text{MeV}, 0.340\,\text{MeV}]$, to compare with the experimental result of

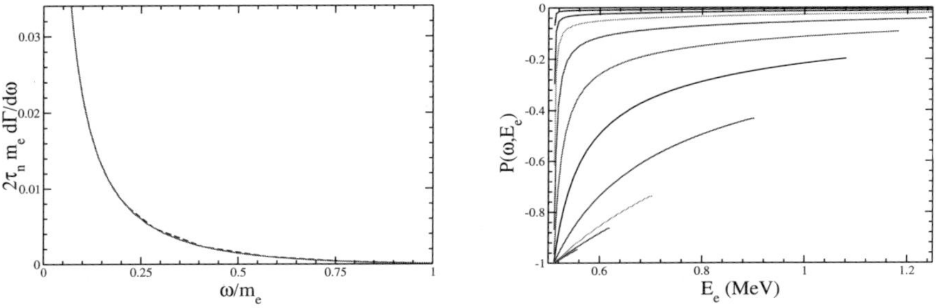

Fig. 1. The photon energy spectrum for radiative neutron β–decay. The dashed line denotes the result to NLO in the SSE, whereas the solid line denotes the leading order result.[3] The photon polarization $P(\omega, E_e)$ in radiative neutron β–decay to NLO in the SSE, as a function of the electron energy E_e for $(E_e^{\max} - E_e)/E_e^{\max} \gtrsim 0.2\%$ and various, fixed photon energy ω. For E_e such that $(E_e^{\max} - E_e)/E_e^{\max} \lesssim 0.2\%$, the polarization plunges to -1 since the electron and photon momenta become anti-parallel. The curves from smallest absolute polarization to largest have $\omega = 0.00539, 0.0135, 0.0265, 0.0534, 0.109, 0.209, 0.390, 0.589, 0.673,$ and 0.736 MeV, respectively.

$(3.13 \pm 0.34) \cdot 10^{-3}$.[5] We also compute the polarization of the emitted photon, where the polarization $P(\omega, E_e) = (T_R - T_L)/(T_R + T_L)$ and $T_{L,R} \equiv d^2\Gamma_{L,R}/d\omega dE_e$. The polarization with ω and E_e is shown in Fig. 1;[1] as ω grows large, the associated electron momentum is pushed towards zero, and the absolute polarization grows larger. In this observable as well the $\mathcal{O}(1/M)$ contributions are $\mathcal{O}(0.5\%)$ or less. Observed departures from this prediction would signify the palpable presence of non-$V - A$ currents; the photon polarization can probe new physics effects to which the correlation coefficients in neutron β–decay are insensitive.[6]

Acknowledgments

S.G. thanks V. Bernard, U.-G. Meißner, and C. Zhang for a very enjoyable collaboration and acknowledges partial support from the U.S. Department of Energy under contract number DE-FG02-96ER40989.

References

1. V. Bernard, S. Gardner, U.-G. Meißner, and C. Zhang, *Phys. Lett.* **B593**, 105 (2004); [*Erratum-ibid.* **B599**, 348 (2004)] and references therein.
2. V. Bernard, T. R. Hemmert, and U.-G. Meißner, *Nucl. Phys.* **A686**, 290 (2001).
3. Y. V. Gaponov and R. U. Khafizov, *Phys. Atom. Nucl.* **59**, 1213 (1996); *Phys. Lett.* **B379**, 7 (1996); *Nucl. Instrum. Meth.* **A440**, 557 (2000).

4. P. C. Martin and R. J. Glauber, *Phys. Rev.* **109**, 1307 (1958) and references therein.
5. J. S. Nico et al., to be published in *Nature*.
6. J. D. Jackson, S. B. Treiman, and H. W. Wyld, Jr., *Phys. Rev.* **106**, 517 (1957).

COMPARISON BETWEEN DIFFERENT RENORMALIZATION SCHEMES FOR COVARIANT BChPT

T. A. Gail*

Physics Department T39, Technische Universität München, 85747 Garching, Germany
* *speaker; E-mail: tgail@ph.tum.de*

In this talk we explore the possibilities of chiral extrapolations using the covariant formulation of chiral effective field theories (EFTs) with baryons. In particular, a study of the convergence of the chiral expansion and of the statistical uncertainties arising from presently unknown parameters in the extrapolation functions is performed for the example of the anomalous magnetic moment of the nucleon. In the calculation of this quantity at next to leading one loop level, i.e. order p^4, we rely on a modified scheme of infrared regularization (IR), which will be discussed in-depth in a forthcoming publication.

Calculating an observable in EFT, one has a freedom of choice which part of the analytic pieces of the result is considered to arise from loop dynamics and which one corresponds to local operators. Taking advantage of this freedom, the IR-scheme[1] was introduced to overcome the long-known problem of power-counting violations in the $\overline{\text{MS}}$-scheme of covariant BChPT. Compared to standard $\overline{\text{MS}}$-results, IR absorbs an infinite sum of terms analytic in the quark mass into low energy constants (LECs). Problems[a] can arise due to the fact, that in EFT at a certain order only a finite number of LECs is included. Therefore, we propose to only consider those analytic terms as part of short distance physics, which can be absorbed into LECs present at the order of the calculation. After this modification the expanded results still reproduce the HBChPT findings at the order of the calculation.

The isovector anomalous magnetic moment of the nucleon in the heavy

[a]For example the unexpanded results can display a non-physical scale dependence or show incorrect thresholds and singularities.

baryon formulation and in modified IR up to $\mathcal{O}(p^4)$ reads:

$$\begin{aligned}
\kappa_V^{\rm HB} &= c_6 - \frac{g_A^2 M_0 m_\pi}{4\pi F_\pi^2} - \frac{g_A^2 m_\pi^2}{8\pi^2 F_\pi^2}\left[2 + c_6 + (7 + 2c_6)\ln\left(\frac{m_\pi}{\lambda}\right)\right] \\
&+ \frac{m_\pi^2}{8\pi^2 F_\pi^2}(4c_4 M_0 - c_6)\ln\left(\frac{m_\pi}{\lambda}\right) - 16 e_{106}^{(r)}(\lambda) M_0 m_\pi^2, \\
\kappa_V^{\overline{\rm IR}} &= c_6 - \frac{g_A^2 M_0 m_\pi}{4\pi F_\pi^2} - \frac{g_A^2 m_\pi^2}{8\pi^2 F_\pi^2}\left[2 + c_6 + (7 + 2c_6)\ln\left(\frac{m_\pi}{\lambda}\right)\right] \\
&+ \frac{m_\pi^2}{8\pi^2 F_\pi^2}(4c_4 M_0 - c_6)\ln\left(\frac{m_\pi}{\lambda}\right) - 16 e_{106}^{(r)}(\lambda) M_0 m_\pi^2 \\
&+ \frac{g_A^2 m_\pi^3}{8\pi F_\pi^2 M_0}(3 + c_6) + \ldots
\end{aligned} \tag{1}$$

The full, unexpanded result for $\kappa_V^{\overline{\rm IR}}$ on which the subsequent discussion is based can be found in forthcoming publication. The resulting quark mass dependence of the anomalous magnetic moment of the nucleon calculated in modified IR has the following properties[b]:

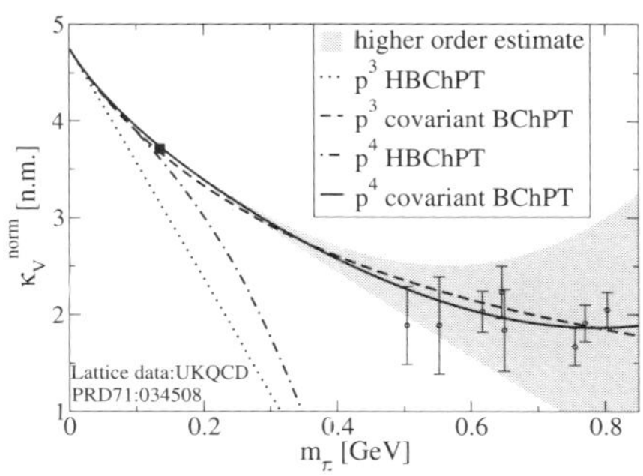

1. A successful chiral extrapolation of currently available lattice data to the physical point can be performed with values for the appearing parameters which are consistent with known information from scattering theory. 2. Statistical uncertainties are small. 3. We do see a clear convergence pattern of the results when going from p^3 to p^4. 4. Estimating higher order effects by adding a typical next order structure and varying its strength

[b]We found the same features in an analogous analysis of the nucleon mass.

within natural size[c], i.e. $\kappa_V^{\overline{\rm IR}} \to \kappa_V^{\overline{\rm IR}} \pm 3\frac{m_\pi^3}{(4\pi F_\pi)^2}$ we finally observe a fast convergence of the expansion up to $m_\pi \approx 500$ MeV.

References

1. T. Becher, H. Leutwyler, Eur.Phys.J.C9:643-671 (1999).

[c]We have also checked, that using realistic quark mass dependencies for $g_A(m_q)$, $F_\pi(m_q)$ and $m_\pi(m_q)$ our $\overline{\rm IR}$-p^4 best-fit curves for κ_V and M_N do not change. This procedure corresponds to an effective inclusion of some fifth order effects.

NON-PERTURBATIVE STUDY OF THE LIGHT PSEUDOSCALAR MASSES IN CHIRAL DYNAMICS

JOSÉ ANTONIO OLLER*

Departamento de Física, Universidad de Murcia,
*E-30071 Murcia, Spain * E-mail: oller@um.es*
www.um.es/fisparts

We perform a non-perturbative chiral study of the masses of the lightest pseudoscalar mesons. The pseudoscalar self-energies are calculated by the evaluation of the scalar self-energy loops with full S-wave meson-meson amplitudes taken from Unitary Chiral Perturbation Theory (UCHPT). These amplitudes, among other features, contain the lightest nonet of scalar resonances σ, $f_0(980)$, $a_0(980)$ and κ. The self-energy loops are regularized by a proper subtraction of the infinities within a dispersion relation formulation of the scattering amplitudes. Values for the bare masses of pions and kaons are obtained, as well as an estimate of the mass of the η_8. We then match to the self-energies from standard Chiral Perturbation Theory (CHPT) to $\mathcal{O}(p^4)$ and resum higher orders from our calculated scalar self-energies. The dependence of the self-energies on the quark masses allows a determination of the ratio of the strange quark mass over the mean of the lightest quark masses, $m_s/\hat{m}$, in terms of the $\mathcal{O}(p^4)$ CHPT low energy constant combinations $2L_8^r - L_5^r$ and $2L_6^r - L_4^r$. In this way, we give a range for the values of these low energy counterterms and for $3L_7 + L_8^r$, once the η meson mass is invoked. The low energy constants are further constraint by performing a fit to the recent MILC lattice data on the pseudoscalar masses. An excellent reproduction of the MILC data is obtained, at the level of 1% of relative error in the pseudoscalar masses, and $m_s/\hat{m} = 25.6 \pm 2.5$ results. This value is consistent with 24.4 ± 1.5 from CHPT and phenomenology and more marginally with the value 27.4 ± 0.5, obtained from Staggered CHPT and the lattice data. We report on the work[1] and more details can be found there.

Keywords: Self-energy, Chiral Lagrangians, Quark Masses.

1. Introduction and Results

The scalar meson-meson S-waves with resonant isospins, I=0, 1/2 and 1, are the source for large corrections in low energy meson-meson dynamics. Well known examples are the scattering length a_0^0, $\Gamma(\eta \to 3\pi)$ or D decays. We are now interested in the effects of the S-waves in the self-energies of the lightest pseudoscalars. We evaluate the diagrams of fig.1. The iterated loops correspond to the generation of the S-waves. One then has for the self-energy of the pseudoscalar P, $\Sigma_P = \Sigma_P^U + \Sigma_P^{tad}$,

$$\Sigma_P^U = - \sum_{Q=\pi^0,\pi^+,\pi^-,K^+,K^-,K^0,\bar{K}^0} \int \frac{d^3k}{(2\pi)^3 2E_Q(\mathbf{k})} T_{P\overline{Q}\to P\overline{Q}}(s_1) \; , \quad (1)$$

$$\Sigma_P^{tad} = - \sum_{Q=\pi^0,\pi^+,K^+,K^0} \lambda_Q \int \frac{d^3k}{(2\pi)^3 2E_Q(\mathbf{k})} T_{P\overline{P}\to Q\overline{Q}}(0) \; , \quad (2)$$

with $s_1 = (M_P - E_Q(\mathbf{k}))^2 - \mathbf{k}^2$ for $p = (M_P, \mathbf{0})$.

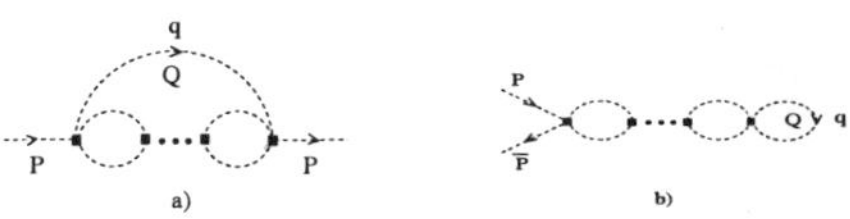

Fig. 1. Diagrams for the calculation of the self-energies.

With this parameterization we fit the data of the MILC Collaboration[2] on the masses of pions and kaons, fig.2. One sees that a good fit to the lattice data results, while obtaining a value for $m_s/\hat{m} = 25.6 \pm 2.5$, that is consistent with the value 24.4 ± 1.5 from CHPT and phenomenology.[3] The central value of the former is much smaller than the value 27.4 ± 0.5 from Staggered CHPT.

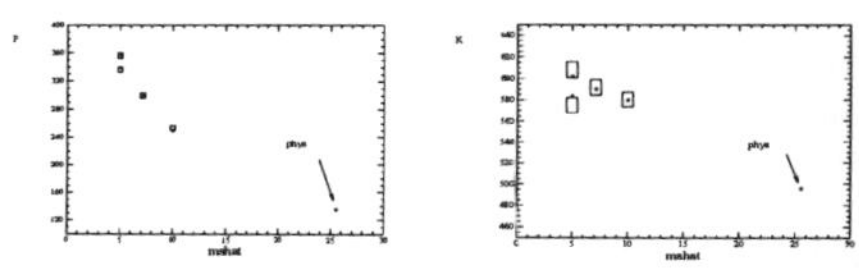

Fig. 2. Lattice data from the MILC Collaboration[2] and their reproduction by our parameterization.

References

1. J. A. Oller and L. Roca, arXiv:hep-ph/0608290. Submitted to Phys. Rev. D.
2. C. Aubin *et al.*, Phys. Rev. **D70**, 094505 (2004).
3. H. Leutwyler, Phys. Lett. **B378**, 313 (1996).

MASSES AND WIDTHS OF HADRONS IN NUCLEAR MATTER

MARTIN KOTULLA

II. Physikalisches Institut, Heinrich-Buff Ring 16, 35392 Giessen, Germany
** E-mail: Martin.J.Kotulla@exp2.physik.uni-giessen.de*

We discuss recent experimental results taken with the Crystal Barrel/TAPS collaboration at the ELSA accelerator on the modification of ω-meson properties in a nuclear medium. The data shows a decreased ω meson mass.

Keywords: omega meson; medium modification; mass

Hadrons are composite systems bound by the strong interaction. However, at larger distances or at energies in the range of the lowest lying hadrons (e.g. the nucleon mass ≈ 1 GeV), the perturbative picture of QCD breaks down and the full complexity manifests itself in a many body structure of gluons, valence quarks and sea quarks. Embedding a hadron into a nucleus, another strongly interacting environment, should necessarily affect (and alter) its properties (e.g. mass). Moreover, chiral symmetry is another player which is at the very heart of QCD. A consequence of chiral symmetry in the hadron spectrum would be the degeneracy of opposite parity states, which is neither realized for baryons nor for mesons. The reason is the spontaneous breakdown of chiral symmetry. However, model calculations suggest a significant temperature and density dependence of the chiral condensate pointing to a partial restoration of chiral symmetry. The experimental observable would be a change of hadron masses towards degeneracy of opposite parity states. Today, the qualitative and quantitative behaviour of the mass of vector mesons is at hand, triggered by a few next generation experiments.[1–3] The general method to measure a modification of a hadron mass is to measure a decay of a sufficiently short lived meson in a nuclear environment. The ω meson is a very well suited probe, its $c\tau = 23$ fm is small enough to have a sufficiently high probability for a decay inside a nucleus and its vacuum width (8.5 MeV) promises a well defined signal. The experiment was performed at the electron stretcher accelerator (ELSA) in

Bonn using the Crystal Barrel and TAPS calorimeters and the ω was reconstructed from its $\pi^0\gamma$ decay.[1] The resulting spectrum of $\pi^0\gamma$ events shows a pronounced ω peak on a smooth background. The background subtracted signal is shown in Fig. 1 for C and Nb targets. Details of the background subtraction can be found in.[1,4] The line-shapes on the nuclear targets shows a significant tail on the low invariant mass side when compared to the reference shapes on LH_2 (vacuum) or a GEANT response simulation. Models are needed to separate quantitatively the integrated mass distribution m(ρ) and to unfold the spectral shape at normal nuclear matter density ρ_0. As an example, Fig. 1 shows the expected experimental line-shape within a BUU calculation with an ω mass lowered by 16% at ρ_0. Although, the experimental resolution of FWHM=55 MeV together with the theoretical models limit the extraction of a precise in-medium signal, evidence for a dropping mass scenario of the ω meson can be concluded. The measurement of the ω absorption as a function of the nuclear size allows to extract the (inelastic) cross section of the ω meson which can be related to its width.[5,6] Preliminary analysis show an increase of the ω meson width. This topic will be elaborated in a forthcoming publication.

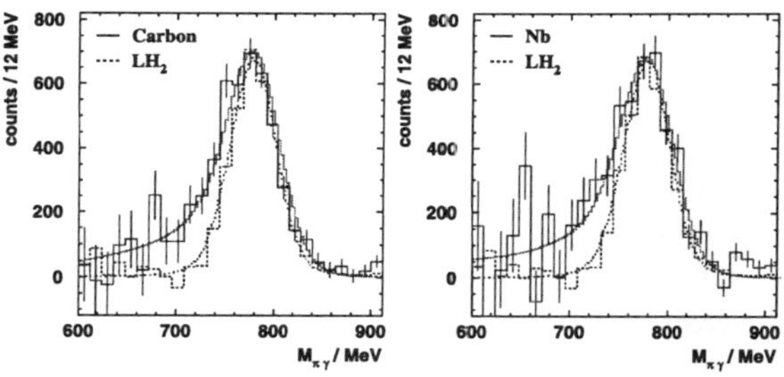

Fig. 1. The ω ($p_\omega \leq 500$ MeV/c) line-shape (solid histogram) on C, Nb compared with a LH_2 reference signal (dashed histogram) and a GEANT simulation (dashed line). A BUU calculation with 16% downward mass shift[7] reproduces the data (solid line).

References

1. D. Trnka *et al.*, *Phys. Rev. Lett.* **94**, p. 192303 (2005).
2. M. Naruki *et al.*, *Phys. Rev. Lett.* **96**, p. 092301 (2006).
3. S. Damjanovic *et al.*, *Phys. Rev. Lett.* **96**, p. 162302 (2006).
4. D. Trnka, *PhD thesis* (2006).
5. P. Muehlich *et al.*, *Nucl. Phys. A* **773**, p. 156 (2006).
6. M. Kaskulov and E. Oset, *priv. communication* (2006).
7. P. Muehlich, *priv. communication* (2005).

CHIRAL EFFECTIVE FIELD THEORY AT FINITE DENSITY

R. J. FURNSTAHL

Department of Physics, The Ohio State University, Columbus, OH 43085, USA
E-mail: furnstahl.1@osu.edu

A path from chiral effective field theory to systematic calculations of nuclear matter and finite nuclei is outlined.

1. Perturbative Chiral EFT for Nuclear Matter

Chiral effective field theory (EFT) has made steady advances in applications to few-body nuclei, which can be solved completely given an EFT Hamiltonian.[1] Complete solutions are necessary, since Weinberg power counting in free-space implies non-perturbative summation even at leading order. Applying the same power counting to medium and heavy nuclei is intractable with present-day wave-function methods.

Lattice calculations might be a possible path.[2] But an alternative power counting at finite density might also apply. The Munich group is pursuing a perturbative chiral EFT approach to nuclear matter and then to finite nuclei through an energy functional.[3] Kaiser et al. construct a loop expansion for the nuclear matter energy per particle, which takes the form:

$$E(k_{\rm F}) = \sum_{n=2}^{\infty} k_{\rm F}^n\, f_n(k_{\rm F}/m_\pi, \Delta/m_\pi) \qquad [\Delta = M_\Delta - M_N \approx 300\,{\rm MeV}]$$

where f_n is not expanded in $k_{\rm F}/m_\pi$ and Δ/m_π.[3] While there remain many open questions in this approach, such as the connection to free-space chiral EFT, the key issue is whether a perturbative loop expansion is justified.

2. Finite Nuclei with Low-Momentum Potentials

The perturbativeness of internucleon interactions depends on the momentum cutoff and the density. Changing the cutoff by renormalization group

(RG) methods leaves observables unchanged by construction (up to approximation errors) but shifts contributions between the potential and the sums over intermediate states in loop integrals. These shifts can weaken or even eliminate sources of non-perturbative behavior such as strong short-range repulsion (e.g., from singular chiral two-pion exchange) or the tensor force. The resulting potentials are generically called $V_{\mathrm{low}\,k}$.[4] An additional bonus is that the corresponding three-nucleon interaction becomes perturbative at lower cutoffs.[5] It is found in practice that few- and many-body calculations can be greatly simplified or converge more rapidly by lowering the cutoff.[6] At present, only the two-body interaction is evolved while the three-body interaction is fit for each cutoff using the N^2LO form.[5] The evolution of consistent chiral three-body interactions to lower momenta is a major goal.

A Weinberg-eigenvalue analysis shows quantitatively the reduced non-perturbativeness with increased density and lower cutoff, which arises from reduced phase space due to Pauli blocking and the cutoff, combined with the favorable momentum dependence of the $V_{\mathrm{low}\,k}$ interaction.[6,7] The saturation mechanism is now dominated by the three-body force contribution, which still remains natural-sized according to chiral EFT power counting.[7] Furthermore, nuclear matter near saturation density is perturbative, at least in the particle-particle channel.

That means that we now have a microscopic basis for nuclear density functional theory (DFT), which is a powerful tool that is widely used in condensed matter and quantum chemistry to treat large many-body systems.[8] DFT can include correlation effects in a Hartree-like framework *exactly* if the correct functional is identified. The effective action formalism provides a framework for which to exploit many-body perturbation theory to construct the functional.[8] In the short term, the density matrix expansion (DME) can be used to derive functionals usable in existing nuclear DFT codes.

Acknowledgments

This work was supported in part by NSF Grant No. PHY–0354916.

References

1. A. Nogga et al., Phys. Rev. C **73**, 064002 (2006), and references therein.
2. D. Lee, B. Borasoy and T. Schafer, Phys. Rev. C **70**, 014007 (2004).
3. N. Kaiser et al., arXiv:nucl-th/0610060, and references therein.
4. S.K. Bogner, T.T.S. Kuo and A. Schwenk, Phys. Rept. **386** (2003) 1.
5. A. Nogga, S.K. Bogner and A. Schwenk, Phys. Rev. **C70** (2004) 061002(R).

6. S.K. Bogner, R.J. Furnstahl, S. Ramanan and A. Schwenk, Nucl. Phys. **A773** (2006) 203; S.K. Bogner et al., nucl-th/0609003 and references therein.
7. S.K. Bogner et al., Nucl. Phys. **A763** (2005) 59.
8. R. J. Furnstahl, J. Phys. G **31**, S1357 (2005) [arXiv:nucl-th/0412093].

THE $\overline{K}$-NUCLEAR INTERACTION: A SEARCH FOR DEEPLY BOUND $\overline{K}$-NUCLEAR CLUSTERS

P. CAMERINI (FINUDA Collaboration)

Physics Department, University of Trieste,
Trieste, Italy
** E-mail: camerini@ts.infn.it*

This talk presents the results of recent experimental searches for antikaon-nuclear bound systems on different nuclei, performed with the FINUDA spectrometer at the DAϕNE ϕ-factory (Frascati,Italy).

Keywords: Kaon; deeply bound; cluster.

1. Experimental searches for $\overline{K}$-nuclear clusters.

The study of $\overline{K}$-nucleus deeply bound systems has been recently proposed as a very promising tool to study the $\overline{K}$-N interaction inside nuclear matter. From the theoretical point of view the opinions on the existence of such aggregates are quite discordant. A general understanding of $\overline{K}$-N and $\overline{K}$-A interactions doesn't foresee the existence of clearly detectable states, as their binding energies should be around $(10 \div 30)$MeV, but their width, $(80 \div 100)$MeV, largely exceeds the level separation, thus preventing their observation. On the other hand, recent phenomenological approaches[1] claim the existence of very strongly bound systems, rather narrow and therefore easily detectable. In this model the strongly attractive $\overline{K}$-N phenomenological interaction in the I = 0 channel allows bindings deep enough to prevent the system to decay into the $\Sigma\pi$ channel.

The first claims of $\overline{K}$-nuclear bounds states is due to two KEK experiments[2] which found bumps in the missing mass spectra in the ^{4}He($_{stopped}K^-$,N) reaction and interpreted them as evidences of the formation of the strange tribarions $S^1(3140)$ and $S^0(3115)$, bound by almost 200 MeV and approximately 20 MeV wide. These results have recently been disputed by new experimental findings and theoretical papers, which point out that the search for such states cannot rely on inclusive measurements only.

Thanks to its large solid angle coverage and to the completeness of the information delivered by its particle identification system, the FINUDA spectrometer is able to reconstruct the invariant mass of the deeply bound cluster from the identification and measurement of its decay products.

The simplest kaon-nucleon aggregate to study beyond the $\Lambda(1405)$ is the $[K^-\text{pp}]$ system which should have[1] a mass of 2232 MeV/c^2 (48 MeV binding) and a width of 61 MeV.

The FINUDA spectrometer was used to study the A(K$^-$,ppπ^-) reaction induced by stopped K$^-$'s in different targets (^6Li,^7Li and ^{12}C). The [K^-pp] system was therefore studied, assuming it can decay into the Λp channel by reconstructing its invariant mass from its decay products. The $\Lambda(1116)$ hyperons were reconstructed on the basis of an invariant mass analysis of the πp pairs. The momentum resolution, mass discrimination capabilities as well as other experimental details can be found in.[3] The analysis of the invariant mass spectrum of the Λp system on a mixture of carbon and litium targets shows a bump[4] at 2.255$\pm$0.009 GeV/c^2. The 67 $\pm$16 MeV wide bump, which is mostly due to Λ's and p's emitted back-to-back cannot be explained by simple $K^-NN \rightarrow \Lambda(\Sigma)N$ absorption and it is actually found at more than 100 MeV below the K^-pp mass threshold. Similarly, the strong back-to-back correlation makes it unlikely to explain the bump as due to final state interactions of the outgoing particles (Λ,p,π) after the K$^-$ absorption. An alternative ongoing analysis based on a background subtraction technique gives similar results: the bump may highlight a ($\sim$100MeV) bound [$\overline{K}$NN] state decaying almost at rest. The values of the bump parameters predicted by[1] for the [K$^-$pp] bound system are $\sim$30% lower than the experimental ones.

An analysis of the [K$^-$ppn] system is also being performed by means of a study of the Λd pairs invariant mass from the $K^-_{stop}A \rightarrow \Lambda d\, A'$ reaction and interesting signals of bound states appear in the invariant mass spectrum. A new campaign of measurements, presently under way, will let us increase the statistics by almost one order of magnitude for both reactions.

References

1. Y.Akaishi,T.Yamazaki, *Phys. Rev.* **C65** (2002),044005 Akaishi, Dote,Yamazaki,*Phys. Rev. Lett.***B613**(2005)140
2. T. Suzuki *et al.*, *Phys. Lett.* **B597** (2004), 263; T.Suzuki *et al.*, *Nucl. Phys.* **A754** (2005), 375c
3. M.Agnello *et al.*, *Phys. Lett.* **B622** (2005) 35
4. M.Agnello *et al.*, *Phys. Rev. Lett.* **94** (2005), 212303

MOMENTS OF GPDs FROM LATTICE QCD

D. G. RICHARDS*

Jefferson Laboratory, 12000 Jefferson Avenue,
Newport News, VA 23606, USA
**E-mail: dgr@jlab.org*

I review the lattice computation of the moments of Generalized Parton Distributions (GPDs), and their chiral extrapolation to the physical quark masses. I illustrate how lattice computations of generalized form factors can provide constraints on phenomenological parameterizations of GPDs, and provide insight into the three-dimensional picture of the nucleon.

Generalized Parton Distributions (GPDs)[1–3] incorporate both the parton distributions measured in DIS, as well as the charge and magnetism distributions probed through form factors. However, the excitement centers on their ability to provide a three-dimensional picture of the hadron, specified by the longitudinal momentum fraction and transverse impact parameter distribution of the partons. Lattice QCD can compute the moments of GPDs and their Q^2 evolution, in a manner analogous to the computation of the moments of parton distributions.

The crucial task for lattice QCD in making contact with experiment is to work at sufficiently light pion masses for reliable chiral extrapolations to be performed. Recently, LHPC has fit several polarized and unpolarized structure-function moments to one-loop chiral perturbation theory, and found good agreement with experiment at the physical quark masses for several benchmark quantities, including the axial-vector charge and the isovector quark momentum fraction.[4,5]

Lattice QCD has an important rôle in constructing pictures of hadron structure. Elsewhere at this conference, insights into the origin of nucleon spin from lattice QCD are described.[5] A further insight is the confirmation that the distribution of the impact parameter of a quark relative to the center of a nucleon becomes narrower as the longitudinal moment fraction approaches unity.[6,7] More recently, LHPC has observed an approach to phenomenological-derived parameterizations of GPDs with decreasing pion mass,[4] illustrated in Figure 1. Finally, QCDSF has studied the transverse

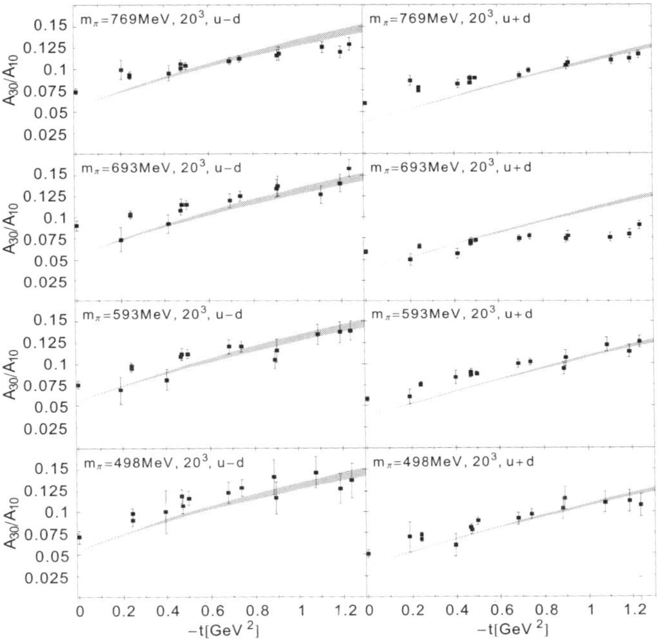

Fig. 1. The ratio of the generalized form factors A_{30}/A_{10} for the combinations $u - d$ and $u + d$; the bands correspond to a phenomenological parametrization.[8]

GPDs, relevant for measurements of single-spin asymmetries.[9] This work was supported by DOE grant DE-AC05-06OR23177 under which Jefferson Science Associates, LLC operates Jefferson Laboratory.

References

1. D. Mueller, D. Robaschik, B. Geyer, F. M. Dittes and J. Horejsi, *Fortschr. Phys.* **42**, p. 101 (1994).
2. X.-D. Ji, *Phys. Rev. Lett.* **78**, 610 (1997).
3. A. V. Radyushkin, *Phys. Rev.* **D56**, 5524 (1997).
4. R. G. Edwards *et al.*, *hep-lat/0610007* .
5. J. N. Negele, *these proceedings* .
6. P. Hagler *et al.*, *Phys. Rev.* **D68**, p. 034505 (2003).
7. M. Gockeler *et al.*, *Phys. Rev. Lett.* **92**, p. 042002 (2004).

8. M. Diehl, T. Feldmann, R. Jakob and P. Kroll, *Eur. Phys. J.* **C39**, 1 (2005).
9. P. Haegler *et al.*, *Proceedings of Lattice 2006, Tucson, Arizona, July 2006* .

GENERALIZED PARTON DISTRIBUTIONS IN EFFECTIVE FIELD THEORY

JIUNN-WEI CHEN
Department of Physics and Center for Theoretical Sciences, National Taiwan University, Taipei 10617, Taiwan
E-mail:jwc@phys.ntu.edu.tw

Generalized parton distributions (GPDs)[1] relate quite different physical quantities, such as Feynman's parton distribution functions (PDFs) and hadron form factors, in the same framework. By generalizing PDFs' one dimensional parton distribution pictures, GPDs provide the three-dimensional pictures.[3] Furthermore, GPDs also give information on highly desirable quantities such as the quark orbital angular momentum contribution to proton spin.[2]

The key to apply ChPT to GPDs is to realize the information of hadronic GPDs is encoded in hadronic twist-2 matrix elements (with the subleading higher twist effects neglected). Since the twist-2 operators are local, color singlet and gauge invariant operators, one can just use the standard techniques of ChPT to match the quark level twist-2 operators to the most general combinations of hadronic operators with the same symmetries.[4] Then by applying the ChPT power counting, one can compute the hadronic twist-2 matrix elements in a systematic chiral expansion. Obviously, one can apply this method to meson, nucleon and nuclear twist-2 matrix elements to get meson, nucleon and nuclear GPDs. ChPT is also a useful tool to help extrapolations of lattice QCD data[5] to infinite volume, zero lattice spacing and physical quark masses.

Several cases of applying ChPT to GPDs have been worked out, e.g., chiral extrapolations of lattice data[4,7,8] including (partially) quenching[9] and finite volume[8] effects, GPDs for quark contribution to proton spin,[10] gravitational form factors,[11] pion GPDs,[12] large N_C relations among nucleon and Δ-resonance distributions[13] (see also earlier work in the large N_C limit[14]), soft pion productions in deeply virtual Compton scattering,[15,16] SU(3) symmetry breaking in the complete set of twist-2[17] and twist-3[18] light cone distribution functions, pion-photon transition distributions[19] and exclusive semileptonic B decays.[20] The method is also generalized to the multi-nucleon case.[16,21] There are also earlier results derived using the soft pion theorem.[22]

References

1. D. Müller *et al.*, Fortschr. Phys. **42**, 101 (1994); A. V. Radyushkin, Phys. Rev. D **56**, 5524 (1997).
2. X. Ji, Phys. Rev. Lett. **78**, 610 (1997);Phys. Rev. D **55**, 7114 (1997).
3. M. Burkardt, Int. J. Mod. Phys. A **18**, 173 (2003); X. Ji, Phys. Rev. Lett. **91**, 062001 (2003); A. V. Belitsky *et al.*, Phys. Rev. D **69**, 074014 (2004).
4. D. Arndt and M. J. Savage, Nucl. Phys. **A697**, 429 (2002); J. W. Chen and X. Ji, Phys. Lett. **B523**, 107 (2001); Phys. Rev. Lett. **87**, 152002 (2001).
5. M. Göckeler *et al.* [QCDSF Collab.], Phys. Rev. Lett. **92**, 042002 (2004); P. Hägler *et al.* [LHPC Collab.], Phys. Rev. D **68**, 034505 (2003).
6. P. Hägler *et al.* [LHPC Collab.], Phys. Rev. Lett. **93**, 112001 (2004);M. Gockeler *et al.* [QCDSF Collaboration], Phys. Lett. B **627**, 113 (2005).
7. W. Detmold *et al.*, Phys. Rev. Lett. **87**, 172001 (2001); Phys. Rev. D **66**, 054501 (2002); *ibid* **68**, 034025 (2003).
8. W. Detmold and C.-J. D. Lin, Phys. Rev. D **71**, 054510 (2005).
9. J. W. Chen and M. J. Savage, Nucl. Phys. A **707**, 452 (2002);Phys. Rev. D **65**, 094001 (2002); J. W. Chen and M. J. Savage, Phys. Rev. D **66**, 074509 (2002); S. R. Beane and M. J. Savage, Nucl. Phys. A **709**, 319 (2002).
10. J. W. Chen and X. Ji, Phys. Rev. Lett. **88**, 052003 (2002).
11. A. V. Belitsky and X. Ji, Phys. Lett. B **538**, 289 (2002).
12. N. Kivel and M. V. Polyakov, arXiv:hep-ph/0203264; M. Diehl, A. Manashov and A. Schafer, Phys. Lett. B **622**, 69 (2005).
13. J. W. Chen and X. Ji, Phys. Lett. **B523**, 73 (2001).
14. D. Diakonov *et al.*, Phys. Rev. D **56**, 4069 (1997); Nucl. Phys. **B480**, 341 (1996); L. L. Frankfurt *et al.*, Phys. Rev. Lett. **84**, 2589 (2000).
15. J. W. Chen and M. J. Savage, Nucl. Phys. A **735**, 441 (2004); N. Kivel *et al.*, nucl-th/0407052; M. C. Birse, J. Phys. G **31**, B7 (2005).
16. S. R. Beane and M. J. Savage, Nucl. Phys. A761, 259 (2005).
17. J. W. Chen and I. W. Stewart, Phys. Rev. Lett. **92**, 202001 (2004).
18. J. W. Chen, H. M. Tsai and K. C. Weng, hep-ph/0511036.
19. B. C. Tiburzi, hep-ph/0508112; B. Pire and L. Szymanowski, Phys. Rev. D **71**, 111501 (2005); Phys. Lett. B **622**, 83 (2005).
20. B. Grinstein and D. Pirjol, Phys. Lett. B **615**, 213 (2005); V. Cirigliano and D. Pirjol, hep-ph/0508095.
21. J. W. Chen and W. Detmold, Phys. Lett. B **625**, 165 (2005).
22. M. V. Polyakov, Nucl. Phys. B **555**, 231 (1999); M. V. Polyakov and C. Weiss, Phys. Rev. D **60**, 114017 (1999); L. Mankiewicz, G. Piller and A. Radyushkin, Eur. Phys. J. C **10**, 307 (1999); M. Penttinen, . V. Polyakov and K. Goeke, Phys. Rev. D **62**, 014024 (2000); M. Diehl, A. Manashov and A. Schafer, arXiv:hep-ph/0608113.

NEAR-THRESHOLD PION PRODUCTION: EXPERIMENTAL UPDATE

M. W. AHMED

Department of Physics, Duke University, and Triangle Universities Nuclear Laboratory, Durham, NC 27708

The near-threshold πN reactions are important to measure since they are a direct consequence of the chiral symmetry breaking and have been explicitly calculated under the framework of Chiral Perturbation Theory (ChPT).[1] The primary focus of the experimental efforts has been the study of π^0 production from the proton near threshold, since it is the lightest pion and the closest approximation to a Goldstone's Boson. These studies have employed two different approaches to produce pions: electro- and photo-production. There have been no new results on pion photo-production since the last Chiral Dynamics Workshop. Recent experiments at Mainz have focused on pion electro-production, and planned experiments at the Jefferson Lab. (JLab) and the High Intensity γ-ray Source (HIγS) will carry out studies using pion electro- and photo-production, respectively.

The recent results from electro-production experiments at Mainz, which explore the Q^2 dependence of the s- and p-wave multipole amplitudes, show a discrepancy between data and the predictions of Chiral Effective Field Theory (ChEFT).[2] This discrepancy is enhanced at the lowest Q^2 of 0.05 $(\mathrm{GeV/c})^2$ which is surprising since at these low momentum transfers the ChEFT should be able to correctly predict the structure functions involved in this process.

The BigBite collaboration at JLab plans to revisit the electro-production studies in the near furture. A wide range of momentum transfers between 0.04 and 0.14 $(\mathrm{GeV/c})^2$ are planned to be studied. The experiment will use the BigBite spectrometer in Hall A of JLab. The data from the BigBite collaboration will be crucial in understanding the exisiting discrepancy between the Mainz data and the ChEFT predictions, especially at low Q^2.

The next generation of pion photo-production experiments are planned to take place at the HIγS facility. Details of the facility and the experiments can be found in this proceedings.[3] The experiments will measure eight structure functions (Rs) as indicated in the cross section below:

$$\frac{d\sigma}{d\Omega^*} = \frac{p_\pi^*}{k_\gamma^*}[R_T^{00} + P_y R_T^{0y} \\ + \Pi_T[(R_{TT}^{00} + P_y R_{TT}^{0y})Cos2\Phi - (P_z R_{TT}^{0z} + P_x R_{TT}^{0x})Sin2\Phi] \\ + \Pi_c[P_z R_{TT'}^{0z} + P_x R_{TT'}^{0x}]],$$

where $\Pi_{T/C}$ is the *Linear* or *Circular* polarization of the beam, and P_{xyz} are target polarizations in x, y, or z directions. The R_T^{00} is the unpolarized transverse differential cross section (σ_T), R_{TT}^{00}/R_T^{00} is the Polarized Photon Asymmetry (Σ), and R_{TT}^{0y}/R_T^{00} is the Polarized Target Asymmetry (T). The measurement of σ_T and Σ will provide the ReE0+ (s-wave) amplitude and ReM1-, ReM1+, ReE1+ (p-wave) amplitudes. The measurement of polarized target asymmetry (T) will provide the ImE0+ (s-wave) amplitude.

The first measurement to be carried out at HIγS will be of the polarized target asymmetry (T) which is sensitive to the charge exchange scattering length a_{cex}. The E0+ amplitude can be expressed as follows:

$$E_{0+}(\gamma p \to \pi^0 p; k) = e^{i\delta_0}[A(k) + i\beta q_+] \text{ where,} \\ \beta = \underbrace{Re[E_{0+}(\gamma p \to \pi^+ n)]}_{28.06\pm0.27\pm0.45\times10^{-3}} \cdot a_{cex}(\pi^+ n \to \pi^0 p).$$

Therefore, a measurement of the the polarized target asymmetry (which in turn is a measurement of the β) can be used to extract the charge exchange scattering length $a_{cex}(\pi^+ n \to \pi^0 p)$ and compared to (-$a(\pi^- p \to \pi^0 n)$). If isospin is conserved in this process then

$$a(\pi^+ n \to \pi^0 p) = \underbrace{-a(\pi^- p \to \pi^0 n)}_{0.1301\pm0.0059}.$$

Hence, a measurement of T (or β) is also a test of isospin conservation. It should be possible to measure the β to a 3% level at HIγS using 100 hours of beam time with 7×10^6 γ/s on target. The measurement will be made using the high resolution Crystal Box and Neutral Meson Spectrometers (NMS). This program at HIγS will start in 2009.

With ongoing programs at Mainz, and planned measurements at JLab and HIγS the hope is to resolve the discrepancies between electroproduction data and ChEFT predictions and to provide valuable data on the dipole amplitudes which dominate the near-threshold pion production.

References

1. V. Bernard, N. Kaiser, and U.G. Meißner, Phys. Rev. Lett., 74, 3752 (1995), and references therein.
2. V. Bernard *et al.*, Nucl. Phys, A607, 379 (1996), and references therein.
3. H. R. Weller, in these proceedings.

PION PHOTOPRODUCTION NEAR THRESHOLD THEORY UPDATE

LOTHAR TIATOR

Institut für Kernphysik, Universität Mainz
D-55099 Mainz, Germany
Email: tiator@kph.uni-mainz.de

Using fixed-t dispersion relations a comparison between experimental information and chiral perturbation theory is possible even at energies below pion threshold. Subtracting the loop corrections from a power series of an expansion of the invariant amplitudes in ν and t gives a unique possibility to derive low-energy constants to any given order.

Keywords: pion photoproduction;low-energy theorems;FFR sumrule

During the last three years no new experiments or re-analysis on threshold pion photoproduction have been performed. The current situation at threshold is given in Ref. 1, where a good agreement of s- and p-wave multipoles was obtained in comparison with Chiral Perturbation Theory.[2] Also dispersion theoretical calculations and a dynamical model calculation with DMT agree quite well with the data.

In a new approach towards low-energy pion photo- and electroproduction, at Mainz we have started to calculate the invariant amplitudes at subthreshold kinematics by using fixed-t dispersion relations.[3,4] This allows a detailed comparison with ChPT in kinematical regions that are not directly accessible to experiments and where the chiral expansion is expected to be even more accurate and reliable.

The transition matrix element for pion photoproduction can be expressed in terms of 4 invariant amplitudes A_i, for which fixed-t dispersion relations (DR) can be formulated. We have evaluated these DR with the Maid2005 imaginary parts and have analyzed the dispersive parts of these amplitudes. For convenience we express the $p\pi^0$ amplitudes by dimensionless functions $\Delta_i(\nu, t)$, in the case of $\Delta_1(\nu, t)$ this is the so-called "FFR discrepancy", which should vanish in the chiral limit,

$$A_1^{\rm disp}(\nu,t) = \frac{e g_{\pi N}}{2M_N^2}\,(\kappa + \Delta_1(\nu,t))\,, \quad A_{2,3,4}^{\rm disp}(\nu,t) \sim \Delta_{2,3,4}(\nu,t)\,.$$

The functions $\Delta_i(\nu,t)$ are regular near the origin of the Mandelstam plane and can be expanded in a power series in ν and t or $\nu_B = (t - M_\pi^2)/(4M_N)$,

$$\Delta_{1,2,4}(\nu,t) = \delta_{00} + \delta_{20}\,\nu^2/M_\pi^2 + \delta_{02}\,\nu_B/M_\pi + \ldots\,,$$
$$\Delta_3(\nu,t) = \delta_{10}\,\nu/M_\pi + \ldots\,.$$

In table 1 we list the leading expansion coefficients, where the vector meson contributions are added to our DR calculations, see Ref. 4 and compare them with the recent covariant Baryon ChPT calculation of Ref. 5. The numbers are in qualitative agreement, differences are mainly due to higher order (h.o.) effects in ChPT and uncertainties in our DR calculations, in particular high-energy contributions above 2.2 GeV that are not included in our approach. While at $\nu = 0, t = 0$ the FFR sum rule is fulfilled, i.e. $\Delta_1 \to 0$, at physical threshold the discrepancy $\Delta_1(\nu_{thr}, t_{thr})$ has a finite value, leading to a violation of the FFR sum rule of about 25% for both proton and neutron, see Ref. 3. This result agrees with the threshold experiments as well as with ChPT calculations. Also for $\Delta_{2,4}(\nu_{thr}, t_{thr})$ we obtain agreement with threshold data derived from the s- and p-wave multipoles of Ref. 1, however, for $\Delta_3(\nu_{thr}, t_{thr})$ we observe a sizeable discrepancy between our DR and the data, which needs further theoretical and experimental investigation.[4]

Table 1. The leading expansion coefficients for the $p\pi^0$ amplitudes from the dispersion integral and ChPT results[5] (in brackets).

	δ_{00}	δ_{10}	δ_{20}	δ_{02}
A_1	−0.04 (0)	—	0.32 (0.53)	1.68 (3.40)
A_2	−6.41 (−6.33)	—	−1.26 (*h.o.*)	1.92 (*h.o.*)
A_3	—	−2.23 (−2.58)	—	—
A_4	21.2 (22.4)	—	2.23 (*h.o.*)	4.50 (*h.o.*)

References

1. A. Schmidt et al., Phys. Rev. Lett. 87 (2001) 232501.
2. V. Bernard, N. Kaiser, and Ulf-G. Meißner, Phys. Lett. B 378 (1996) 337.
3. B. Pasquini, D. Drechsel, and L. Tiator, Eur. Phys. J. A 23 (2005) 279.
4. B. Pasquini, D. Drechsel, and L. Tiator, Eur. Phys. J. A 27 (2006) 231.
5. V. Bernard, B. Kubis, and U.-G. Meißner, Eur. Phys. J. A 25 (2005) 419.

Name	Institution	Email
Mohammad Ahmed	Duke University/TUNL	mohammad.ahmed@duke.edu
Constantia Alexandrou	University of Cyprus	alexand@ucy.ac.cy
Silas Beane	University of New Hampshire	silas@physics.unh.edu
Paulo Bedaque	University of Maryland	bedaque@umd.edu
Paolo Beltrame	Univeristy of Karslruhe	paolo.beltrame@lnf.infn.it
Fatiha Benmokhtar	University of Maryland	fatiha@jlab.org
Claude Bernard	Washington University, St. Louis	cb@lump.wustl.edu
Aron Bernstein	MIT	bernstein@lns.mit.edu
Johan Bijnens	Lund University	bijnens@thep.lu.se
Matthew Blackston	Duke University/TUNL	blackston@tunl.duke.edu
Brian Bunton	Duke University	bunton@phy.duke.edu
Paolo Camerini	University of Trieste-INFN	camerini@ts.infn.it
Irinel Caprini	National Institute of Physics and Nuclear Engineering	caprini@theory.nipne.ro
Oscar Cata	University of Washington	ocata@phys.washington.edu
Shailesh Chandrasekharan	Duke University	sch@phy.duke.edu
Wei Chen	Duke University	wc19@phy.duke.edu
Jiunn-Wei Chen	National Taiwan U.	jwc@phys.ntu.edu.tw
Jian-ping Chen	Jefferson Lab	jpchen@jlab.org
Wei Chen	Duke University/TUNL	weichen@tunl.duke.edu
Deepshikha Choudhury	Ohio University	choudhur@phy.ohiou.edu

Name	Institution	Email
Vincenzo Cirigliano	Los Alamos National Laboratory	cirigliano@lanl.gov
Thomas Clegg	UNC/TUNL	clegg@tunl.duke.edu
Gilberto Colangelo	University of Bern	gilberto@itp.unibe.ch
Christopher Crawford	University of Tennessee	chris2@lns.mit.edu
Timothy Daniels	UNC/TUNL	tdaniels@physics.unc.edu
Kees de Jager	Jefferson Lab	kees@jlab.org
Paul Debevec	University of Illinois	debevec@uiuc.edu
Achim Denig	University of Karlsruhe	achim.denig@iekp.fzk.de
Alexandre Deur	Jefferson Lab	deurpam@jlab.org
Marina Dorati	Universita' degli Studi di Pavia	marina.dorati@pv.infn.it
Evgeny Epelbaum	Forschungszentrum Juelich	e.epelbaum@fz-juelich.de
James Esterline	Duke University/TUNL	jhe@tunl.duke.edu
Renee Fatemi	MIT	rfatemi@mit.edu
Jerry Feldman	George Washington University	feldman@gwu.edu
Kevin Fissum	Lund University and MAX-lab	kevin.fissum@nuclear.lu.se
Andreas Fuhrer	University of Bern	afuhrer@itp.unibe.ch
Richard Furnstahl	Ohio State University	furnstahl.1@osu.edu
Tobias Gail	Technische Universität München	tgail@ph.tum.de
Liping Gan	University of North Carolina Wilmington	ganl@uncw.edu
Haiyan Gao	Duke University	gao@phy.duke.edu
Anders Grdestig	University of South Carolina	anders@physics.sc.edu

Name	Institution	Email
Susan Gardner	University of Kentucky	gardner@pa.uky.edu
Ashot Gasparian	NC A&T State University	gasparan@jlab.org
Simona Giovannella	INFN	simona.giovannella@lnf.infn.it
Leonardo Giusti	CERN	leonardo.giusti@cern.ch
Jose Goity	Hampton University/Jefferson Lab	goity@jlab.org
Maarten Golterman	San Francisco State University	maarten@stars.sfsu.edu
Silvia Goy Lopez	Universita di Torino	silvia.goy.lopez@cern.ch
Harald Griesshammer	George Washington University	hgrie@gwu.edu
Hans Hammer	Bonn University	hammer@itkp.uni-bonn.de
Christoph Hanhart	IKP/Forschungszentrum Jülich	c.hanhart@fz-juelich.de
Thomas Hemmert	TU Muenchen	themmert@ph.tum.de
Seth Henshaw	TUNL Graduate Student	henshaw@tunl.duke.edu
Renato Higa	Jefferson Lab	higa@jlab.org
Barry Holstein	University of Massachusetts	holstein@physics.umass.edu
David Hornidge	Mount Allison University	dhornidge@mta.ca
Calvin Howell	Duke University/TUNL	howell@tunl.duke.edu
Haiming Hu	BES Collaboration	huhm@ihep.ac.cn
Charles Hyde-Wright	Old Dominion University	chyde@odu.edu
Jan Stern	Universite de Paris XI. Sud	stern@ipno.in2p3.fr
Nasser Kalantar	KVI	nasser@kvi.nl
Peter Kammel	UIUC	pkammel@uiuc.edu

Name	Institution	Email
Karol Kampf	Inst. of Particle and Nuclear Physics	kampf@ipnp.troja.mff.cuni.cz
Viktor Kashevarov	P.N. Lebedev Physical Institute	kashev@kph.uni-mainz.de
Paul Kingsberry	MIT/Duke U.	kingsberry@tunl.duke.edu
Martin Kotulla	University of Giessen	Martin.J.Kotulla@exp2.physik.uni-giessen.de
Bastian Kubis	University of Bonn	kubis@itkp.uni-bonn.de
Andrzej Kupsc	Uppsala University	kupsc@tsl.uu.se
Frank Lee	George Washington University	fxlee@gwu.edu
Dean Lee	North Carolina State University	dean_lee@ncsu.edu
Winfried Leidemann	University of Trento	leideman@science.unitn.it
Vadim Lensky	Inst. for Theo. & Exp. Phys.	v.lensky@fz-juelich.de
Heinrich Leutwyler	University of Bern	leutwyler@itp.unibe.ch
Richard Lindgren	University of Virginia	ral5q@virginia.edu
Diane Markoff	NC Central University/TUNL	dmarkoff@nccu.edu
Matteo Martini	LNF and INFN	matteo.martini@lnf.infn.it
Vicent Mateu	Universitat de Valencia-CSIC	mateu@ific.uv.es
Judith McGovern	University of Manchester	judith.mcgovern@manchester.ac.uk
Dustin McNulty	MIT	mcnulty@jlab.org
Thomas Mehen	Duke University	mehen@phy.duke.edu
Ulf Meissner	University of Bonn and Hemholtz Institute	meissner@itkp.uni-bonn.de

Name	Institution	Email
Johan Messchendorp	KVI	messchendorp@kvi.nl
Robert Michaels	Jefferson Lab	rom@jlab.org
Rory Miskimen	University of Massachusetts	miskimen@physics.umass.edu
Matthew Moulson	INFN/Frascati	moulson@lnf.infn.it
Bachir Moussallam	IPN, Université Paris-Sud	moussall@ipno.in2p3.fr
Bernhard Musch	Technische Univ. Muenchen	Bernhard.Musch@ph.tum.de
Fred Myhrer	University of South Carolina	myhrer@physics.sc.edu
Yasuki Nagai	OSAKA UNIVERSITY	nagai@rcnp.osaka-u.ac.jp
Satoshi Nakamura	TRIUMF	snakamura@triumf.ca
John Negele	Massachusetts Institute of Technology	negele@mitlns.mit.edu
Andreas Nogga	Forschungszentrum Juelich	a.nogga@fz-juelich.de
Blaine Norum	University of Virginia	ben@virginia.edu
Jose Oller	Universidad de Murcia	oller@um.es
Giuseppina Orlandini	University of Trento	orlandin@science.unitn.it
Shelley Page	University of Manitoba	spage@cc.umanitoba.ca
Vladimir Pascalutsa	College of William & Mary	vlad@jlab.org
Barbara Pasquini	University of Pavia	pasquini@pv.infn.it
Manuel Pavon	Universidad de Granada	mpavon@ugr.es
Ronald Pedroni	NCATSU/TUNL	pedroni@tunl.duke.edu
Jose R. Pelaez	Universidad Complutense	jrpelaez@fis.ucm.es
Brent Perdue	Duke University/TUNL	perdue@tunl.duke.edu

Name	Institution	Email
Daniel Phillips	Ohio University	phillips@phy.ohiou.edu
Lucas Platter	Ohio University	lplatter@phy.ohiou.edu
Dinko Pocanic	University of Virginia	pocanic@virginia.edu
Richard Prior	North Georgia College & State University	rprior@ngcsu.edu
Xin Qian	Duke University	xqian@jlab.org
Andrea Quadri	University of Milan	andrea.quadri@mi.infn.it
Michael Ramsey-Musolf	Caltech/U. Wisconsin-Madison	mjrm@caltech.edu
Robert Redwine	MIT	redwine@mit.edu
David Richards	Jefferson Science Associates	dgr@jlab.org
Craig Roberts	Argonne National Laboratory	cdroberts@anl.gov
Michael Ronquest	University of Virginia	mr7y@virginia.edu
martin savage	University of Washington	savage@phys.washington.edu
Gerrit Schierholz	DESY	gerrit.schierholz@desy.de
Martin Schmid	University of Bern	schmidm@itp.unibe.ch
Kimiko Sekiguchi	RIKEN	kimiko@rarfaxp.riken.jp
Silvano Simula	INFN Sezione Roma III	simula@roma3.infn.it
Christopher Smith	Bern University	chsmith@itp.unibe.ch
Roxanne Springer	Duke University	rps@phy.duke.edu
Sean Stave	Duke/TUNL	stave@mit.edu
Lothar Tiator	Universitt Mainz	tiator@kph.uni-mainz.de
Brian Tiburzi	University of Maryland	bctiburz@phy.duke.edu

Name	Institution	Email
Werner Tornow	Duke University/TUNL	tornow@tunl.duke.edu
Ubirajara van Kolck	University of Arizona	vankolck@physics.arizona.edu
Marc Vanderhaeghen	College of William & Mary and Jefferson Lab	marcvdh@jlab.org
Steven Vigdor	Indiana University	vigdor@indiana.edu
Thomas Walcher	Mainz University	walcher@kph.uni-mainz.de
Andre Walker-Loud	University of Maryland	walkloud@u.washington.edu
Henryk Witala	Jagiellonian University	witala@if.uj.edu.pl
Henry Weller	Duke University and TUNL	hrweller@msn.com
Charles Whisnant	James Madison University	whisnacs@jmu.edu
Qiang Ye	Duke Univeristy	yq@tunl.duke.edu
Ross Young	Jefferson Lab	young@jlab.org
Fabien Zehr	Universitaet Basel	fabien.zehr@unibas.ch
Oliver Zimmer	TU Munich	oliver.zimmer@ph.tum.de
Xing Zong	Duke University	xz12@phy.duke.edu

AUTHOR INDEX